Principles and Applications of Electrochemistry

Principles and Applications of Electrochemistry

Fourth edition

D. R. CROW
Professor of Electrochemistry and Dean of Research
University of Wolverhampton

Stanley Thornes (Publishers) Ltd

First edition published by Chapman & Hall 1974
Second edition published by Chapman & Hall 1979
Third edition published by Chapman & Hall 1988
Fourth edition published by Chapman & Hall 1994

Reprinted in 1998 by:
Stanley Thornes (Publishers) Ltd
Ellenborough House
Wellington Street
CHELTENHAM
GL50 1YW
United Kingdom

98 99 00 01 02 / 10 9 8 7 6 5 4 3 2 1

A catalogue record for this book is available from the British Library

ISBN 0–7487–4378–2

Typeset by Pure Tech Corporation, Pondicherry, India
Printed and bound in Great Britain by T.J. International Ltd., Padstow, Cornwall

Preface

It has been my good fortune to have had the various editions of this book in print for rather more than twenty years. Those who have used it have been kind enough to comment on its appropriateness for them as either teacher or student, and their identification of errors and obscurities has always been communicated to me with diplomacy. There are now many more books, both general and specific, which deal with the subject of Electrochemistry: in some of them I flatter myself that there is evidence of the influence of earlier editions of this text in regard to structure and treatment. There is always something to learn from different approaches and I acknowledge their help to me in formulating my own ideas for presenting this update.

This fourth edition represents a considerable revision. Structurally it is different from its predecessors in that principles and applications are separated. Several derivations which at first (and maybe even subsequent) readings of their context, may be omitted without in any way detracting from the main theme have been placed in Appendices. Other areas have been re-ordered, expanded or given greater prominence. An example is the section on Electrochemical Sensors which now forms a chapter on its own; another area is that of Electro-analytical Techniques which again has a chapter devoted to it and which includes a number of modern methods not included in earlier editions.

Maintenance of a proper balance between selectivity and detail is always a problem in writing a text such as this one: however, in producing this new edition, which is both re-structured and modestly expanded, the aim has been to bring the essentials into focus and context, but to point the way to further study, expanded treatment and greater intellectual depth. Above all, it has been my concern that the *shape* of the subject of Electrochemistry should be stated at the outset and not lost or obscured during the reading of particular parts, that its coherence should be evident and that its wider application should be appreciated.

There are a number of people whose help I would wish to acknowledge. In terms of this edition I am indebted to my typist Ms Tracey Simcox whose speed and accuracy of word-processing, combined with an ability to decipher my drafts, contributed to my keeping to the schedule agreed with the publishers. Which leads me to express my gratitude to the various editors and staff of Chapman and Hall with whom I have worked over the years: their courtesy, understanding and help have always been so supportive. My immediate colleagues and my own students are among those already mentioned whose use of earlier editions and association with various undergraduate courses have provided me with feedback on the effectiveness of the book.

Finally, I am thankful for the inestimable benefit of a happy and settled home where my wife Margaret, despite a busy academic life of her own, is a constant source of support and encouragement.

D. R. C.

Contents

1 The development and structure of electrochemistry

1.1 The ubiquitous nature of electrochemistry

The subject of electrochemistry is concerned with the study and exploitation of the transference of electrical charges across interfaces and through solution. Transfer of material across biological membranes, storage of electricity in batteries, the production of nylon, electroplating, nerve action, corrosion—these are some areas in which few people would argue that the phenomena are electrochemical.

Yet within the more academic tradition, electrochemistry has been seen as a large and important area of physical chemistry—in many respects rightly so. The subject finds its place in courses of physical chemistry and it appears in chapters of books of that subdivision of the subject of chemistry. Still, it is difficult to define the limits of electrochemistry, for its influence permeates so much of wider chemistry.

The Periodic Table of the Elements, the foundation of systematic inorganic chemistry, is complemented by the Electrochemical Series. Any discussion of periodicity quickly introduces such terms as *electronegativity* and *ionic* character—and the language has turned inexorably to that of electrochemistry. Metal extraction and chemical analysis require electrochemical principles for their understanding and effective exploitation, both on an industrial and on a laboratory scale. Organic synthesis increasingly sees electrochemistry put to use; modern development owes much to the control of electrochemical parameters made possible by modern instrumentation although the Kolbe synthesis was established in the mid-nineteenth century.

1.2 The historical dimension

Publication of the laws of electrochemical conversion some 160 years ago by Michael Faraday established quantitative electrochemistry: direct application of those laws is seen in coulometry.

The nineteenth century saw the building and emergence of the discipline of *Thermodynamics*—a tool of enormous power and an intellectual insight of awesome proportions. Electrochemistry played its part in this development: formulation of the Third Law in particular owed much to investigations of the temperature dependence of cell emf's. Largely under the influence of Nernst, electrochemistry was itself given, and seen as having, a thermodynamic foundation. Indeed, the dominance of Nernst and electrochemical thermodynamics was to have something of a retarding influence on earlier twentieth century progress in Western electrochemistry. Development of electrochemical kinetics owed more to the researches of Central and Eastern Europe. In retrospect, the work of Heyrovsky, starting in Prague in the late twenties, bridges and ultimately leads to a proper regard for the thermodynamic and kinetic principles in electrochemistry.

1.3 The domains of electrochemistry

Electrochemistry is concerned with charges and with their movement and transfer from one medium to another. The ultimate unit of charge is that carried by the electron; electrons are important in electrochemistry and their functions here are similar to some of those which they exhibit in related disciplines more usually regarded as the province of physics. The science of *Thermionics* is built on the exchange of electrons between a solid and a vacuum; that of *transistor electronics* is based on their transfer between one solid phase and another.

When electrons exchange between metals (or other *electronic* conductors), in this context usually termed *electrodes*, and species in solution within which that metal is placed, *Electrodics* is a suitable name to give to the discipline which emerges.

The behaviour of species in solution with an excess or deficiency of electrons, so that they form the negatively or positively charged entities termed *ions*, is the interest of the science which may be called *Ionics.* Electrodics, concerned with electrode processes, and ionics, concerned with the behaviour of ions in solution, constitute much of the fabric ofelectrochemistry.

Separating the solid (or sometimes liquid) material of an electrode and a solution there exists an interfacial region of great importance. The interplay of these three regions is implied by Figure 1.1.

The extremely large field gradients at electrode/solution interfaces caused by imposed potential differences, induce gross distortions in the positions (orbitals) which electrons may occupy in solute ions and in electrodes. Such constraints affect the transfer of electrons between the solid and solution species.

Systems of the sort shown in Figure 1.1 only take on practical significance when combined in pairs to form *electrochemical cells* (Figure 1.2). Differing chemical characteristics of systems labelled 1 and 2 give rise to voltages across terminals connected to the electrodes. This is the arrangement in the wide range of batteries commercially available, except that pastes often replace solutions, render the arrangement stationary and lead to the description *dry*. In cases where the processes which generate the cell voltage

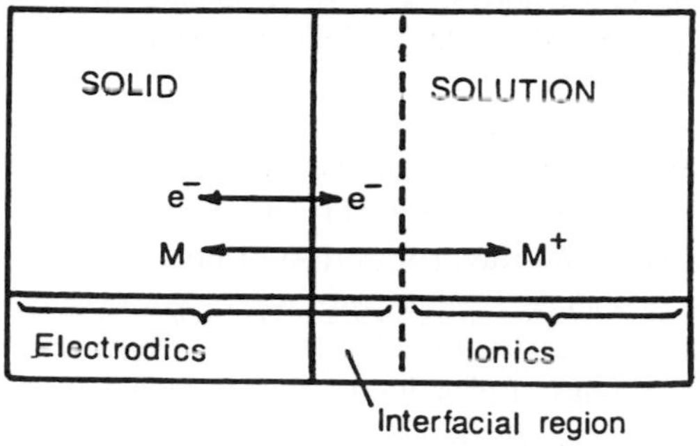

Figure 1.1 Electrodics is concerned with the exchange of electrons between electrodes and species in solution (these are frequently ions). Ionics is the discipline concerned with the behaviour of ions in solution.

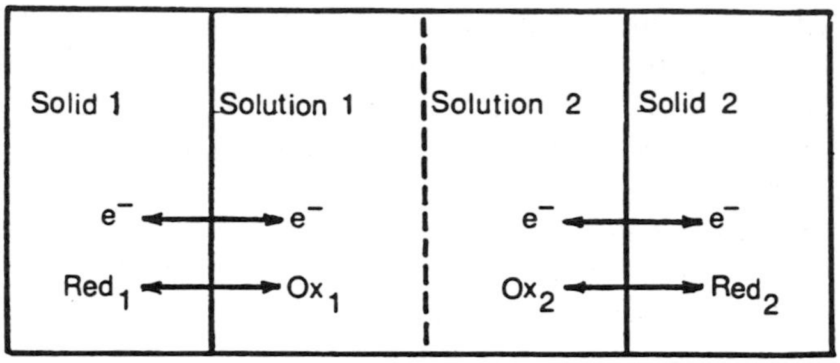

Figure 1.2 General arrangement of an electrochemical cell. Spontaneous generation of a voltage occurs by interfacing of certain electrode/solution pairs of differing chemical characteristics. Imposition of external voltage can reverse the discharge process in rechargeable cells. Supply of 'fuels' to solution sectors under conditions that electron exchange may occur between them and the electrodes can generate an exploitable voltage.

can be reversed, after discharge, by imposition of an external voltage, the battery is rechargeable. The efficiency of such processes is largely governed by the absence of side reactions, the lead–acid cell installed in our motor cars being particularly effective. In other variations, a voltage is generated by the transfer of electrons between the electrodes and *fuels* continuously supplied to the adjacent solution sectors. This is easier said than done and it is necessary to have appropriate catalysts incorporated to ensure rapid electron transfer. Manned spacecraft use hydrogen and oxygen in such *fuel cells* to provide a major part of the electricity requirements, and to supplement the water supply by this product of the process.

The arrangement shown in Figure 1.2 is quite general and of particular practical importance is the variant where one *half-cell* has its properties maintained constant so that relative changes in the other half may be investigated. This approach is the basis of much of electroanalytical chemistry.

The nature, behaviour and structure of species in solution are closely linked with the way in which they undergo electron exchange reactions. Changing the *form* of these species, such as by complexation, often produces a change in electrode behaviour. The nature of metallic deposits produced in electroplating is, for example, significantly affected by the presence of additives to the solution. Even today the reasons for this are far from completely understood and the production of many plated surfaces remains as much an art as a science.

In many arrangements, the magnitudes of currents and voltages are related to the *concentrations* of dissolved solutes. Measurement of such quantities, under carefully controlled experimental conditions, constitutes the very large area of electroanalytical chemistry. Much recent effort has been devoted to the development of electrochemical *sensors*. These are systems which miniaturize and confine complex solution chemistry and electrochemistry to small probes whose sensitive and reversible responses to specific solution components allows their concentration to be determined. A particularly attractive type of sensor is that based on the specific reactions of enzymes.

Implicit in the demonstration and implementation of Faraday's laws are the concepts of atomicity and the nature of ions and electrons. Evidently charges transferred through solutions and across electrode/solution interfaces are 'atomic' in nature.

The two laws are tersely expressed as follows:

1. In electrolytic processes the amount of chemical decomposition is proportional to the *quantity of electricity* passed.
2. The masses of different species dissolved from or deposited at electrodes *by the same quantity of electricity* are in direct proportion to M_R/z for each species. M_R is the relative molar mass, z the change in charge number which occurs in the electrode reaction which may be represented in general terms by

$$A^{n+} \pm ze \rightleftharpoons A^{(n\mp z)+}$$

Thus, when a given current is caused to pass for a given time through a series of electrolyte solutions, the extent of decomposition is always the same in terms of $1/z$ moles. In this statement lies the definition of the Faraday constant as the amount of electricity required to deposit a mole of any species from solution. It has the value $96\,487\,\mathrm{C\,mol^{-1}}$ and its units emphasize the fact that 96 487 coulombs is the amount of electricity associated with '1 mole of unit charges, or electrons'. For practical purposes throughout the following chapters the value of the constant is rounded to $96\,500\,\mathrm{C\,mol^{-1}}$.

Faraday's laws, while allowing a measure of *how much* electrochemical change occurs, can say nothing about *how* (or why) such changes take place. Further, the electron exchange reactions involve species which originate in the solution and which must travel to the electrode to become involved in the process there. In the bulk of the solution electroactive solutes will behave in a way dependent on their structure, their interaction with the solvent and the prevailing conditions. Such factors will influence the transfer of an ion through the solution to the edge of the interfacial region and its subsequent negotiation of that region. Each of these circumstances constitutes an area (or areas) of study in their own right. Their influence is augmented in the *overall* electrochemical processes quantitatively expressed in terms of Faraday's laws. In short, the behaviour of ions may be considered in the bulk of a solution, at interfaces and at electrodes—three domains of electrochemistry. In Table 1.1 important features of the subject, upon which subsequent chapters focus, are listed within these domains.

The sequence of Chapters 2–7, concerned with *principles*, reflects directly the journey of an ion in solution to an electrode and its reaction there. Chapters 8–11 are concerned with the exploitation of those principles in a range of *applications*.

Table 1.1 Features of electrochemistry

Ionics	Interfacial phenomena	Electrodics
Nature and behaviour of ions in solution and fused state	Double layer theory	Kinetics and mechanism of electrode reactions
Ionic equilibria	Adsorption	
Acid–base theory	Zeta potential	Electron-transfer processes
Transport processes	Electrokinetic phenomena	
Potential-determining ion reactions and reversible half-cell potentials	Colloidal systems	Electrocatalytic processes
Sensor systems	Ion-exchange processes	

PART I PRINCIPLES

2 Ionic interaction: the ways in which ions affect each other in solution

2.1 The nature of electrolytes

Electrolytes are species giving rise to ions to a greater or lesser extent, strong electrolytes being completely ionized even in the solid and fused states. In the latter case, and also when dissolved in a solvent, the ions become free to move and the highly ordered lattice structure characteristic of crystals is destroyed. Weak electrolytes, on the other hand, are ionized to only a small extent in solution, ionization increasing with dilution according to the well-known Ostwald Law considered in Chapter 3.

2.1.1 Ion–ion and ion–solvent interactions

Although strong electrolytes are completely ionized, their ions are not entirely free to move independently of one another through the body of a solution, except when this is infinitely dilute. A fairly realistic picture of the situation in a solution containing the oppositely charged ions of an electrolyte is as follows. Ions will move randomly with respect to one another due to fairly violent thermal motion. Even in this condition, however, coulombic forces will exert their influence to some extent with the result that each cation and anion is surrounded on a time average by an 'ion atmosphere' containing a relatively higher proportion of ions carrying charge of an opposite sign to that on the central ion.

Movement of ions under the influence of an applied field will be very slow and subject to disruption by the thermal motion. Under the influence of such a field, movement of the atmosphere occurs in a direction opposite to that of the central ion, resulting in the continuous breakdown and reformation of the atmosphere as the ion moves in one direction through the solution. The time lag between the restructuring of the atmosphere and the movement of the central ion causes the atmosphere to be asymmetrically distributed around the central ion causing some attraction of the latter in a direction opposite to that of its motion. This is known as the *asymmetry*, or *relaxation* effect. In addition, central ions experience increased viscous hindrance to their motion on account of solvated atmosphere ions which, on account of the latter's movement in the opposite direction to the central ion, produce movement of solvent in this opposing direction as well. This is known as the *electrophoretic* effect.

Such interactions must obviously increase in significance with increasing concentration of the electrolyte. In the extreme condition of infinite dilution, all interionic effects are eliminated, ions move without the above restrictions, current may pass freely and conductivity reaches a maximum value. In another extreme situation, interionic attraction may become so great that the formation of discrete ion pairs may occur. The most favourable conditions for such behaviour are high electrolyte concentration and high charges on the ions. Ion pairs consist of associated ions, the formation of which must be regarded as a time-averaged situation since in any such system there will be a continual interchange of ions amongst the pairs. For a species to be regarded as an ion pair, it must be a 'kinetically distinct' species. That is, although it is an unstable and transient entity, it nevertheless has a lifetime of such duration that it can experience a number of kinetic collisions before exchanging an ion partner.

2.1.2 Dissolution, solvation and heats of solution

Very many salts are known to dissolve readily in solvents with heats of solution that are usually fairly small in magnitude and which may be exothermic or endothermic. At first sight this is a phenomenon rather difficult to account for, since crystal structures have high lattice energies. A lattice energy is the large-scale analogue of the dissociation energy of an individual ionic 'molecule'. In a crystal, the energies of a large number of component ion formula units contribute to the total lattice energy which is effectively the energy evolved when the lattice is built up from free ions. Since such energies are large, we are led to suppose that a large amount of energy is required to break down the ordered structure and liberate free ions. A way in which the observed easy dissolution can be explained is by the simultaneous occurrence of another process which produces sufficient energy to compensate for that lost in the rupture of the lattice bonds. Exothermic reactions of individual ions with the solvent—giving rise to the heat of solvation—provide the necessary energy. From the First Law of Thermodynamics, the algebraic sum of the lattice and solvation energies is the heat of solution. This explains both why the heats of solution are usually fairly small and why they may be endothermic or exothermic—depending upon whether the lattice energy or the solvation energy is the greater quantity.

A great difficulty when dealing with electrolytes is to ascribe individual properties to individual ions. Individual thermodynamic properties cannot be determined, only mean ion quantities being measurable. Interionic and ion–solvent interactions are so numerous and important in solution that, except in the most dilute cases, no ion may be regarded as behaving independently of others. On the other hand, there is no doubt that certain dynamic properties such as ion conductances, mobilities and transport numbers may be determined, although values for such properties are not absolute but vary with ion environment.

2.2 Ion activity

Since the properties of one ion species are affected by the presence of other ions with which it interacts electrostatically, except at infinite dilution, the concentration of a species is an unsatisfactory parameter to use in attempting to predict its contribution to the bulk properties of a solution. What is rather required is a parameter similar to, and indeed related to, concentration, i.e. the actual number of ions present, but which expresses the availability of the species to determine properties, to take part in a chemical reaction or to influence the position of an equilibrium. This parameter is know as activity (a) and is related to concentration (c) by the simple relationship

$$a_i = \gamma_i c_i \tag{2.1}$$

γ_i is known as an activity coefficient which may take different forms depending on the way in which concentrations for a given system are expressed, i.e. as molarity, molality or mole fraction. For instance, the chemical potential (μ_i) of a species i may be expressed in the form

$$\mu_i = \mu_i^{\ominus} + RT \ln x_i \gamma_i \tag{2.2}$$

where x_i is the mole fraction and γ_i is known as the rational activity coefficient. Similar expressions in terms of molar or molal concentrations may be used.

2.2.1 Chemical and electrochemical potential

Before going any further it is necessary to be quite clear what chemical potential (μ_i) really means; somewhat formal definitions, while necessary, tend to obscure the mental picture and experiment required to understand the practical relevance of such parameters.

In essence, μ_i is the *change in free energy* of a system when 1 mole of *uncharged* species i is added to it (in fact to such a large amount of the system under fixed conditions that no other changes of significant size occur).

In a solution where individual particles do *not* interact with one another (an *ideal* solution), μ_i is given by

$$\mu_i = \mu_i^{\ominus} + RT \ln x_i \tag{2.3}$$

In a (non-ideal) solution where particles interact with one another equation (2.3) must be modified to

$$\mu_i = \mu_i^{\ominus} + RT \ln a_i$$

$$= \mu_i^{\ominus} + RT \ln x_i \gamma_i \qquad \text{(See equation (2.2))}$$

or

$$\mu_i = \mu_i^{\ominus} + RT \ln x_i + RT \ln \gamma_i \tag{2.4}$$

In equation (2.4) the term $RT \ln \gamma_i$ may be seen as a modifying function to correct the ideal equation for the effects of interaction.

If solute species are ions, particularly strong interactions occur because of the electrostatic forces between charges. Equation (2.4) will no longer take account of their energy contribution to the system.

Charges on ions already present in the system will generate an electrical potential, ϕ. Addition of 1 mole of further changes, of magnitude z_i, will induce an *extra* change in free energy of magnitude $z_i F \phi$. Thus, the free energy of the system is increased by both the transfer of matter and by the transfer of charge. The sum of these contributions is known as the *electrochemical* potential $\tilde{\mu}_i$. Thus

$$\tilde{\mu}_i = \mu_i + z_i F \phi \tag{2.5}$$

or, in terms of equation (2.4)

$$\tilde{\mu}_i = \mu_i^{\ominus} + RT \ln a_i + z_i F \phi$$

and

$$\tilde{\mu}_i = \mu_i^{\ominus} + RT \ln x_i + RT \ln \gamma_i + z_i F \phi \tag{2.6}$$

2.2.2 *Mean ion activity*

Until about 1923, activity coefficients were purely empirical quantities in that when concentrations were modified by their use, correct results could be predicted for the properties of a system. On the basis of the Debye–Hückel theory, to be discussed shortly, activity coefficients become rationalized and theoretically predictable quantities.

For the purposes of deriving relationships in which activity coefficients occur it is very convenient to make use of the idea of individual ion activities and activity coefficients. However, as already stressed, such quantities are incapable of measurement and so are meaningless in a practical sense. One ion species, deriving from dissolved electrolyte, cannot on its own determine properties of a system; it will always do so in concert with an equivalent number of oppositely charged ions. It is therefore only possible to use a form of activity or activity coefficient which takes account of both types of ions characteristic of an electrolyte. Such forms are known as mean ion activities ($a_\pm$) and mean ion activity coefficients ($\gamma_\pm$) and are defined by

$$(a_\pm)^\nu = a_+^{\nu_+} \cdot a_-^{\nu_-} \tag{2.7a}$$

and

$$(\gamma_+)^{\nu} = (\gamma_+)^{\nu_+} \cdot (\gamma_-)^{\nu_-} \tag{2.7b}$$

where $\nu = \nu_+ + \nu_-$, the latter being the number of cations and anions respectively deriving from each formula unit of the electrolyte.

2.3 The Debye–Hückel equation

Equation (2.6) in many ways sets the scene for this present treatment of electrochemistry. Three things should be emphasized about it:

1. Its origin is in thermodynamics, with its starting point the Gibbs definition of chemical potential.
2. The third term, $RT \ln \gamma_i$, is the one of immediate interest and provides the starting point for attempting to predict values of γ_i from know physical parameters.
3. The fourth term, $z_i F\phi$, will feature prominently in later chapters and is the foundation for interpreting interfacial behaviour and electrode potentials.

2.3.1 A theoretical model for calculating activity coefficients

In this section the outline thinking of Debye and Hückel is given: detailed derivations of important relationships required are given in Appendix I.

A potential ϕ exists in the vicinity of an ion by virtue of the charge which it carries: this may be resolved into *two* parts. One contribution is due to the ion itself, the other due to its atmosphere of appropriately signed net charge.

Thus

$$\phi = \phi_0 + \phi_i \tag{2.8}$$

where ϕ_0 has been used to represent the contribution of the ion itself and ϕ_i that due to its atmosphere.

It is clear that the term $RT \ln \gamma_i$ has the units of energy per mole: the crucial step in the derivation of Debye and Hückel is the identification of $RT \ln \gamma_i$ with the contribution that the *ion atmosphere* makes to the total (molar) energy of ion species i. In the case of an individual ion, the contribution is $kT \ln \gamma_i$ and this may be equated to the work which must be performed to give an ion its charge, $z_i\epsilon$, ϵ being the electronic charge.

The work done, dw, in charging an ion by an increment of charge, $d\epsilon$, is given by

$$dw = \phi \, d\epsilon \tag{2.9}$$

Therefore

$$dw = \phi_0 d\epsilon + \phi_i d\epsilon$$

Thus $kT \ln \gamma_i$ may be equated to that part of the work, w_i, providing the ion with its charge, which is associated with ϕ_i, i.e. to

$$w_i = \int_0^{z_i \epsilon} \phi_i \, d\epsilon \tag{2.10}$$

The laws of electrostatics (Appendix I) may be used to rationalize the terms in equation (2.8) as follows

$$\phi = \pm \left(\frac{z_i \epsilon}{4\pi\varepsilon_0 \varepsilon} \right) \frac{1}{a} \mp \left(\frac{z_i \epsilon}{4\pi\varepsilon_0 \varepsilon} \right) \left(\frac{\kappa}{1 + \kappa a} \right) \tag{2.11}$$

where ε_0 is the permittivity of a vacuum and ε is the relative permittivity (dielectric constant) of the solvent. The two terms of equation (2.11) have the general form of the potential at the *surface of a charged sphere*: the signs for the two contributions are seen to be consistent with the opposite charges associated with ion and atmosphere. The term a represents the distance of closest approach of an ion and another from the atmosphere, both regarded as spheres (Figure 2.1).

The expression

$$\frac{\kappa}{1 + \kappa a}$$

in the second term on the right-hand side of equation (2.11) corresponds to $1/a$ in the first term. Thus the expression $(1 + \kappa a)/\kappa$ is an effective radius—that of the ion atmosphere. Thus, the effect of the atmosphere on the potential of a given ion is equivalent to the effect of the same charge distributed over a sphere of radius $(1 + \kappa a)/\kappa$ or $(1/\kappa + a)$. In fact, it is the quantity $1/\kappa$ which is usually defined as the thickness of the ion atmosphere or Debye length. This is reasonable in dilute solution where $1/\kappa \gg a$ but becomes unrealistic in more concentrated solutions to the extent that a calculated value of $1/\kappa$ may become less than a, implying that the edge of the atmosphere resides at a distance less than the distance of closest approach. It is seen that κa is the ratio of the ion diameter to the atmosphere radius.

Combining equations (2.10) and (2.11) gives w_i as

$$w_i = -\int_0^{z_i \epsilon} \left(\frac{z_i \epsilon}{4\pi\varepsilon_0 \varepsilon} \right) \left(\frac{\kappa}{1 + \kappa a} \right) d\epsilon$$

(assuming i to be a positive ion), therefore

$$w_i = -\frac{z_i^2 \epsilon^2 \kappa}{8\pi\varepsilon_0 \varepsilon (1 + \kappa a)}$$

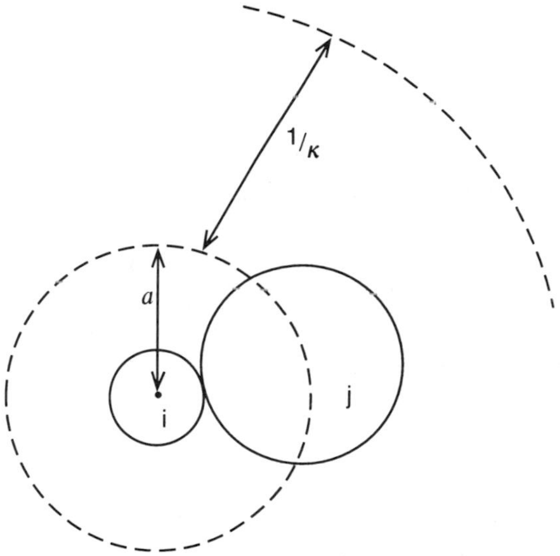

Figure 2.1 The distance of closest approach, a, for ion i and atmosphere ion j, and the Debye length, $1/\kappa$. Radius of ion atmosphere = $1/\kappa + a \sim 1/\kappa$ when $1/\kappa \ll a$ as is the case for dilute solutions.

so that

$$kT \ln \gamma_i = -\left(\frac{z_i^2 \epsilon^2}{8\pi\varepsilon_0\varepsilon}\right)\left(\frac{\kappa a}{1+\kappa a}\right)$$

or, in terms of the mean ion activity coefficient of the electrolyte

$$\ln \gamma_{\pm} = -\frac{\epsilon^2}{8\pi\varepsilon_0\varepsilon kT}\,|z_+z_-|\left(\frac{\kappa}{1+\kappa a}\right) \tag{2.12}$$

It is shown in Appendix II that κ may be expressed in the form

$$\kappa = \left(\frac{2\times 10^3 \epsilon^2 N}{\varepsilon_0\varepsilon kT}\right)^{\frac{1}{2}} \sqrt{I} \tag{2.13}$$

where N is the Avogadro constant and I is the ionic strength of the solution given by

$$I = \tfrac{1}{2}\sum m_i z_i^2 \tag{2.14}$$

m_i being the concentration of the electrolyte in the units $\mathrm{mol\,kg^{-1}}$.

Substitution of equation (2.13) into equation (2.12), after conversion to logarithms to base 10 gives

$$\log \gamma_{\pm} = -\frac{\epsilon^2 N}{2\cdot 3(8\pi\varepsilon_0\varepsilon RT)}\,|z_+z_-|\,\frac{\left(\dfrac{2\times 10^3\epsilon^2N^2}{\varepsilon_0\varepsilon RT}\right)^{\frac{1}{2}}\sqrt{I}}{1+\left(\dfrac{2\times 10^3\epsilon^2N^2}{\varepsilon_0\varepsilon RT}\right)^{\frac{1}{2}}a\sqrt{I}}$$

(Note that $k = R/N$.) Or, more simply, in a form which collects constants,

$$-\log \gamma_{\pm} = \frac{|z_+z_-|\,A\sqrt{I}}{1+Ba\sqrt{I}} \qquad (2.15)$$

This is the Debye–Hückel equation in which A and B are constant for a particular solvent at a given temperature and pressure and which are given explicity by

$$A = \frac{\epsilon^2 N}{2\cdot 3RT(8\pi\varepsilon_0\varepsilon)}\left(\frac{2\times 10^3N^2\epsilon^2}{\varepsilon_0\varepsilon RT}\right)^{\frac{1}{2}}$$

$$B = \left(\frac{2\times 10^3N^2\epsilon^2}{\varepsilon_0\varepsilon RT}\right)^{\frac{1}{2}}$$

The above constants may be calculated for water at 298 K by substitution of the data.

$$\epsilon = 1.6021\times 10^{-19}\,\mathrm{As};\ \mathrm{N} = 6.023\times 10^{23}\,\mathrm{mol}^{-1}$$
$$\varepsilon_0 = 8.8542\times 10^{-12}\,\mathrm{kg}^{-1}\,\mathrm{m}^{-3}\,\mathrm{s}^4\,\mathrm{A}^2;\ \varepsilon = 78.54$$
$$R = 8.314\,\mathrm{J\,K}^{-1}\,\mathrm{mol}^{-1};\ T = 298\,\mathrm{K}$$

Thus

$$A = 0.509\,\mathrm{mol}^{-\frac{1}{2}}\,\mathrm{kg}^{\frac{1}{2}}$$
$$B = 3.290\times 10^9\,\mathrm{m}^{-1}\,\mathrm{mol}^{-\frac{1}{2}}\,\mathrm{kg}^{\frac{1}{2}}$$

In view of the fact that the product Ba in equation (2.15) is frequently of the order of unity, it is common practice for activity coefficients to be calculated via the simplified form

$$-\log \gamma_{\pm} \sim \frac{|z_+z_-|\,A\sqrt{I}}{1+\sqrt{I}} \qquad (2.16)$$

2.3.2 Limiting and extended forms of the Debye–Hückel equation

For very dilute solutions the denominator of equation (2.15) is very little different from unity. Under such conditions the 'Limiting Law' expression holds, viz.

$$-\log \gamma_{\pm} = |z_+z_-|\,A\sqrt{I} \qquad (2.17)$$

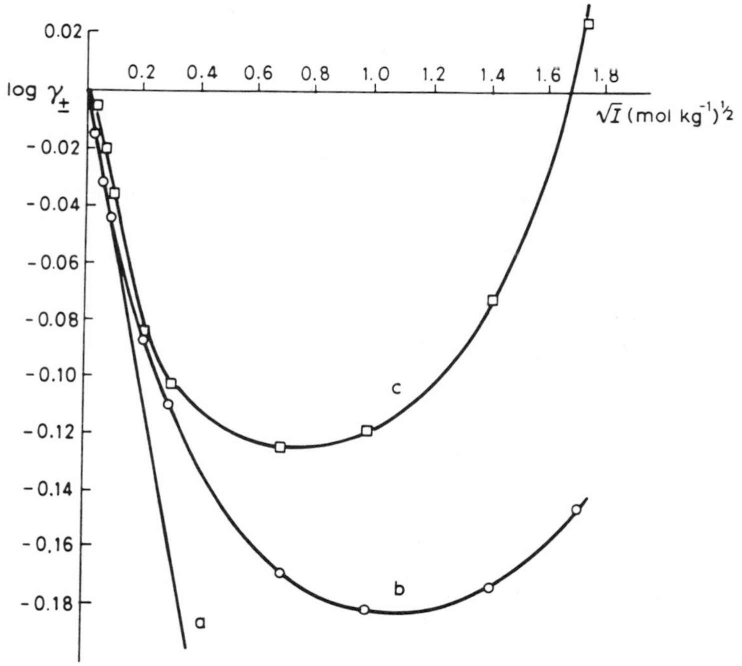

Figure 2.2 Variation of mean ion activity coefficients with ionic strength ($\log \gamma_\pm$ versus $\sqrt{I}$). (a) Limiting law line for a 1:1 electrolyte, (b) experimental graph for NaCl, (c) experimental graph for KOH.

An experimental test of the Debye–Hückel model as represented by equation (2.15) may be undertaken by rearranging this expression to the form

$$-\frac{A\,|\,z_+z_-\,|\,\sqrt{I}}{\log \gamma_\pm} = 1 + Ba\sqrt{I} \tag{2.18}$$

The left-hand side of equation (2.18) may be determined experimentally under conditions where the limiting law holds. Then, if B for the solvent is known, the left-hand side of the equation may be plotted as a function of $\sqrt{I}$ to give a value of the distance of closest approach, a, from the slope of the graph. Here a drawback is encountered since very often the values obtained for a are physically meaningless in that they are far too small or even negative. Nevertheless, with the assumption of a reasonable magnitude for a ($\sim 0.4\,\text{nm}$) equation (2.15) does hold for many electrolytes up to $I \sim 0.1\,\text{mol kg}^{-1}$.

In practice activity coefficients initially decrease with increasing concentration of electrolyte (Figure 2.2). Such behaviour is entirely consistent with both the Debye–Hückel equation and its limiting form. However, in

(*a*) [+ −]
(*b*) [+ − +]$^+$, [− + −]$^-$
(*c*) [+ − + −], [− + − +]

Figure 2.3 Possible associations in a 1:1 electrolyte. (a) ion pairs, zero net charge; (b) triple ions, unit charge; (c) quadruple ions, zero net charge.

practice, activity coefficients show a turning point at some value of I, after which they progressively increase. It is thus seen to be necessary to modify equation (2.15) by the addition of a further term which is an increasing function of I, i.e.

$$\log \gamma_{\pm} = -\frac{A\,|z_+z_-|\sqrt{I}}{1 + Ba\sqrt{I}} + bI \qquad (2.19)$$

This last relationship has become known as the Hückel equation.

2.4 Ion association

Mean ion activity coefficients cannot always be realistically predicted from the Debye–Hückel equation: this suggests that the model used to describe the distribution of ions about other ions does not apply. This idea is supported by the observation that for many cases experimental values of conductances are lower than those predicted by the Onsager equation (see Chapter 4).

2.4.1 Ionization, dissociation and association

Strong electrolytes are fully *ionized*: this applies to the solid, solution and fused states. Dilution of a solution of a strong electrolyte does not change the state of ionization unlike the case for a weak electrolyte where undissociated molecular species are in equilibrium with ions

$$BA \rightleftharpoons B^+ + A^-$$

Here the proportion of ions to undissociated species increases with dilution.

In the solid state, strong electrolytes are fully *associated* within a rigidly geometrical distribution of discrete ions.

It was suggested by Bjerrum that, under certain conditions, a *degree of association* of oppositely charged ions in solution is feasible. High electrolyte concentration, strong interionic forces (favoured by solvents of *low* dielectric constant since coulombic forces are inversely proportional to ε), small ions and high charges are the factors which should encourage such behaviour.

There is experimental evidence for the formation of ion pairs, triple ions and even quadruple ions (Figure 2.3).

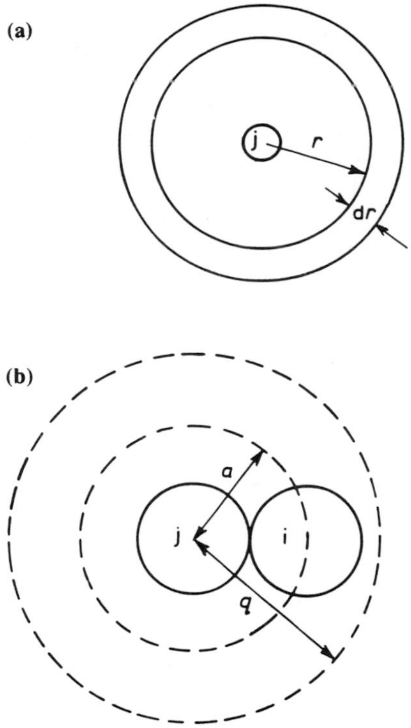

Figure 2.4 (a) Model for determination of distribution of i ions within shells of specified dimensions about j ions (b) Relation of a to q for ion pairing.

Such evidence is provided by rationalization of the discrepancies relating to such parameters as conductivity referred to above and by experimental measurement of association constants for processes such as

$$B^+ + A^- \rightleftharpoons [B^+A^-]$$

2.4.2 The Bjerrum equation

Association leads to a smaller number of particles in a system and associated species have a lower charge than non-associated ones. This will obviously serve to diminish the magnitudes of properties of a solution which are dependent on the number of solute particles and the charges carried by them.

Bjerrum's basic assumption was that the Debye–Hückel theory holds so long as the oppositely charged ions of an electrolyte are separated by a distance q greater than a certain minimum value given by

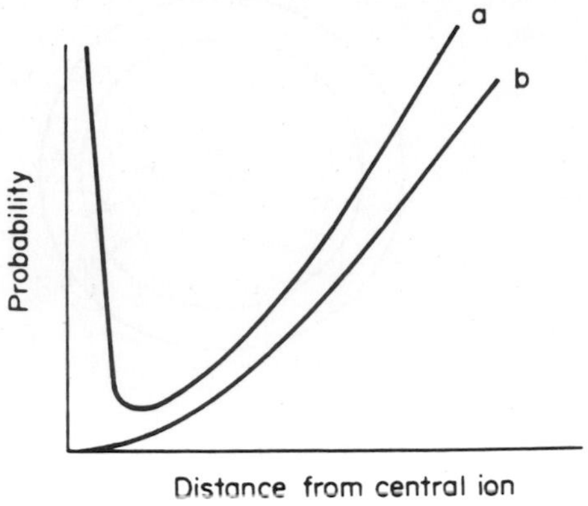

Figure 2.5 Shapes of probability curves for distribution of (a) i ions about j ions; (b) j ions about j ions.

$$q = \frac{z_i z_j \epsilon^2}{8\pi\varepsilon_0\varepsilon kT} \tag{2.20}$$

When the ion separation is less than q, ion pairing is regarded as taking place. Equation (2.20) may be derived from consideration of the Boltzmann distribution of i-type ions in a thin shell of thickness dr at a distance r from a central j-type ion (Figure 2.4)

The number of i-type ions in such a shell is given by

$$\mathrm{d}N_i = N_i \exp\left[\pm\frac{z_i\epsilon\phi_j}{kT}\right] 4\pi r^2\,\mathrm{d}r \tag{2.21}$$

The potential at a small distance from the central j-ion may be assumed to arise almost entirely from that ion and is given by equation (2.11) as

$$\phi_j = \pm\frac{z_j\epsilon}{4\pi\varepsilon_0\varepsilon r} \tag{2.22}$$

Thus,

$$\mathrm{d}N_i = 4\pi N_i \exp\left(-\frac{z_i z_j \epsilon^2}{4\pi\varepsilon_0\varepsilon kTr}\right) r^2\,\mathrm{d}r \tag{2.23}$$

We expect to find decreasing probability of finding i-type ions per unit volume at increasing values of r. However, the volumes of the concentric shells increase outwards from a j-ion so that, in fact, the probability passes through a minimum at some critical distance. A plot of calculated probability versus r shows such a minimum (Figure 2.5).

It may be seen from the minimum condition, $\mathrm{d}N_i/\mathrm{d}r = 0$, that this distance is given by equation (2.20).

For a 1:1 electrolyte in aqueous solution at 298 K, q has the value 0.357 nm. Should the sum of the respective ionic radii be less than this figure then ion-pair formation will be favoured.

It is evident that for a given electrolyte at constant temperature, lowering of dielectric constant will encourage association. For tetraisoamylammonium nitrate the sum of ion radii is of the order of 0.7 nm. This gives a value of about 42 for ε by substitution into equation (2.20) and implies that, for solvents of greater dielectric constant than 42, there should be no association but rather complete dissociation, i.e. the Debye–Hückel theory should hold good. Conductance measurements have verified that, in fact, virtually all ion pairing has ceased for $\varepsilon > 42$.

Problems

2.1 Estimate the radius of the ion atmosphere surrounding a given ion at 298 K in an aqueous solution of (i) a 1:1 electrolyte, (ii) a 1:2 electrolyte for the concentrations 0.1, 0.01, 0.001 and 0.0001 mol kg^{-1}. Note the dependence of the value of the atmosphere radius upon ionic strength and also upon the relative permittivity (dielectric constant) of a medium by comparing values of $1/\kappa$ for water (ε = 78.54) and for *N,N*-dimethylformamide (ε = 36.70).

2.2 Calculate the ionic strength of an aqueous solution of magnesium chloride of concentration 0.0015 mol kg^{-1} at 298 K. In terms of the Debye–Hückel limiting law, calculate (i) the activity coefficients of the Mg^{2+} and Cl^{-} ions in the solution, (ii) the mean ion activity coefficient of the ions, $\gamma_{\pm}$.

2.3 Calculate the ionic strength of (i) a solution of lanthanum nitrate, $La(NO_3)_3$ at 298 K at a concentration of 0.001 mol kg^{-1}, (ii) a solution containing lanthanum nitrate and sodium chloride at concentrations of 0.001 and 0.002 mol kg^{-1}, respectively.

2.4 Determine the mean activity coefficient of $La(NO_3)_3$ in the environments of the previous question (i) from the Debye–Hückel Limiting Law, (ii) from the extended form of the law.

2.5 Estimate the values of the Debye–Hückel constants A and B for water as solvent at 298 K.

2.6 The following values of mean ion activity coefficients are reported (R. H. Stokes and R. A. Robinson (1955) *Electrolyte Solutions*, Butterworths, London, p. 234) for sodium chloride in water at 298 K for the range of molalities shown.

m mol kg^{-1}	0.001	0.002	0.005	0.01	0.02	0.05	0.10
$\gamma_{\pm}$	0.9649	0.9519	0.9273	0.0922	0.8706	0.8192	0.7784

By a suitable graphical procedure, calculate the distance of closest approach of the ions.

2.7 Calculate the Bjerrum critical distance for ion pairing in water at 298 K for 1:1, 2:2 and 3:3 electrolytes.

3 Ionic equilibria: the behaviour of acids and bases

3.1 Classical theory. The Arrhenius dissociation model

Arrhenius put forward the idea of incomplete dissociation of weak electrolytes on the basis of observed anomalies in the values of colligative properties for electrolytes. Thus, the depression of freezing point of water containing a non-electrolyte is less than that with the same molar proportion of a weak electrolyte. Except at very high dilutions, however, the depression does not approach that to be expected for the total number of ions to which the complete dissociation of the electrolyte gives rise. The classical dissociation theory, concerned essentially with aqueous solutions, rests on the concept of incomplete dissociation of weak electrolytes which increases with dilution. The Arrhenius theory considers the parent electrolyte and ions produced from it as behaving in their own right, any chemical interaction with solvent molecules being ignored, the solvent being simply regarded as a medium within which dissociation and dispersal of ions may occur.

The Arrhenius theory similarly attempts to define acids and bases as isolated species in solution giving rise, respectively, to hydrogen ions and hydroxyl ions. While this view accounts for some properties of acids and bases and their reactions, it cannot begin to explain acidic and basic characteristics in non-aqueous media. Acidic and basic properties are, in fact, *consequent upon* interaction with the solvent and until such interaction has taken place the properties are not shown.

Despite the shortcomings of the classical theory, the relations derived from its use do allow the calculation of reliable equilibrium data for weak electrolytes, including acids and bases, in aqueous solution.

Consider a solution of a binary electrolyte, MX, at a concentration C mol dm^{-3}. If only a fraction α, of MX ionizes, there will result an equilibrium mixture of $C(1-\alpha)$ mol dm^{-3} of MX in company with $C\alpha$ of both M^+ and X^- ions. The equilibrium for the dissociation process

$$MX \rightleftharpoons M^+ + X^-$$

is then given by

$$K' \sim \frac{[M^+][X^-]}{[MX]} = \frac{\alpha^2 C}{(1-\alpha)} \tag{3.1}$$

The approximation sign accounts for the use of concentrations in place of activities. Equation (3.1) has become known as the Ostwald Dilution Law and, for a very weak electrolyte where $\alpha \ll 1$, takes the form $K' \sim \alpha^2 C$.

3.2 The Brønsted–Lowry concept of acids and bases

In studies of catalysis of solution reactions by acids and bases, Brønsted observed that catalytic activity was shown by species which at the time were not regarded as acids or bases. An extension of the classical view of acids and bases became desirable and a general definition of such substances to include both aqueous and non-aqueous media was necessary.

Earlier definitions of acids and bases as species producing, respectively, hydrogen ions and hydroxyl ions are only valid in aqueous solution. The concept of Brønsted and Lowry, while extending the classical definition, does not exclude the treatment of Arrhenius for aqueous media. The extended view regards acids as proton donors and bases as proton acceptors regardless of whether substances concerned are ionic or neutral. A terse summary of this definition may be given as follows

$$\text{acid} \rightleftharpoons \text{base} + \text{proton} \tag{3.2}$$

As examples, the following equilibria may be cited

$$\begin{aligned} H_2SO_4 &\rightleftharpoons HSO_4^- + H^+ \\ HSO_4^- &\rightleftharpoons SO_4^{2-} + H^+ \\ H_3O^+ &\rightleftharpoons H_2O + H^+ \\ H_2O &\rightleftharpoons OH^- + H^+ \\ OH^- &\rightleftharpoons O^{2-} + H^+ \end{aligned} \tag{3.3}$$

The species on the right-hand side of equation (3.3), along with the protons, are in each case what are known as the conjugate bases of the various acids. In a similar way we may define conjugate acids of appropriate bases. It is apparent that some species, known as ampholytes, are capable of behaving as both acids and bases. In this respect the nature of water when acting as a solvent is of particular significance, as it means that all acidic and basic properties in aqueous solution will set up equilibria involving the solvent.

3.2.1 The importance of solvent in generating acid–base properties

Owing to the fact that unsolvated protons cannot exist in solution, equations (3.3) are meaningless as written. They may, however, prove useful when considering coupled acid–base systems provided that it is borne in mind that H^+ refers to the solvated proton. Thus, the reaction of sulphuric acid and ammonia may be considered as

$$H_2SO_4 \rightleftharpoons HSO_4^- + H^+$$

$$\frac{NH_3 + H^+ \rightleftharpoons NH_4^+}{H_2SO_4 + NH_3 \rightleftharpoons HSO_4^- + NH_4^+} \tag{3.4}$$

Or, in general,

$$\text{acid}_1 + \text{base}_2 \rightleftharpoons \text{base}_1 + \text{acid}_2 \tag{3.5}$$

where base_1 is the conjugate base of the original acid_1 and acid_2 is the conjugate acid of the original base_2.

3.2.2 *Relative strengths of conjugate pairs*

The position of equilibrium in systems represented by equation (3.5) is determined by the relative strengths of the two acids and bases. If acid_1 is stronger than acid_2 or, which comes to the same thing, if base_2 is stronger than base_1, the equilibrium is displaced to the right. To such reversible systems the Law of Mass Action may be applied to give for the equilibrium constant

$$K \sim \frac{[\text{base}_1][\text{acid}_2]}{[\text{acid}_1][\text{base}_2]} \tag{3.6}$$

where K is dependent only upon temperature and the nature of the solvent. Equation (3.6) may be used for dilute solutions where the effects of ion interaction are minimized.

Relationships so far used show no involvement of solvent and obscure the fact that acidic and basic properties are *consequent upon* interaction of species with solvent.

3.2.3 *Types of solvent and general acid–base theory*

Equation (3.5) may be applied quite generally to all dissociation processes, e.g. to acid dissociations such as

$$HNO_3 + H_2O \rightleftharpoons NO_3^- + H_3O^+ \tag{3.7}$$

and to self-ionization (or autoprotolytic) reactions such as

$$\begin{aligned} H_2O + H_2O &\rightleftharpoons H_3O^+ + OH^- \\ NH_3 + NH_3 &\rightleftharpoons NH_4^+ + NH_2^- \\ C_2H_5OH + C_2H_5OH &\rightleftharpoons C_2H_5O^- + C_2H_5OH_2^+ \\ H_2SO_4 + H_2SO_4 &\rightleftharpoons HSO_4^- + H_3SO_4^+ \end{aligned} \tag{3.8}$$

The last processes are important for almost all solvents.

There are four types of solvent which may be distinguished in terms of acid–base theory:

1. Acidic or protogenic solvents. These provide protons, common examples being sulphuric and ethanoic acids.
2. Basic or protophilic solvents. These have the ability to bind protons, liquid ammonia being typical.
3. Amphoteric or amphiprotic solvents such as water and ethanol which behave as either acids or bases.
4. Aprotic solvents. Such solvents show no self-ionization. Most, like benzene, do not take part in protolytic reactions although some, such as dimethylformamide and dimethylsulphoxide show basic properties.

The term lyonium ion is often given to the ion species resulting from proton solvation (e.g. H_3O^+, NH_4^+) while the solvent residue (less the proton) is called the lyate ion (e.g. $C_2H_5O^-, NH_2^-$).

The proton affinities of the solvent and its lyate ion decide the acid–base characteristics of a particular solute in that solvent. Thus, if a solute has a greater proton affinity than that of the lyate ion deriving from the solvent, it will behave as a base in that solvent. On the other hand, if the conjugate base to a solute has a smaller proton affinity than the solvent itself, the solute behaves as an acid.

For example, the conjugate base of benzoic acid, $C_6H_5COO^-$, has less affinity for protons than water so that the reaction

$$C_6H_5COOH + H_2O \rightleftharpoons C_6H_5COO^- + H_3O^+ \tag{3.9}$$

occurs in aqueous solution which is in accord with experience, benzoic acid behaving as a weak acid in aqueous solution. This, however, is not the case in concentrated sulphuric acid. Here benzoic acid has a greater proton affinity than the lyate ion from sulphuric acid, HSO_4^-, so that the reaction

$$C_6H_5COOH + H_2SO_4 \rightleftharpoons C_6H_5COOH_2^+ + HSO_4^- \tag{3.10}$$

occurs in this medium, i.e. benzoic acid is now behaving as a base. Aniline in liquid ammonia behaves as a weak acid because the solvent has a rather greater affinity for protons that does the species $C_6H_5NH^-$

$$C_6H_5NH_2 + NH_3 \rightleftharpoons C_6H_5NH^- + NH_4^+ \tag{3.11}$$

On the other hand, in glacial ethanoic acid, aniline has a much greater proton affinity than does the lyate ion CH_3OO^- and so behaves as a strong base

$$C_6H_5NH_2 + CH_3COOH \rightleftharpoons C_6H_5NH_3^+ + CH_3COO^- \tag{3.12}$$

3.3 Strengths of acids and bases in aqueous solution

Somewhat vague reference was made above to the relative strengths of acids and bases. It is necessary to give quantitative significance to such terms.

The strengths of acids for a particular solvent are measured and expressed with respect to a chosen standard for that solvent: for water, the standard is the acid–base pair H_3O^+/H_2O. The strength of some other acid, HA, may then be defined with respect to the reaction

$$HA + H_2O \rightleftharpoons A^- + H_3O^+ \tag{3.13}$$

For example

$$CH_3COOH + H_2O \rightleftharpoons CH_3COO^- + H_3O^+ \tag{3.14}$$

A^- is the conjugate base of HA, while the ethanoate ion, CH_3COO^-, is the conjugate base of ethanoic acid, CH_3COOH.

3.3.1 Dissociation constants of acids and the self-ionization constant of water

The equilibrium constant for reaction (3.13) gives a measure of the strength of HA in that it defines the extent to which the reaction proceeds to the left or right, i.e.

$$K = \frac{[A^-][H_3O^+]}{[HA][H_2O]}$$

or

$$K_a = \frac{[A^-][H_3O^+]}{[HA]} \tag{3.15}$$

where K_a is the dissociation constant of HA; $K_a = K[H_2O]$, the term $[H_2O]$ being omitted from the denominator and absorbed into the constant, K_a, since it is very large and approximately constant for dilute solution. If HA is the water molecule itself then the equilibrium constant refers to the self-ionization of water, viz.

$$H_2O + H_2O \rightleftharpoons H_3O^+ + OH^- \tag{3.16}$$

for which

$$K = \frac{[H_3O^+][OH^-]}{[H_2O]^2}$$

It is usual in this case to absorb $[H_2O]^2$ into the constant K to give

$$K_w = [H_3O^+][OH^-] \tag{3.17}$$

where K_w is known as the ionic product of water which has the value 10^{-14} at 298 K.

3.3.2 Dissociation constants of bases

The strength of a base, B, is defined in terms of the hydroxyl ions produced in water, i.e. by the reaction

$$B + H_2O \rightleftharpoons BH^+ + OH^- \tag{3.18}$$

For example

$$NH_3 + H_2O \rightleftharpoons NH_4^+ + OH^- \tag{3.19}$$

BH^+ is the conjugate acid of B, while the ammonuim ion is the conjugate base of NH_3. The dissociation constant, K_b of the base B may now be expressed as

$$K_b = \frac{[BH^+][OH^-]}{[B]} \tag{3.20}$$

Equation (3.15) and equation (3.20) will now be developed a little further in the examples provided by ethanoic acid and ammonia. For ethanoic acid equation (3.15) becomes

$$K_a = \frac{[CH_3COO^-][H_3O^+]}{[CH_3COOH]} \tag{3.21}$$

If both sides of equation (3.21) are multiplied by K_w, it may be rewritten as

$$K_a = \frac{[CH_3COO^-][H_3O^+]}{[CH_3COOH]} \cdot \frac{K_w}{[H_3O^+][OH^-]}$$

Therefore

$$K_a = \frac{[CH_3COO^-]}{[CH_3COOH][OH^-]} \cdot K_w \tag{3.22}$$

Now

$$\frac{[CH_3COO^-]}{[CH_3COOH][OH^-]} = \frac{1}{K'_b} = \frac{1}{K_h} \tag{3.23}$$

where K'_b is the dissociation constant of the *conjugate base* of CH_3COOH, that is the ethanoate ion. K'_b takes its form from the reaction.

$$CH_3COO^- + H_2O \rightleftharpoons CH_3COOH + OH^- \tag{3.24}$$

K'_b is alternatively expressed as K_h, the *hydrolysis constant.* In a similar way equation (3.20) may be considered as applied to ammonia for which it becomes

$$K_b = \frac{[NH_4^+][OH^-]}{[NH_3]} \tag{3.25}$$

and

$$K_b = \frac{[NH_4^+][OH^-]}{[NH_3]} \cdot \frac{K_w}{[H_3O^+][OH^-]}$$

Therefore

$$K_b = \frac{[NH_4^+]}{[NH_3][H_3O^+]} \cdot K_w \tag{3.26}$$

Now

$$\frac{[NH_4^+]}{[NH_3][H_3O^+]} = \frac{1}{K_a'} = \frac{1}{K_h} \tag{3.27}$$

where K_a' is the dissociation constant of the *conjugate acid* of NH_3, that is the ammonium cation, K_a' taking its form from the reaction

$$NH_4^+ + H_2O \rightleftharpoons NH_3 + H_3O^+ \tag{3.28}$$

Again, an alternative designation of K_a' is K_h.

Thus for both of the above cases

$$K_a = \frac{K_w}{K_b'} \text{ or } K_b = \frac{K_w}{K_a'} \tag{3.29}$$

i.e. the dissociation constant of an acid or base is given by K_w divided by the dissociation constant of the corresponding conjugate base or acid.

It should be clear that the foregoing only relates to weak acids and bases. Dissociation constants have no meaning for the cases of completely ionized strong acids and bases.

3.3.3 Zwitterions

Amino acids carry both amino and carboxyl groups and belong to a class of species known as 'zwitterions' carrying both positively and negatively charged centres. Glycine, the simplest with the basic formula NH_2CH_2CO OH, exists in aqueous solution as the zwitterion $^+H_3NCH_2COO^-$. Such species clearly combine the acidic and basic properties described above.

In the presence of base, glycine ($G^\pm$) will form a salt of the anion $H_2NCH_2COO^-$, (G^-), while with an acid it produces a salt of the cation $^+H_3NCH_2COOH$, (G^+). It is apparent that the acid–base properties of glycine will be characterized by two acid dissociation constants which may be defined in terms of the equilibria.

$$\underset{(G^+,\ \text{acid solution})}{^+H_3NCH_2COOH} \rightleftharpoons \underset{(G^\pm,\ \text{neutral solution})}{^+H_3NCH_2COO^- + H^+} \rightleftharpoons \underset{(G^-,\ \text{basic solution})}{H_2NCH_2COO^- + 2H^+}$$

$$K_{a1} \cong \frac{[H^+][G^\pm]}{[G^+]};\ K_{a2} \cong \frac{[H^+][G^-]}{[G^\pm]} \tag{3.30}$$

3.3.4 *The values of dissociation constants*

Experimental values of K_a, K_b vary over many orders of magnitude. Rather than deal with such an unwieldy range of values, it is more convenient to express strengths of acids or bases on a logarithmic scale. The logarithmic exponent (pK) of a dissociation constant is defined as

$$pK = -\log_{10} K \tag{3.31}$$

the negative sign indicating that a high pK value is to imply a weaker acid and stronger base while a low value implies a stronger acid and weaker base.

3.4 Extent of acidity and the pH scale

A consideration of the self-ionization of water,

$$2H_2O \rightleftharpoons H_3O^+ + OH^-$$

makes it clear that this solvent is neutral owing to equal concentrations of H_3O^+ and OH^- ions. At 298 K each of these is equal to 10^{-7} mol dm^{-3}. It must be remembered that, since K_w is temperature dependent, so also is the concentration of H_3O^+ and OH^- corresponding to the condition of neutrality. Owing to the vast range of acidity possible, concentrations of H_3O^+ and OH^- are better expressed in logarithmic form in the same way as dissociation constants. Again, negative exponents are considered, thus,

$$\text{pH} = -\log a_{H_3O^+} \sim -\log[H_3O^+] \tag{3.32}$$

and

$$\text{pOH} = -\log[OH^-]$$

Also

$$\log[H_3O^+] + \log[OH^-] = \log K_w \tag{3.33}$$

or,

$$\text{pH} + \text{pOH} = \text{p}K_w \text{ (at 298 K)} \tag{3.34}$$

so that for most practical purposes the scale may be regarded as extending from 0 to 14.

These simple relationships may be used to calculate the pH values of solutions of acids, bases and various types of salt.

3.4.1 *Calculation of pH for solutions of strong acids and bases*

It should be clear from equation (3.32) that solutions of hydrochloric acid at concentration 0.1, 0.01 and 0.001 mol dm^{-3} have pH values which are respectively 1, 2 and 3. In a complementary way, application of equation

(3.34) shows that solutions of sodium hydroxide at concentrations 0.1, 0.01 and 0.001 mol dm^{-3} have pH values of 13, 12 and 11, respectively.

Similarly, it is readily calculated that a solution of 0.05 mol dm^{-3}HCl has a pH of 1.30 while that of a 0.003 mol dm^{-3} solution of this strong monobasic acid is 2.52: a solution of 0.005 mol dm^{-3} NaOH may be calculated to have a pOH value of 2.30 and a pH of 11.70. At these concentrations, the contributions of H_3O^+ and OH^- from the solvent are negligible by comparison with their concentrations deriving from dissolved acid or base. However, at concentrations much lower than those considered above, this simple process cannot be continued. It only holds so long as the contributions to $[H_3O^+]$ and $[OH^-]$ by water itself are very small relative to those of added acids or bases. In view of the magnitude of the ionic product of water, as soon as the concentrations of H_3O^+ or OH^- deriving from added acid or base become of the same order as the water equilibrium values of 10^{-7} mol dm^3, the contributions to pH from *both* solvent and solute must be taken account of.

Values of pH calculated for some hydrochloric acid solutions of low concentration are given in Table 3.1.

Table 3.1 Calculation of pH values of solutions of low concentration of strong acid

Concn. of added acid (mol dm^{-3})	Total concn. of H_3O^+ (mol dm^{-3})	pH
10^{-5}	1.01×10^{-5}	4.996
10^{-6}	1.10×10^{-6}	5.95
10^{-7}	2.00×10^{-7}	6.70
10^{-8}	1.10×10^{-7}	6.96

Values for a similar range of solutions of sodium hydroxide are collected in Table 3.2.

Table 3.2 Calculation of pH values of solutions of low concentrations of strong base

Concn. of added base (mol dm^{-3})	Total concn. of OH^- (mol dm^{-3})	pOH	pH
10^{-5}	1.01×10^{-5}	4.996	9.004
10^{-6}	1.10×10^{-6}	5.95	8.05
10^{-7}	2.00×10^{-7}	6.70	7.30
10^{-8}	1.10×10^{-7}	6.96	7.04

3.4.2 Calculation of pH for solutions of weak acids and bases

For the case of a weak acid equation (3.15) contains the quantity $[H_3O^+]$ which is required to calculate pH. This expression may be manipulated as follows

$$K_a = \frac{[A^-][H_3O^+]}{[HA]} \sim \frac{[H_3O^+]^2}{C} \qquad (3.35)$$

since $[A^-] = [H_3O^+]$ and $[HA] \sim C$, the analytical concentration of acid if the extent of dissociation is small. Thus

$$[H_3O^+] = K_a^{\frac{1}{2}} C^{\frac{1}{2}}$$

So that in terms of $pH = -\log_{10}[H_3O^+]$

$$pH = -\tfrac{1}{2}\log K_a - \tfrac{1}{2}\log C$$

which, in terms of equation (3.31) becomes

$$pH = \tfrac{1}{2}(pK_a - \log C) \qquad (3.36)$$

Similarly, for the case of a weak base equation (3.20) contains the quantity $[OH^-]$ which provides a value of pOH. This expression may be approximated as follows

$$K_b = \frac{[BH^+][OH^-]}{[B]} \sim \frac{[OH^-]^2}{C} \qquad (3.37)$$

C now being the concentration of base.
Thus

$$[OH^-] = K_b^{\frac{1}{2}} C^{\frac{1}{2}}$$

So that in terms of $pOH = -\log_{10}[OH^-]$

$$pOH = -\tfrac{1}{2}\log K_b - \tfrac{1}{2}\log C$$
$$= \tfrac{1}{2}(pK_b - \log C)$$

Therefore the pH of the solution is given in terms of equation (3.34) by

$$pH = pK_w - \tfrac{1}{2}(pK_b - \log C) \qquad (3.38)$$

Equation (3.36) and equation (3.38) provide a rapid route to calculation of pH when the values of dissociation constants are known.

3.5 Hydrolysis. Salt solutions showing acid–base properties

Reactions such as

$$CH_3COO^- + H_2O \rightleftharpoons CH_3COOH + OH^- \qquad \text{(see(3.24))}$$

and

$$NH_4^+ + H_2O \rightleftharpoons NH_3 + H_3O^+ \qquad \text{(see (3.28))}$$

are *hydrolysis* reactions for salts of a weak acid and base, respectively.

In the first case it is also necessary to have cations and in the second anions present in solution since individual ion species cannot exist on their own. If these other ions do not show acidic or basic character they play no part in the hydrolysis equilibria. Thus a solution of sodium ethanoate shows a *basic* reaction in aqueous solution, sodium ions showing no hydrolysis while the ethanoate ions undergo reaction (3.24). Similarly a solution of ammonium chloride is *acidic* in aqueous solution: chloride ions are not hydrolysed and the acidity is produced by reaction (3.28).

A salt derived from both a weak acid and a weak base will show hydrolysis reactions for both cation and anion.

3.6 Calculation of the pH of salt solutions

The procedure here is very much like that adopted in the case of solutions of weak acids and bases: the equation representing the reaction of the hydrolysable components with water is manipulated via approximations to give an explicit relationship between hydrogen ion concentration and a dissociation constant.

3.6.1 Salts derived from weak acids and strong bases

For the case of sodium ethanoate, the hydrolysis reaction is (3.24) for which K_h(or K_b') is

$$K_h = \frac{[CH_3COOH][OH^-]}{[CH_3COO^-]} \qquad \text{(see equation (3.23))}$$

Since $[CH_3COOH] = [OH^-]$, it follows that

$$K_h = \frac{[OH^-]^2}{[CH_3COO^-]} \tag{3.39}$$

If the degree of hydrolysis is small, which in practical terms means a K_h value less that 0.01, $[CH_3COO^-]$ is approximately the concentration of sodium ethanoate, C, originally dissolved. Thus,

$$K_h \sim \frac{[OH^-]^2}{C} \tag{3.40}$$

and

$$[OH^-] \sim \sqrt{K_h C}$$

therefore,

$$[H_3O^+] = \frac{K_w}{[OH^-]} \sim \frac{K_w}{\sqrt{K_h C}}$$

and

$$[H_3O^+] = \left(\frac{K_w K_a}{C}\right)^{\frac{1}{2}}$$

Or, taking logarithms

$$pH = \tfrac{1}{2}(pK_w + pK_a + \log C) \tag{3.41}$$

In equation (3.41), pK_a refers to the *parent weak acid*, CH_3COOH.

3.6.2 *Salts derived from weak bases and strong acids*

For the case of ammonium chloride, whose hydrolysis reaction is (3.28), K_h (or K'_a) is given by

$$K_h = \frac{[NH_3][H_3O^+]}{[NH_4^+]} \qquad \text{(see equation (3.27))}$$

Since $[NH_3] = [H_3O^+]$ it follows that

$$K_h = \frac{[H_3O^+]^2}{[NH_4^+]} \tag{3.42}$$

Also, if K_h is small, $[NH_4^+] \sim C$, the concentration of ammonium chloride taken, so that

$$[H_3O^+] \sim \sqrt{K_h C} = \left(\frac{K_w C}{K_b}\right)^{\frac{1}{2}}$$

or, taking logarithms

$$pH = \tfrac{1}{2}(pK_w - pK_b - \log C) \tag{3.43}$$

where pK_b refers to the *parent weak base*, NH_3.

3.6.3 *Salts derived from weak acids and weak bases*

In the case of the salt ammonuim ethanoate both hydrolysis reactions (3.24) and (3.28) occur and it is readily shown that

$$K_h = \frac{K_w}{K_a K_b} \tag{3.44}$$

Let it be supposed that the concentration of dissolved salt is C and that $[NH_3] \sim [CH_3COOH]$, i.e. the two hydrolysis reactions have proceeded to the same extent. If K_h is small, $[NH_4^+]$ and $[CH_3COO^-]$ are both approximately C. In terms of these approximations, K_h is given by

$$K_h = \frac{[NH_3][CH_3COOH]}{[NH_4^+][CH_3COO^-]} \sim \frac{[CH_3COOH]^2}{C^2}$$

or,

$$[CH_3COOH] = CK_h^{\frac{1}{2}} = C\left(\frac{K_w}{K_a K_b}\right)^{\frac{1}{2}} \tag{3.45}$$

Now K_a for the parent acid is given by

$$K_a = \frac{[CH_3COO^-][H_3O^+]}{[CH_3COOH]}$$

so that

$$[CH_3COOH] = \frac{C[H_3O^+]}{K_a} \tag{3.46}$$

Combining equations (3.45) and (3.46) provides an explicit relationship between K_a, K_b, K_w and $[H_3O^+]$, viz.

$$[H_3O^+] = \left(\frac{K_a K_w}{K_b}\right)^{\frac{1}{2}} \tag{3.47}$$

or, in terms of pH,

$$\text{pH} = \tfrac{1}{2}(\text{p}K_w + \text{p}K_a - \text{p}K_b) \tag{3.48}$$

3.7 Buffer systems

Bufffer systems consist of weak acids or bases dissolved in company with one of their completely ionized salts. The purpose of such systems is to maintain an almost constant pH which is only slightly affected by the addition of acids or bases.

3.7.1 The Henderson–Hasselbalch equation

Consider the system ethanoic acid/sodium ethanoate in which the state of ionization of the various species may be represented as follows:

$$CH_3COOH + H_2O \rightleftharpoons CH_3COO^- + H_3O^+$$
$$CH_3COO^- + Na^+$$

The presence of the fully dissociated ethanoate salt will suppress the protolysis of the acid so that this remains virtually undissociated. This means that the concentration of free ethanoic acid, $[CH_3COOH]$, in the system is to all intents and purposes the original concentration, $[\text{acid}]_0$. Similarly, the concentration of ethanoate ion $[CH_3COO^-]$ is very close to that of the salt, $[\text{salt}]_0$.

Thus, the acid dissociation constant may be given in the form

$$K_a \sim \frac{[\text{salt}]_0}{[\text{acid}]_0} \cdot [H_3O^+]$$

or

$$[H_3O^+] \sim K_a \frac{[\text{acid}]_0}{[\text{salt}]_0} \tag{3.49}$$

or, taking logarithms,

$$\text{pH} = \text{p}K_a + \log \frac{[\text{salt}]_0}{[\text{acid}]_0} \tag{3.50}$$

Equation (3.50) is one form of the Henderson–Hasselbalch equation which states that the pH of a buffer solution is a function of the dissociation constant of the weak acid and the acid: salt ratio.

For a weak base in company with one of its completely ionized salts, the argument is similar. Consider the case of ammonia in the presence of ammonium chloride:

$$\begin{aligned} NH_3 + H_2O &\rightleftharpoons NH_4^+ + OH^- \\ & NH_4^+ + Cl^- \end{aligned}$$

Since the ammonium chloride is almost completely dissociated, $[NH_4^+] \sim [\text{salt}]_0$ and $[NH_3] \sim [\text{base}]_0$ so that

$$K_b = \frac{[\text{salt}]_0}{[\text{base}]_0} \cdot [OH^-]$$

therefore,

$$[OH^-] = \frac{[\text{base}]_0}{[\text{salt}]_0} \cdot K_b \tag{3.51}$$

or

$$\text{pH} = \text{p}K_w - \text{p}K_b - \log \frac{[\text{salt}]_0}{[\text{base}]_0} \tag{3.52}$$

This is a further form of the Henderson–Hasselbalch equation (3.52) being complementary to equation (3.50).

Addition of small amounts of strong acid or base serve only to interconvert small amounts of the weak acid or base and the salt comprising the buffer so that the concentration ratio remains almost constant. The most effective buffers have this ratio of the order of unity.

Amino acids such as glycine may form acidic and basic buffers. In terms of equation (3.30) it may be readily seen that

$$[H^+]^2 \sim K_{a1}K_{a2}\frac{[G^+]}{[G^-]}$$

or,

$$pH \cong \tfrac{1}{2}pK_{a1} + \tfrac{1}{2}pK_{a2} - \tfrac{1}{2}\log\frac{[G^+]}{[G^-]}$$

A solution of pure glycine thus shows a pH (the isoelectric point) of $\frac{1}{2}(2.34 + 9.78) = 6.06$, the pK_a values used being the accepted ones for 298 K. In the presence of strong acid, only the $G^+/G^\pm$ equilibrium is of significance so that

$$pH \cong pK_{a1} + \log\frac{[G^\pm]}{[G^+]} \tag{3.53}$$

Correspondingly, in the presence of base, only the $G^\pm/G^-$ equilibrium is of significance and

$$pH \cong pK_{a2} + \log\frac{[G^-]}{[G^\pm]} \tag{3.54}$$

3.7.2 *Efficiency of buffer systems: buffer capacity*

The efficiency of a buffer system is measured in terms of its 'buffer capacity', β, which, for an acidic buffer may be defined by

$$\beta = \frac{d[B]}{dpH} \tag{3.55}$$

Equation (3.55) gives β as the change of concentration of strong base, d[B], which is required for a given pH change. For a basic buffer, the buffer capacity is given by

$$\beta = -\frac{d[A]}{dpH} \tag{3.56}$$

where d[A] is now the change of concentration of strong acid required to produce the change dpH. Acidic buffer solutions may be prepared by mixing a weak acid, HA and a strong base, BOH, so that the following interaction occurs

$$HA + BOH \rightleftharpoons B^+ + A^- + H_3O^+$$

the pH of the resulting solution being given by

$$pH = pK_a + \log\frac{[A^-]}{[HA]}$$

or

$$pH = pK_a + \log \frac{x}{a - x} \tag{3.57}$$

where $a = [HA] + [A^-]$ and $x = [B^+]$.

β is obtained from this expression by differentiation as

$$\beta = \frac{d[B]}{dpH} = \frac{dx}{dpH} = 2.3x\left(1 - \frac{x}{a}\right) \tag{3.58}$$

Basic buffer solutions may be prepared by mixing a weak base and strong acid so that the following interaction occurs

$$BOH + HA \rightleftharpoons BH^+ + A^- + OH^-$$

the pH of the resulting solution now being given by

$$pH = pK_w - pK_b + \log \frac{[BOH]}{[BH^+]}$$

or,

$$pH = pK_w - pK_b + \log \frac{(a - x)}{x} \tag{3.59}$$

where $a = [BOH] + [BH^+]$ and $x = [BH^+]$. Here

$$\beta = -\frac{dx}{dpH} = 2.3x\left(\frac{x}{a} - 1\right) \tag{3.60}$$

Buffering is only satisfactory when the salt:acid or salt:base ratios lie between 0.1 and 10 so that the effective buffer range of any system, in terms of the dissociation constants of the acid or base involved, is given by

$$pH = pK_a \pm 1$$

and

$$pOH = pK_b \pm 1 \tag{3.61}$$

The variation of buffer capacity with variation of the salt:acid ratio is shown in Figure 3.1.

It is seen from equations (3.58) and (3.60) that for β to be a maximum, $d\beta/dx = 0$, i.e. $x_{max} = \frac{1}{2}a$, which corresponds to a 1:1 ratio of acid or base to salt. It is clear from equation (3.57) that, under the above conditions, $pH = pK_a$. The variation of β with pH is shown in Figure 3.2.

Calculations of pH from the given forms of the Henderson–Hasselbalch equation are only valid when the buffer ratios may be calculated from the quantities of components used to make the solution. Such procedure is

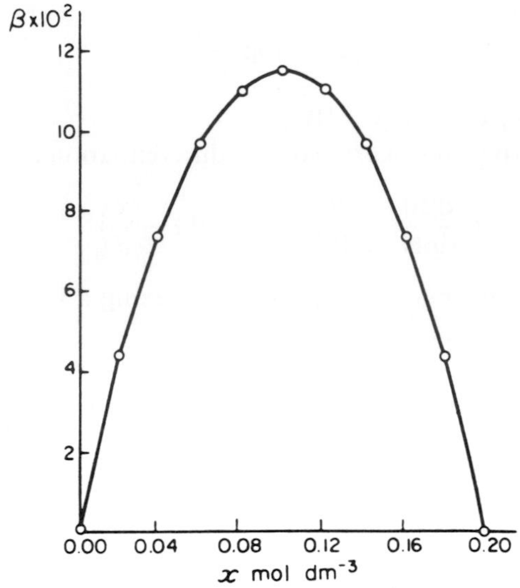

Figure 3.1 Variation of buffer capacity for an acid buffer with the proportion of salt to acid. Points calculated from equation (3.58) using $a = 0.2\,\text{mol dm}^{-3}$; $x = 0\text{–}0.2\,\text{mol dm}^{-3}$.

only justified for the range $4 < \text{pH} < 10$; outside this range, the values of concentrations involved in the buffer ratio will differ considerably from those initially dissolved. Consider firstly the case of a solution with pH < 4 containing a buffer system represented by HA/A^-. It is now no longer true to say that very little acid will ionize; it is of sufficient strength that the amount of free unionized acid is significantly less than that originally dissolved. There is correspondingly more of the anion, A^-, than can be accounted for by the total ionization of the salt, alone. In order to establish the amounts of free acid and anion that will be present in the buffer system, It is necessary to consider the equilibrium

$$HA + H_2O \rightleftharpoons H_3O^+ + A^-$$

from which it is readily seen that

$$[HA] = [HA]_0 - [H_3O^+]$$

and

$$[A^-] = [A^-]_0 + [H_3O^+]$$

The Henderson–Hasselbalch equation now takes the form

$$\text{pH} = \text{p}K_a + \log \frac{[A^-]_0 + [H_3O^+]}{[HA]_0 - [H_3O^+]} \tag{3.62}$$

Similar arguments hold for a buffer system of pH greater than 10, which we

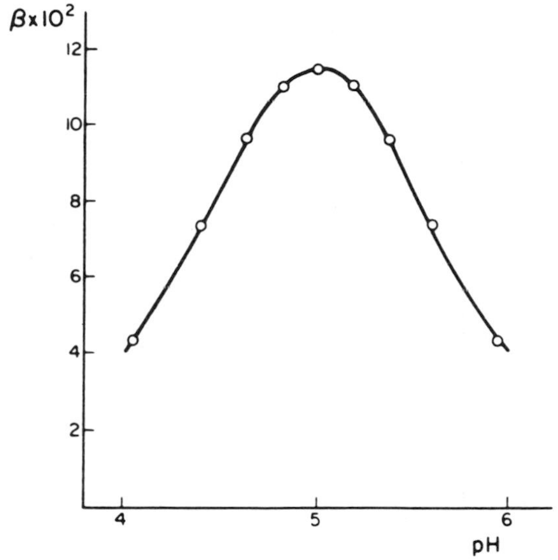

Figure 3.2 Variation of buffer capacity with pH for an acid buffer. Points calculated from equations (3.58) and (3.57) using $pK_a = 5; a = 0.2\,\text{mol dm}^{-3}$.

may represent by B/BH^+. It is now necessary to make suitable allowance for the equilibrium

$$B + H_2O \rightleftharpoons BH^+ + OH^-$$

The base now ionizes to a far from negligible extent so the $[B] < [B]_0$ and $[B] = [B]_0 - [OH^-]$. Further, $[BH^+] = [BH^+]_0 + [OH^-]$. Thus,

$$pH = pK_w - pK_b - \log \frac{[BH^+]_0 + [OH^-]}{[B]_0 - [OH^-]} \tag{3.63}$$

3.8 Operation and choice of visual indicators

In an acid–base titration there will only be observed a pH of 7 at the equivalence point if both titrant and titrand are strong electrolytes. If one is weak, the salt formed will undergo hydrolysis and the solution at the equivalence point will be either slightly acid or alkaline.

Acid-base indicators show differing colours with varying hydrogen ion concentration in a solution. The change in colour occurs in general over a 'colour change interval' of some two pH units. It is necessary to select indicators for particular titrations which show clear colours at pH values close to those known to hold at the equivalence point.

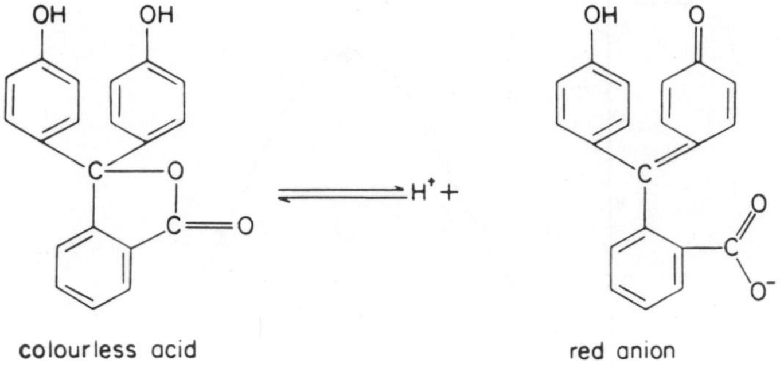

Figure 3.3 The acidic and anionic forms of phenolphthalein.

3.8.1 Functioning of indicators

Indicators are themselves weak organic acids or bases whose undissociated forms differ in colour from the ionic forms due to their considerably different electronic structure and hence absorption spectra.

The dissociation of an acid indicator molecule in water occurs according to

$$\underset{\text{colour 1}}{HIn} + H_2O \rightleftharpoons H_3O^+ + \underset{\text{colour 2}}{In^-} \tag{3.64}$$

In dilute solution the equilibrium constant, K_i, for this reaction is

$$K_i = \frac{[H_3O^+][In^-]}{[HIn]} \tag{3.65}$$

from which

$$[H_3O^+] = K_i \frac{[HIn]}{[In^-]} \tag{3.66}$$

or,

$$pH = pK_i + \log \frac{[In^-]}{[HIn]} \tag{3.67}$$

The titration exponent, pK_i, may be defined as the pH at which the concentrations of basic (In^-) and acidic (HIn) forms of the indicator are equal.

Phenolphthalein is an acid indicator of this type with acidic and basic species given in Figure 3.3.

Equation (3.66) may alternatively be expressed in terms of the fraction, x, of indicator in the alkaline form

$$[H_3O^+] = K_i \frac{1-x}{x} \tag{3.68}$$

The latter is sometimes known as the indicator equation: it is quite general in its application to both indicator acids and bases.

3.8.2 *Titrimetric practice*

It is found in practice that an observer can normally detect no change from a full acid or full alkaline colour until at least 9% of the indicator is in the alkaline- or acid-coloured form, respectively. Thus in passing from an acid to an alkaline solution, no change in colour is apparent until 9% of the indicator assumes the alkaline form, i.e. $x = 0.09$ and $1 - x = 0.91$. Insertion of these values into equation (3.68) gives

$$[H_3O^+] = \frac{0.91}{0.09} K_i \sim 10K_i$$

or

$$pH = pK_i - 1 \tag{3.69}$$

Similarly, in passing from alkaline to acid solution,

$$[H_3O^+] = \frac{0.09}{0.91} K_i \sim 0.1K_i$$

or

$$pH = pK_i + 1 \tag{3.70}$$

The effective range of pH over which colour changes can be detected is therefore given by

$$pH = pK_i \pm 1 \tag{3.71}$$

Problems

(Use concentrations rather than activities except where indicated otherwise)

3.1 Calculate the pH of the following solutions: (i) 0.001 mol dm^{-3} hydrochloric acid, (ii) 0.001 mol dm^{-3} sulphuric acid, (iii) 0.002 mol dm^{-3} sodium hydroxide, (iv) 0.0015 mol dm^{-3} calcium hydroxide, (v) 0.01 mol dm^{-3} ethanoic acid, (vi) 0.02 mol dm^{-3} ammonium hydroxide.

3.2 The pH of a 0.01 mol dm^{-3} aqueous solution of adipic acid is observed to be 3.22. Calculate the pH of a 0.15 mol dm^{-3} solution of adipic acid at the same temperature.

3.3 Calculate the volume of 0.2 mol dm^{-3} sodium hydroxide which must be added to 200 cm^3 of 0.2 mol dm^{-3} ethanoic acid at 298 K($K_a = 1.75 \times 10^{-5}$) to produce a solution with a pH of 5.50.

3.4 Calculate the pH of the following salt solutions at 298 K: (i) 0.2 mol dm^{-3} ammonium chloride, (ii) 0.1 mol dm^{-3} sodium ethanoate, (iii) 0.1 mol dm^{-3} ammonium ethanoate. For ammonia, $pK_b = 4.75$; for ethanoic acid $pK_a = 4.76$.

3.5 (i) Estimate the pH of the solution containing 0.02 mol dm^{-3} ethanoic acid and 0.03 mol dm^{-3} sodium ethanoate (pK_a of ethanoic acid = 4.76). (ii) Estimate the pH of the solution containing 0.02 mol dm^{-3} ammonium hydroxide and 0.01 mol dm^{-3} ammonium chloride (pK_b of ammonia = 4.75).

3.6 To 25 cm^3 of a solution of isopropylamine of concentration 0.15 mol dm^{-3} are added 10 cm^3 of 0.12 mol dm^{-3} hydrochloric acid. The total volume is made up to 250 cm^3 with distilled water. What is the pH of the resulting solution? (pK_b for isopropylamine = 4.03.)

3.7 Calculate the pH of the following glycine-based buffer solutions: (i) 25 cm^3 0.02 mol dm^{-3} glycine +10 cm^3 0.02 mol dm^{-3} NaOH, (ii) 25 cm^3 0.02 mol dm^{-3} glycine +18 cm^3 0.02 mol dm^{-3} NaOH, (iii) 25 cm^3 0.02 mol dm^{-3} glycine +8 cm^3 0.02 mol dm^{-3} HCl, (iv) 25cm^3 0.02 mol dm^{-3} glycine +16 cm^3 0.02 mol dm^{-3} HCl. (The pK_{a1}, pK_{a2} values of glycine are 2.34 and 9.78 respectively.)

3.8 Calculate the pH of an aqueous phosphate buffer solution at 298 K which has been prepared by mixing equal volumes of 0.1 mol dm^{-3} solutions of Na_2HPO_4 and NaH_2PO_4. The second dissociation constant of phosphoric acid may be taken as $K_{a2} = 6.34 \times 10^{-8}$. (Use the Debye–Hückel equation for calculation of ion activity coefficients.)

3.9 Consider the titration of 10 cm^3 of 0.1 mol dm^{-3} lactic acid (K_a = 1.38 × 10^{-4}) with 0.1 mol dm^{-3} sodium hydroxide. Calculate the pH of the solution at the following stages during the titration: (i) at the start (no NaOH added), (ii) at quarter neutralization (2.5 cm^3 NaOH added), (iii) at half neutralization (5.0 cm^3 NaOH added), (iv) at three-quarters neutralization (7.5 cm^3 NaOH added), (v) at neutralization (10 cm^3 NaOH added), (vi) when a total of 20 cm^3 of NaOH has been added.

3.10 The pH of an aqueous solution of sodium benzoate at 298 K is 8.32 for a concentration of 0.04 mol dm^{-3}. Given that the ionization constant of benzoic acid at 298 K is 6.31×10^{-5}, calculate a value for the ionic product of water at this temperature. Use the Debye-Hückel equation for the estimation of the mean ion activity coefficient.

3.11 A 0.05 mol dm^{-3} solution of ammonium chloride in water at 273 K has a pH of 6.08. The thermodynamic dissociation constant of ammonium hydroxide at 273 K is 1.227×10^{-4}. Calculate the value of the thermodynamic ionic product of water at 273 K.

4 The conducting properties of electrolytes

4.1 The significance of conductivity data

Experimental determinations of the conducting properties of electrolyte solutions are important essentially in two respects. Firstly, it is possible to study quantitatively the effects of interionic forces, degrees of dissociation and the extent of ion pairing. Secondly, conductance values may be used to determine quantities such as solubilities of sparingly soluble salts, ionic products of self-ionizing solvents, dissociation constants of weak acids and to form the basis for conductimetric titration methods.

The resistance of a portion of an electrolyte solution may be defined in the same way as for a metallic conductor by

$$R = \rho\left(\frac{l}{A}\right) \tag{4.1}$$

ρ being the specific resistance or resistivity and l and A the length (m) and area (m^2) respectively of the portion of solution studied. $1/R$ is known as the conductance of the material. Of greater importance here is the reciprocal of ρ, known as the conductivity, κ, i.e.

$$\kappa = \frac{1}{\rho} = \frac{l}{RA} \tag{4.2}$$

κ has the units $\Omega^{-1}\,m^{-1}$ or Sm^{-1}, S (Siemens) being the unit of reciprocal resistance.

4.1.1 Measurement of conductivity

A Wheatstone bridge arrangement may be used to measure the resistance of a portion of a solution bounded by electrodes of fixed area held at a fixed separation from each other. These electrodes are usually of platinum with a coating of finely divided platinum (platinum black) electrodeposited on their surfaces. Two complications are immediately apparent: firstly, application of a direct voltage across the electrodes is likely to cause electrolysis and the electrodes are said to become polarized: secondly, it is difficult to measure the area of the electrodes and the distance between them (maintaining these parameters at fixed values can also present some problems). To overcome the first difficulty, it is essential to use an alternating voltage

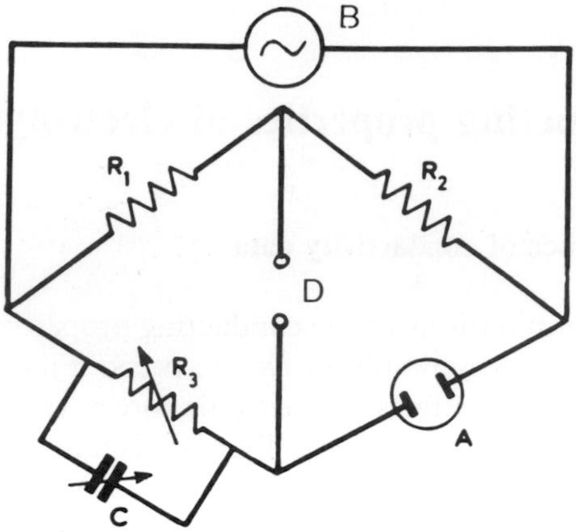

Figure 4.1 A.c. bridge circuit for conductivity determination. (A) Conductivity cell. (B) Oscillator. (C) Capacitance (variable) to balance capacitance of cell. (D) Detector.

source so that no significant accumulation of electrolysis products can occur at the electrodes, the changes occurring in one half-cycle being reversed in the next half-cycle. The catalytic properties of the platinum black ensure that the electrode reactions occur rapidly and stay in phase with the applied alternating voltage. The second problem is solved by determining a cell constant with a solution of accurately known κ. What is actually measured, of course, is the resistance R, with κ determined from

$$R = \left(\frac{1}{\kappa}\right)\left(\frac{l}{A}\right) \tag{4.3}$$

If κ for a standard solution of a reference electrolyte is known, l/A (the cell constant) may be calculated from an observed resistance using the cell in question and the standard electrolyte. Potassium chloride is the accepted standard for which accurately determined values of κ at different concentrations and temperature in aqueous solution are available. Once the cell constant is known the conductivity of any electrolyte may be determined from its measured resistance using equation (4.3). The essential circuit is shown in Figure 4.1.

The variable condenser connected in parallel with the variable resistance, R_3, serves to balance the capacity effects of the conductance cell. Adjustment of C and R_3 is made until the detector indicates zero voltage difference between the opposite junctions of the network. In this condition of bridge balance, the resistance R_C may be found from the expression $R_1/R_2 = R_3/R_C$. The position of balance may be indicated by a minimum signal on an oscilloscope.

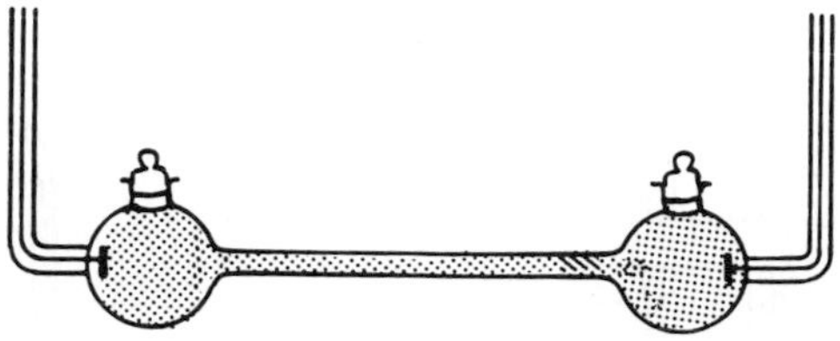

Figure 4.2 Conductivity cell with high cell constant; suitable for precision measurement of high conductivities.

For maximum sensitivity in measuring high conductivities, a high cell constant is required. Here the cell should be of the type shown in Figure 4.2 with small electrodes separated by a large distance. Conversely, for the measurement of small conductivities, l/A should be as small as possible as in Figure 4.3. This latter design is very adaptable as a 'dip-type' cell.

4.1.2 Molar conductivity

Conductivity as a practical quantity has restricted use since it is not possible to compare values for different electrolyte concentrations owing to the section of solution investigated containing different numbers of ions. Molar conductivity, given the symbol Λ, on the other hand, is defined in such a way that at any concentration the conductivity of one mole of any electrolyte may be determined.

Imagine two electrodes held at a separation of 1 m. If a solution has a concentration $C\,\text{mol m}^{-3}$ then the volume of solution containing one mole $= 1/C\,\text{m}^3$ and both electrodes would have to have an area $1/C\,\text{m}^2$, for a separation of 1 m, if one mole of electrolyte were to be held between them at concentration C. The molar conductivity is thus given by the conductivity, κ, multiplied by the volume which contains one mole of electrolyte, i.e. by

$$\Lambda(\Omega^{-1}\,\text{m}^2\,\text{mol}^{-1}) = \frac{\kappa(\Omega^{-1}\,\text{m}^{-1})}{C(\text{mol m}^{-3})} \tag{4.4a}$$

In terms of the centimetre as the unit of length, it is acceptable to express κ in the units $\Omega^{-1}\,\text{cm}^{-1}$ and C in mol cm^{-3} and to obtain Λ as

$$\Lambda(\Omega^{-1}\,\text{cm}^2\,\text{mol}^{-1}) = \frac{\kappa(\Omega^{-1}\,\text{cm}^{-1})}{C(\text{mol cm}^{-3})} \tag{4.4b}$$

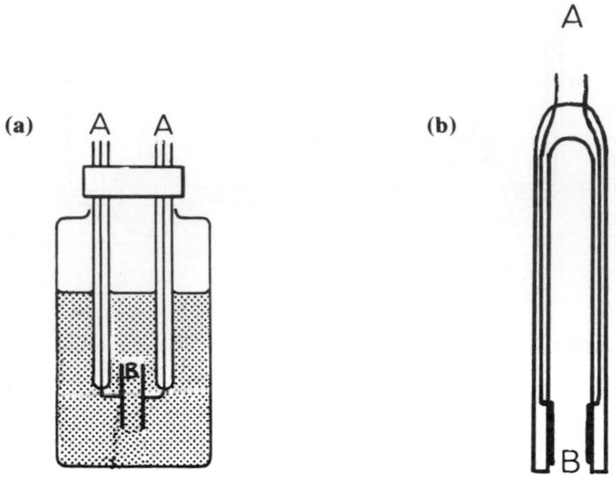

Figure 4.3 Conductivity cells with low cell constant. (a) Bottle-type (b) Dip-type: A. Contact wires sealed into insulating glass envelope. B. Exposed surfaces of platinised platinum electrodes.

Indeed, to many people the latter units produce values for Λ which are more convenient (Table 4.1) than those obtained with SI units (Table 4.2).

However, the more usual units of concentration are $\mathrm{mol\,dm^{-3}}$ which, when combined with κ in the units $\Omega^{-1}\,\mathrm{cm}^{-1}$, require modification of equation (4.4b) to

$$\Lambda(\Omega^{-1}\,\mathrm{cm}^2\,\mathrm{mol}^{-1}) = \frac{1000\kappa(\Omega^{-1}\,\mathrm{cm}^{-1})}{C(\mathrm{mol\,dm}^{-3})} \tag{4.4c}$$

To correct molar conductivities in the units $\Omega^{-1}\,\mathrm{cm}^2\,\mathrm{mol}^{-1}$ to $\Omega^{-1}\,\mathrm{m}^2\,\mathrm{mol}^{-1}$ it is necessary to divide the former by 10^4.

The usually accepted numerical values of equilibrium constants are obtained by the use of concentrations in the units $\mathrm{mol\,dm^{-3}}$ or $\mathrm{mol\,kg^{-1}}$ and the values for the dissociation constant of ethanoic acid given in Table 4.3, and estimated via equation (4.8) are those corresponding to the latter units.

Clearly, it is necessary to be familiar with the units which may be used for the various quantities and, in particular, to exercise the greatest care and consistency in combining them.

4.1.3 Empirical variation of molar conductivity of electrolyte solutions with concentration

In respect of the variation of their molar conductivities with concentration, strong and weak electrolytes show distinct characteristics. For *strong* electrolytes Kohlrausch established an empirical relationship between Λ and $\sqrt{C}$, viz.

$$\Lambda_C = \Lambda_0 - k\sqrt{C} \qquad (4.5)$$

which holds up to concentrations in the region of $5\,\mathrm{mol\,m^{-3}}$, extrapolation of a plot of Λ_C versus $\sqrt{C}$ giving Λ_0 the molar conductivity at infinite dilution.

Weak electrolytes show no such linear relationship. Their dissociation is considered to increase with increasing dilution of the solution so that the dissociation constant of ethanoic acid may be expressed in the manner already considered in Chapter 3.

$$\mathrm{CH_3COOH} \rightleftharpoons \mathrm{CH_3COO^-} + \mathrm{H^+}$$

Equilibrium concentrations	$(1-\alpha)C$	αC	αC

$$K = \frac{[\mathrm{CH_3COO^-}][\mathrm{H^+}]}{[\mathrm{CH_3COOH}]} = \frac{\alpha^2 C}{(1-\alpha)} \qquad (4.6)$$

For such electrolytes, Arrhenius showed that α, the degree of dissociation, may be given quite well by the following

$$\alpha = \frac{\Lambda_C}{\Lambda_0} \qquad (4.7)$$

Λ_C being the measured molar conductivity at a finite electrolyte concentration. Such an expression is obviously meaningless for strong electrolytes where ionization is complete at all concentrations. A combination of equations (4.6) and (4.7) gives

$$K = \frac{C\Lambda_C^2}{\Lambda_0^2\left(1 - \dfrac{\Lambda_C}{\Lambda_0}\right)} = \frac{C\Lambda_C^2}{\Lambda_0(\Lambda_0 - \Lambda_C)} \qquad (4.8)$$

In order to apply equation (4.8) to the determination of dissociation constants it is necessary to have independent means of determining Λ_0. This is made possible by the Independent Migration Principle considered in the next section.

4.1.4 *The independent migration of ions*

Kohlrausch demonstrated that each ion of an electrolyte makes its own unique contribution to the total molar conductivity of the electrolyte which is independent of the other ion(s). The phenomenon has expression as the Law of Independent Migration of Ions in the form, for a 1:1 electrolyte

$$\Lambda_0 = \lambda_+^0 + \lambda_-^0 \tag{4.9}$$

λ_0^+, λ_0^- being the ionic conductivities at infinite dilution. Equation (4.9) has been written for infinite dilution since it is only under such conditions, when ion–ion interactions are at a minimum, that the law strictly holds. It is then applicable to both strong and weak electrolytes. Its validity is demonstrated in the data of Table 4.1.

Table 4.1 Infinite dilution values of molar conductivities of some strong electrolytes at 298 K (Ω^{-1} cm^2 mol^{-1})

Salt	Λ_0	Salt	Λ_0	$\Delta\Lambda_0$
KCl	130	NaCl	108.9	21.1
KNO_3	126.3	$NaNO_3$	105.2	21.1
$\Delta\Lambda_0$	3.7		3.7	

In quoting the Λ or Λ_0 value for a given electrolyte, it is necessary to be very careful to specify the formula unit to which the value applies. The molar conductivity of an electrolyte solution reflects the amount of current that it can carry, i.e. the rate of transfer of charge through it. When comparing values for different electrolytes, it is essential to define 1 mole in all cases as the amount associated with 1 mole of unit charges (i.e. 6.023×10^{23} elementary units).

There is no problem in the case of uni-univalent electrolytes such as potassium chloride where the mole is specified as KCl but for magnesium sulphate or sodium sulphate a mole would be specified as $\frac{1}{2}MgSO_4$ and $\frac{1}{2}Na_2SO_4$, respectively (rather than $MgSO_4$ or Na_2SO_4).

Thus,

$$\Lambda_0(MgSO_4) = 2.662 \times 10^{-2}\,\Omega^{-1}\,\mathrm{m}^2\,\mathrm{mol}^{-1}$$

but

$$\Lambda_0(\tfrac{1}{2}MgSO_4) = 1.331 \times 10^{-2}\,\Omega^{-1}\,\mathrm{m}^2\,\mathrm{mol}^{-1}$$

and

$$\Lambda_0(Na_2SO_4) = 2.604 \times 10^{-2}\,\Omega^{-1}\,\mathrm{m}^2\,\mathrm{mol}^{-1}$$

but

$$\Lambda_0(\tfrac{1}{2}Na_2SO_4) = 1.302 \times 10^{-2}\,\Omega^{-1}\,\mathrm{m}^2\,\mathrm{mol}^{-1}$$

We clearly have to be equally careful in the case of ion conductivities. There is, for example, no significance in comparing the values of $\lambda^0(Na^+)$, $\lambda^0(Mg^{2+})$ and $\lambda^0(Fe^{3+})$ but there is real significance in a comparison of $\lambda^0(Na^+)$, $\lambda^0(1/2Mg^{2+})$ and $\lambda^0(1/3Fe^{3+})$ in that molar values thus defined are referred in each case to the amount of species associated with 1 mole of electrons. Some values of limiting ion conductivities are listed in Table 4.2.

Table 4.2 Limiting molar conductivities of some ion species at 298 K

Cation	$\lambda_0^+ (\Omega^{-1} m^2 mol^{-1})$	Anion	$\lambda_0^- (\Omega^{-1} m^2 mol^{-1})$
H_3O^+	3.499×10^{-2}	OH^-	1.976×10^{-2}
Na^+	0.502×10^{-2}	MnO_4^-	0.613×10^{-2}
Ag^+	0.619×10^{-2}	Cl^-	0.764×10^{-2}
$1/2\,Ca^{2+}$	0.595×10^{-2}	$1/2\,SO_4^{2-}$	0.800×10^{-2}
$1/3\,Fe^{3+}$	0.680×10^{-2}	$1/3\,Fe(CN)_6^{3-}$	0.991×10^{-2}

In the application of the Kohlrausch law, the values of molar ion conductivities used must refer to the quantity of ions contained in the specified amount of electrolyte. Thus

$$\begin{aligned}\Lambda_0(\tfrac{1}{2}Na_2SO_4) &= \lambda^0(Na^+) + \lambda^0(\tfrac{1}{2}SO_4^{2-}) \\ &= (0.502 + 0.800) \times 10^{-2}\,\Omega^{-1}\,m^2\,mol^{-1} \\ &= 1.302 \times 10^{-2}\,\Omega^{-1}\,m^2\,mol^{-1}\end{aligned}$$

It is, of course, equally true to write

$$\begin{aligned}\Lambda_0(Na_2SO_4) &= \lambda^0(2Na^+) + \lambda^0(SO_4^{2-}) \\ &= (1.004 + 1.600) \times 10^{-2}\,\Omega^{-1}\,m^2\,mol^{-1} \\ &= 2.604 \times 10^{-2}\,\Omega^{-1}\,m^2\,mol^{-1}\end{aligned}$$

A more general form of the Kohlrausch law is thus seen to be

$$\Lambda_0 = \nu_+\lambda_+^0 + \nu_-\lambda_-^0 \tag{4.10}$$

where ν_+, ν_- are the numbers of moles of cation and anion, respectively, to which 1 mole of the electrolyte gives rise in solution.

The Kohlrausch law is required to establish the value of Λ_0 for weak electrolytes. For example Λ_0 for ethanoic acid may be calculated from experimentally determined values of Λ_0 for hydrochloric acid, sodium ethanoate and sodium chloride (all strong electrolytes). From the Kohlrausch relation we may write

$$\begin{aligned}{}_{HCl}\Lambda_0 &= (\lambda_{H^+})^0 + (\lambda_{Cl^-})^0 && \text{(i)} \\ {}_{NaEtO}\Lambda_0 &= (\lambda_{Na^+})^0 + (\lambda_{EtO^-})^0 && \text{(ii)} \\ {}_{NaCl}\Lambda_0 &= (\lambda_{Na^+})^0 + (\lambda_{Cl^-})^0 && \text{(iii)}\end{aligned} \tag{4.11}$$

Addition of (i) and (ii) followed by subtraction of (iii) gives

$$({}_{HCl}\Lambda_0 + {}_{NaEtO}\Lambda_0) - {}_{NaCl}\Lambda_0 = (\lambda_{H^+})^0 + (\lambda_{EtO^-})^0 \tag{4.12}$$

Experimental values at 298 K for hydrochloric acid, sodium ethanoate and sodium chloride are 4.262×10^{-2}, 0.910×10^{-2} and $1.265 \times 10^{-2}\,\Omega^{-1}\,m^2\,mol^{-1}$, respectively. Insertion of these data into equation (4.12) gives a value of $3.907 \times 10^{-2}\,\Omega^{-1}\,m^2\,mol^{-1}$ for the limiting molar conductivity of ethanoic acid at 298 K. This value, combined with measured values at infinite concentrations, was used to calculate the dissociation constant of ethanoic acid according to equation (4.8) and the results are shown in Table 4.3.

Table 4.3 Values of dissociation constant of ethonaic acid calculated from experimental values of Λ at 298 K

C(mol dm^{-3})	$K \times 10^5$
0.001	1.851
0.005	1.851
0.010	1.846
0.050	1.771
0.100	1.551

The results may be taken to verify the validity of equation (4.8). Values of K show fairly satisfactory constancy but a pronounced trend towards a limiting value at the lowest concentrations. This confirms the approach of activity coefficients to unity in this region, extrapolation to $C = 0$ providing the limiting (thermodynamic) value of K.

4.2 Conductivity and the transport properties of ions

Transport is the general term used for the movement of matter of various sorts; here it is the movement of ions through solution which is of concern. There are *four* factors which encourage ions to move through solution namely

(i) concentration gradients;
(ii) potential gradients;
(iii) temperature gradients;
(iv) mechanical stirring.

All of these have significance in various electrochemical contexts; the convective effects represented by (iii) and (iv) above will be of greater concern in later chapters dealing with electrode processes and their exploitation. For the present it is the effects of concentration and potential gradients which are important and in the treatment which follows the convective processes are assumed to be excluded. In practical terms they are easily controlled via thermostatic control and protection of apparatus from vibration and shock.

In Chapter 2, the electrochemical potential, $\tilde{\mu}_i$ of ion species i was defined by

$$\tilde{\mu}_i = \mu_i + z_i F\phi \tag{4.13}$$

where μ_i is the chemical potential while ϕ is an electrical potential experienced by the ion. It has been shown that in an electrolyte solution, ϕ arises from both the charge on the ions and from that on the atmospheres by which they are surrounded.

In terms of the conventional thermodynamic definition of μ_i, equation (4.13) may be expressed in a form involving the concentration of the ion, thus

$$\tilde{\mu}_i \sim \mu_i^{\ominus} + RT \ln c_i + z_i F\phi \tag{4.14}$$

This is seen to be of similar form to equation (2.4) except that molar concentration $c_i(\mathrm{mol\,dm^{-3}})$ is more conveniently used than mole fraction x_i. The expression is applicable in dilute solution where concentration may be approximated for activity.

Spontaneous movement of i will take place from regions in solution of higher to those of lower electrochemical potential. The rate of such movement, v_i, is proportional to the gradient $(\partial\tilde{\mu}_i/\partial x)$ in which x is the distance of i from a reference plane so that

$$v_i = k_i \left(\frac{\partial\tilde{\mu}_i}{\partial x}\right) \tag{4.15}$$

where k_i is a proportionality constant.

Then, in terms of equation (4.14), the rate of movement becomes

$$v_i = k_i \left[RT\left(\frac{\partial \ln c_i}{\partial x}\right) + z_i F\frac{\partial\phi}{\partial x}\right] \tag{4.16}$$

It is now necessary to introduce the idea of the flux, j_i, of ion i. This is defined as the quantity which passes normally across a plane of unit area in unit time, its units being either $\mathrm{mol\,s^{-1}m^{-2}}$ or $\mathrm{mol\,s^{-1}\,cm^{-2}}$. Thus, in terms of dn_i, moles crossing the plane in time dt we can write

$$j_i = \frac{dn_i}{dt} = c_i v_i \tag{4.17}$$

so that, in terms of equation (4.16) the flux becomes

$$j_i = k_i c_i \left[RT\left(\frac{\partial \ln c_i}{\partial x}\right) + z_i F\left(\frac{\partial\phi}{\partial x}\right)\right] \tag{4.18}$$

emphasizing that it is a function of both the concentration and the potential gradients.

4.2.1 Diffusion and conductivity: the Nernst–Einstein equation

It is useful to consider the significance of the two gradient terms separately. Thus, when $(\partial\phi/\partial x) = 0$, equation (4.18) becomes

$$j_i = k_i c_i RT\left(\frac{\partial \ln c_i}{\partial x}\right)$$

$$= k_i RT\left(\frac{\partial c_i}{\partial x}\right)$$

or

$$j_i = D_i\left(\frac{\partial c_i}{\partial x}\right) \tag{4.19}$$

Equation (4.19) gives expression to Fick's first law of diffusion and introduces the diffusion coefficient of i, defined as $D_i = k_i RT$, and having the units $m^2\,s^{-1}$ or $cm^2\,s^{-1}$.

Under conditions of zero concentration gradient, the flux may be expressed by

$$j_i = k_i c_i z_i F\left(\frac{\partial\phi}{\partial x}\right)$$

i.e.

$$j_i = c_i\left(\frac{D_i}{RT}\right) z_i F\left(\frac{\partial\phi}{\partial x}\right) \tag{4.20}$$

or, in terms of the passage of charge

$$z_i F_{j_i} = i = c_i\left(\frac{D_i}{RT}\right) z_i^2 F^2\left(\frac{\partial\phi}{\partial x}\right) \tag{4.21}$$

i ($C\,s^{-1}\,m^2$ or $A\,m^{-2}$) being the current density.

In the simplest formulation conductivity, κ, is given by equation (4.2) as

$$\kappa = \frac{l}{RA}$$

$$= \frac{lI}{EA} \quad \text{(Substituting } \tfrac{1}{R} = \tfrac{I}{E} \text{ by Ohm's Law where } I = \text{current, } E = \text{potential)}$$

or

$$\kappa = \frac{I/A}{E/l} = \frac{\text{current density}}{\text{potential gradient}}$$

Therefore

$$\kappa = \frac{i}{\left(\dfrac{\partial\phi}{\partial x}\right)} = \frac{i}{\vec{F}} \tag{4.22}$$

using the symbol $\vec{F}$ for field strength or potential gradient. By inserting the expression for i (equation (4.21)) into equation (4.22) the conductivity, κ_i, of ion i becomes

$$\kappa_i = \frac{c_i D_i z_i^2 F^2}{RT} \tag{4.23}$$

or,

$$\lambda_i = \frac{\kappa_i}{c_i} = \frac{D_i z_i^2 F^2}{RT} \tag{4.24}$$

Introducing the general Kohlrausch law as expressed in equation (4.10) it is possible to write for a general electrolyte

$$\Lambda_0 = \frac{F^2}{RT}\left[\nu_+ D_+^0 z_+^2 + \nu_- D_-^0 z_-^2\right]$$

and since $\nu_+ z_+ = \nu_- z_-$

$$\Lambda_0 = \frac{\nu_+ z_+ F^2}{RT}\left[z_+ D_+^0 + z_- D_-^0\right] \tag{4.25}$$

or, in the case that $\nu_+ = \nu_- = 1$; $z_+ = z_- = z$ and equation (4.25) becomes

$$\Lambda_0 = \frac{z^2 F^2}{RT}\left[D_+^0 + D_-^0\right] \tag{4.26}$$

Equations (4.24), (4.25) and (4.26) are all forms of the Nernst–Einstein equation.

4.2.2 Ion speeds and conductivity: the Einstein and Stokes–Einstein equations

Since the conductivity of an electrolyte is a measure of the current it can carry, and therefore of the rate of charge transfer, it is also a function of the rate with which the constituent ions carry their charge through a solution. This rate depends upon the concentration and valency of the ions as well as upon their speeds.

Movement of ions through a solution may be induced by the imposition of an electric field—a consequence of an applied potential between the electrodes. The electric field force experienced by an ion causes it to accelerate. This acceleration, however, is opposed by the retarding forces of the asymmetry and electrophoretic effects as well as by the solvent viscosity (Chapter 2), so that an ion ultimately moves with a uniform velocity determined by a balance of these opposing forces. For a strong electrolyte at concentration c, giving rise to cations with charge z_+ at concentration c_+ and anions with charge z_- at concentration c_-, the amount of charge crossing unit area of solution in unit time is given by

$$c_+\nu_+z_+F + c_-\nu_-z_-F = i_+ + i_- = i \tag{4.27}$$

where i is the current density, i_+, i_- are the current density contributions from cation and anion while ν_+, ν_- are the speeds of the two ionic species.

In the case of a weak electrolyte, where each molecule produces ν_+ cations and ν_- anions, with a degree of ionization α, then

$$c_+ = \alpha\nu_+c \quad \text{and} \quad c_- = \alpha\nu_-c$$

Substitution for c_+, c_- into equation (4.27) gives

$$i = \alpha cF(\nu_+v_+z_+ + \nu_-v_-z_-) \tag{4.28}$$

Now, the speeds with which ions move are linear functions of the field strength, $\vec{F}$, so we may write,

$$v_+ = u_+\vec{F} \quad \text{and} \quad v_- = u_-\vec{F} \tag{4.29}$$

Here the proportionality constants u_+, u_- are the *mobilities* of the respective ionic species, i.e. their speeds under a unit field strength at a specified concentration.

Introducing the expressions v_+, v_- from equations (4.29) into equation (4.28) gives

$$i = \alpha cF\vec{F}(\nu_+u_+z_+ + \nu_-u_-z_-) \tag{4.30}$$

Therefore, in terms of equation (4.22) the conductivity becomes

$$\kappa_c = \frac{1}{\vec{F}} = \alpha cF(\nu_+u_+z_+ + \nu_-u_-z_-)$$

or

$$\Lambda_c = \frac{\kappa_c}{c} = \alpha F(\nu_+u_+z_+ + \nu_-u_-z_-) \tag{4.31}$$

Now, at infinite dilution $\alpha = 1$ and u_+, u_- will reach limiting values u_+^0, u_-^0; Equation (4.31) then becomes

$$\Lambda_0 = F(\nu_+u_+^0z_+ + \nu_-u_-^0z_-) \tag{4.32}$$

and since

$$\Lambda_0 = \nu_+\lambda_+^0 + \nu_-\lambda_-^0 \qquad \text{(equation (4.10))}$$

then

$$\Lambda_0 = \nu_+u_+^0z_+F + \nu_-u_-^0z_-F \tag{4.33}$$

where

$$\lambda_+^0 = u_+^0z_+F \quad \text{and} \quad \lambda_-^0 = u_-^0z_-F \tag{4.34}$$

Equations (4.32), (4.33) and (4.34) are clearly valid for both strong and weak electrolytes at infinite dilution.

It should be noted with caution that some workers have used the term 'mobility' in place of molar ion conductivity. Such practice is misleading and should be avoided; the distinction between molar ion conductivity, as a measure of the amount of current that an ion can carry, and mobility, which is a speed in a field of unit potential gradient, should be clearly understood.

Evidently equations (4.34) make possible the calculation of the speeds with which ions move under the influence of an applied field when their limiting molar conductivities are known.

Consider the case of a singly charged cation for which $\lambda_+^0 = 0.5 \times 10^{-2}$ $\Omega^{-1}\,\mathrm{m}^2\,\mathrm{mol}^{-1}$

$$u_+^0 = \frac{\lambda_+^0}{F} = \frac{0.5 \times 10^{-2}\,\Omega^{-1}\mathrm{m}^2\,\mathrm{mol}^{-1}}{96\,500\,\mathrm{C\,mol}^{-1}}$$
$$= 5.2 \times 10^{-8}\frac{\Omega^{-1}\,\mathrm{m}^2}{\mathrm{A\,s}}$$
$$= 5.2 \times 10^{-8}\frac{\mathrm{m}^2}{\mathrm{V\,s}}$$
$$= 5.2 \times 10^{-8}\frac{\mathrm{m\,s}^{-1}}{\mathrm{V\,m}^{-1}}$$

i.e. at infinite dilution the cation moves with a speed of $5.2 \times 10^{-8}\,\mathrm{m\,s}^{-1}$ in a potential gradient of $1\,V\,\mathrm{m}^{-1}$. Or, as the mobility would more usually be expressed,

$$u_+^0 = 5.2 \times 10^{-8}\,\mathrm{m}^2\,\mathrm{s}^{-1}\mathrm{V}^{-1}$$

Table 4.4 Ion mobilities at 298 K in aqueous solution

Ion	u^0 ($\mathrm{m^2 s^{-1} V^{-1}}$)
H_3O^+	36.3×10^{-8}
OH^-	20.5×10^{-8}
Li^+	4.0×10^{-8}
Na^+	5.2×10^{-8}
K^+	7.6×10^{-8}
Ag^+	6.4×10^{-8}
Mg^{2+}	5.5×10^{-8}
Zn^{2+}	5.5×10^{-8}
Cl^-	7.9×10^{-8}
Br^-	8.1×10^{-8}
NO_3^-	7.4×10^{-8}
SO_4^{2-}	8.3×10^{-8}

A list of ion mobilities is given in Table 4.4. It is seen that H_3O^+ and

OH^- ions in aqueous solution are exceptional in having extremely high mobilities. In view of what has become known about the extent of association of water molecules ('iceberg' structures) such large values cannot be accounted for on the basis of the independent migration of H_3O^+ and OH^- species. In fact, a somewhat unique transport mechanism operates whereby protons are exchanged between neighbouring solvent molecules producing movement of charge and causing continuous destruction and reformation of the species. A very similar mechanism operates in the case of $H_3SO_4^+$ and HSO_4^- in systems where concentrated sulphuric acid is used as solvent. For the other ions listed it is to be noted that the speeds are very small indeed, even in the extreme condition of infinite dilution when they can move unencumbered by other ions. It follows that electrical migration should be interpreted as a very slow drift of ions towards electrodes superimposed on their much more rapid thermal motions.

A combination of the general form of equation (4.34) with equations (4.17) and (4.19) allows a relationship between diffusion coefficient and mobility to be derived. If the migration of ions induced by an applied field exactly balanced their diffusion then

$$D_i\left(\frac{\partial c_i}{\partial x}\right) - c_i u_i \vec{F} = 0 \qquad (4.35)$$

The Boltzmann distribution law may be used to express the concentration of ion species i, $c_i(x)$ at a point where the applied potential is $\phi(x)$ in terms of the concentration c_i^0 at $\phi(x) = 0$, i.e.

$$c_i(x) = c_i^0 \exp\left[-\frac{z_i F\phi(x)}{RT}\right]$$

Therefore

$$\left(\frac{\partial c_i}{\partial x}\right) = c_i^0\left[\frac{\partial}{\partial\phi}\exp\left(-\frac{z_i F\phi(x)}{RT}\right)\right]\frac{\partial\phi}{\partial x}$$

$$= c_i^0\left(-\frac{z_i F}{RT}\right)\exp\left(-\frac{z_i F\phi(x)}{RT}\right)\frac{\partial\phi}{\partial x}$$

$$= c_i(x)\left[-\frac{z_i F}{RT}\right]\frac{\partial\phi}{\partial x}$$

$$= \left[\frac{c_i(x) z_i F}{RT}\right]\vec{F}$$

Therefore

$$D_i \left(\frac{c_i z_i F}{RT}\right) \vec{F} = c_i u_i \vec{F}$$

or

$$D_i = \frac{u_i RT}{z_i F} \tag{4.36}$$

Equation (4.36) is the Einstein equation.

The mobility of an ion species may be related to the viscosity of the solvent medium by combining a form of equation (4.15) with the Stokes equation. The latter may be written in terms of the viscous force on a single solvated ion of radius r_i under terminal velocity v_i and may be equated to the diffusion force experienced by that ion, viz.

$$6\pi r_i v_i \eta = \frac{1}{N}\left(\frac{\partial \mu_i}{\partial x}\right) \tag{4.37}$$

The right-hand side of equation (4.37) requires the factor $1/N$ for consideration of a single ion and has been expressed in terms of μ_i rather than $\tilde{\mu}_i$ if no external field is applied. Under these circumstances equation (4.15) becomes

$$v_i = k_i \left(\frac{\partial \mu_i}{\partial x}\right) = \frac{D_i}{RT}\left(\frac{\partial \mu_i}{\partial x}\right)$$

Therefore,

$$\left(\frac{\partial \mu_i}{\partial x}\right) = \frac{v_i RT}{D_i} = 6\pi r_i v_i \eta N$$

and

$$D_i = \frac{RT}{6\pi r_i \eta N} \tag{4.38}$$

Equation (4.38) is the Stokes–Einstein equation. The right-hand side may be equated to the corresponding expression in the Einstein equation so that

$$\frac{RT}{6\pi r_i \eta N} = \frac{u_i RT}{z_i F}$$

or

$$u_i = \frac{z_i F}{6\pi r_i \eta N}$$

Substitution of $\lambda_i^0 = F z_i u_i^0$ yields, after rearrangement

$$\lambda_i^0 \eta = \frac{z_i^2 F^2}{6\pi r_i N} \tag{4.39}$$

Equation (4.39) is the theoretical expression of Walden's rule which in its earlier empirical form suggested that the product of λ_i^0 and η should be approximately constant for the same solute species in different solvents. Differing degrees of solvation in different solvents cause effective ionic radii as well as viscosity to vary with solvent; in consequence, gross departures from the 'rule' may be observed.

4.3 Rationalization of relationships between molar conductivity and electrolyte concentration

Three classes of behaviour may be distinguished. In the case of strong, completely dissociated electrolytes any variation of conductivity with concentration can be traced to the varying interaction between ions as their proximity is varied by concentration changes. For weak, incompletely dissociated electrolytes changes in conductivity may be expected to occur as their degree of dissociation is forced to increase by increasing dilution. The conditions under which the ions deriving from strong electrolytes may associate into ion pairs have already been considered. Since such associations reduce the effective number of conducting species, occurrence of such phenomena may be expected to reduce the conductivity below values which would be obtained were all the ions to be unassociated.

4.3.1 Strong, completely dissociated electrolytes

Onsager attempted formulation and quantitative assessment of the relaxation and electrophoretic effects (Chapter 2). Since these are functions of the nature of ion atmospheres, it is to be expected that expressions resulting from the Debye–Hückel theory will have some significance in such formulation. There has developed some variation in the form of the original expression derived by Onsager: in part this has arisen because of changes in, and rationalization of, units. The ion conductivity λ_+, of a cation species in a very dilute solution of a strong electrolyte may be expressed as

$$\lambda_+ = \lambda_+^0 - \left[|z_+z_-| \left(\frac{\epsilon^2 N}{12\pi\varepsilon\varepsilon_0 RT} \right) \left(\frac{q}{1+\sqrt{q}} \right) \lambda_+^0 + \left(\frac{F^2 z_+}{6\pi\eta N} \right) \right] \frac{\kappa}{1+\kappa a} \qquad (4.40)$$

where

$$q = \frac{|z_+z_-|(\lambda_+^0 + \lambda_-^0)}{(|z_+| + |z_-|)(|z_+|\lambda_-^0 + |z_-|\lambda_+^0)} \qquad (4.41)$$

so that, if $\kappa a < 1$ and κ is replaced by its expression in equation (2.13)

$$\lambda_+ = \lambda_+^0 - \left(\frac{2 \times 10^3 N^2 \epsilon^2}{\varepsilon\varepsilon_0 RT}\right)^{\frac{1}{2}}$$
$$\times \left[|z_+z_-|\left(\frac{\epsilon^2 N}{12\pi\varepsilon\varepsilon_0 RT}\right)\left(\frac{q}{1+\sqrt{q}}\right)\lambda_+^0 + \left(\frac{F^2 z_+}{6\pi\eta N}\right)\right]\sqrt{I} \qquad (4.42)$$

Since a complementary expression to equation (4.40) may be written for λ_- in terms of λ_-^0 and z_-, addition of λ_+ and λ_- gives the molar conductivity according to the Kohlrausch principle, viz.

$$\Lambda_+ = \Lambda_0 - B\left[|z_+z_-|\left(\frac{\epsilon^2 N}{12\pi\varepsilon\varepsilon_0 RT}\right)\left(\frac{q}{1+\sqrt{q}}\right)\Lambda_0 + \frac{F^2(|z_+|+|z_-|)}{6\pi\eta N}\right]\sqrt{I} \qquad (4.43)$$

where B is the factor in the Debye–Hückel equation. For a 1:1 electrolyte $q = 0.5$ $|z_+| = |z_-| = 1$, so that for water as solvent at 298 K, using $\varepsilon = 78.5$ and $\eta = 8.937 \times 10^{-4}\,\text{kg m}^{-1}\text{s}^{-1}$ (0.008937 poise) equation (4.43) may be expressed in terms of the values for the various constants as

$$\Lambda = \Lambda_0 - (3.290 \times 10^9)[2.381 \times 10^{-10}(0.294)\Lambda_0 + 1.842 \times 10^{-12}]C^{\frac{1}{2}}$$

therefore

$$\Lambda = \Lambda_0 - (0.230\Lambda_0 + 6.060 \times 10^{-3})C^{\frac{1}{2}} \qquad (4.44)$$

for Λ in the units $\Omega^{-1}\,\text{m}^2\,\text{mol}^{-1}$.

Or, in general,

$$\Lambda_C = \Lambda_0 - (B_1\Lambda_0 + B_2)C^{\frac{1}{2}} \qquad (4.45)$$

which is of the same form as the empirical square-root law (equation (4.5)) established by Kohlrausch.

In equation (4.45) the constants have had the subscripts 1 and 2 attached to distinguish them from the Debye–Hückel constant B, but to emphasize their relationship with that constant.

In Figure 4.4 is shown an example of the variation of experimentally determined Λ_C values as a function of $C^{\frac{1}{2}}$ for a 1:1 electrolyte. The Onsager relation is a limiting one in that it only holds good for 1:1 electrolytes at concentrations less than 0.001 M, deviations occurring as shown at higher concentrations due to the neglect of higher terms in the limiting equation. The theoretical Onsager slope may be calculated from equation (4.44) as

$$0.23\Lambda_0 + 60.60 = 89.68\left(\frac{\Omega^{-1}\,\text{cm}^2\,\text{mol}^{-1}}{\text{mol}^{1/2}\,\text{dm}^{-3/2}}\right) \qquad (4.46)$$

which is in good agreement with the value indicated for the limiting slope in Figure 4.4. Deviations for electrolytes with higher valency products occur at

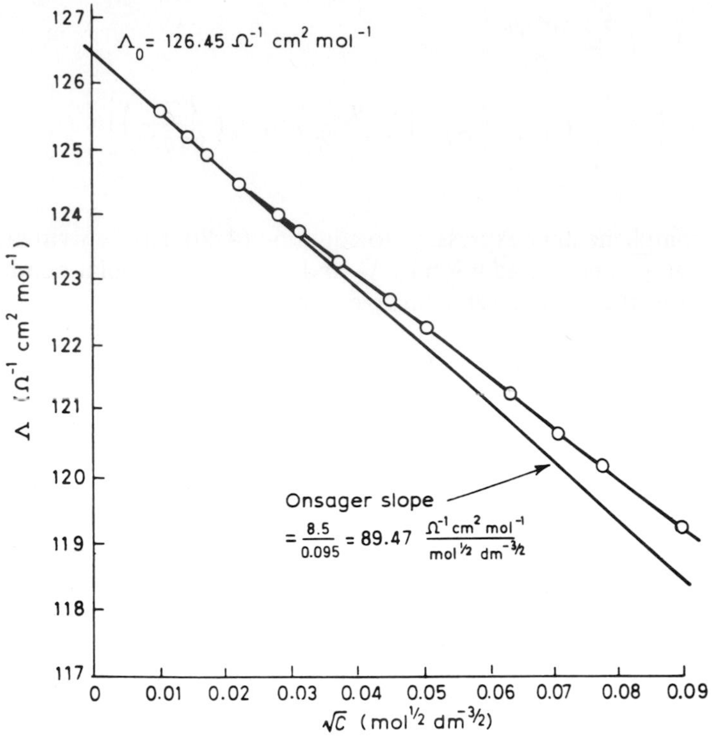

Figure 4.4 Experimental plot of molar conductivity versus square root of concentration for sodium chloride in water at 298 K.

even lower concentrations. Electrolytes with a valency product equal to or greater than 4 show marked negative deviations. These are attributed to ion association and are to be distinguished from positive deviations associated with shortcomings of the Onsager equation.

Rearrangement of equation (4.45) gives

$$\Lambda_0 = \frac{\Lambda_C + B_2 C^{1/2}}{1 - B_1 C^{1/2}} \tag{4.47}$$

Shedlovsky observed that the value of Λ_0 as calculated from equation (4.47) was not constant, but showed almost linear variation with concentration. The linear extrapolation function

$$\Lambda_0' = \frac{\left(\Lambda_C + B_2\sqrt{C}\right)}{\left(1 - B_1\sqrt{C}\right)}$$

when plotted against concentration yields a further value of Λ_0 when extrapolated to zero concentration. This value can be defined in terms of an empirical relationship

$$\Lambda_0 = \frac{\left(\Lambda_C + B_2\sqrt{C}\right)}{\left(1 - B_1\sqrt{C}\right)} - bC \tag{4.48}$$

where b is an empirical constant.

Rearrangement of equation (4.48) yields

$$\Lambda_C = \Lambda_0 - (B_1\Lambda_0 + B_2)\sqrt{C} + bC\left(1 - B_1\sqrt{C}\right) \tag{4.49}$$

This equation holds good for a number of electrolytes up to a concentration of 0.1 M.

In conclusion, we can see that in an ideal solution of a strong electrolyte Λ is independent of concentration. An approach to this condition is made in very dilute solution as seen in the portion BC of the graph of Λ versus dilution given in Figure 4.5 for sodium chloride solutions in water at 298 K. Over the region AB extreme departures from ideality occur and, with increasing concentration, ion–ion and ion–solvent interactions become more and more significant. Over this region a conductivity coefficient of may be defined by

$$\frac{\Lambda}{\Lambda_0} = g_\Lambda \tag{4.50}$$

4.3.2 Weak, incompletely dissociated electrolytes

Experimental use of equations (4.5) and (4.7) from the Arrhenius theory does make possible the determination of dissociation constants even though both equations are based upon erroneous assumptions. Ions, even in dilute solution, do not behave as ideal solutes and their conductivities are functions of concentration. The reasons that equilibrium constants determined by the Arrhenius equations are fairly good are that, firstly, interactions between ions are less numerous than for a strong electrolyte and secondly, and more important, corrections to α by the use of the Onsager equation and introduction of activity coefficients from the Debye–Hückel theory almost compensate one another. The Onsager equation in the case of a weak electrolyte becomes

$$\Lambda_C = \alpha\left[\Lambda_0 - (B_1\Lambda_0 + B_2)\sqrt{(\alpha C)}\right] \tag{4.51}$$

and equation (4.50) now takes the form

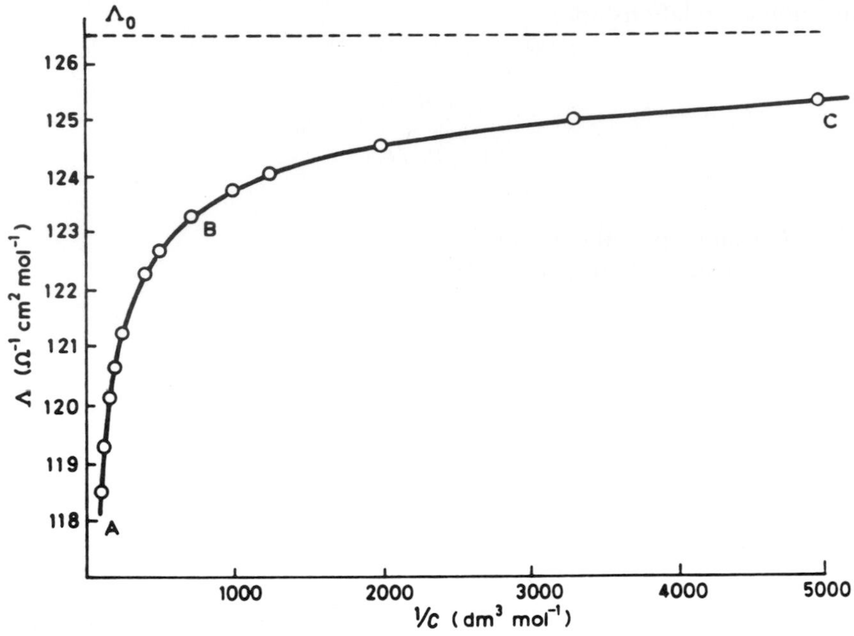

Figure 4.5 Plot of experimental molar conductivity versus dilution for a strong, completely ionized electrolyte (NaCl) in water at 298 K.

$$\frac{\Lambda}{\Lambda_0} = \alpha g_\Lambda \tag{4.52}$$

which may be regarded as the precise form of equation (4.7).

4.3.3 Electrolyte systems showing ion pairing

It has been seen that deviations from the Onsager equation in its limiting form occur for uni-univalent electrolytes at higher concentrations where observed molar conductivities are somewhat higher than predicted so that the slope of the Λ versus $C^{1/2}$ graph is somewhat lower than the theoretical Onsager slope.

Electrolytes with a valency product < 4 show fair to good agreement with the Onsager equation at low concentrations, although the upper concentration limit at which deviation begins to occur becomes progressively lower as the valency product increases. When this product is > 4 it is no longer possible to perform experiments at the extremely high dilutions where the Onsager equation might be expected to hold. Even for the lowest concentrations accessible the deviations are now extreme (Figure 4.6).

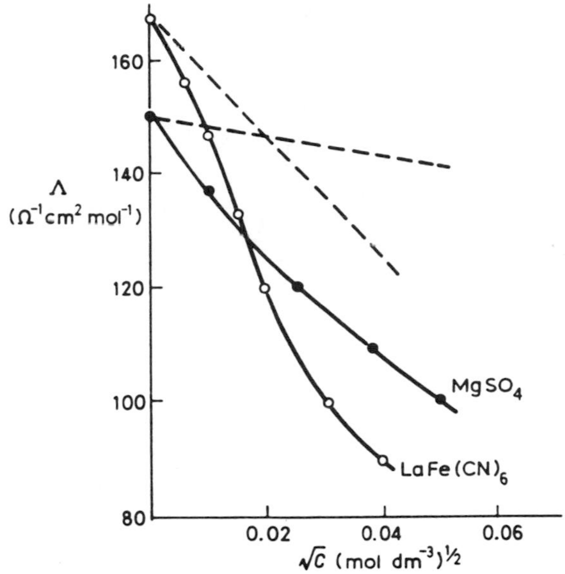

Figure 4.6 Deviations from the Onsager equation, shown by electrolytes with valency products 4 and 6, indicating ion pairing. The dashed lines are the Onsager slopes.

However, such deviations occur in the opposite sense to those obtained with low-valency product electrolytes, i.e. observed conductivities are now substantially lower than those predicted. Such deviations indicate a drastic reduction in the number of conducting species in solution, i.e. association to form ion pairs has taken place. Such deviations become more marked in solvents of low dielectric constant. In such cases the molar conductivity versus $C^{1/2}$ graph may show a minimum (Figure 4.7) and this is attributed to the formation of triple ions which, unlike ion pairs, carry a net charge.

4.4 Conductivity at high field strengths and high frequency of alternation of the field

In normal fields of the order of a few volts per centimetre, conductivities show no measurable variation with field. Wien, however, using fields of the other of 100 kV cm^{-1} observed an increase in conductivity whose magnitude is a function of both the concentration and the charge on the ions of the electrolyte. For a given concentration of a particular electrolyte, a limiting value of conductivity is reached at higher field strengths.

Under the influence of such high field strengths, the ions move very rapidly indeed (up to m s^{-1}). Since the rearrangement time of the atmosphere

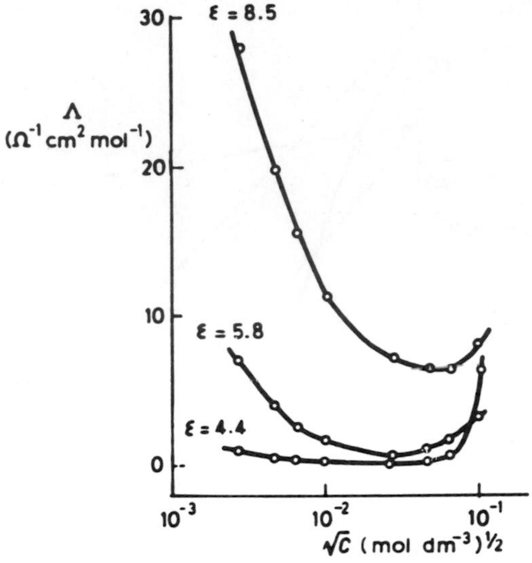

Figure 4.7 Plots of molar conductivity versus $\sqrt{C}$ for tetraisoamylammonium nitrate at various values of dielectric constant.

about an ion is slow by comparison, the retardation of the ion's motion by the electrophoretic and relaxation effects becomes progressively smaller as the field strength becomes larger. This effect is only observed for strong electrolytes.

The conductivity of weak electrolytes, e.g. weak acids, is also increased under the influence of high fields. This dissociation field effect, or second Wien effect, is caused quite differently from that described for strong electrolytes. The high field in this case changes the values of the dissociation constants of weak electrolytes. For an acid dissociation

$$HA \underset{k_{-1}}{\overset{k_1}{\rightleftharpoons}} H^+ + A^-$$

where k_1,k_{-1} are the forward and backward rate constants, the dissociation constant, K_a, is given by

$$K_a = \frac{k_1}{k_{-1}}$$

Since the H–A bond has some electrostatic character, the acid molecule has, to a limited extent, the properties of an ion pair. While the rate of formation of an ion pair is independent of an external field, the dissociation rate is increased, i.e. k_1 increases relative to k_{-1} with consequent increase in K_a. The equilibrium shifts to a new position corresponding to a higher concentration of H^+ and A^-.

An increase in conductivity of strong electrolytes is also observed with high-frequency fields operating at frequencies greater than 5 megacycles per second. At such frequencies, a central ion oscillates at a frequency comparable to the relaxation time of the atmosphere. The relaxation effect thus becomes smaller the greater the frequency and the conductivity rises. The electrophoretic effect remains unchanged.

4.5 Electrical migration and transport numbers

Section 4.2 showed that electrolyte conductivities are dependent upon the transport properties of the constituent ions in the solvent considered. The phenomenon of electrical migration, i.e. the movement of cations and anions under the influence of an applied electrical field, may be used to identify properties of individual ions.

In order to determine λ_+, λ_- from measurements of the conductance of the electrolyte, it is necessary to know the fraction of the total current passed which is carried by each ion type. Such fractions are known as transport numbers, t. By definition, the sum of the transport numbers of all ion species in an electrolyte solution is unity.

Equation (4.27) gave the anion and cation current density contributions,

$$i_+ = c_+ v_+ z_+ F; \quad i_- = c_- v_- z_- F$$

and

$$i = i_+ + i_- = c_+ v_+ z_+ F + c_- v_- z_- F$$

So that the fraction, t_+, of the current density carried by the cation is

$$t_+ = \frac{i_+}{i_+ + i_-} = \frac{c_+ v_+ z_+ F}{F(c_+ v_+ z_+ + c_- v_- z_-)} \tag{4.53}$$

For a uni-univalent electrolyte, $c_+ = c_- = c$ and $|z_+| = |z_-| = 1$, so that equation (4.53) becomes

$$t_+ = \frac{v_+}{v_+ + v_-} \tag{4.54}$$

Similarly, for the anion,

$$t_- = \frac{v_-}{v_+ + v_-} \tag{4.55}$$

The concentration dependence of transport numbers is implicit in equations (4.54) and (4.55) owing to the concentration dependence of ion speeds.

For a solution containing several electrolytes, the transport number of an individual species is defined similarly, viz.

$$t_+ = \frac{v_+}{\Sigma v}; \quad t_- = \frac{v_-}{\Sigma v} \tag{4.56}$$

From such expressions it is evident that transport number values are very much dependent on the nature and concentration of other ion species present. The greater the number of other ion species, the smaller will be the fraction of the total current carried by the ion under consideration and hence the smaller its transport number. This phenomenon is made use of in a number of electroanalytical techniques to be described later; if the transport number of a particular species can be made so small that it approaches zero, a condition has been reached where that species ceases to migrate and no current is carried by it.

Now, from equation (4.29)

$$v_+ = u_+\vec{F} \quad \text{and} \quad v_- = u_-\vec{F}$$

so that equation (4.53) may equally well be written

$$t_+ = \frac{c_+u_+z_+}{c_+u_+z_+ + c_-u_-z_-}$$

and since $c_+ = \alpha\nu_+ c$ and $c_- = \alpha\nu_- c$

$$t_+ = \frac{\alpha\nu_+cu_+z_+}{\alpha\nu_+cu_+z_+ + \alpha\nu_-cu_-z_-}$$

Therefore

$$t_+ = \frac{\nu_+u_+z_+}{\nu_+u_+z_+ + \nu_-u_-z_-}$$

Also for electroneutrality, $|\nu_+z_+| = |\nu_-z_-|$.

Therefore

$$t_+ = \frac{u_+}{u_+ + u_-} \quad \text{and} \quad t_- = \frac{u_-}{u_+ + u_-} \tag{4.57}$$

Now $\lambda_+^0 = u_+^0 z_+ F$; $\lambda_-^0 = u_-^0 z_- F$ (see equation (4.34))

So that for the special case of infinite dilution,

$$t_+^0 = \frac{u_+^0}{u_+^0 + u_-^0} \quad \text{and} \quad t_-^0 = \frac{u_-^0}{u_+^0 + u_-^0}$$

Therefore

$$t_+^0 = \frac{\lambda_+^0/z_+F}{\lambda_+^0/z_+F + \lambda_-^0/z_-F}; \quad t_-^0 = \frac{\lambda_-^0/z_-F}{\lambda_+^0/z_+F + \lambda_-^0/z_-F} \tag{4.58}$$

and remembering again that $|\nu_+z_+| = |\nu_-z_-|$ the last relationships become

$$t_+^0 = \frac{\nu_+\lambda_+^0}{\nu_+\lambda_+^0 + \nu_-\lambda_-^0}; \quad t_-^0 = \frac{\nu_-\lambda_-^0}{\nu_+\lambda_+^0 + \nu_-\lambda_-^0}$$

or, in terms of the modified Kohlrausch law given in equation (4.10)

$$t_+^0 = \frac{\nu_+\lambda_+^0}{\Lambda_0}; \; t_-^0 = \frac{\nu_-\lambda_-^0}{\Lambda_0} \tag{4.59}$$

Equations (4.59) enable λ_+^0, λ_-^0 to be determined from a knowledge of t_+^0, t_-^0 and Λ_0, the transport numbers at infinite dilution being calculated by extrapolating to zero concentration a series of values obtained by measurements made over a range of electrolyte concentration.

Equations (4.59) hold for both strong and weak electrolytes at infinite dilution. At finite concentrations the same *form* of equations (4.59) holds approximately for strong electrolytes, the quantities having values corresponding to the concentration used. In the case of a weak electrolyte the approximate relationships must include the degree of ionization corresponding to its concentration.

It is apparent from the above relationships that t_+, t_- are related by the expression

$$t_+ + t_- = 1 \tag{4.60}$$

The experimental determination of transport numbers is considered in Chapter 8.

Problems

4.1 The resistance of a 0.005 mol dm^{-3} solution of sodium chloride at 298 K was found to be 2.619×10^3 Ω. In a separate experiment (using the same conductivity cell) the resistance of 0.1 mol dm^{-3} potassium chloride was determined as 122.6 Ω. Given that the conductivity of 0.1 mol dm^{-3} potassium chloride at 298 K is 0.01289 Ω^{-1} cm^{-1}, calculate the molar conductivity of the sodium chloride solution.

4.2 Given that the molar conductivities at infinite dilution at 298 K for sodium propionate, sodium nitrate and nitric acid are 0.859×10^{-2}, 1.2156×10^{-2} and 4.2126×10^{-2} Ω^{-1} m^2 mol^{-1} respectively, calculate the molar conductivity of propionic acid at infinite dilution at 298 K.

4.3 If the mobility of the silver ion in aqueous solution at 298 K is 6.40×10^{-8} m^2 s^{-1} V^{-1}, calculate (i) the diffusion coefficient of the silver ion, (ii) its molar ion conductivity and (iii) its effective radius. The viscosity of water at 298 K is 8.94×10^{-4} kg m^{-1} s^{-1}.

5 Interfacial phenomena: double layers

5.1 The interface between conducting phases

At the interface between any pair of conducting phases a potential difference exists whose magnitude is a function of both the composition and nature of the phases. There are many types of interface for which this phenomenon is of practical importance, for example,

(i) metal/electrolyte solution;
(ii) metal/metal;
(iii) electrolyte solution/electrolyte solution;
(iv) solution of lower concentration/semipermeable membrane/solution of higher concentration.

The observed potentials are produced by the electrical double layer whose structure is responsible for many of the properties of a given system; the double layer itself arises from an excess of charges at the interface which may be ions, electrons or oriented dipoles.

To understand why this is so, it is necessary to consider the nature of a metallic conductor and how its introduction to an electrolyte solution disturbs the ion distribution considered in Chapter 2. Metals consist of an ordered arrangement of positive nuclei surrounded by mobile electrons which occupy closely spaced levels, culminating in that of the highest energy at what is known as the Fermi level. Transfer of electrons to and from this level is involved in electrode processes but these are not the concern for the moment. It is the electrostatic effect of the surface accumulation of negative charges on other charged species and dipoles in solution which is of immediate importance here.

5.2 The electrode double layer

The expression 'double layer' is something of a misnomer and arose from the original simple view of an organized arrangement of positive ions from solution, to compensate for the negative charges in the surface, to form an interfacial region similar to a parallel plate condenser (Figure 5.1). The terminology, with qualifications, has persisted. Although double layers are a general interfacial phenomena, some attention will be paid here to electrode–electrolyte interfaces because of their importance in electrode kinetics. Fur-

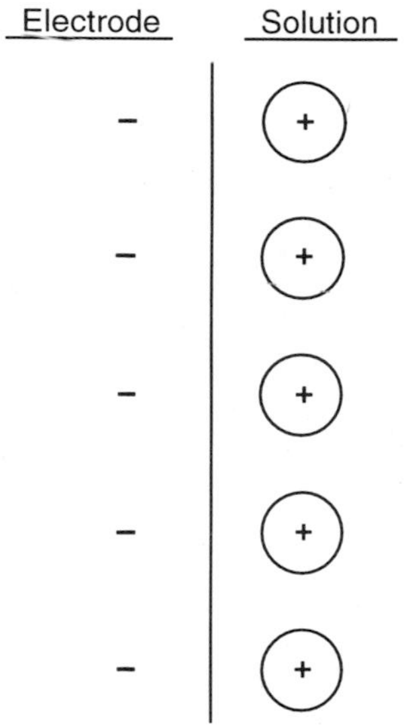

Figure 5.1 Simple condenser (Helmholtz) model of the double layer.

ther, the theory leads to a proper interpretation of electrokinetic phenomena, to an understanding of the factors affecting colloid stability and to the elucidation of cell membrane and ion-exchange processes.

For the case of an electrode dipping into a solution of an electrolyte, it is clear that, for electroneutrality, the excess charge residing on the electrode surface must be exactly balanced by an equal charge of opposite sign on the solution side. It is the *distribution* of this latter charge that is of interest. When only electrostatic interaction operates, ions from the solution phase may approach the electrode only so far. The surface array of ions is 'cushioned' from the electrode surface by a layer of solvent (in this case water) molecules (Figure 5.2). The line drawn through the centre of such ions at this distance of closest approach marks a boundary known as the 'outer Helmholtz plane' (OHP). The region within this plane constitutes the compact part of the double layer or the Helmholtz layer.

The size of the ions forming up at the outer Helmholtz plane are such that sufficient of them to neutralize the charge on the electrode cannot all fit here. The remaining charges are held with increasing disorder as the distance from the electrode surface increases, where electrostatic forces become weaker and where dispersion by thermal motion is more effective. This less-ordered arrangement of charges of sign opposite to that on the electrode constitutes the *diffuse part of the double layer*. Thus, all the charge which neutralizes that on the electrode is held in a region between the outer Helmholtz plane and the bulk of the electrolyte solution. The variation of potential, ϕ, with distance from the electrode surface is shown in Figure 5.3.

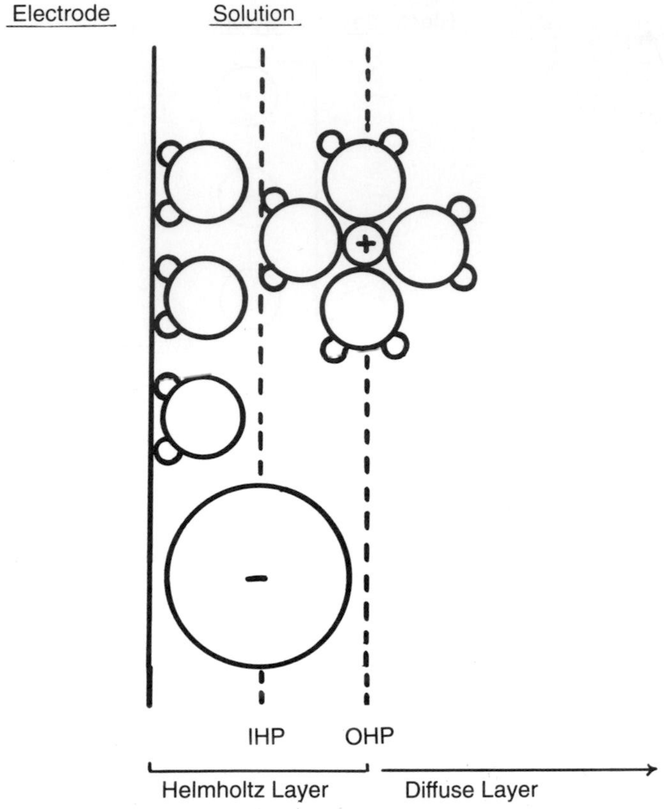

Figure 5.2 Relative positions of inner and outer Helmholtz planes of electrode double layer.

This is only the case for purely electrostatic interaction between the electrode and ions in solution. In other cases, specific adsorption of ions may occur in which van der Waals and chemical forces participate. Most anions are specifically adsorbed, thereby losing most, if not all, of their inner hydration shell. This behaviour contrasts with that of most cations, which retain their hydration molecules. Specifically adsorbed species can evidently approach much closer to the electrode surface as has been shown in Figure 5.2.

A line drawn through the centres of such species aligned at the electrode surface defines a further boundary within the Helmholtz layer—the so-called inner Helmholtz plane. The extent to which specific adsorption occurs is controlled by the nature of ions in solution as well as by the nature of the electrode material and any potential applied to it.

Uncharged species, if they are less polar than the solvent or are attracted to the electrode material by van der Waals or chemical forces, will accumulate at the interface. Such species are known as *surfactants*. Where specific

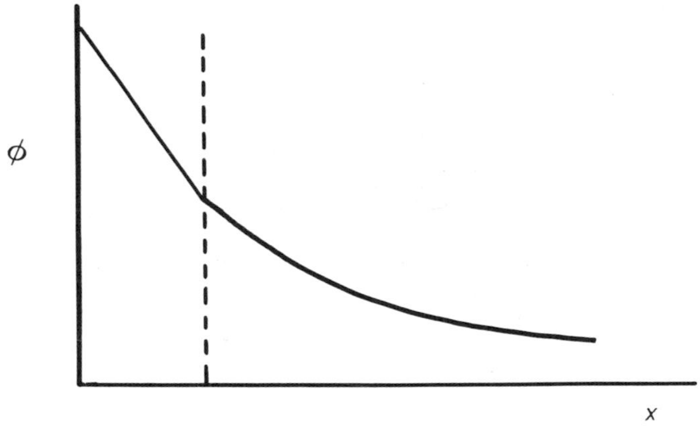

Figure 5.3 Varition of potential with distance from electrode surface.

adsorption occurs the charge distribution in the diffuse layer will change to maintain electroneutrality.

5.3 Polarized and non-polarized electrodes

These terms sometimes cause difficulty, but this is not really necessary: to polarize something is quite simply to cause a charge separation as in the classical Helmholtz model. If the metallic conductor considered above behaves as an *ideal polarized electrode* it allows no movement of charge between itself and the solution phase. In such cases it is possible to control the charge distribution by the imposition of an externally applied potential and to effectively change the character of the surface 'condenser'. In contrast, an ideal non-polarized electrode allows free and unimpeded exchange of electrons across the interface.

There is something which should be clearly understood in advance of later chapters concerned with non-polarized electrodes. Although a double layer is identifiable with the structure and behaviour of a polarized electrode, it is also present in non-polarized electrodes: this is why it is an important influence on electrode processes. However, the origin in the latter case is different and involves *charge transfer* across the interface.

5.4 Electrocapillarity: the Lippman equation

It has been noted that the potential of an ideal polarized electrode may be varied at will by altering its charge distribution through variation of applied potential, but without altering the equilibrium position at the interface.

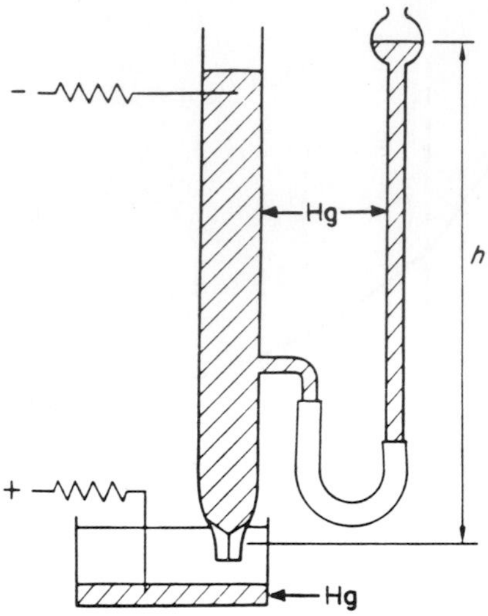

Figure 5.4 Principle of the Lippmann electrometer.

Such a system is equivalent to a perfect condenser without leakage and, like a perfect condenser, it may be charged by connection to a reference electrode and source of emf. It will then retain the imposed potential when the source is removed.

In practice, this behaviour is difficult to attain. However, the mercury electrode in potassium chloride and similar electrolyte solutions approaches very closely to the behaviour required of an ideal polarized electrode. *Within a certain range of potentials*, ions do not react at its surface nor does the metal dissolve, i.e. neither electrochemical reactions nor the establishment of electrochemical equilibria occur.

5.4.1 Variation of charge with applied potential at a mercury/solution interface

Interfacial tension measurements on liquid metal electrodes, such as mercury, have provided a great deal of information about double layer structure and have indicated the factors governing adsorption at a charged interface. The interfacial tension, γ, of a mercury/solution interface may be observed with the Lippmann electrometer, which is shown schematically in Figure 5.4. The mercury meets the solution in a fine-bore capillary, the meniscus being observed by means of a microscope. If a potential is applied to the mercury, the position of the meniscus is seen to change but may be restored to its original position by changing the reservoir height, h. The

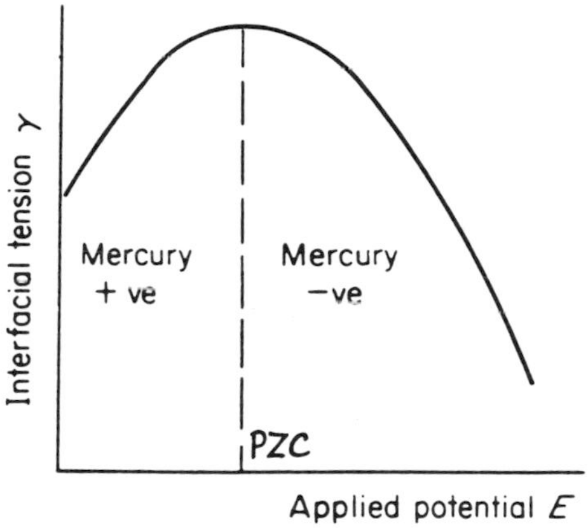

Figure 5.5 Schematic variation of interfacial tension with applied potential (electrocapillary curve) for a mercury/aqueous electrolyte solution interface.

amount by which h is required to be changed is a function of the change in interfacial tension caused by the applied potential. Mercury takes up a positive charge with respect to an aqueous solution; when a small negative potential is applied, some of this charge is neutralized and the interfacial tension rises. A point is reached where all the positive charge is neutralized and γ will then reach its maximum value. As an excess of negative charge is added with further increase in negative potential, γ will again decrease (Figure 5.5).

The dependence of interfacial tension on applied potential may be derived by application of the Gibbs adsorption isotherm to the system of phases in equilibrium in an electrochemical cell incorporating an almost ideal polarized electrode.

The Gibbs adsorption isotherm is expressed in the form

$$d\gamma + \Sigma\Gamma_i \, d\mu_i = 0 \tag{5.1}$$

where γ is the interfacial tension between two phases and Γ_i is the interfacial concentration of adsorbed species, i, this latter being neutral so that its chemical potential, μ_i, is a function of the pressure, temperature and composition of the phase.

The concern here, however, is with *charged* species. When i carries charge its chemical potential is also a function of the electrical potential, ϕ, of the phase in which it exists. Thus, for ions of charge z_i, the electrochemical potential, $\tilde{\mu}_i$ (see equation (2.5)) has been defined by

$$\tilde{\mu}_i = \mu_i + z_i F\phi \qquad (5.2)$$

So that, for ions, the Gibbs adsorption isotherm is express in terms of electrochemical potential by

$$d\gamma + \Sigma\Gamma_i d\tilde{\mu}_i = 0 \qquad (5.3)$$

Equation (5.3) clearly relates the interfacial tension to the electrical potential difference between the phases.

Application of equation (5.3) to the series of interfaces involved in the cell incorporating the mercury/electrolyte solution interface of the Lippmann electrometer allows derivation of the following expression

$$\left(\frac{\partial\gamma}{\partial E}\right)_{P,T,\mu} = -\sigma \qquad (5.4)$$

This is known as the Lippmann equation and relates the variation of interfacial tension with applied potential to σ, the number of charges per unit area at the interface.

Derivation of equation (5.4) is given in Appendix III and, since this involves a consideration of the series of interfaces in an electrochemical cell, its detailed scrutiny is probably better left until Chapter 6 has been read.

Expressing σ as the product CE, where C is the capacitance of the double layer regarded as a condenser and E the applied potential, then at constant T, P and μ,

$$d\gamma = -CE\,dE \qquad (5.5)$$

which, on integration, gives

$$\gamma = -\frac{C}{2}E^2 + \text{constant} \qquad (5.6)$$

This last expression is a form of the equation to a parabola. At the maximum of the curve

$$\left(\frac{\partial\gamma}{\partial E}\right) = 0$$

a situation corresponding to the potential of zero charge (PZC) where the Helmholtz distribution of charges is destroyed.

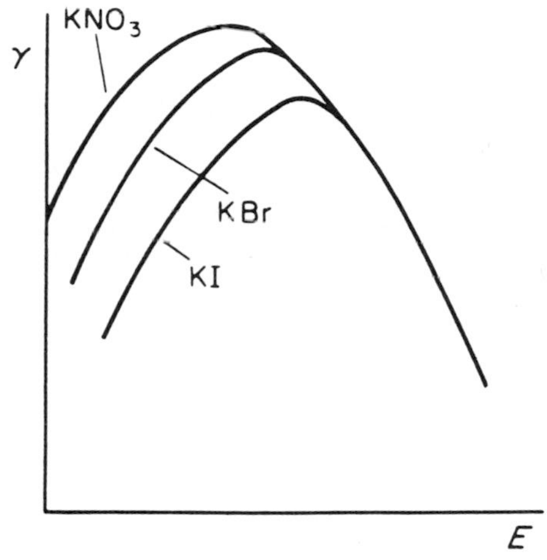

Figure 5.6 Effect on electrocapillary curves of specific adsorption of anions.

5.4.2 Specific adsorption

Electrocapillary curves obtained with the Lippmann electrometer are not usually parabolic. For a few electrolytes, such as potassium nitrate (and even then within a limited concentration range) parabolic curves are found but more usually the curves show varying degrees of distortion. Such behaviour is always found with cations and anions which are specifically adsorbed.

The capacity of the double layer formed at the interface may be found by differentiating the Lippmann equation.

$$\left(\frac{\partial^2 \gamma}{\partial E^2}\right)_{P,T,\mu} = -\frac{\partial \sigma}{\partial E} = C \tag{5.7}$$

Were C to be a constant for a given electrode, identical electrocapillary curves would be obtained whatever the electrolyte dissolved in solution. Alkali metal nitrates do show almost identical parabolas, but other salts of given alkali metals each give their own characteristic curves (Figure 5.6.)

It is seen that the variations in electrocapillary curves for such salts occur only on the part of the curves which correspond to a positive charge being carried by the mercury electrode: they are regarded as arising from specific adsorption of the various anions into the double layer. Surface active cations similarly affect the side of the curve which corresponds to mercury adopting a negative charge. Non-electrolytes which are surface active, such as gelatin, also modify the shape of electrocapillary curves (often quite drastically) and this will be seen to be of significance in the technique of polarography to be considered in Chapter 9.

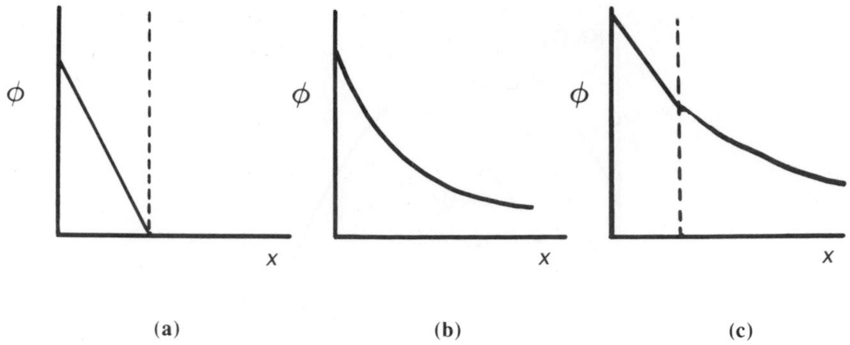

Figure 5.7 Potential variations at electrode/solution interfaces according to (a) Helmholtz, (b) Gouy and Chapman, (c) Stern.

5.5 Models for the double layer

The nature of the interfacial region considered in section 5.2 and represented schematically in Figure 5.2 has evolved via a number of theoretical models. All of these explain some experimental phenomena and do so in terms of electrostatic forces.

5.5.1 Distribution of charge according to Helmholtz, Gouy and Chapman, and Stern

Helmholtz considered the interfacial region to be limited to the condenser model described by Figure 5.1, the potential variation across the region being linear as in Figure 5.7(a). Gouy and Chapman, by contrast, appreciated that a large charged plane surface presented to the ions of an electrolyte in solution would, rather like a 'giant ion', induce a one-dimensional charge distribution similar to that proposed in three dimensions for the Debye–Hückel model of electrolyte behaviour. In this approach the potential variation is non-linear (Figure 5.7(b)) and the interfacial region extends further into the electrolyte solution; in fact, so far as to a position where the ions may be considered to behave in the way they would before the introduction of the metal.

In experimental terms these two models represent extremes of behaviour in terms of electrolyte concentration and modern interpretations are based on the compromise model proposed by Stern. This distinguishes the Helmholtz and the diffuse regions as parts of the whole interfacial region and identifies corresponding linear and non-linear variations of ϕ with distance according to Figure 5.7(c).

Modern interpretation of the interfacial region as represented by Figure 5.2 is based on the Stern model but includes the effects of specific adsorption (particularly associated with the work of Graham) and takes account of solvation of electrolyte ions and of the electrode surface.

5.5.2 *The diffuse double layer*

The similarity of the three-dimensional Debye–Hückel treatment for the distribution of ions in solution and that possible for a one-dimensional distribution normal to the electrode surface has been mentioned. But what, it may be asked, does this achieve? It provides an expression for the potential at any point within the diffuse, non-linear region and, particularly important, the potential at the OHP as a limiting condition. Derivation of this expression is given in Appendix IV and it may be written as

$$\phi = \frac{\sigma}{\varepsilon_0 \varepsilon \kappa} e^{\kappa(a-x)} \tag{5.8}$$

where σ is the charge density at the electrode surface i.e. the charge per unit area at the position where $x = 0$. Distance from the electrode surface is represented by x while a now stands for the distance of closest approach of ions to the surface (defined in terms of the position of the OHP identified in Figure 5.2).

The constant κ has similar significance to its use in Debye–Hückel theory: in the latter $1/\kappa$ was the radius of the ion atmosphere, here it may be identified with δ, the thickness of the diffuse layer.

5.5.3 *The zeta potential*

In the condition that x approaches a, i.e. the outer limit of the Helmholtz layer, $(a - x) \rightarrow 0$. Under these conditions ϕ will be designated ϕ_0; i.e.

$$\phi_0 = \frac{\sigma}{\varepsilon_0 \varepsilon \kappa} = \zeta \tag{5.9}$$

ϕ_0 may, for present purposes, be identified approximately with ζ (the 'zeta' potential), i.e. the potential at the point where the potential difference across the interface ceases to be uniform, viz. the edge of the outer Helmholtz layer where the diffuse layer begins.

The capacity, C_D, of the (electrolyte concentration dependent) diffuse layer is given by

$$C_D = \frac{\sigma}{\zeta} = \varepsilon_0 \varepsilon \kappa$$

therefore,

$$\zeta = \frac{\sigma \delta}{\varepsilon_0 \varepsilon} \tag{5.10}$$

This equation forms the basis for the explanation and description of all electrokinetic phenomena.

There are effectively two components which make up the total potential drop across the interface; viz. ϕ_0 across the diffuse part, and $(\phi - \phi_0)$ across the fixed part. The total capacitance of the double layer, C, is made up of that due to the inner (adsorption) layer, which we may designate C_H, and that due to the diffuse layer, C_D. Since these capacitances are connected in series

$$\frac{1}{C} = \frac{1}{C_H} + \frac{1}{C_D}$$

or

$$C = \frac{C_H C_D}{C_H + C_D} \tag{5.11}$$

Now, if the electrolyte solution is very dilute, $C_D \ll C_H$ and $C \sim C_D$. The double layer is now essentially all diffuse, and this was the model adopted by Gouy and Chapman in their work on the double layer. On the other hand, when the solution is very concentrated $C_D \gg C_H$ and $C \sim C_H$ which defines the earliest model of the double layer due to Helmholtz.

The relationship between ϕ, ϕ_0 and ζ is summarized in Figure 5.8.

5.6 Electrokinetic phenomena

Electrokinetic properties are associated with phases in contact with each other and are of particular significance for colloidal systems, although by no means restricted to these. Imposition of an emf across such interfaces causes movements of the phases with respect to one another while forced movement of the phases produces a characteristic emf. Thus cause and effect are readily interchangeable. Electrokinetic effects may be summarized as in Table 5.1.

Table 5.1 Electrokinetic phenomena

Motion caused by imposed emf	Emf produced by movement of phases
Electro-osmosis—liquid caused to move through a static diaphragm	*Streaming potential*—potential produced by liquid being forced through a diaphragm
Electrophoresis—solid particles caused to move through a stationary liquid	*Sedimentation potential*—potential produced by the free fall of particles through a liquid (the Dorn effect)

When motion, by whichever means, occurs it does so along a shear plane separating those ions or molecules intimately attached to the solid surface and others moving relative to them. This plane must be close to *yet outside* the OHP and it follows that while the potential at his plane of movement, the ζ potential, will often be close to ϕ_0 it will *not* identify exactly with it.

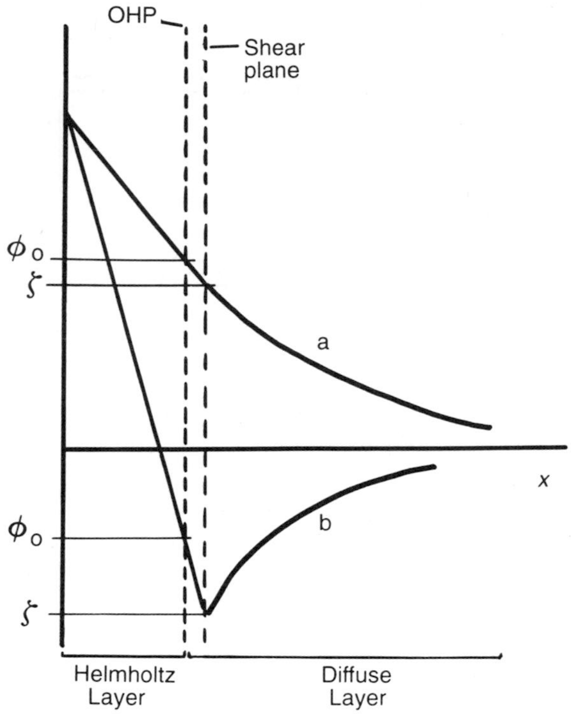

Figure 5.8 Potential variations across electrode/solution interface showing relationship between ϕ_0 and the ζ potential (a) without specific adsorption and (b) with specific adsorption.

5.6.1 Electro-osmosis

A diaphragm through which liquid is forced by imposition of an emf may be regarded as comprising a series of capillaries around the internal surface of which there exists a double layer of separated charges (Figure 5.9).

Let it be assumed that, during the movement of liquid through such a capillary, the fall of velocity is confined to the double layer by frictional forces. The velocity gradient in the layer is then v/δ, while the potential gradient down the length of the tube is $E/l = V(\mathrm{V\,m^{-1}})$.

If the surface charge per unit area is σ, then the electrical force per unit area $= V\sigma$.

The viscous force per unit area $= \eta(v/\delta)$, η being the coefficient of viscosity of the liquid. If liquid flows through the capillaries at a constant rate the electrical force balances the viscous force, i.e.

$$\eta = \left(\frac{v}{\delta}\right) = V\sigma \tag{5.12}$$

therefore,

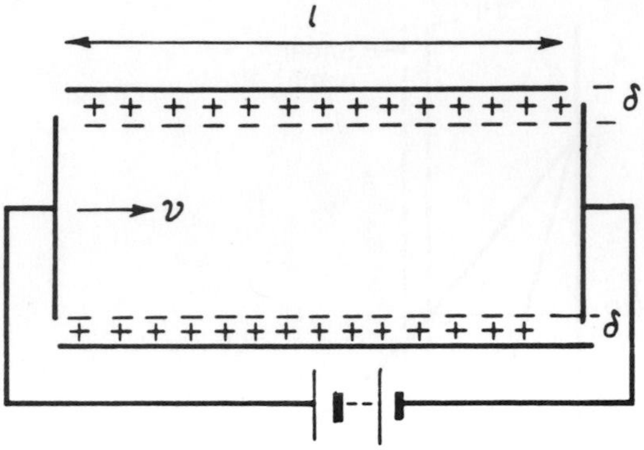

Figure 5.9 Cylindrically symmetrical double layer around the surface of a capillary.

$$\sigma = \left(\frac{\eta}{V}\right)\left(\frac{v}{\delta}\right) \tag{5.13}$$

Now

$$\zeta = \left(\frac{\delta}{\varepsilon_0\varepsilon}\right)\sigma \qquad \text{(see equation (5.10))}$$

therefore,

$$\zeta = \frac{\eta v}{\varepsilon_0\varepsilon V} \quad \text{or} \quad v = \frac{\zeta\varepsilon_0\varepsilon V}{\eta} \tag{5.14a}$$

For a potential gradient of $1\,\mathrm{V\,m^{-1}}$, v is identified with u_0 the electro-osmotic mobility. Therefore,

$$\zeta = \frac{\eta u_0}{\varepsilon_0\varepsilon} \tag{5.14b}$$

or,

$$u_0 = \frac{\zeta\varepsilon_0\varepsilon}{\eta} \tag{5.14c}$$

Equations (5.14a)–(5.14c) are all forms of the Smoluchowski equation. If the volume flow per unit time and the cross-sectional area of all capillaries are Φ and q respectively, then

$$v = \frac{\Phi}{q} \tag{5.15}$$

and

$$\zeta = \frac{\eta\Phi}{\varepsilon_0\varepsilon V q} \tag{5.16}$$

and, since for a single capillary, $q = \pi r^2$

$$\Phi = \frac{\zeta \varepsilon_0 \varepsilon V \pi r^2}{\eta} \tag{5.17}$$

or

$$\zeta = \frac{\eta \Phi}{\varepsilon_0 \varepsilon V \pi r^2} \tag{5.18}$$

Equation (5.18) may be used to determine the zeta potential from measurements of Φ, r and V. Since it is often by no means easy to measure V precisely, it is better to measure the current flowing, I, and the conductivity of the liquid, κ, and to replace V in equation (5.18) by $I/q\kappa$. Thus,

$$\zeta = \frac{\eta \Phi}{\varepsilon_0 \varepsilon V q} = \frac{\eta \Phi \kappa}{\varepsilon_0 \varepsilon I} \tag{5.19}$$

It is also possible to determine ζ from measurements on a single capillary, for under these conditions the Poiseuille equation may be used, viz.

$$\Phi = \frac{\pi P r^4}{8 \eta l}$$

where P is the driving pressure. This equation may be substituted into equation (5.18), the significance of P now being the difference in pressure at the ends of the capillary resulting from electro-osmotic flow, i.e.

$$P = \frac{8 \varepsilon_0 \varepsilon V l \zeta}{r^2} \tag{5.20}$$

5.6.2 Streaming potential

The velocity of a liquid flowing in a capillary varies with the distance from the centre of the tube as shown in Figure 5.10. The liquid at the surface of the tube is stationary so that the double layer at the interface consists of a stationary and a moving part. It is the relative movement of these two planes of the double layer which gives rise to the streaming potential. The velocity of the liquid at any point on the parabolic front distance x from the wall is given by

$$u = \frac{P(r^2 - x^2)}{4 \eta l} \tag{5.21}$$

Thus, the moving part of the double layer, at a distance $(r - \delta)$ from the centre of the tube, moves with a velocity u_δ given by

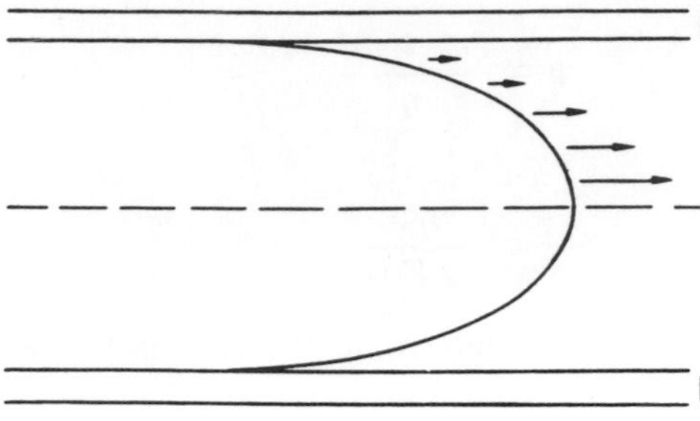

Figure 5.10 Variation of liquid velocity through a capillary with distance from the centre.

$$u_\delta = \frac{P}{4\eta l}\left[r^2 - (r-\delta)^2\right]$$
$$= \frac{P}{4\eta l}(2r\delta - \delta^2)$$

therefore

$$u_\delta \sim \frac{Pr\delta}{2\eta l} \quad \text{since } \delta^2 \ll 2r\delta \tag{5.22}$$

As the movement of the front of liquid forces one layer of charges past the other, a current is produced which must be given by the product of the total charge around a unit length of tube and the velocity of the moving part of the layer, i.e.

$$I = 2\pi r \sigma u_\delta \tag{5.23}$$

which, on substitution for u_δ from equation (5.22) becomes

$$I = \frac{\pi r^2 \sigma \delta P}{\eta l} \tag{5.24}$$

If the liquid has conductivity κ, the conductance of the liquid in the capillary is $\pi r^2 \kappa / l$ and its resistance $l/\pi r^2 \kappa$. Thus, by Ohm's Law,

$$E_s = \frac{Il}{\pi r^2 \kappa}$$

where E_s is the streaming potential.

So

$$E_s = \frac{\sigma \delta P}{\eta \kappa} \qquad \text{(by substitution for } I \text{ from equation (5.24))}$$

but

$$\zeta = \frac{\sigma\delta}{\varepsilon_0\varepsilon} \qquad \text{(see equation (5.10))}$$

from which

$$\sigma\delta = \zeta\varepsilon_0\varepsilon$$

so that

$$E_s = \frac{\zeta\varepsilon_0\varepsilon P}{\eta\kappa} \tag{5.25}$$

It is seen from a comparison of equations (5.19) and (5.25) that

$$\frac{E_s}{P} = \frac{\Phi}{I} \tag{5.26}$$

Equation (5.26) has indeed been experimentally verified.

5.6.3 Electrophoresis

Here solid particles, which may be of colloidal dimensions or even larger, are caused to move through a static solvent under the influence of an electric field. The electrophoretic velocity, v, in a field V is given by

$$v = \frac{\zeta\varepsilon_0\varepsilon V}{\eta} \qquad \text{(see equation (5.14a))}$$

where it is assumed that the thickness of the double layer is small in comparison with the size of the particles.

The velocity attained by an ion moving in the field is

$$v = \frac{z_i\epsilon V}{R} \tag{5.27}$$

where R is the viscous resistance to the motion of the ion given by Stokes' Law for a spherical particle of radius r by

$$R = 6\pi\eta r$$

Therefore

$$v = \frac{z_i\epsilon V}{6\pi\eta r} \tag{5.28}$$

Now, such particles experience the electrophoretic and relaxation effects. To discuss the electrophoretic effect, it is assumed that each particle is surrounded by a diffuse double layer of a thickness dependent upon the concentrations of ions in the solution. During migration through a solution a particle drags with it a layer of liquid of thickness d (say): d is usually less than δ. Within this layer there are a number of ions and their presence modifies the effective charge on the moving particle. If the effective charge is $\Delta z\epsilon$, equation (5.28) may be written as

$$v = \frac{\Delta z\epsilon V}{6\pi\eta(r + d)} \tag{5.29}$$

For a spherical condenser we have, from equation (5.10)

$$C = \frac{\Delta z\epsilon}{\zeta}$$

also,

$$C = (r + d)4\pi\varepsilon_0\varepsilon$$

for the spherical condenser of radius $(r + d)$; therefore,

$$\Delta z\epsilon = 4\pi\varepsilon_0\varepsilon\zeta(r + d) \tag{5.30}$$

From equations (5.29) and (5.30)

$$v = \frac{\varepsilon_0\varepsilon\zeta V}{\left(\frac{3}{2}\right)\eta} \tag{5.31}$$

It is seen that equation (5.31) is of the same form as equation (5.14a) but contains a different numerical factor.

In solutions of high ionic concentration, i.e. the condition where a very thin double layer is formed equation (5.14a) is expected to apply. On the other hand, in very dilute solutions equation (5.31) will be the more likely form. Corrections for the relaxation effect are rather difficult to assess for large particles on account of the Onsager correction only being applicable to cases where a central ion is *small* in comparison with its atmosphere.

The determination of electrophoretic velocities may be carried out experimentally by the use of methods suitable for transport number measurements. Moving boundary techniques have proved useful despite the problem of a difficulty in selecting suitable indicator ions. Reliable estimates of electrophoretic velocities make possible the determination of zeta potentials. Since colloids migrate at characteristic rates under the influence of an electric field, electrophoresis provides an important means of separation. Coatings, such as rubber or graphite, may be deposited on metal electrodes by this means and additives to these may be co-deposited.

Of particular importance is the separation and purification of proteins. Ampholytic protein particles migrate in an electric field at a rate which is characteristic, not only of their surface properties and charge and composition of the solution, but also of the pH. This is because the charge which the particles acquire by their own loss or gain of protons is a function of pH. At a given pH, different proteins are thus dissociated to different extents and have characteristic mobilities. An originally sharp boundary between a buffer solution containing various proteins and another buffer solution without proteins, splits into several boundaries corresponding to each species. A purified protein shows a characteristic variation of migration velocity with pH (the so-called 'mobility curve'). The velocity has a different sign on either side of the isoelectric point and is zero at this point. The slope of the velocity versus pH plot at the isoelectric point is a characteristic of a given protein.

5.7 Behaviour of colloidal systems

5.7.1 Stability of colloidal dispersions

The stability of colloid particles is attributed to the nature of the double layer which exists at the interface between their surfaces and the solution in which they are dispersed. Breakdown of colloidal systems, i.e. aggregation for flocculation, is similarly caused by changes in the double layer structure.

The concern here is with lyophobic colloids. Lyophilic colloid systems, on account of their affinity for the solvent, form thermodynamically stable solvated systems. Lyophobic colloids, on the other hand, are in a state of unstable equilibrium with the medium and are susceptible to irreversible breakdown when the equilibrium is subjected to even small disturbances. Lyophobic colloid particles carry similar charges, as may be confirmed by the direction of their migration in an electric field, and these usually originate from the preferential adsorption of ions from the solution. For example, negatively charged hydrosulphide ions are adsorbed at the surface of colloidal particles of arsenic(III) sulphide. Consequently, the diffuse double layer, surrounding each particle, must contain an equivalent number of positively charged (hydrogen) ions—the 'counterions' or 'gegenions'.

The repulsive forces between such particles are usually large due to the fairly large number of unit charges which each carries. Not only are the forces of repulsion large, they are of long range in comparison with the short-range attractive (dispersion) forces. The overriding repulsive forces prevent the particles aggregating, and their magnitude controls the stability of the colloid system. The magnitude of these forces may be changed by changing the number of charges carried by the particles. Adsorption of oppositely charged ions, leading to partial or complete neutralization of those on the particles, reduces the repulsive forces, allows more free play of the attractive van der Waals forces and, if occurring to a sufficiently large extent, results in aggregation. The observed effects of adding electrolytes to lyophobic colloid systems are qualitatively in agreement with this interpretation, so that, for instance, the higher the charges carried by the ions of an added electrolyte, the more efficient it is found to be and the lower the concentration required to induce aggregation.

5.7.2 Colloidal electrolytes

Some electrolytes containing large ions, particularly soaps, dyes and many synthetic detergents, behave as normal electrolytes only in very dilute solution. At higher concentrations they show unusually low osmotic pressures and their conductivities show large deviations from the Onsager relationship. Such behaviour may be attributed to the formation of micelles by aggregation of similarly charged ions. This process of micelle formation occurs at a critical concentration for each system and is encouraged by large ion size. For example, cetylpyridinium salts show the Onsager dependence of molar conductivity on the square root of electrolyte concentration at very low concentrations (10^{-3} M). Beyond a critical concentration, however, the conductivity declines very rapidly and ultimately assumes a minimum almost constant value. Micelle formation, caused by the aggregation of about 68 cetylpyridinium ions, causes the observed drop in conductivity. In these micelles the cations are arranged with the cationic groups facing the solvent and with the hydrocarbon chains pointing inwards. The gegenions contained in the double layer surrounding the particles then reduce the latter's effective charge and mobility and give rise to the sharp drop in conductivity. With a knowledge of the transport numbers of cation and anion over the concentration range, the ion conductivities may be calculated. The conductivity of the cetylpyridinium ion increases sharply beyond the critical concentration due to the increased mobility of the hydrocarbon chain constituent. There is lower frictional resistance offered to the movement of the micelle than to that of the total number of original individual particles, and this more than outweighs the effects of the more dense ion atmosphere due to the increased concentration of charges in the micelle.

After the critical concentration the conductivity of the anion is observed to drop very sharply to zero and to pass into negative values. This is the same type of transport number behaviour as that shown by cadmium iodide (Chapter 8) and indicates that the anions are preferentially transported to the cathode rather than to the anode. It is evident that ion association between anions and positively charged micelles is the cause. Hartley calculated that in cetylpyridinium bromide about 53 bromide ions are associated with each micelle of 68 cationic species to give a net charge of +15.

5.7.3 Polyelectrolytes

Micelle particles are usually spherical in shape due to the fact that the constituent ions tend to orientate themselves with lyophobic fragments pointing inwards away from the solvent. Consequently the distribution of charge will tend to be spherically symmetrical. Polyelectrolytes, by contrast, are long-chain polymeric species which carry ionizable groups along the chain. Depending on their proximity to one another, charged groups along such a chain interact and the extent of interaction may be affected by changing the conditions of the system. Thus, in a dilute solution of sodium polymethacrylate, the repulsion between neighbouring carboxylate groups causes almost complete extension of the chain. In the parent acid, however, which is weak with ionization of the acid groups occurring to only a small extent, the chain is coiled. The coils open when alkali is added to neutralize the acid groups and increase the number of carboxylate repulsions. The addition of other salts to solutions of the full extended poly salt causes the latter to recoil due to the increased ionic strength which reduces the intergroup repulsions.

Measurements of conductance and transport numbers similar to those used for micelles confirm the importance of association of counterions for polyelectrolytes. An interesting feature is that in an electric field such polyions exhibit abnormally high induced dipoles. It is apparent that the associated counterions have considerable mobility along the length of the chain so that the field causes polarization and orientation of the chain along the field direction.

Problems

5.1 Calculate the approximate thickness (δ) of the diffuse double layer which is established at a negatively charged plane solid surface in contact with a 0.001 mol dm^{-3} aqueous solution of sodium sulphate at 298 K. Relative permittivity of water = 78.54.

5.2 Particles of the colloidal dispersion of a noble metal have have an effective mean radius of 0.25 μm in a 0.04 mol dm^{-3} aqueous solution of a 1:1 electrolyte and have an electrophoretic mobility of 3.50×10^{-8} m^2 V^{-1} at 298 K. Calculate an approximate value for the zeta (ζ) potential. Viscosity of water at 298 K= 8.94×10^{-4} kg m^{-1} s^{-1}, relative permittivity of water at 298 K = 78.5. Comment upon the magnitude of the quantity κa in relation to the range of values calculated for the examples to Chapter 2.

5.3 A glass tube 15 cm long and mean internal diameter 1.2 mm is filled with water from a static source while a potential difference of 250 V is applied between its ends. A temperature of 298 K is maintained throughout the time of imposition of the potential difference. Calculate the rate of electro-osmotic flow of the water at 298 K given that the zeta (ζ) potential for a glass/water interface is -40 mV, the viscosity of water is 8.9×10^{-4} kg m^{-1} s^{-1}, and its relative permittivity is 78.5 at this temperature.

6 Electrode potentials and electrochemical cells

6.1 Comparison of chemical and electrochemical reactions

In this chapter the transfer of electrons through the electrode/solution interface is considered. Since such transfer requires electrons to exchange between species in solution and the Fermi level in the electrode, those species become chemically changed in the process. In this sense the phenomena considered here are fundamentally different to those shown under conditions of polarization considered in Chapter 5. A rather different model is now required for the electrode/solution interface.

Electrode reactions are oxidation–reduction processes of a somewhat unique type which obey the scheme:

$$\text{oxidant} + n\text{e} \rightleftharpoons \text{reductant} \tag{6.1}$$

ne representing a transfer of n unit charges, i.e. electrons.

The difference between chemical and electrochemical reactions lies in the different sources of electrons. A chemical oxidation–reduction system is made up of two individual systems,

$$\text{Ox}_1 + n\text{e} \rightleftharpoons \text{Red}; \text{ e.g. } Fe^{3+} + e \rightleftharpoons Fe^{2+}$$
$$\text{Red}_2 - n\text{e} \rightleftharpoons \text{Ox}_2; \text{ e.g. } Ce^{3+} - e \rightleftharpoons Ce^{4+}$$

Overall

$$\text{Ox}_1 + \text{Red}_2 \rightleftharpoons \text{Red}_1 + \text{Ox}_2$$

e.g.

$$Fe^{3+} + Ce^{3+} \rightleftharpoons Fe^{2+} + Ce^{4+} \tag{6.2}$$

It is not normally possible to isolate the two contributing processes, since it is only possible to observe changes in one system by coupling it with the second.

Electrochemically, the individual processes may be separated, however. It is, for example, often possible for a metallic conductor, dipping into a solution of an oxidizing or reducing agent, to exchange electrons with such species and effectively bring about their reduction or oxidation. Such reactions are in some ways simpler in that if the electrode is only providing or taking up electrons, it may otherwise be regarded as inert.

It is possible to control precisely the rate with which electrons are provided (or taken up) by an electrode by variation of a potential applied to it via an externally connected emf source (the circuit being completed by the inclusion of a suitable reference electrode).

There are two ways in which such electron exchange reactions may occur. They may be forced to occur, which is the situation encountered in electrolysis or they may occur spontaneously as happens in batteries and galvanic cells.

When two electrochemical redox systems are coupled together, one electrode providing and the other taking up electrons, the net effect is similar to that of the chemical scheme. Such is the situation observed with electrochemical cells for which there are associated overall chemical reactions. Owing to the precision with which electrochemical measurements may be made for such systems, it is often possible to use them to obtain precise thermodynamic data characteristic of the reactions occurring within cells.

In considering electron exchange reactions *at* electrodes it must not be forgotten that an oxidant or reductant in solution has to have some means of *reaching* the electrode so that the electrochemistry can take place. There are a number of ways in which mass transfer can occur and the interplay between the relative rates of mass transfer and of electron exchange processes will be important in later chapters. For the present it is possible to arrive at preliminary conclusions by assuming virtually infinite rates for *all* processes.

6.2 Electrode potentials: their origin and significance

A metal dipping into a solution of its ions has an equilibrium such as

$$M^{n+} + ne \rightleftharpoons M \qquad (6.3)$$

eventually established at its surface. For many such systems equilibrium is established rapidly; in other cases the approach to equilibrium is slow, at least at room temperature.

Such an electrode will adopt a potential difference with respect to the solution whose value is a function of the position of equilibrium (6.3). If this is established rapidly, the potential difference (the electrode potential) may be easily measured by means of a potentiometer device which compares it with another (reference) electrode. If the process is a slow one, a continuously variable potential will be observed and no steady value may be determined experimentally.

Other electrodes involve gases in equilibrium with ions in solution, e.g. hydrogen and chlorine electrodes function through operation of the following equilibria

$$\tfrac{1}{2}H_2 \rightleftharpoons H^+ + e \qquad (6.4)$$

and

$$\tfrac{1}{2}\,Cl_2 + e \rightleftharpoons Cl^- \qquad (6.5)$$

These require the gas to be bubbled over the surface of some inert electrode material dipping into a solution of the ions of the gas. Surface adsorbed gas molecules enter into equilibrium with ions in solution and cause the electrode to adopt a potential characteristic of the position of equilibrium.

For the hydrogen electrode it is seen that the oxidized form is in solution; for the chlorine electrode the oxidized form is adsorbed at the surface. In redox electrodes both oxidized and reduced forms are present in solution, electrons being exchanged at an inert conductor immersed in the solution. A platinum wire placed in the Fe(III)–Fe(II) system constitutes such an electrode operating on the equilibrium

$$Fe^{3+} + e \rightleftharpoons Fe^{2+} \qquad (6.6)$$

Each electrode system described constitutes what is known as a 'half-cell'; it is necessary to couple two such half-cells to form a complete electrochemical cell. When all the equilibrium components of a half-cell are in their standard states of unit activity, the electrode is said to be a standard electrode and to adopt its standard potential.

6.2.1 Types of potential operating at the electrode/solution interface

It is necessary to clarify what is measured by the potentiometric means referred to above and to understand which of a number of potentials applicable to interfaces are capable of experimental measurement.

On the face of it the problem is a simple one in which it is required to determine the difference in potential of an electrode material and that of the solution with which it is in contact. In terms of the model represented in Figure 6.1, it is required to measure $\Delta\phi$ given by

$$\Delta\phi = \phi_M - \phi_S \qquad (6.7)$$

where ϕ_M, ϕ_S are the *inner*, or Galvani, potentials of the two phases.

Up until now the symbol ϕ has been used somewhat indiscriminately to represent a potential: it is now necessary to be more circumspect and to introduce new symbols to represent a number of types of potential which need to be distinguished.

This model is rather different to that considered in Chapter 5. Implicit in equilibrium (6.3) is the *polarizability* of the electrode, i.e. transfer of charge can occur. It must be remembered, however, that charges transferred between the two phases have to pass through the non-homogeneous sections of the interfacial region. A consideration of the journey of a charge from

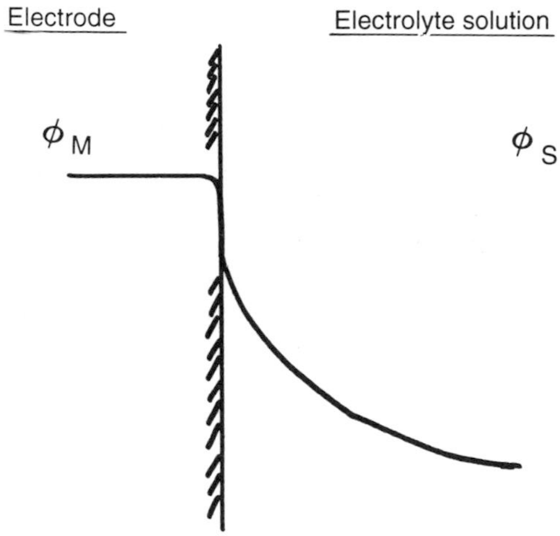

Figure 6.1 Relationship between Galvani potentials of electrode and solution phases.

solution to electrode and vice versa brings into better focus the differing regions of potential which the charge must negotiate.

It is necessary to perform a mental experiment whereby the electrode and solution are separated from one another but with the charge distributions considered to be the same as when combined. It is then possible to imagine the influence of the electrode on a charge approaching it from the solution side and passing through its surface and similarly to imagine the influence of the solution on a charge approaching it from the electrode side and emerging through the solution surface.

Consider first the transfer of charge from solution to electrode: at a position some 100 nm from its surface a potential now to be designated ψ, due to the electrode, will be experienced. This is analogous to the potential arising in the vicinity of an ion considered in Chapter 2.

ψ is the *outer*, or Volta, potential. The charge, moving from this position through the diffuse region of rising ψ, will eventually experience a sharp variation of potential *at the surface* designated χ.

After penetration of the surfacc the charge experiences the inner region of the electrode where the constant (Galvani) potential is ϕ. This process is shown in Figure 6.2(a).

Figure 6.2(b) represents the complementary mental experiment for the electrode-to-solution journey which requires the introduction of inner (ϕ_S), outer (ψ_S), and surface (χ_S) potentials for the solution. It is clear that for both media and in general the Galvani potential may be expressed by

$$\phi = \chi + \psi \tag{6.8}$$

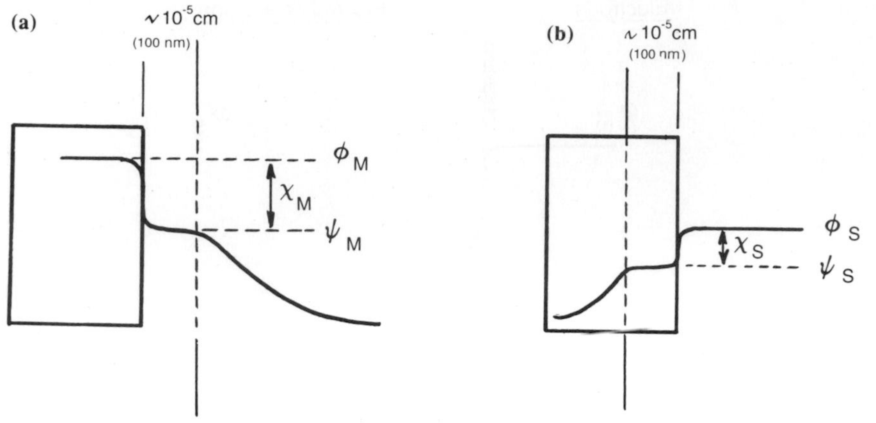

Figure 6.2 Relationship between inner, outer and surface potentials. (a) solution-to-electrode charge transfer; (b) electrode-to-solution charge transfer.

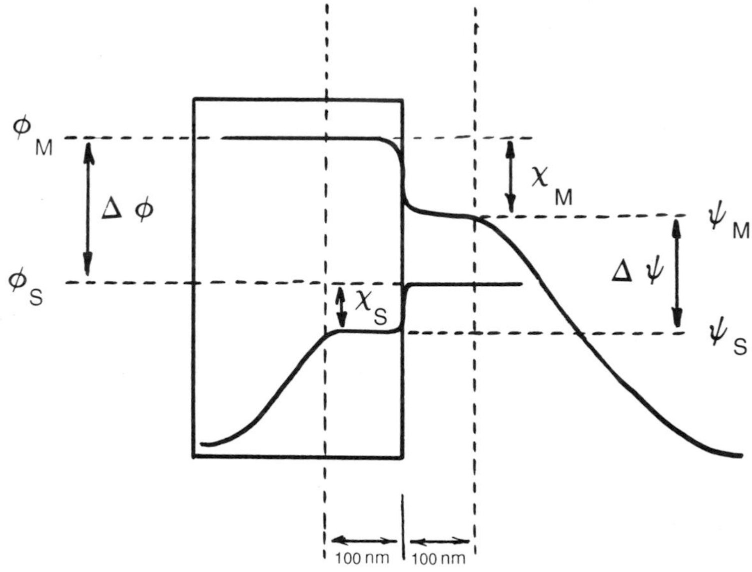

Figure 6.3 Combination of the components of Figure 6.2 to show the relationship between ϕ, ψ and χ for electrode and solution phases.

The relationship between ϕ_M and ϕ_S and that between ψ_M and ψ_S is better understood when the two curves of Figure 6.2 are brought together as shown in Figure 6.3.

6.2.2 *Measurable and non-measurable quantities*

Outer (Volta) potentials can be measured; surface potentials cannot. This is apparently a considerable problem for it means that ϕ cannot be measured. This implies that $\Delta\phi$ cannot be measured which at first sight may seem to be somewhat at odds with the comments made about *experimental* potentiometric measurement in section 6.2. What *can* be measured is a value of $\Delta\phi$ *relative to a reference* $\Delta\phi$: such relative values are known as *electrode potentials* (E). It is perhaps appropriate now to summarize these important concepts.

(i) electrode potential, $E = \Delta\phi = \phi_M - \phi_s$, is expressed in terms of non-measurable quantities;
(ii) a reference electrode potential, $E_{ref} = \Delta\phi_{ref}$ may be defined, quite arbitrarily, for convenience;
(iii) the difference, $E - E_{ref} = \Delta\phi - \Delta\phi_{ref}$, can be measured by potentiometric means.

If E_{ref} is designated zero and is the common reference for *all* measurements of $E - E_{ref}$, then it is possible to draw up an internationally agreed series of values of E. This series, within the IUPAC system, is considered in more detail in a later section.

6.3 Electrode potentials and activity: the Nernst equation

Consider a redox half-reaction of the type given by equation (6.3), e.g.

$$Cu^{2+} + 2e \rightleftharpoons Cu \tag{6.9}$$

This equilibrium is established at room temperature when a piece of metallic copper is immersed in a solution of copper(II) sulphate.

When equilibrium is established, the electrochemical potentials to the left must equal those to the right, thus

$$\tilde{\mu}_{Cu^{2+}} + 2\tilde{\mu}_e = \tilde{\mu}_{Cu} \tag{6.10}$$

Now, for *each phase*, $\tilde{\mu}$ is given by

$$\tilde{\mu}_i = \mu_i^{\ominus} + RT\ln a_i + z_iF\phi \qquad \text{(see equation (2.5))} \tag{6.11}$$

Equation (6.11) may be substituted for each phase into equation (6.9) to obtain the equality:

$$[\mu^{\ominus}_{Cu^{2+}} + RT\ln a_{Cu^{2+}} + 2F\phi_{Cu^{2+}}] + [2\mu^{\ominus}_e + 2RT\ln a_e - 2F\phi_e]$$
$$= [\mu^{\ominus}_{Cu} + RT\ln a_{Cu} + 0F\phi_{Cu}] \tag{6.12}$$

It should be noted that a negative sign is associated with the term $2F\phi_e$ since electrons carry negative charge, and that the term involving ϕ_{Cu} becomes zero due to the neutrality of copper atoms.

Thus, since $\Delta\phi = \phi_{electrode} - \phi_{solution} = \phi_{Cu} - \phi_S$ is required.

$$[0F\phi_{Cu} - 2F\phi_{Cu^{2+}} + 2F\phi_e] = [\phi_e - \phi_{Cu^{2+}}]2F$$
$$= \mu^{\ominus}_{Cu^{2+}} + 2\mu^{\ominus}_e - \mu^{\ominus}_{Cu} + RT\ln a_{Cu^{2+}} - RT\ln a_{Cu}$$

or,

$$\Delta\phi = \frac{\mu^{\ominus}_{Cu^{2+}} + 2\mu^{\ominus}_e - \mu^{\ominus}_{Cu}}{2F} + \frac{RT}{2F}\ln\frac{a_{Cu^{2+}}}{a_{Cu}}$$

or

$$\Delta\phi = \Delta\phi^{\ominus} + \frac{RT}{2F}\ln\frac{a_{Cu^{2+}}}{a_{Cu}} \tag{6.13}$$

Equation (6.13) is known as the Nernst equation which may be expressed in the general form applicable to the general half-reaction process

$$\text{oOx} + ne \rightleftharpoons \text{r Red} \tag{6.14}$$

$$E = E^{\ominus} + \frac{RT}{nF}\ln\frac{(a_{ox})^o}{(a_{red})^r} \tag{6.15}$$

In equations (6.13) and (6.14), $\Delta\phi^{\ominus}$ and $E^{\ominus}$ are symbols for the standard electrode potential. Although equation (6.13) has been expressed in terms of $\Delta\phi$ and $\Delta\phi^{\ominus}$ to emphasize that electrode potentials are Galvani potential differences, the following sections will adopt the more familiar notation of E and $E^{\ominus}$.

A shortened, and for many purposes an adequate alternative, derivation of the Nernst expression may be effected as follows.

The van't Hoff reaction equation expresses the free energy change for a chemical reaction in the form

$$\Delta G = \Delta G^{\ominus} + RT\ln\frac{\prod(\text{activities of products})}{\prod(\text{activities of reactants})} \tag{6.16}$$

So that for the electrode reaction (6.3) as written, equation (6.16) takes the form

$$\Delta G = \Delta G^{\ominus} + RT\ln\frac{a_M}{a_{M^{n+}}} \tag{6.17}$$

Now, the free energy change of a reversible electrode reaction is related to the electrode potential by

$$\Delta G = -nEF \tag{6.18}$$

and, for the standard state

$$\Delta G^{\ominus} = -nE^{\ominus}F \tag{6.19}$$

Relationships (6.18) and (6.19) follow from simple reasoning: reduction of one mole of M^{n+} to M requires the passage of n Faradays, or a quantity of electricity nF coulombs. Passage of charge nF through a potential difference of E volts constitutes electrical work of nFE joules. This work, done by the system, at constant temperature and pressure, is equal to the decrease in free energy of the system, $-\Delta G$. Hence the equality (6.18) and, under standard conditions (6.19).

Substitution of equations (6.18) and (6.19) into equation (6.17) gives

$$nEF = nE^{\ominus}F + RT \ln \frac{a_{M^{n+}}}{a_M}$$

or,

$$E = E^{\ominus} + \frac{RT}{n\mathrm{F}} \ln a_{M^{n+}} \tag{6.20}$$

where a_M has been omitted as the activity of the metal may be regarded as constant and unity. The logarithmic term always involves a ratio of terms characteristic of the oxidized form to those characteristic of the reduced form. Thus, for a redox electrode, e.g. the Fe(III)–Fe(II) system

$$E = E^{\ominus} + \frac{RT}{F} \ln \frac{a_{Fe^{3+}}}{a_{Fe^{2+}}} \tag{6.21}$$

and for a chlorine electrode,

$$\begin{aligned} E &= E^{\ominus} + \frac{RT}{F} \ln \frac{(a_{Cl_2})^{\frac{1}{2}}}{a_{Cl^-}} \\ &= E^{\ominus} + \frac{RT}{F} \ln \frac{1}{a_{Cl^-}} \end{aligned} \tag{6.22}$$

since $a_{Cl_2}^{\frac{1}{2}} = 1$ at 1 atmosphere pressure.

Thus, in general,

$$E_{eq} = E^{\ominus} + \frac{RT}{nF} \ln \frac{a_{Ox}}{a_{red}} \sim E^{\ominus} + \frac{RT}{nF} \ln \frac{[\mathrm{Ox}]}{[\mathrm{Red}]} \tag{6.23a}$$

At 298 K, with the value of R as 8.314 J K^{-1}mol^{-1} and F as 96 500 C mol^{-1}, equation (6.23a) may be expressed as

$$E_{eq} \sim E^{\ominus} + \frac{0.02567}{n} \ln \frac{[\mathrm{Ox}]}{[\mathrm{Red}]} \tag{6.23b}$$

or as

$$E_{eq} \sim E^{\ominus} + \frac{0.05913}{n} \log \frac{[\mathrm{Ox}]}{[\mathrm{Red}]} \tag{6.23c}$$

These are forms of the Nernst equation, in which E_{eq} has been used to emphasize that it is an equilibrium potential referring to the position of dynamic equilibrium between oxidized and reduced forms which is established rapidly at the electrode surface. Only to such a system can this—a thermodynamic equation—be applied.

6.4 Disturbance of the electrode equilibrium

The equilibrium at the electrode may be disturbed by making it more oxidizing or reducing by superimposing an external emf. Thus, if a potential, E, is applied such that $E < E_{eq}$, some of the oxidized form is reduced until a new equilibrium position is reached where $E'_{eq} = E$. Conversely, if $E > E_{eq}$, some of the reduced form is oxidized. Such changes may only be made within limits which are consistent with equilibrium being maintained at the electrode. While such an approach may be useful, it is much oversimplified; electron exchanges proceed with finite rates which vary widely and mass transfer processes occur at finite speeds. It is, however, convenient to consider cases where (i) the electron exchange rate > mass transfer rate and (ii) mass transfer rate > electron transfer rate.

6.4.1 Why electrons transfer

Clearly if electrons are to exchange between the material of an electrode and a species in solution it is required that the two participants approach to within a minimum distance of each other. To achieve this there will be rearrangements of hydration molecules in the surface, interface and solute species. Such processes will be considered later: for the present it is the energetic requirements for electron transfer when the two media may be assumed to be appropriately oriented which are of interest.

Electron exchanges will occur between the highest available energy level in the electrode (the Fermi level) and the lowest energy orbitals of the solute species. These energy levels are related to the various potentials considered in section 6.2, in terms of the schematic plan given in Figure 6.4.

If $U_F \sim U_{solute}$ it is easy for electrons to transfer in either direction. Shift of U_F by imposition of an external potential will disturb the equilibrium.

6.4.2 The distinction between fast and slow systems

If equilibrium between metal M and its ion M^+ is established rapidly a characteristic potential will be adopted by M relative to that of the constant

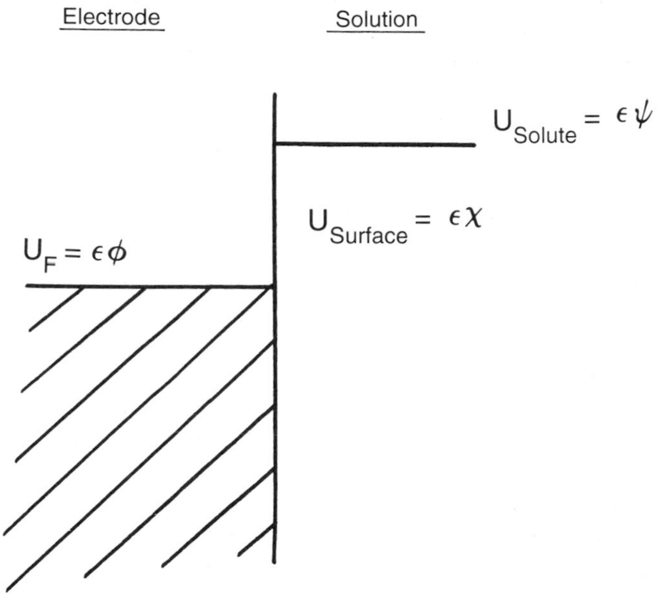

Figure 6.4 Energy levels relating to the various potentials at the electrode–solution interface.

potential of a reference electrode. A plausible model for this state of affairs is given in Figure 6.5 where an approximate coincidence is shown of the energy of the Fermi level U_F and the energy U_{M^+} of the maximum of the electronic energy distribution curve of the ion.

No net current will be observed since identical current components flow in either direction due to identical movements of charge. As soon as a potential is applied more negative (more reducing, or more cathodic) than the equilibrium value, E_{eq}, a net flow of electrons occurs from electrode to solution, i.e. from U_F' to U_{M^+} and the process $M^+ + e \rightarrow M$ is accelerated relative to the reverse reaction. A net cathodic current flows (Figure 6.6).

If a potential is applied more positive (more oxidizing, or more anodic) than E_{eq} a net current flows in the opposite direction since the process $M \rightarrow M^+ + e$ now occurs with electrons flowing from U_{M^+} to U_F''. The current–voltage graph takes the form shown in Figure 6.7.

Similar considerations apply to oxidation and reduction processes at inert electrodes. Let us consider the same metal ion M^+ and its reduction at such an inert electrode, i.e. one with which it does not react or enter into equilibrium. In this case no current will be observed until E_{eq} is reached, since only the oxidized form, M^+, is present with none of the reduced form, M. If the reduced form, M, can be dissolved in the electrode, e.g. in the form of an amalgam in the case of a mercury electrode, the corresponding anodic current–voltage curve may be obtained. To do this the amalgam

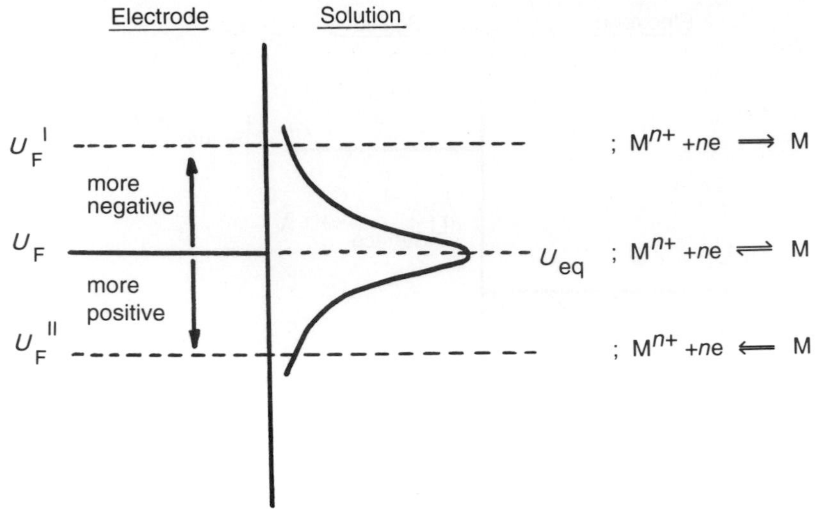

Figure 6.5 Dependence of the nature of an electrode reaction on the Fermi level of the electrode.

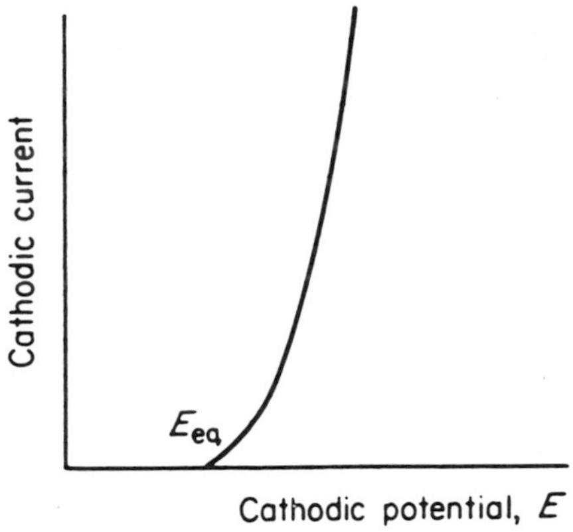

Figure 6.6 Current–potential curve for a cathodic process.

electrode and reference electrode would have to be placed in a solution containing no M^+ but some indifferent electrolyte to act as current carrier. At potentials with respect to the reference electrode more negative than E_{eq} no net current would be observed, but at more positive values a net anodic current would appear (Figure 6.8).

Consider now a redox system, e.g. M^{3+}/M^{2+}, and let both forms be present in a solution into which are placed an inert electrode and a suitable

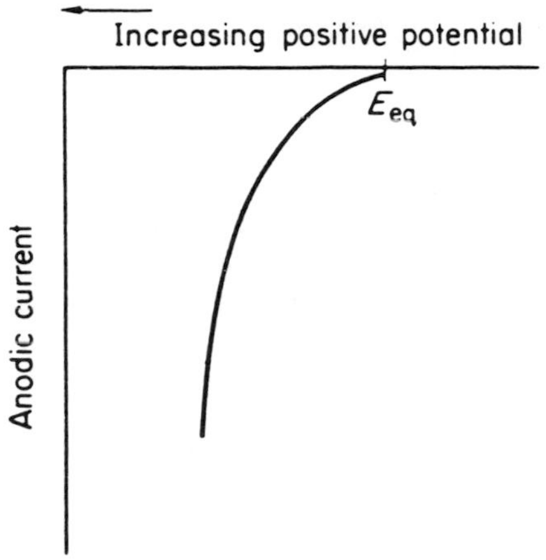

Figure 6.7 Current–potential curve for an anodic process.

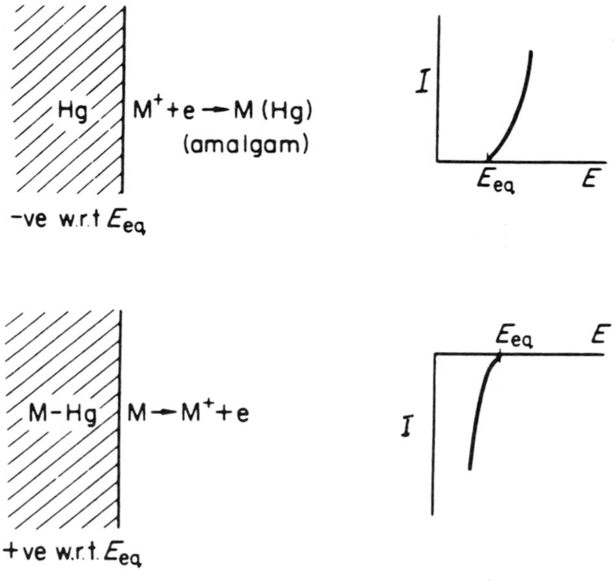

Figure 6.8 Cathodic and anodic reactions, with corresponding current–potential curves, at an inert electrode in which the reduced form is soluble.

reference electrode. If the equilibrium

$$M^{3+} + e \rightleftharpoons M^{2+}$$

is established instantaneously at the electrode surface then, with no exter-

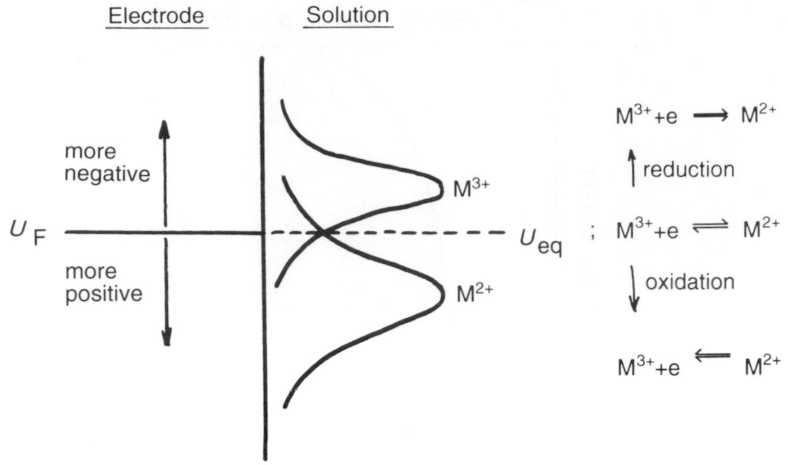

Figure 6.9 The nature of the net redox process shown as a function of the interaction of the Fermi level of the (inert) electrode and the energy distribution curves of the oxidized and reduced forms of the couple.

nally applied potential, the inert electrode adopts a characteristic potential, E_{eq} corresponding to $U_F = U_{redox}$ (Figure 6.9).

The distribution of electronic energy levels in M^{3+} and M^{2+} will differ to a small extent due to differing degrees of solvation consequent upon differing ion charges.

At potentials more negative than E_{eq} the current–voltage curve for the process $M^{3+} + e \rightarrow M^{2+}$ may be developed while at more positive potentials the curve for the process $M^{2+} - e \rightarrow M^{3+}$ occurs in a complementary way, the one passing smoothly into the other through E_{eq} (Figure 6.10).

All systems considered so far in this section may be classed as fast or reversible, the term 'fast' referring to the rapid attainment of equilibrium between an electrode and species in solution.

Figure 6.10 shows the variation of *net* current with applied potential (dashed curve) in relation to the currents of the individual forward and backward processes. This shows more clearly the flow of equal and opposite currents at E_{eq}; these being denoted by I_0, the exchange current. For such a system at potentials only slightly removed from E_{eq}, a net oxidation or reduction may be made to occur under almost reversible conditions with the Nernst equation applying.

In the case of slow or 'irreversible' systems, equilibrium is established so slowly that the condition is never seen to be reached. No significant current is seen near to E_{eq} and applied potentials well removed to both cathodic and anodic sides of this value are often required to produce currents of the same order of magnitude as those obtained for a fast system. Evidently

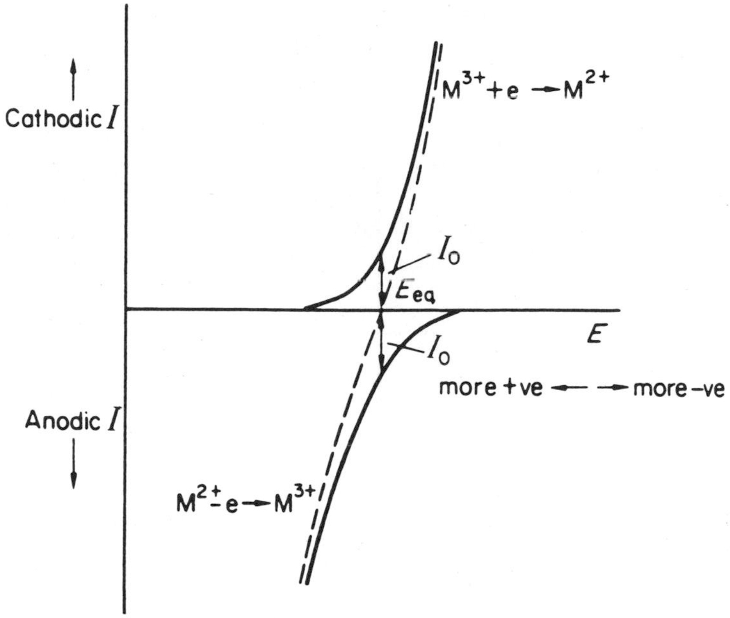

Figure 6.10 Anodic–cathodic current–potential curve for a 'fast' system.

the average energies of the electronic levels are both significantly removed from U_F (Figure 6.11).

In Figure 6.12 are shown schematically the shapes of current–voltage curves to be expected for a slow anodic–cathodic reaction. It is seen that, not only is a considerably more negative potential required for the slow system in order to obtain the same current as for the fast one, but the rate of increase of current is much less for the slow process. For the fast system the current rises almost vertically. In practice this would show some small deviation from the vertical and this reflects the influence of the comparative slowness of mass transfer processes. Slow electrochemical systems are said to require large overvoltage, $(E - E_{eq})$ or $(E_{eq} - E)$, in order to produce a significant net current and take place under irreversible conditions.

When an electrochemical reaction occurs the concentration of oxidizable or reducible species at the electrode surface is depleted and if fresh material is not provided electrolysis stops. In fact, mass transfer occurs, which tends to maintain surface concentrations constant. If the electrode reaction occurs much more rapidly than mass transfer processes, the latter are rate determining and control the magnitude of current which flows. This gives rise finally to a stationary state in which material reaching the electrode is oxidized or reduced as fast as it arrives at the surface. Later chapters

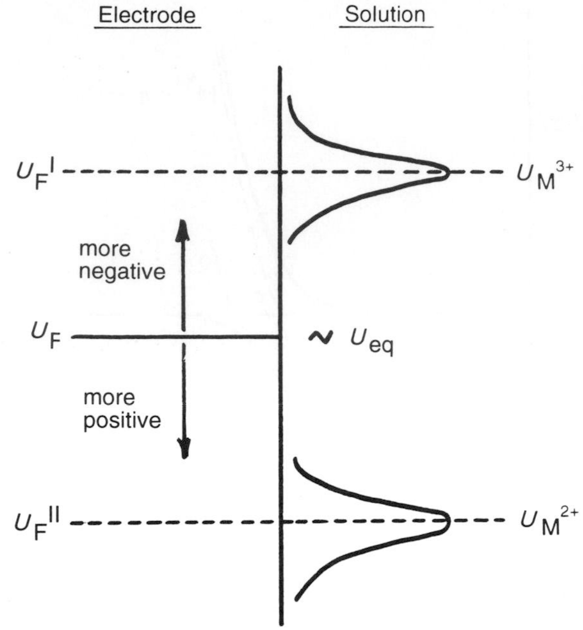

Figure 6.11 The energies of electronic levels of oxidized and reduced forms are both well removed from U_F and substantial additional energy, in the form of extra applied potential, is required to induce net reaction.

will consider the important implications of such situations for analytical electrochemistry.

6.5 The hydrogen scale and the IUPAC convention

In the last section variation of the relative value of the potential of an electrode with respect to some reference value was considered. It has been established that it is impossible to determine absolute values of potentials adopted by electrodes. What *can* be done is to measure the value of $\Delta\phi$ for one redox system relative to that for another reference value. In more colloquial terms, the emf of a cell obtained by coupling the electrode (half-cell) in question with another electrode can be measured. If the latter is so constructed that it maintains an almost constant potential *whatever the potential difference* between the two electrodes, electrode potentials can be measured with respect to an arbitrarily chosen standard and given physical significance. The internationally agreed standard is the hydrogen electrode; this device operating with hydrogen gas at 1 atmosphere pressure, in contact with platinized platinum, in a solution of hy-

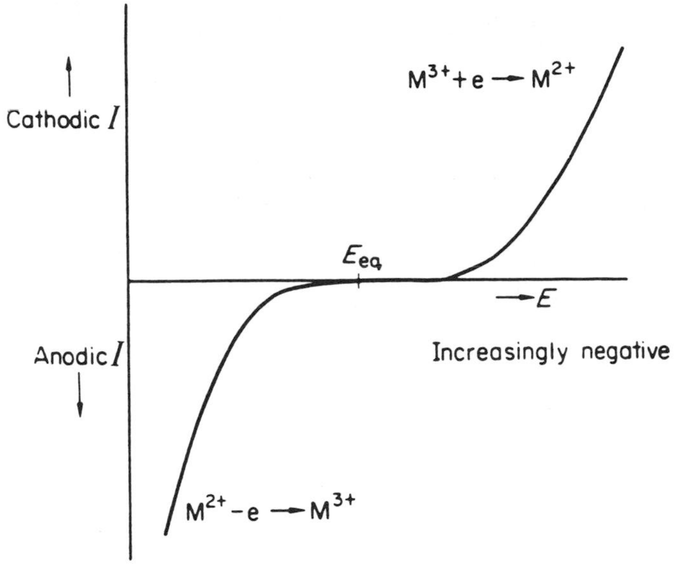

Figure 6.12 Anodic–cathodic current–potential curve for a 'slow' system.

drogen ions of unit activity, is assigned a potential of zero *at all temperatures.*

6.5.1 The standard hydrogen electrode

The hydrogen electrode, depending for its operation on establishment of the equilibrium

$$\frac{1}{2}H_2 \rightleftharpoons H^+ + e$$

requires that gaseous hydrogen and hydrogen ions in solution be brought together at the surface of a suitable catalyst. Equilibrium is attained rapidly at platinum black. The finely divided platinum is supported on platinum foil acting as an electronic conductor. This part of the electrode is made by first cleaning the foil in chromic acid and then plating it in a solution of 1% platinum(IV) chloride with an anode of platinum. The necessary thin layer of the catalytic material is in this way cathodically deposited on the foil. Unfortunately, this process causes the occlusion of some chlorine in the deposit and this must be removed by making the electrode the cathode for a short-term electrolysis in dilute sulphuric acid. Chlorine is swept away in the stream of electrolytically generated hydrogen.

In operation this electrode dips into a solution of hydrogen ions of constant activity while hydrogen gas passes over its surface. In Figure 6.13

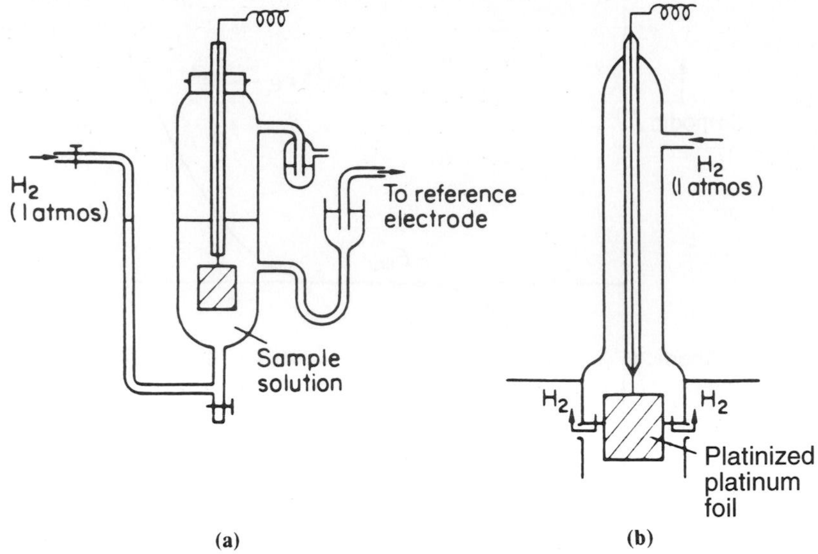

Figure 6.13 Two forms of hydrogen electrode. (a) Lindsey type: hydrogen bubbles upwards over the platinum foil surface. (b) Hildebrand type: a series of holes are blown in the glass envelope surrounding the platinum foil. The gas flow rate is adjusted so that hydrogen only escapes through these. The level of solution inside the envelope fluctuates so that part of the foil is exposed alternately to solution and gas.

are shown two types of hydrogen electrode: the one has the electrolyte solution enclosed and protected from possible air contamination, the other may be dipped into a solution whose hydrogen ion concentration is to be determined. The former is more desirable for the determination of electrode potentials; the latter is suited for following changes in hydrogen ion concentrations as in titrations.

It is vital that the hydrogen gas is pure and for the most accurate work even commercially pure hydrogen should be bubbled through alkaline pyrogallol or passed over heated palladized asbestos to remove the last traces of oxygen. Since the passage of gas through the electrode solution can cause a change of its concentration, prior passage of the purified gas through a sample of this solution is desirable.

Satisfactory functioning of the hydrogen electrode is, above all, dependent on the complete absence of catalyst poisons such as mercury and arsenic and particularly sulphur compounds. Consequently, rubber tubing should not be used for connections; PVC or polythene are much more reliable materials. Before using a hydrogen electrode to determine the emf of a cell formed by coupling it with another half-cell, it should be checked against a duplicate electrode. The steady emf of a cell made up of two identical hydrogen half-cells should, of course, be zero. In operation hydrogen electrode should be found to assume a steady potential within 20 min and this should be independent of the rate of bubbling. Dependence of the potential on bubbling rate is characteristic of poisoning as are also slowness in reaching equilibrium and general variability of potential. Under these conditions it is necessary to clean and replatinize the electrode surface.

By applying the Nernst equation to the hydrogen electrode equilibrium it is seen that the potential of a hydrogen electrode is given by

$$E = E^{\ominus} + \frac{RT}{F}\ln\frac{a_{H^+}}{(a_{H_2})^{\frac{1}{2}}} = E^{\ominus} + \frac{RT}{F}\ln a_{H^+} - \frac{RT}{2F}\ln P_{H_2} \qquad (6.24)$$

where P_{H_2} is the pressure of hydrogen gas.

When both the activities of hydrogen and hydrogen ion are unity, $E = E^{\ominus}$, which is arbitrarily given the value zero at all temperatures. Potentials of all other electrodes may then be given values relative to this standard.

6.5.2 Electrode potential and cell emf sign conventions

The signs which are to be given to electrode potentials and cell emf follow the convention adopted by the International Union of Pure and Applied Chemistry (IUPAC).

By the electrode potential of the half-cell M^{n+}/M is implied the emf of the cell formed by coupling the latter with a hydrogen half-cell. Since the potential of the latter is zero under standard conditions, the emf determined is the electrode potential of the M^{n+}/M couple. The cell may be represented by

$$\text{Pt}, H_2(a = 1)|H^+(a = 1)||M^{n+}|M \qquad (6.25)$$

the electrode potential of the couple M^{n+}/M then being defined by

$$E_{cell} = E_{right} - E_{left} \qquad (6.26)$$

with $E_{left} = 0$ for a standard hydrogen electrode.

In equation (6.25) a number of symbols commonly used in such cell representations are given. A vertical line usually implies a phase boundary between the species or components brought into equilibrium, e.g. an electrode and an electrolyte. A double vertical line usually represents the introduction of a salt bridge (see section 6.7) while a dotted vertical line may be used to represent the interface between two solutions in which electrical contact is maintained without short-term mixing. Commas usually link the components of an electrode or electrolyte system.

The sign of the electrode potential is decided very easily by whether hydrogen gas is evolved or hydrogen ionizes at the left-hand electrode. If on closing the circuit of cell (6.25) the hydrogen electrode becomes the positive pole by giving up electrons to hydrogen ions to give gaseous hydrogen, the unknown electrode potential is negative and equal to the emf developed by the cell. On the other hand, if the hydrogen electrode becomes the negative pole, taking up electrons as hydrogen ionizes, the unknown electrode potential is positive and again equal to the emf developed by the cell. A table of International Standard Electrode Potentials on the hydrogen scale is given in Table 6.1.

Table 6.1 Standard electrode potentials, $E^\ominus$ (volts with respect to the Standard hydrogen electrode), at 298 K

Electrode	$E^\ominus$	Electrode	$E^\ominus$
Li^+/Li	−3.05	Cu^{2+}/Cu^+	+0.16
Ca^{2+}/Ca	−2.87	Bi^{3+}/Bi	+0.23
Na^+/Na	−2.71	Cu^{2+}/Cu	+0.34
Mg^{2+}/Mg	−2.37	O_2/OH^-	+0.40
Al^{3+}/Al	−1.66	Fe^{3+}/Fe^{2+}	+0.76
Zn^{2+}/Zn	−0.76	Ag^+/Ag	+0.80
Cd^{2+}/Cd	−0.40	Hg^{2+}/Hg	+0.80
Ni^{2+}/Ni	−0.25	Hg^{2+}/Hg_2^{2+}	+0.92
Pb^{2+}/Pb	−0.13	Cl_2/Cl^-	+1.36
H^+/H	0.00	Ce^{4+}/Ce^{3+}	+1.61

When an electrochemical cell is formed from two half-cells, one of which is not the hydrogen half-cell, the emf may be calculated from rule (6.26) using the individual half-cell potentials determined with respect to hydrogen. The electrode which has the more negative potential is always written on the left as the negative pole of the cell, while that with the more positive electrode potential is always written on the right as the positive pole of the cell. When this rule is followed, no ambiguity as to signs arises. It is seen that only in the *determination* of the sign of a half-cell potential is there a possibility of having a negative pole on the right; but here we are seeking to establish in which direction a particular cell reaction (involving the standard hydrogen electrode) is spontaneous. For a half-cell with established sign and magnitude of *E*, the direction of the spontaneous cell reaction is specified. Consider the Daniell cell in which one electrode is zinc dipping into a zinc sulphate solution, the other being copper dipping into copper sulphate solution. The two solutions are prevented from mixing by a porous membrane separating them. In accordance with the convention this cell may be represented by

$$Zn|ZnSo_4 \vdots CuSO_4|Cu$$

i.e.

$$Zn|Zn^{2+} \vdots Cu^{2+}|Cu$$

It is instructive to investigate such a cell more analytically in terms of

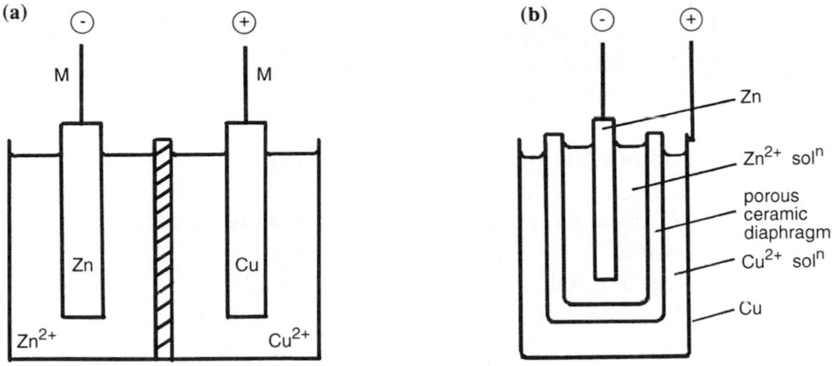

Figure 6.14 The Daniell cell. (a) Schematic form; (b) practical arrangement.

the potential differences at the various interfaces which it involves. For these purposes it is useful to visualize the cell in the schematic form of Figure 6.14(a), although the practical arrangement is as shown in Figure 6.14(b).

Combining equation (6.26) with the general expression for the potential of the two half-cells, viz.

$$E_{\text{half-cell}} = \Sigma\Delta\phi_{\text{interfaces}} \tag{6.27}$$

the expression for the emf of the cell becomes:

$$E_{\text{cell}} = ({}_{\text{R}}\phi_{\text{M}} - \phi_{\text{Cu}}) + (\phi_{\text{Cu}} - \phi_{\text{Cu}^{2+}}) + (\phi_{\text{Cu}^{2+}} - \phi_{\text{Zn}^{2+}}) + (\phi_{\text{Zn}^{2+}} - \phi_{\text{Zn}}) + (\phi_{\text{Zn}} - {}_{\text{L}}\phi_{\text{M}}) \tag{6.28}$$

There are *two* ways of using equation (6.28). Firstly, it may be expressed in terms of various $\Delta\phi$ terms as follows:

$$E_{\text{cell}} = \Delta\phi^{\text{M}}_{\text{Cu}} + \Delta\phi^{\text{Cu}}_{\text{Cu}^{2+}} \pm (\Delta\phi^{\text{Cu}^{2+}}_{\text{Zn}^{2+}}) - \Delta\phi^{\text{Zn}}_{\text{Zn}^{2+}} - \Delta\phi^{\text{M}}_{\text{Zn}} \tag{6.29}$$

Equation (6.29) emphasizes the existence of the small but significant liquid junction potential, $\Delta\phi^{\text{Cu}^{2+}}_{\text{Zn}^{2+}}$ a quantity difficult to measure accurately but easily minimized. Secondly, cancellation of like terms in equation (6.28) yields the simple expression

$$E_{\text{cell}} = {}_{\text{R}}\phi_{\text{M}} - {}_{\text{L}}\phi_{\text{M}} \tag{6.30}$$

Equation (6.30) emphasizes that the cell emf is the difference in inner potentials of the two connecting wires.

6.5.3 Calculation of cell emf values from tabulated data

It is a useful exercise to write down all the information, including electrode reactions, which is implied by the use of data from Table 6.1 applicable to the half-cell components to left and right. Thus, for the Daniell cell, knowing only the values of $E^{\ominus}_{Cu^{2+}|Cu}$ and $E^{\ominus}_{Zn^{2+}|Zn}$ it is possible to summarize its behaviour and orientation in the following way

Left-hand electrode	*Right-hand electrode*
ANODE	CATHODE
Negative pole	Positive pole
Oxidation occurs	*Reduction* occurs
$Zn \rightarrow Zn^{2+} + 2e$	$Cu^{2+} + 2e \rightarrow Cu$
$E^{\ominus} = -0.76\,V$	$E^{\ominus} = +0.34\,V$

Overall

$$Zn + Cu^{2+} \rightarrow Zn^{2+} + Cu$$

$$\begin{aligned} E^{\ominus}_{cell} &= +0.34 - (-0.76) \\ &= +1.10\,V \end{aligned}$$

It can be seen that when the cell is represented correctly there is no ambiguity in regard to the direction in which the reaction proceeds spontaneously.

The meaning and significance of the terms *anode* and *cathode* which have been introduced above should be clearly understood. It is particularly important to distinguish them from *polarities*. The term *cathode* describes the electrode at which *reduction* occurs: in a spontaneous cell, as discussed here, this will be the *more positive* pole of the cell, but in an electrolysis cell where such processes are *induced*, such an electrode will be the more negative pole. In a complementary way, the term *anode* describes the electrode at which *oxidation* occurs: for a spontaneous cell this will be the more negative pole but the more positive one in an electrolysis cell.

6.6 Other reference electrodes

It is often more convenient to use subsidiary or secondary reference electrodes whose potentials have been accurately determined with respect to that of the hydrogen electrode. A number of useful systems give reproducible potentials over a long period. Quite apart from the experimental inconveniences of the hydrogen electrode, secondary reference electrodes may be chosen which do not show the major disadvantages of the primary standard. The latter cannot be used in solutions containing chemically reducible species and is susceptible to poisoning.

The most useful reference systems are those described as electrodes of the second kind. These are quite different to systems such as Cu/Cu^{2+} and Zn/Zn^{2+} which are electrodes of the first kind.

Electrodes of the second kind have the following form: metal, in contact with one of its sparingly soluble salts, placed in a solution containing a strongly ionized salt with a common anion. The calomel electrode is a case in point and is represented by

$$Hg|Hg_2Cl_2(s),\ KCl\,aq.$$

The term *calomel* deriving from the earlier trivial name for mercury(I) chloride.

Other examples are the silver/silver chloride and mercury/mercury(I) sulphate electrodes

$$Ag|AgCl(s), KCl\,aq.$$

and

$$Hg|Hg_2SO_4(s), K_2SO_4\,aq.$$

It is important to be very clear in regard to the reactions upon which the operation of such electrodes depends. For the calomel half-cell the reaction is

$$Hg_2Cl_2 + 2e \rightleftharpoons 2Hg + 2Cl^-$$

and similarly for the silver–silver chloride electrode

$$AgCl + e \rightleftharpoons Ag + Cl^-$$

In particular it should be noted that the structure and reaction of the silver/silver chloride electrode is quite distinct from those of the silver/silver ion electrode.

The potential adopted by each of these reference electrodes is controlled by the activity of the *anion* in solution. This may be shown quite simply by considering a calomel electrode in which the chloride ion activity is a_{Cl^-}. The potential adopted by the *mercury* depends upon the activity of the Hg_2^{2+} ion, so that

$$E = E^{\ominus}_{Hg} + \frac{RT}{2F} \ln a_{Hg_2^{2+}} \qquad (6.31)$$

Now, $a_{Hg_2^{2+}} \times a^2_{Cl^-} = K_{Hg_2Cl_2}$, the solubility product of calomel, i.e.

$$a_{Hg_2^{2+}} = \frac{K_{Hg_2Cl_2}}{a^2_{Cl^-}} \qquad (6.32)$$

Substitution of equation (6.32) into equation (6.31) gives

$$E = E^{\ominus}_{Hg} + \frac{RT}{2F} \ln \frac{K_{Hg_2Cl_2}}{a^2_{Cl^-}} \qquad (6.33)$$

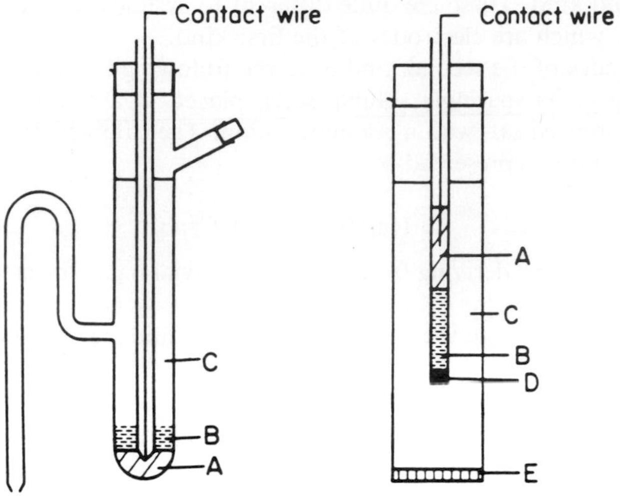

Figure 6.15 Two forms of calomel electrode. A. Mercury. B. Mercury–calomel paste. C. Potassium chloride solution. D. Asbestos or glass wool plug. E. Sintered glass.

or

$$E = \left[E^{\ominus}_{Hg} + \frac{RT}{2F}\ln K_{Hg_2Cl_2}\right] - \frac{RT}{2F}\ln a^2_{Cl^-} \tag{6.34}$$

Since the terms in brackets form a constant, which we may denote by $E^{\ominus}_{Hg_2Cl_2}$, equation (6.34) may take the form

$$E = E^{\ominus}_{Hg_2Cl_2} - \frac{RT}{F}\ln a_{Cl^-} \tag{6.35}$$

Three types of calomel have been commonly used, distinguishable by different concentration of potassium chloride. Their characteristics are summarized in Table 6.2.

Table 6.2 Potentials adopted by calomel electrodes with different concentrations of KCl

Electrode	Potential (V, versus Standard hydrogen electrode)
0.1 mol dm^{-3} KCl	+0.336
1 mol dm^{-3} KCl	+0.283
Saturated KCl	+0.242

plus1.5bp

Two forms of calomel electrode are shown in Figure 6.15.

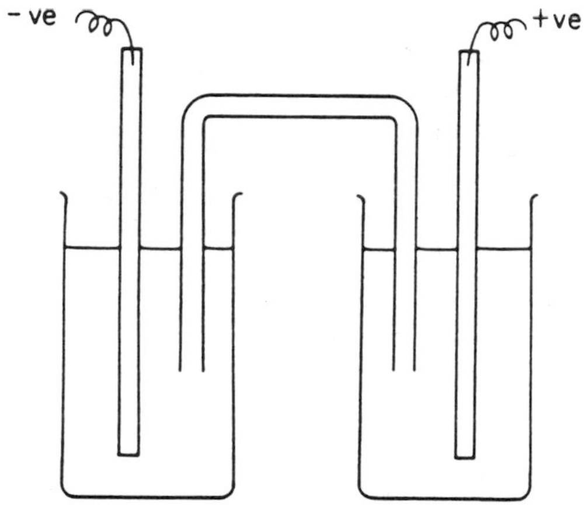

Figure 6.16 Connection of half-cells by means of a salt bridge containing a suitable electrolyte to eliminate the junction potential. For KCl and NH_4NO_3, $t_+ \sim t_- \sim 0.5$.

6.7 Concentration cells and emf measurements

It has been seen that the coupling of two half-cells produces an electrochemical cell which may be used to produce an emf. In the Daniell cell the two half-cell components are brought into electrical contact by a porous membrane separating the copper sulphate and zinc sulphate solutions. There will inevitably be interdiffusion of zinc and copper ions. Usually the ions from such solutions in contact will diffuse at different rates, leading to a charge separation across the interface which will give rise to a potential difference in this region which ultimately becomes steady. Any measurement of the cell emf will under these conditions include a contribution from this diffusion or liquid junction potential. Liquid junction potentials are extremely difficult to reproduce in practice and, even though their magnitudes do not normally exceed 100 mV, it is wisest for them to be eliminated if at all possible. This may be achieved by connecting solutions in two half-cells by means of a salt bridge (Figure 6.16). This is either a glass or flexible tube containing a saturated solution of either potassium chloride or ammonium nitrate. To prevent excessive diffusion, the ends of the tubes are often plugged with porous material, such as filter paper or glass wool, and the electrolytes are frequently set in agar gel. Transport numbers of cation and anion in solutions of potassium chloride and ammonium nitrate are approximately equal and if such species serve to carry current in a salt bridge between two half-cells the rates of movement of charge in either direction will be approximately equal. Such minimization of charge separation may serve to reduce liquid junction potentials to a few millivolts.

Cell emf's may be measured potentiometrically by comparison with a standard cell of known reproducible emf. A Weston Standard Cadmium

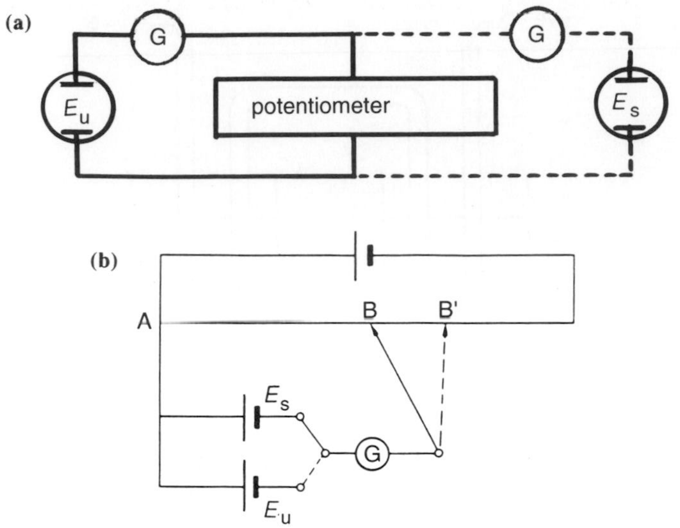

Figure 6.17 (a) Schematic circuit for determination of cell emf. **(b)** Poggendorf potentiometric circuit for determination of cell emf. E_s = emf of standard, E_u = emf of unknown. $E_u/E_s = AB'/AB$.

cell is frequently used for such purposes to calibrate coils or, more historically, the slide-wire of a potentiometer (Figure 6.17). By means of such a device the natural emf of a cell is balanced by an equal and opposite applied emf. Balance in this 'null-point' method is identified by approach to the equilibrium potential E_{eq} along the path identified in Figure 6.9 and recognized by zero deflection in the galvanometer.

It is clearly necessary for the two emfs to act *in opposition* : if the polarity of a cell is unknown this will be soon remedied for only if connection is made in accordance with this property will balance be obtainable.

6.8 Concentration cells without liquid junctions

Concentration cells are made up of two half-cells which are similar chemically but which differ in the activity of some common component: the difference gives rise to an emf because of the difference in potential of the two half-cells. The activity difference may be either between the solutions or between the electrode materials.

6.8.1 Cells with amalgam electrodes

Such a cell is formed by two metal amalgam electrodes of different metal activity dipping into a common solution of a soluble salt of the metal, e.g.

Anode (−ve pole) Hg, Tl	$\mid TlNO_3 \mid$	Tl, Hg (+ve pole) *cathode*		
a_2		a_1	$(a_2 > a_1)$	
E_2		E_1		

activity of Tl^+ ions in solution $= a_+$.

Oxidation

$$Tl(a_2) \rightarrow Tl^+ + e$$

$$E_2 = E^{\ominus}_{Tl} + \frac{RT}{F} \ln \frac{a_+}{a_2}$$

Reduction

$$Tl^+ + e \rightarrow Tl(a_1)$$

$$E_1 = E^{\ominus}_{Tl} + \frac{RT}{F} \ln \frac{a_+}{a_1}$$

Overall $Tl(a_2) \rightarrow Tl(a_1)$. Since $a_2 > a_1, E_1 > E_2$. Therefore,

$$E_{cell} = E_1 - E_2 = \frac{RT}{F} \ln \frac{a_2}{a_1} \qquad (6.36)$$

It should be noted that equation (6.36) gives the instantaneous cell emf which will fall as the ratio a_2/a_1 decreases due to transfer of material. Overall, the cell reaction involves the passage of thallium from the higher to the lower activity. When these activities become equal, the potentials of both electrodes are the same and the cell ceases to operate. For the passage of 1 Faraday, the free energy change accompanying the movement of 1 mole of thallium from a_2 to a_1 is

$$\Delta G = -RT \ln \frac{a_2}{a_1}$$

and since $\Delta G = -nEF$ (and $n = 1$)

$$E_{cell} = \frac{RT}{F} \ln \frac{a_2}{a_1} \qquad \text{(see equation (6.36))}$$

6.8.2 Cells with gas electrodes operating at different pressures

Here we may consider a cell consisting of two hydrogen electrodes operating at different pressures dipping into a common solution of hydrochloric acid, e.g.

Anode	(−ve pole)		(+ve pole)	*Cathode*
	Pt, H_2	$\mid HCl \mid$	H_2, Pt	
	P_2		$P_1 (P_2 > P_1)$	
	E_2		E_1	

Oxidation

$$\tfrac{1}{2}H_2(P_2) \rightarrow H^+ + e$$
$$E_2 = E^{\ominus}_{H_2} + \frac{RT}{F}\ln\frac{a_{H^+}}{(a_{H_2})^{1/2}}$$
$$= \frac{RT}{F}\ln a_{H^+} - \frac{RT}{F}\ln(P_2)^{1/2}$$

Reduction

$$H^+ + e \rightarrow \tfrac{1}{2}H_2(P_1)$$
$$E_1 = E^{\ominus}_{H_2} + \frac{RT}{F}\ln\frac{a_{H^+}}{(a_{H_2})^{1/2}}$$
$$= \frac{RT}{F}\ln a_{H^+} - \frac{RT}{F}\ln(P_1)^{1/2}$$

Overall $\tfrac{1}{2}H_2(P_2) \rightarrow \tfrac{1}{2}H_2(P_1)$.

Again, $E_1 > E_2$. Therefore,

$$E_{cell} = E_1 - E_2 = \frac{RT}{2F}\ln\frac{P_2}{P_1} \tag{6.37}$$

6.8.3 *Concentration cells without transference*

One can form such a cell by connecting two cells of the Harned type in opposition to one another; in this way a composite cell is formed.

The simplest type of Harned cell is

Anode (−ve pole)		*Cathode* (+ve pole)
H_2(1 atm), Pt	HCl	AgCl(s), Ag

Oxidation	*Reduction*
$\tfrac{1}{2}H_2 \rightarrow H^+ + e$	AgCl(s) + e Ag(s) + Cl^-

Overall cell reaction:

$$H_2(1\ \text{atm}) + AgCl(s) \rightarrow Ag(s) + Cl^- + H^+$$

Thus, the free energy change for the passage of 1 Faraday is

$$\Delta G = -nFE = \Delta G^{\ominus} + RT\ln\frac{a_{Ag}a_{H^+}a_{Cl^-}}{(a_{H_2})^{\frac{1}{2}}a_{AgCl}}$$

which, since the activities of silver, hydrogen and silver chloride are constant and unity, becomes

$$\Delta G = \Delta G^{\ominus} + RT\ln a_{H^+}a_{Cl^-}$$

or

$$E = E^{\ominus} - \frac{RT}{F}\ln a_{H^+}a_{Cl^-} = E^{\ominus} - \frac{2RT}{F}\ln(a_{\pm})_{HCl} \tag{6.38}$$

Such cells are useful for determining mean ion activity coefficients of acid in the central solution (see Chapter 8).

A composite of two such cells may be represented as

$$\underset{E_2}{\text{Ag, AgCl(s)}|\underset{(a_\pm)_2}{\text{HCl}}|\text{Pt, H}_2\text{(1 atm)}} \quad \underset{E_1}{\text{Pt, H}_2\text{(1 atm)}|\underset{(a_\pm)_1}{\text{HCl}}|\text{AgCl(s), Ag}}$$

The left-hand cell has electrodes reversed with respect to the Harned cell considered above, i.e.

$$E_{\text{left cell}} = \frac{2RT}{F}\ln(a_\pm)_2 - E^\ominus = -E_2 \qquad \text{(according to equation (6.38))}$$

The right-hand cell has electrodes in the conventional arrangement, i.e.

$$E_{\text{right cell}} = E^\ominus - \frac{2RT}{F}\ln(a_\pm)_1 = E_1$$

The resultant emf of the composite cell is thus given by

$$E_{\text{cell}} = E_{\text{left}} + E_{\text{right}} = -E_2 + E_1 = \frac{2RT}{F}\ln\frac{(a_\pm)_2}{(a_\pm)_1} \tag{6.39}$$

In order for this to be positive, it is necessary that $(a_\pm)_2 \geq (a_\pm)_1$ so that $E_1 > E_2$. No physical transfer of material occurs from one side to the other but, if $(a_\pm)_2 > (a_\pm)_1$, the net effect during the working of the cell is the deposition of H^+ and Cl^- at their respective reversible electrodes in the left-hand cell and dissolution of H^+ and Cl^- at appropriate electrodes on the right. If the cell is allowed to discharge, the resultant effect is a spontaneous decline in $(a_\pm)_2$ and increase in $(a_\pm)_1$.

If the cell is rewritten with the hydrogen electrodes at extreme left and right, the section of the composite cell containing the higher mean activity of HCl, $(a_\pm)_2$, must now appear on the right as follows:

$$\underset{E_1}{\text{Pt, H}_2\text{(1 atm)}|\underset{(a_\pm)_1}{\text{HCl}}|\text{AgCl(s), Ag}} \quad \underset{E_2}{\text{Ag, AgCl(s)}|\underset{(a_\pm)_2}{\text{HCl}}|\text{Pt, H}_2\text{(1 atm)}}$$

The left-hand cell now has the conventional Harned orientation so that

$$E_{\text{left cell}} = E^\ominus - \frac{2RT}{F}\ln(a_\pm)_1 = E_1$$

Conversely, the right-hand cell is now reversed and its emf is given by

$$E_{\text{right cell}} = \frac{2RT}{F}\ln(a_\pm)_2 - E^\ominus = -E_2$$

so that the resultant emf of the composite cell is given by

$$E_{\text{cell}} = E_{\text{left}} + E_{\text{right}} = E_1 - E_2 = \frac{2RT}{F}\ln\frac{(a_\pm)_2}{(a_\pm)_1}$$

which is seen to be identical to equation (6.39).

The electrode at extreme right adopts a relatively positive potential (due to deposition of H^+ from $(a_\pm)_2$) with respcct to the electrode at extreme left which adopts a relatively negative potential (due to dissolution of H^+ into $(a_\pm)_1$). The overall effect in a discharging cell remains a spontaneous decline in $(a_\pm)_2$.

In the general case with an electrolyte comprising ν_+ cations and ν_- anions ($\nu_+ + \nu_- = \nu$) equation (6.39) becomes

$$E = \frac{\nu}{\nu_\pm} \cdot \frac{RT}{F} \ln \frac{(a_\pm)_2}{(a_\pm)_1} \tag{6.40}$$

ν_+ or ν_- being used in the latter expression according to whether the outer electrodes are reversible with respect to cations or anions.

Consider the connection of two Harned cells of the type considered above with HCl at molality $m_1 = 0.01\,\text{mol kg}^{-1}$ and $m_2 = 0.1\,\text{mol kg}^{-1}$ respectively and $E = 0.2225\,\text{V}$. Calculation of $\gamma_\pm$ for both cases via the extended Debye–Hückel equation yields $(a_\pm)_1 = 0.0090$ and $(a_\pm)_2 = 0.0755$. Individual cell emf values are, according to equation (6.38), $E_1 = 0.4643\,\text{V}$ and $E_2 = 0.3551\,\text{V}$.

Thus, by equation (6.39) $E_{\text{cell}} = 0.1092\,\text{V}$ whether the connection of the two cells is via the hydrogen or silver components: in both cases the extreme right-hand electrode is seen to be positive relative to the extreme left-hand electrode. In the former case the silver electrode on the extreme right adopts a potential of +0.4643 V while that at the extreme left adopts +0.3551 V; in the latter case the platinum contact on the extreme right adopts a potential of −0.3551 V while that at the extreme left adopts −0.4643 V. Calculations of E_{cell}, based upon individual cells, are seen to be consistent with the general expression (6.40) involving the ratio of the mean activities of HCl, i.e.

$$E_{\text{cell}} = 2 \times 0.02567 \ln \frac{0.0755}{0.0090} = 0.1092\,\text{V}$$

6.9 Concentration cells with liquid junctions

Cells within this category may be conveniently divided into two classes according to whether a liquid junction potential is present or eliminated by connection of the two half-cells by means of a salt bridge.

6.9.1 Cells with a liquid junction potential

Consider two half-cells having identical electrodes dipping into their respective solutions containing the same electrolyte but at different mean ion activities. Electrical contact between the half-cells is made by the two solutions meeting at a junction. Such a cell may be represented by

$$\begin{array}{cccc} \mathrm{M} & |\ \mathrm{MX} & \vdots\ \mathrm{MX} & |\ \mathrm{M} \\ (a_+)_1, & (a_-)_1 & (a_+)_2, & (a_-)_2 \end{array}$$

where

$$(a_+)_2 > (a_+)_1, (a_-)_2 > (a_-)_1$$

with the electrodes reversible with respect to M^+ cations.

At the right-hand electrode, for the passage of 1 Faraday, 1 mole of M^+ ions will be deposited. However, migration of t_+ moles of M^+ across the junction will to some extent make good the loss of M^+ by the deposition process. Similarly, at the left-hand electrode, although 1 mole of M dissolves as M^+ ions, t_+ moles migrate out of the region towards the positive pole. This behaviour is summarized below.

Anode (−ve pole)	$(a_+)_1, (a_-)_1$	$(a_+)_2, (a_-)_2$	*Cathode* (+ve pole)
	$M \rightarrow M^+ + e$		$M^+ + e \rightarrow M$
	t_+ moles of M^+ migrate out	t_+ moles of M^+ migrate in	
	net gain of $(1 - t_+)$ moles of M^+ = t_- moles at $(a_+)_1$	net loss of $(1 - t_+)$ moles of M^+ = t_- moles at $(a_+)_2$	
	gain of t_- moles of X^- at $(a_-)_1$	loss of t_- moles of X^- at $(a_-)_2$	

t_+, t_- are the average transport numbers for the two activities of the electrolyte involved. It is clear that the overall process involves the transfer of material from the higher to the lower activity, viz.

$$\begin{array}{ccccc} t_-M^+ & + t_-X^- & \rightarrow & t_-M^+ & + t_-X^- \\ (a_+)_2 & (a_-)_2 & & (a_+)_1 & (a_-)_1 \end{array}$$

For the process the free energy change per Faraday is

$$\Delta G = RT \ln \frac{(a_+)_1^{t_-} (a_-)_1^{t_-}}{(a_+)_2^{t_-} (a_-)_2^{t_-}} \qquad (6.41)$$

since, by definition

$$a_+a_- = (a_\pm)^2; a_+^{t_-} a_-^{t_-} = (a_\pm)^{2t_-}$$

therefore,

$$\Delta G = 2t_{-}RT \ln \frac{(a_{\pm})_1}{(a_{\pm})_2}$$

therefore

$$E_{\text{cell}} = 2t_{-}\frac{RT}{F} \ln \frac{(a_{\pm})_2}{(a_{\pm})_1} \tag{6.42}$$

It should be stressed that the transport number which appears in the equation for the cell emf is that of the ionic species with respect to which the electrodes are not reversible. For the general case where the electrolyte species provides ν ions of which there are ν_+ cations and ν_- anions, equations (5.31) takes the forms

$$E = t_{-}\left(\frac{\nu}{\nu_+}\right)\left(\frac{RT}{nF}\right) \ln \frac{(a_{\pm})_2}{(a_{\pm})_1} \tag{6.43}$$

or,

$$E = t_{+}\left(\frac{\nu}{\nu_-}\right)\left(\frac{RT}{nF}\right) \ln \frac{(a_{\pm})_2}{(a_{\pm})_1} \tag{6.44}$$

according to whether the electrodes are reversible with respect to cations (equation (6.43)) or anions (equation (6.44)).

6.9.2 Cells with eliminated liquid junction potentials

Consider now the same half-cells as used in the previous section but joined via a salt bridge. This cell is represented by

$$\underset{}{\text{M}} \;\Big|\; \underset{(a_+)_1,\,(a_-)_1}{\text{MX}} \;\Big\|\; \underset{(a_+)_2,\,(a_-)_2}{\text{MX}} \;\Big|\; \text{M}$$

where

$$(a_+)_2 > (a_+)_1, \quad (a_-)_2 > (a_-)_1$$

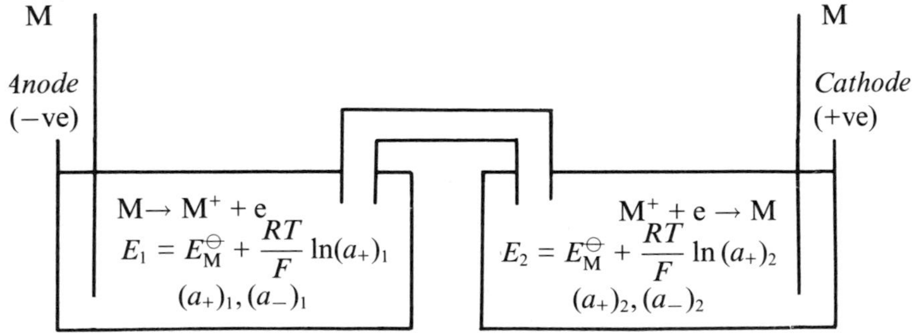

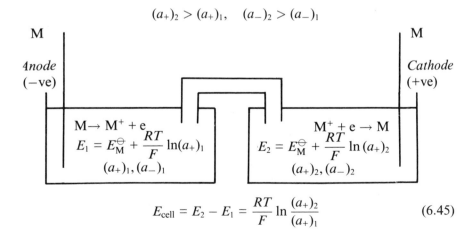

$$E_{\text{cell}} = E_2 - E_1 = \frac{RT}{F} \ln \frac{(a_+)_2}{(a_+)_1} \tag{6.45}$$

Now, since individual ion activity coefficients are inaccessible to measurement, the cell emf must be related to determinable mean ion activities. By definition,

$$\left(\frac{(a_\pm)_2}{(a_\pm)_1}\right)^2 = \frac{(a_+)_2(a_-)_2}{(a_+)_1(a_-)_1}$$

If it is assumed that

$$\frac{(a_+)_2}{(a_+)_1} \sim \frac{(a_-)_2}{(a_-)_1}$$

then

$$\frac{(a_\pm)_2}{(a_\pm)_1} \sim \frac{(a_+)_2}{(a_+)_1}$$

and

$$E = \frac{RT}{F} \ln \frac{(a_\pm)_2}{(a_\pm)_1} \tag{6.46}$$

6.9.3 Calculation of liquid junction potentials

It is apparent that the difference between the emf values of the cells considered in sections 6.9.1 and 6.9.2 gives the liquid junction potential (E_{lj}) involved in the former. Thus

$$E_{\text{lj}} = 2t_- \frac{RT}{F} \ln \frac{(a_\pm)_2}{(a_\pm)_1} - \frac{RT}{F} \ln \frac{(a_\pm)_2}{(a_\pm)_1} \tag{6.47}$$

Therefore,

$$E_{\text{lj}} = (2t_- - 1)\frac{RT}{F} \ln \frac{(a_\pm)_2}{(a_\pm)_1} \tag{6.48}$$

Now

$$t_+ + t_- = 1$$

Therefore,

$$(2t_- - 1) = t_- - t_+$$

Therefore,

$$E_{\text{lj}} = (t_+ - t_-)\left(\frac{RT}{F}\right) \ln \frac{(a_\pm)_2}{(a_\pm)_2} \tag{6.49}$$

The general form of equation (6.48) is

$$E_{\text{lj}} = \left(t_\mp \left(\frac{\nu}{\nu_\pm}\right) - 1\right) \frac{RT}{nF} \ln \frac{(a_\pm)_2}{(a_\pm)_1} \tag{6.50}$$

6.10 Membrane equilibria

When two solutions of the same salt at different concentrations are separated by a membrane which is permeable to both ion species, the potential across the membrane is identifiable with the junction potential between the solutions given by equation (6.50). If the membrane has a pore size that shows *restricted* permeability to one ion species, the transport number appearing in equation (6.50) is that of this species across the membrane.

Such phenomena are of great significance in biological cell membrane systems. Cell solutions usually contain a higher proportion of potassium salts than sodium salts while external solutions usually show the reverse. The cell surface may therefore be treated as a membrane separating solutions of potassium and sodium ions and which exhibits considerably lower permeability to sodium than to potassium ions. Potassium ions may therefore pass through the membrane from inside to outside the cell at a faster rate than sodium ions pass into the cell. The greater tendency for small, positively charged ions to pass to the outside of the cell leads to a charge distribution in which the interior tends to be negatively charged with respect to the exterior. Although a somewhat crude description, it does in fact summarize the essential condition at the membrane in nerve and muscle cells where electrical impulses are passed from one cell to another.

It is found that the above polarity may be reversed upon addition of species which may disturb the structure of the membrane and open it for transmission of larger ions. Such is the case if a quaternary ammonium salt is introduced near the membrane surface. It would appear that the organic cations are able to penetrate the membrane faster than the inorganic ions, so that they (temporarily at least) open the membrane structure and allow freer passage of inorganic ions. Such a mechanism has been postulated as a contributory factor in the control of the transmission of electrical impulse from one nerve cell to another and from nerve to muscle cells in an organism.

6.10.1 Membrane potentials

Consider two solutions with different concentrations of electrolyte and non-electrolyte species separated by a membrane. Suppose that the membrane allows passage of solvent molecules and at least one ionic species. After a time the concentrations of species in solutions either side of the membrane will reach equilibrium values. The concentration of any species which *cannot* pass through the membrane will remain constant.

When equilibrium is established, an *osmotic pressure difference* is set up across the boundary between the two solutions associated with an unequal distribution of diffusible material. Additionally, a *potential difference* (the membrane potential) is established across the membrane.

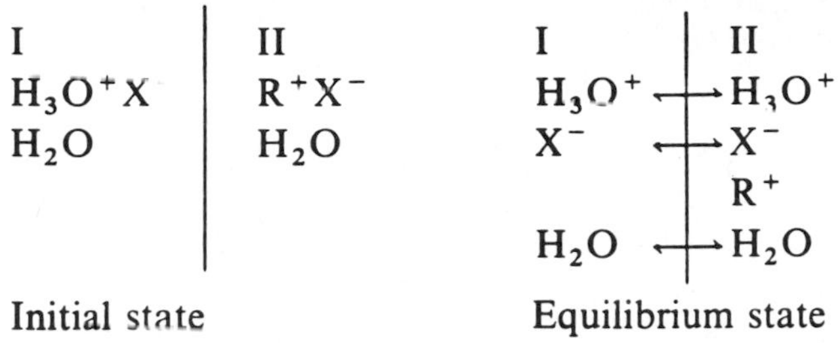

(a) **(b)**

Figure 6.18 Distribution of ions within a system of solutions of electrolyte, RX, and acid, HX, separated by a membrane impermeable to R^+. (a) Initial state, (b) equilibrium state.

Consider the simple system in which an aqueous solution of a salt comprising R^+ and X^- ions is separated from an aqueous solution of the acid HX by a membrane which is *impermeable to* R^+. If the two solutions are labelled I and II, the initial and equilibrium situations may be represented by Figure 6.18.

It is assumed that R^+ on account of its size cannot diffuse through the membrane but that H_3O^+ and X^- will pass through until equilibrium is achieved. In this state the concentrations of H_3O^+ and X^- to the left of the membrane must be equal. In solution II to the right of the membrane X^- must electrically balance *both* R^+ and H_3O^+.

At equilibrium the following relationships must hold for electrochemical and chemical potentials of the three diffusible species.

$$
\begin{aligned}
{}^{\mathrm{I}}\tilde{\mu}_{\mathrm{H_3O^+}} &= {}^{\mathrm{II}}\tilde{\mu}_{\mathrm{H_3O^+}} \\
{}^{\mathrm{I}}\tilde{\mu}_{\mathrm{X^-}} &= {}^{\mathrm{II}}\tilde{\mu}_{\mathrm{X^-}} \\
{}^{\mathrm{I}}\mu_{\mathrm{H_2O}} &= {}^{\mathrm{II}}\mu_{\mathrm{H_2O}}
\end{aligned}
\qquad (6.51)
$$

Since the membrane is not permeable to R^+, the pressure of the two phases will differ and it is necessary to allow for the variation of chemical potential with pressure, $(\partial\mu_{\mathrm{i}}/\partial P)_{\mathrm{T}} = V_i$. The chemical potential of species i is related to its activity a_{i} by

$$\mu_{\mathrm{i}} = \mu_{\mathrm{i}}^{\ominus} + RT \ln a_{\mathrm{i}} \qquad (6.52)$$

Now, $\mu_{\mathrm{i}}^{\ominus}$, which occurs in both electrochemical and chemical potentials, may be expressed as

$$\mu_{\mathrm{i}}^{\ominus} = \mu_{\mathrm{i}}^{*} + \int_0^P V^{\ominus}\,\mathrm{d}P \qquad (6.53)$$

Where μ_{i}^{*}, is the value corresponding to $P = 0$ and $V^{\ominus}$ is the molar volume in the standard state. If, as an approximation we neglect the compressibility of the solution, we may write $V^{\ominus} = V =$ constant. Thus

$$\mu_{\mathrm{i}}^{\ominus} = \mu_{\mathrm{i}}^{*} + VP \qquad (6.54)$$

and

$$\mu_i = \mu_i^* + VP + RT \ln a_i \tag{6.55}$$

Therefore, for water, acting as solvent in this case,

$${}^{I}\mu_{H_2O} = {}^{I}\mu_{H_2O}^* + V_{H_2O}P^{I} + RT \ln {}^{I}a_{H_2O} \tag{6.56}$$

and

$${}^{II}\mu_{H_2O} = {}^{II}\mu_{H_2O}^* + V_{H_2O}P^{II} + RT \ln {}^{II}a_{H_2O} \tag{6.57}$$

Therefore

$$V_{H_2O}(P^{I} - P^{II}) = RT \ln \frac{{}^{II}a_{H_2O}}{{}^{I}a_{H_2O}} \tag{6.58}$$

and since for charged species the following holds

$$\tilde{\mu}_i = \mu_i + z_i F\phi \qquad \text{(see equations (2.5), (5.2))}$$

it is possible to arrive at expressions for the ionic diffusible species which complement equation (6.58) viz.

$$(P^{I} - P^{II})V_{H_3O^+} = RT \ln \frac{{}^{II}a_{H_3O^+}}{{}^{I}a_{H_3O^+}} + F(\phi^{II} - \phi^{I}) \tag{6.59}$$

and

$$(P^{I} - P^{II})V_{X^-} = RT \ln \frac{{}^{II}a_{X^-}}{{}^{I}a_{X^-}} - F(\phi^{II} - \phi^{I}) \tag{6.60}$$

where $(\phi^{II} - \phi^{I}) = \Delta\phi$ (the membrane potential).

Elimination of $\Delta\phi$ between equations (6.59) and (6.60) gives

$$(P^{I} - P^{II}) = \left(\frac{RT}{V_{H_3O^+} + V_{X^-}}\right) \ln \left(\frac{{}^{II}a_{H_3O^+}\,{}^{II}a_{X^-}}{{}^{I}a_{H_3O^+}\,{}^{I}a_{X^-}}\right) \tag{6.61}$$

Elimination of $(P^{I} - P^{II})$ between equations (6.58) and (6.61) gives

$$\left(\frac{1}{V_{H_2O}}\right) \ln \left(\frac{{}^{II}a_{H_2O}}{{}^{I}a_{H_2O}}\right) = \left(\frac{1}{V_{H_3O^+} + V_{X^-}}\right) \ln \left(\frac{{}^{II}a_{\pm}^2}{{}^{I}a_{\pm}^2}\right) \tag{6.62}$$

or,

$$\frac{({}^{I}a_{\pm})^2}{({}^{I}a_{H_2O})^x} = \frac{({}^{II}a_{\pm})^2}{({}^{II}a_{H_2O})^x} \tag{6.63}$$

where

$$x = \frac{V_{H_3O^+} + V_{X^-}}{V_{H_2O}}$$

If $a_{H_2O} \sim 1$ in both phases, equation (6.63) becomes

$$^{I}a_{\pm} \sim {}^{II}a_{\pm} \tag{6.64}$$

The expression for $\Delta\phi$ may now be obtained by eliminating $(P^{I} - P^{II})$ between equation (6.58) and either equation (6.59) or equation (6.60)

$$\Delta\phi = \frac{RT}{F} \ln\left(\frac{^{II}a_{H_3O^+}}{^{I}a_{H_3O^+}}\right)\left(\frac{(^{I}a_{H_2O})^{x+}}{(^{II}a_{H_2O})^{x+}}\right) = \frac{RT}{F} \ln\left(\frac{^{I}a_{X^-}}{^{II}a_{X^-}}\right)\left(\frac{(^{II}a_{H_2O})^{x-}}{(^{I}a_{H_2O})^{x-}}\right) \tag{6.65}$$

where

$$x^{+} = \frac{V_{H_3O^+}}{V_{H_2O}}; \quad x^{-} = \frac{V_{X^-}}{V_{H_2O}}$$

If again

$$^{I}a_{H_2O} \sim {}^{II}a_{H_2O}$$

$$\Delta\phi = \frac{RT}{F} \ln\left(\frac{^{II}a_{H_3O^+}}{^{I}a_{H_3O^+}}\right) = \frac{RT}{F} \ln\left(\frac{^{I}a_{X^-}}{^{II}a_{X^-}}\right) \tag{6.66}$$

The general expression for the membrane potential is

$$\Delta\phi = \frac{RT}{nF} \ln\frac{^{II}a_{i}}{^{I}a_{i}} = \frac{RT}{F} \ln\lambda \tag{6.67}$$

Here λ is known as the Donnan distribution coefficient, expressed by the membrane equilibrium condition in the presence of various ion types by

$$\left(\frac{^{II}a_{+}}{^{I}a_{+}}\right) = \left(\frac{^{II}a_{2+}}{^{I}a_{2+}}\right)^{1/2} = \left(\frac{^{II}a_{3+}}{^{I}a_{3+}}\right)^{1/3} = - - -$$

$$- - - = \left(\frac{^{I}a_{-}}{^{II}a_{-}}\right) = - - -$$

$$\left(\frac{^{I}a_{3-}}{^{II}a_{3-}}\right)^{1/3} = - - - - = \lambda \tag{6.68}$$

Although the individual ion activities contained in λ cannot be measured, they may be replaced, for experimental purposes, by mean ion activities provided that solutions of such dilution are used that the Debye–Hückel limiting law holds.

The Donnan equilibrium has frequently been cited as controlling the movement of water and electrolytes into and out of biological cells. While chemical processes within the living cell can control the permeability of cell membranes to various species, explanations of the origin of steady natural cell membrane potentials cannot successfully be built on a *thermodynamic* equation. Expressions for such potentials of the *form* of the Donnan equation may be established from *kinetic* principles which involve *steady-state* concentration differences of species within and external to natural cells.

6.10.2 Dialysis

Membrane equilibria are made use of in the separation by dialysis of inorganic ions from solutions of biologically important polyelectrolytes such as nucleic acids and proteins. Dialysis is based on the principle that a membrane allows free passage of small particles in true solution through it while retaining particles of colloidal dimensions. If the solvent on the exit side of the membrane is continuously renewed, the particles escaping through the membrane are removed, further transference through the membrane encouraged and separation of the colloid species made feasible. In the technique of electrodialysis, removal of ions is made easier by an electric field. The solution containing the macroparticles is placed between two membranes with pure solvent on either side and an emf imposed between electrodes placed in the solvent compartments.

It is sometimes advantageous to take advantage of membrane hydrolysis and this is used to convert proteins into acidic forms without recourse to conventional chemical means which might interfere with the system. Consider dialysis into pure water of a salt NaR from a solution through a membrane which allows passage of Na^+ but is impermeable to R^-

I		II
H_2O	|	H_2O, NaR
	Na^+ $\longleftarrow$ OH^-	
	|	RH

Sodium ions from II diffuse into I along with an equivalent number of hydroxyl ions. These latter arise from the dissociation of water which is necessary to maintain electroneutrality of I. The hydrogen ions produced by this process then associate with anions R^- to form the weak acid RH and maintain the electroneutrality of II. The initial and final equilibrium concentrations for the two solutions on either side of the membrane are as follows:

	I	II
Initial State	$[Na^+] = 0$	$[Na^+] = [R^-] = c$
Equilibrium condition	$[Na^+] = x = [OH^-]$	$[Na^+] = c - x$ $[R^-] = c$ $[H^+] = x$

At equilibrium, in accordance with equation (6.64)

$$({}^{\mathrm{I}}a_{Na^+}\,{}^{\mathrm{I}}a_{OH^-}) = ({}^{\mathrm{II}}a_{Na^+}\,{}^{\mathrm{II}}a_{OH^-})$$

therefore,

$$x^2 \sim (c - x)\left(\frac{K_w}{x}\right)$$

therefore,

$$x = (K_w c)^{\frac{1}{3}} \tag{6.69}$$

Although the number of sodium ions passing into I to meet the equilibrium conditions is not large, continuous replacement of solution I by pure water forces the process to continue by encouraging a continuous movement towards equilibrium. In this way hydrolysis of the species NaR may be effected to a significant extent.

6.10.3 Ion-exchange resins

Typical cation exchange resins possess open three-dimensional structures with sulphonic acid groups attached in a regular manner throughout the network. For electroneutrality, there are required to be cations contained within the network (e.g. hydrogen or sodium ions) equal in number to the acid groups. If such a resin is placed in an acid or salt solution, water enters the free space in the network and causes swelling. We now have the situation where anions and cations of the dissolved species can move between the external solution and that inside the resin. The sulphonic acid groups, however, are fixed—not in this case by a membrane impermeable to their motion through it, but by chemical bonding. The effect will nevertheless be the same as for the membrane system. The solution inside the resin will show a larger osmotic pressure and the resin will continue to swell until a balance is achieved with the restoring forces of the extended structure. There will be an unequal distribution of electrolyte ions between the resin solution and the external solution.

In the case of a hydrogen ion-exchange resin placed in a solution of a 1:1 acid HA, while the ratio $[H^+]/[A^-]$ in the external solution must be 1:1, the ratio internally will be found to be up to several orders of magnitude greater than this, i.e. the hydrogen ions are allowed ready access to the

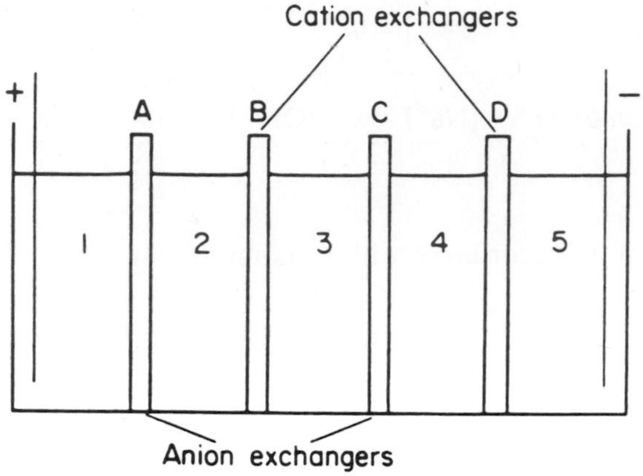

Figure 6.19 Schematic desalination plant.

interior of the resin—in fact they pass almost unhindered through the resin with a transport number very close to unity—while the resin presents an almost impermeable barrier to the A^- anions. Anion-exchange resins work on the same principle with cathodic groups distributed through the interior network of the resin. Free passage of anions is now possible with almost total restriction on the entry and passage of cations through the resin.

Combinations of cation- and anion-exchange resins are used in electrolytic desalination plants to produce fresh water from brackish water or even sea water. The salt water is placed in a series of compartments separated alternately by anion and cation exchangers (Figure 6.19).

An emf applied between electrodes placed in the extreme cells constrains the ions to move in opposite directions through the solution in the field produced. Free movement is not possible, since it is restricted by the ion exchangers. Thus the anion exchanger A allows free passage of anions from solution 2 to solution 1 but cations cannot pass from left to right through A to solution 2. Similarly, the cation exchanger B allows free passage of cations into solution 3 but does not allow anions through from 3 to 2. Thus, solution 1 becomes more concentrated in ions while solution 2 becomes more dilute. Similarly, solutions 3 and 5 become more concentrated and solution 4 more dilute. Such separation can continue until the desalination of solutions 2 and 4 is about 95% complete.

Problems

6.1 Write the half-cell and overall cell reactions for the following cells:

(i) $Zn|ZnSo_4(aq., 0.01 mol\,kg^{-1})||CuSO_4(aq., 0.001 mol\,kg^{-1})|Cu$;

(ii) $Pb|Pb(NO_3)_2$ (aq., 0.01 mol kg^{-1})||KCl (aq., satd.), $Hg_2Cl_2(s)|Hg$;
(iii) $Zn|ZnSO_4$ (aq., 0.01 mol kg^{-1})||$FeSO_4$, $Fe_2(SO_4)_3$, H_2SO_4, aq.|Pt
(each species at 0.001 mol kg^{-1}).

Estimate the emf values of the cells at 298 K, using the Debye–Hückel equation to estimate activity coefficients and the $E^{\ominus}$ data from Tables 5.1 and 5.2.

6.2 Devise electrochemical cells suitable for studying the following equilibria and calculate their standard emf values at 298 K.

(i) $Ce^{4+} + Fe^{2+} \rightleftharpoons Fe^{3+} + Ce^{3+}$;
(ii) $Zn + 2AgCl \rightleftharpoons Zn^{2+} + 2Ag + 2Cl^-$.

(The standard potential of the silver/silver chloride electrode is +0.222 V.)

6.3 For the cell Pt|H_2(1 atm)|HCl(0.0256 mol kg^{-1}|AgCl|Ag, the emf at 298 K is 0.4182 V. Estimate the standard emf of the cell and the standard potential of the silver/silver chloride electrode.

6.4 For the following cell

$$Hg|Hg_2Cl_2(s), KCl\,(aq., satd.)||KCl\,(aq., 0.08\ mol\,kg^{-1}), AgCl(s)|Ag,$$

the emf was observed to be 0.051 V at 298 K. If the potential of the saturated calomel electrode at 298 K is 0.242 V, and the standard potential of the silver/silver chloride electrode is 0.222 V, obtain an estimate of the activity coefficient of the chloride ions in 0.08 mol kg^{-1} aqueous potassium chloride. Compare this value with those obtained from the Debye–Hückel limiting and extended laws.

6.5 Calculate the change in standard free energy at 298 K for the reaction

$$Sn(s) + Pb^{2+} = Sn^{2+} + Pb(s)$$

given that $E^{\ominus}_{Sn^{2+}/Sn} = -0.140$ V and $E^{\ominus}_{Pb^{2+}/Pb} = -0.126$ V.

6.6 The emf of the cell

$$Ag|AgNO_3\,(aq., 0.01\,mol\,kg^{-1})||AgNO_3\,(aq., 0.1\,mol\,kg^{-1}|Ag$$

is 0.054 V, while that of the cell

$$Ag|AgNO_3\,(aq., 0.01\,mol\,kg^{-1})\vdots AgNO_3\,(aq., 0.1\,mol\,kg^{-1})|Ag$$

is 0.058 V. If the mean ion activity coefficients of aqueous 0.01 and 0.10 mol kg^{-1} silver nitrate are 0.898 and 0.735, respectively, obtain a mean value for the transport number of the silver ion.

6.7 Consider the situation where a solution of sodium chloride of initial concentration C_1 is brought into Donnan equilibrium with a solution of the sodium salt of a protein, NaR, at initial concentration C_2, the two solutions being separated by a membrane permeable to Na^+ and Cl^- ions but impermeable to R^-. Show that at equilibrium the number of moles, x, of NaCl which have entered the compartment containing the protein is given by $x = C_1^2/(C_2 + 2C_1)$.

6.8 A series of experiments was conducted in which 0.01 mol dm^{-3} aqueous solutions of colloidal electrolytes of general formula Na_yX were placed on one side of a membrane and equilibrated with an equal volume of 0.1 mol dm^{-3} sodium chloride solution on the other side. For the cases of $y = 10$, 15 and 20, calculate the fraction of sodium chloride which would have entered the colloid-containing section when Donnan equilibrium was established.

6.9 A cellophane bag containing 100 cm^3 of 0.1 mol dm^{-3} NaR solution was placed in 100 cm^3 of pure water and the whole system brought to equilibrium at 298 K. Calculate (i) the pH inside and outside the bag, and (ii) the potential difference across the bag wall. Assume that NaR is completely dissociated.

6.10 An approximately 250 cm^3 capacity rectangular perspex cell was divided into two equal compartments by a mechanically supported leak-proof membrane. In the left hand compartment were placed 100 cm^3 of very dilute hydrochloric acid containing 1.5 g of a completely dissociated monobasic colloidal acid. The membrane was impermeable to the anion of the acid. The right-hand compartment was filled with 100 cm^3 distilled water and the whole system was thermostated at 298 K and allowed to equilibrate. After equilibration, the pH of the solution to the right of the membrane was found to be 3.37 while that to the left was 2.72. Estimate the relative molecular mass of the colloidal acid.

6.11 Show that if, in the Donnan equilibration of the sodium salt of a protein Na_zR, the concentration of the protein is very much lower than that of other salts, the membrane potential may be expressed by

$$\Delta\phi \cong \left(\frac{RT}{2F}\right)\left(\frac{z\,{}^{\mathrm{I}}m_{\mathrm{R}^{z-}}}{{}^{\mathrm{II}}m}\right)$$

where ${}^{\mathrm{I}}m_{\mathrm{R}^{z-}}$ is the molality of the non-permeable protein solution and ${}^{\mathrm{II}}m$ is the molality of the electrolyte solution with which it is equilibrated.

6.12 The following values of membrane potential were obtained when the sodium salt of haemoglobin (Na_zHb) was equilibrated with sodium chloride, the pH of both solutions being maintained at 7.8 with phosphate buffer:

${}^{\mathrm{I}}m_{\mathrm{Hb}^{z-}}$ (mol kg^{-1})	$\Delta\phi$(mV)
1.0×10^{-3}	-0.245
2.0×10^{-3}	-0.485
3.0×10^{-3}	-0.720
4.0×10^{-3}	-0.960
5.0×10^{-3}	-1.204

Calculate the average number of sodium ions combined with the haemoglobin molecule if ${}^{\mathrm{II}}m = 0.453$ mol kg^{-1}.

7 Electrode processes

7.1 Equilibrium and non-equilibrium electrode potentials

In Chapter 6 the reversible potential adopted by a metal electrode M when placed in a solution of ions M^+ was considered. The steady potential resulted from the rapid establishment of the equilibrium

$$M \rightleftharpoons M^+ + e \tag{7.1}$$

no net current flowing when the forward and backward rates of the reaction are equal. The further such an equilibrium lies to the *right* the more *negative* is the electrode potential.

If a potential rather more oxidizing or reducing than the equilibrium value is imposed upon such an electrode a small net current can be made to flow but with small disturbance of the electrode potential. This is because, although the applied excess potential causes a net reaction in one direction, the equilibrium re-asserts itself so rapidly that the electrode potential hardly alters. This process is of course used as the basis of the experimental determination of the electrode potential: exact balance of an applied potential from a potentiometer and that of the electrode system under study is recognized by zero net current in the galvanometer (see Figure 6.17).

7.1.1 Current–potential relationships for fast and slow systems

An idealized graph of current versus potential for a fast system takes the form shown in Figure 7.1. The small slope on the linear part of the curve is due to the effects of mass transfer processes which are slow relative to the electron exchange rate. The dashed regions represent the condition that these mass transfer processes control the overall rate of reaction.

It has been seen that some electrodes never show reversible potentials under experimental conditions because the rate of attainment of equilibrium is low: such processes are more realistically called *slow* rather than irreversible. In order for even a small cathodic or anodic current to flow a potential well in excess of the equilibrium value must be applied to the respectively negative or positive side of this value. For such cases a current voltage curve will take the form shown in Figure 7.2.

Chapter 6 introduced the equal and opposite currents flowing at E_{eq}. This *exchange current*, I_0, features importantly in electrode kinetics considered in the next section. It is, however, useful to introduce the concept here since

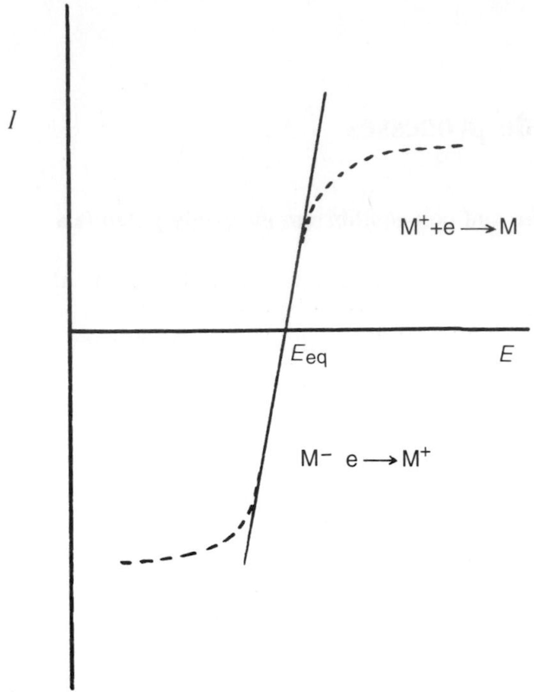

Figure 7.1 Idealized current-potential relationship for the system $M \rightleftharpoons M^+ + e$ in which equilibrium is established rapidly. The dashed regions correspond to the condition that mass transfer processes control the overall rate of the net anodic or cathodic reaction.

the values of *net* currents, consequent upon imposition of applied potentials, differ markedly relative to I_0 for reversible and irreversible systems. The two extremes of behaviour are contrasted in Figure 7.3.

For reversible systems I_0 is large: a small net change ΔI is sustainable without undue departure from reversibility (Figure 7.3(a)) and Nernstian conditions still apply. For irreversible processes (Figure 7.3(b)) ΔI is large with respect to I_0 (which is very small) and disturbance to the equilibrium is significant.

7.1.2 Mass transfer and electron-exchange processes

In considering electron-exchange reactions at electrodes, the layer of solution very close to the electrode surface comes under scrutiny. When conditions exist or are imposed such that the electronic levels in the electrode material and an electroactive solute material are compatible, electron transfer may take place. It will do so according to the Frank–Condon Principle

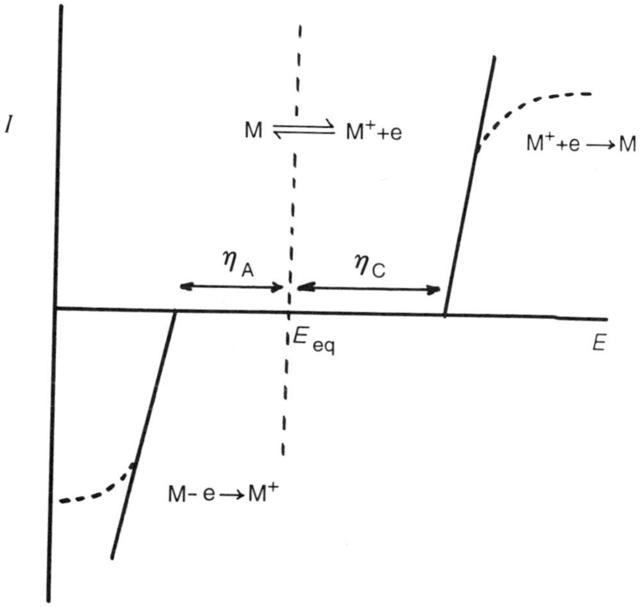

Figure 7.2 Idealized current–potential curve for a slow or irreversible system which requires an overvoltage, η_C, for the process $M^+ + e \rightarrow M$ and η_A for $M \rightarrow M^+ + e$. Here the equilibrium $M \rightleftharpoons M^+ + e$ is not established under the experimental conditions used and E_{eq} is not realizable. The dashed curves correspond to control of the rates of the net anodic and cathodic reactions by mass-transfer processes.

according to which the rates of electron transfer take place very much more rapidly than molecular rearrangements. The interplay of solvent interactions, applied potential and the very large field gradient within the double layer (of the order of 10^9 V m^{-1}) serves to produce a solute structure to (or from) which electrons may be transferred. When this happens a current flows through the electrode, through the external circuit and through the complementary electrode.

In order to sustain a current for a given potential it is necessary for the supply of material at the electrode surface to be sustained and the movement of ions through the solution will effectively constitute the *electrolytic current* which complements the *electronic current* flowing in the external circuit. Such movement cannot be increased indefinitely and a point must be reached where solute species react with the electrode as fast as they reach it. The current then approaches the limiting values indicated by the dashed curves in Figures 7.1 and 7.2. Before proceeding much further, the nature of the various mass transfer processes should be considered.

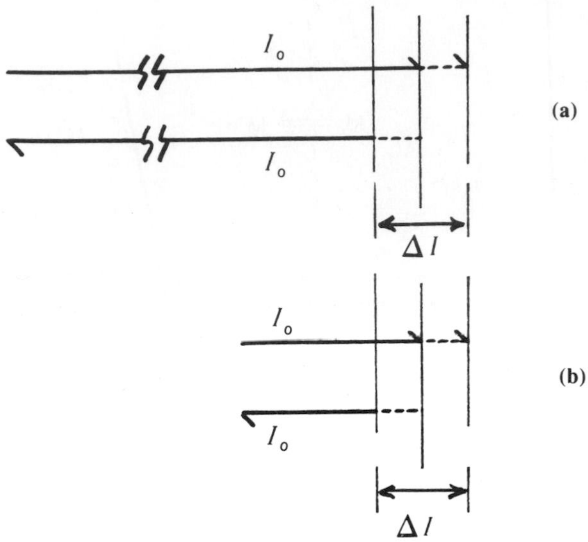

Figure 7.3 Diagrammatic representation of relative magnitudes of a net current ΔI and the exchange current I_0 for (a) a reversible system where $\Delta I \ll I_0$ and (b) an irreversible system where $\Delta I \sim I_0$ or $\Delta I > I_0$.

7.1.3 Types of mass transfer

There are three types of mass transfer.

1. *Migration.* The movement of cations and anions through a solution under the influence of an applied potential between electrodes placed in that solution. This phenomenon has been met with already and the significance of transport numbers has been discussed.
2. *Diffusion.* An electrode reaction depletes the concentration of oxidant or reductant at an electrode surface and produces a concentration gradient there. This gives rise to the movement of species from the higher to the lower concentration. Unlike migration, which only occurs for charged particles, diffusion occurs for both charged and uncharged species.
3. *Convection.* This includes thermal and stirring effects which can arise extraneously through vibration, shock and temperature gradients. More importantly, some electro-analytical methods are based upon controlled convection at electrodes. Considerable variation in the relative rates of mass transfer and electron-exchange processes is possible: the simplest interpretation of electrochemical reactions assumes that the rates of all processes occur rapidly.

7.2 The kinetics of electrode processes: the Butler–Volmer equation

The equilibrium

$$\mathrm{M} \underset{v_c}{\overset{v_a}{\rightleftharpoons}} \mathrm{M}^{n+} + ne \qquad \text{(see equation (7.1))}$$

may be considered from a kinetic viewpoint: v_a is the rate of the ionization (dissolution) or anodic process while v_c represents the rate of the discharge or cathodic process.

There will be an activation energy barrier to both processes which can be visualized as in Figure 7.4. Here the reaction profile involving the transition state (Figure 7.4(a)) may be seen as reflecting the intersection of reactant and use product energy parabolas (Figure 7.4(b)).

The rates for the processes may be expressed in terms of the Arrhenius equation, viz.

$$v_a = k_a e^{-\Delta G_a^{\ddagger}/RT} \tag{7.2}$$

and

$$v_c = k_c[\mathrm{M}^{n+}]e^{-\Delta G_c^{\ddagger}/RT} \tag{7.3}$$

where k_a, k_c are corresponding rate constants and $[\mathrm{M}^{n+}]$ is the concentration of metal ions (strictly the value at the electrode surface which, for present purposes, may be regarded as the outer Helmholtz plane).

At equilibrium $v_a = v_c$ so that

$$k_a \exp\left[-\frac{\Delta G_a^{\ddagger}}{RT}\right] = k_c[\mathrm{M}^{n+}]\exp\left[-\frac{\Delta G_c^{\ddagger}}{RT}\right]$$

Therefore

$$\frac{\exp\left[-\Delta G_a^{\ddagger}/RT\right]}{\exp\left[-\Delta G_c^{\ddagger}/RT\right]} = \frac{k_c}{k_a}[\mathrm{M}^{n+}]$$

so that, in terms of $\Delta G = \Delta G_c^{\ddagger} - \Delta G_a^{\ddagger}$

$$\exp\left[-\frac{\Delta G}{RT}\right] = \frac{k_c}{k_a}[\mathrm{M}^{n+}] = \exp[nF\Delta\phi RT]$$

Therefore

$$\frac{nF\Delta\phi}{RT} = \ln\left(\frac{k_c}{k_a}\right) + \ln[\mathrm{M}^{n+}]$$

or

$$\Delta\phi = \frac{RT}{nF}\ln\left(\frac{k_c}{k_a}\right) + \frac{RT}{nF}\ln[\mathrm{M}^{n+}]$$

or

$$\Delta\phi = \Delta\phi^{\ominus} + \frac{RT}{nF}\ln[\mathrm{M}^{n+}] \tag{7.4}$$

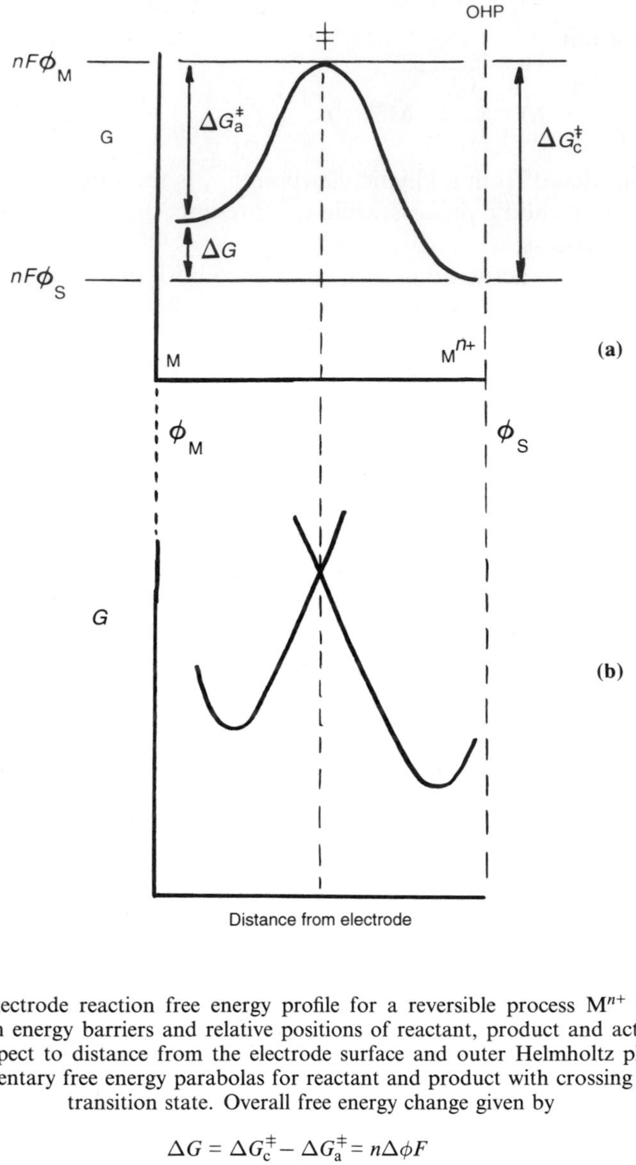

Figure 7.4 Electrode reaction free energy profile for a reversible process $M^{n+} + ne \rightleftharpoons M$. (a) Activation energy barriers and relative positions of reactant, product and activated complex with respect to distance from the electrode surface and outer Helmholtz plane (OHP). (b) Complementary free energy parabolas for reactant and product with crossing point at the transition state. Overall free energy change given by

$$\Delta G = \Delta G_c^\ddagger - \Delta G_a^\ddagger = n\Delta\phi F$$

$$\Delta\phi = \phi_M - \phi_S$$

which, in terms of the more commonly used symbols for electrode potentials becomes

$$E = E^\ominus + \frac{RT}{nF}\ln[M^{n+}] \tag{7.5}$$

Equation (7.5) clearly identifies with equation (6.20), i.e. the Nernst equation which has now been derived kinetically. A rather different insight is now given into $E^{\ominus}$ as a function of the rate constants of the two processes occurring at the electrode.

In irreversible systems no perceptible reaction occurs at E_{eq} or $\Delta\phi^{\ominus}$. The transition state can now only be formed by application of extra (cathodic or anodic) potential—the overvoltage, η. Thus to effect reduction of M^{n+} for this case the magnitudes of $\Delta G_c^{\ddagger}$ and $\Delta G_a^{\ddagger}$ become

$$\Delta G_c^{\ddagger} = (\Delta G_c^{\ominus})^{\ddagger} + \alpha nF\eta \tag{7.6}$$

and

$$\Delta G_a^{\ddagger} = (\Delta G_a^{\ominus})^{\ddagger} - (1 - \alpha)nF\eta \tag{7.7}$$

In this view η is seen as serving *two* functions: (i) part of it (a fraction α) assists the cathodic process; (ii) a fraction $(1 - \alpha)$ retards the anodic process. α is known as the *transfer coefficient*. This interpretation may be represented by Figure 7.5(a). Figure 7.5(b) shows a view of the effect of the overvoltage η imposed across the region between the electrode surface and the outer Helmholtz plane (OHP). Having reached the OHP a metal ion must then cross the interfacial region where it will experience increasing influence by η: if it reached as far as the surface proper it would experience the full value, but only a fraction of it, α, is required to produce the transition state which occurs at an intermediate distance.

It is seen that the idea is restricted in that it is not easy to represent both the cathodic and anodic functions on the same figure. However, the same argument as given above can be applied to the anodic overvoltage required to produce a net dissolution.

In terms of the new activation energies the rates of cathodic and anodic processes now become.

$$v_c' = k_c[M^{n+}]\exp\left[-\frac{\{(\Delta G_c^{\ominus})^{\ddagger} + \alpha nF\eta\}}{RT}\right] \tag{7.8}$$

and

$$v_a' = k_a \exp\left[-\frac{\{(\Delta G_a^{\ominus})^{\ddagger} - (1 - \alpha)nF\eta\}}{RT}\right] \tag{7.9}$$

Since for this case the condition of equal rates is not realized, a net cathodic or anodic reaction occurs for which the corresponding net current may be calculated. Respective rates may be expressed in terms of cathodic and anodic current densities as follows:

$$v_c' = \frac{i_c}{nF} \tag{7.10}$$

and

$$v_a' = \frac{i_a}{nF} \tag{7.11}$$

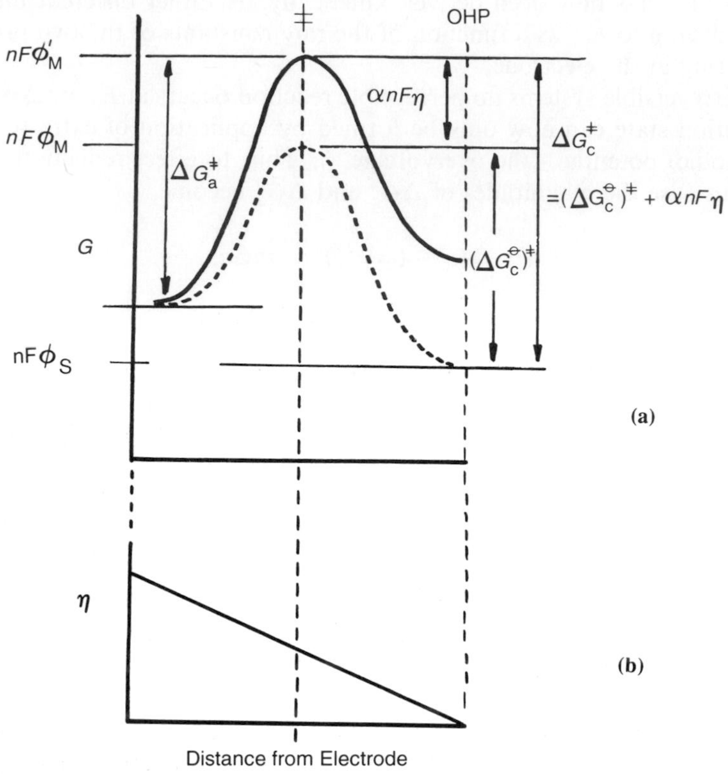

Figure 7.5 Electrode reaction free energy profile for an irreversible process. (a) Formation of transition state by imposition of cathodic overvoltage $\eta = \Delta\phi - \Delta\phi^{\ominus}$ where $\Delta\phi = \phi'_M - \phi_S$ and $\Delta\phi^{\ominus} = \phi'_M - \phi_S$. (b) Diagrammatic representation of the effect of the imposition of η across the region between electrode surface and outer Helmholtz plane.

Therefore,

$$i_c = nFk_c[M^{n+}]\exp\left[-\frac{(\Delta G_c^{\ominus})^{\ddagger}}{RT}\right]\exp\left[-\frac{\alpha nF\eta}{RT}\right] \tag{7.12}$$

and

$$i_a = nFk_a\exp\left[-\frac{(\Delta G_a^{\ominus})^{\ddagger}}{RT}\right]\exp\left[\frac{(1-\alpha)nF\eta}{RT}\right] \tag{7.13}$$

The forms of equations (7.12) and (7.13) may be simplified to

$$i_c = i_0\exp\left[-\frac{\alpha nF\eta}{RT}\right] \tag{7.14}$$

and

$$i_a = i_0 \exp\left[\frac{(1-\alpha)nF\eta}{RT}\right] \tag{7.15}$$

or, for currents I_c, I_a, I_0.

$$I_c = I_0 \exp\left[-\frac{\alpha nF\eta}{RT}\right] \tag{7.16}$$

and

$$I_a = I_0 \exp\left[\frac{(1-\alpha)nF\eta}{RT}\right] \tag{7.17}$$

i_0 and I_0 being the exchange current density and exchange current respectively.

It is seen that equations (7.16) and (7.17) are consistent with the condition

$$I_c = I_a = I_0 \text{ when } \eta = 0.$$

Clearly for the hypothetical reaction scheme proposed, a net cathodic current density $i = i_c - i_a$ is given by

$$i = i_0\left\{\exp\left[-\frac{\alpha nF\eta}{RT}\right] - \exp\left[\frac{(1-\alpha)nF\eta}{RT}\right]\right\} \tag{7.18}$$

Equation (7.18) is known as the Butler–Volmer equation. This expression only strictly holds in the above form for processes involving a single electron. When electrochemical reactions involving more than one electron are considered it is more rigorous to replace α by α_c and to replace $(1-\alpha)$ by α_a since $\alpha_c + \alpha_a \neq 1$ except for $n = 1$, thus

$$i = i_0\left\{\exp\left[-\frac{\alpha_c nF\eta}{RT}\right] - \exp\left[\frac{\alpha_a nF\eta}{RT}\right]\right\} \tag{7.19}$$

Application of equation (7.19) may be demonstrated by consideration of an electrode process with the following characteristics:

$$n = 1;\ \alpha_c = 0.65;\ \alpha_a = 0.35,\ I = 1\,\text{mA};\ A = 1\,\text{cm}^2$$

For a cathodic overvoltage $\eta = 0.1$ V a net cathodic current flows given by (noting that $F/RT = 38.95\,\text{C J}^{-1}$ at 298 K)

$$\begin{aligned}(I_c)_{\text{net}} &= \exp(-0.65 \times 38.95 \times (-0.1)) - \exp(0.35 \times 38.95 \times (-0.1))\\ &= 12.575 - 0.256\\ &= 12.319\,\text{mA}.\end{aligned}$$

If an anodic overvoltage $\eta = +0.1\,\text{V}$ is imposed the net anodic current is given by

$$\begin{aligned}(I_a)_{net} &= \exp(-0.65 \times 38.95 \times 0.1) - \exp(0.35 \times 38.95 \times 0.1)\\ &= 0.080 - 3.909\\ &= -3.829\,\text{mA}.\end{aligned}$$

Conversely, with an electrode process where α_a favours the anodic component (e.g. for the characteristics $n = 1$; $\alpha_c = 0.35$; $\alpha_a = 0.65$; $I_0 = 1\,\text{mA}$; $A = 1\,\text{cm}^2$), then for a cathodic overvoltage $\eta = -0.1\,\text{V}$, the net cathodic current is now given by

$$\begin{aligned}(I_c)_{net} &= \exp(-0.35 \times 38.95 \times (-0.1)) - \exp(0.65 \times 38.95 \times (-0.1))\\ &= 3.909 - 0.080\\ &= 3.829\,\text{mA}.\end{aligned}$$

While for an overvoltage $\eta = +0.1\,\text{V}$, the magnitude of the net anodic current is

$$\begin{aligned}(I_c)_{net} &= \exp(-0.35 \times 38.95 \times (0.1)) - \exp(0.65 \times 38.95 \times (0.1))\\ &= 0.256 - 12.575\\ &= -12.319\,\text{mA}.\end{aligned}$$

It has been common practice in the United Kingdom and the United States to represent current–voltage curves with cathodic currents increasing in the positive vertical direction and with voltages increasing negatively from left to right. This indeed has been the practice adopted here and the signs of the currents in the above calculations are consistent with this. However, many people consider it more logical to assign negative signs to cathodic currents deriving from a negative overvoltage and vice versa. It is a simple matter to rearrange signs in the above examples to conform with this alternative convention.

7.3 The relationship between current density and overvoltage: the Tafel equation

If the overvoltage is small, the exponential terms of equation (7.18) may be expanded and all terms except the first two neglected. The expression then reduces to

$$i = \frac{i_0 nF\eta}{RT} \tag{7.20}$$

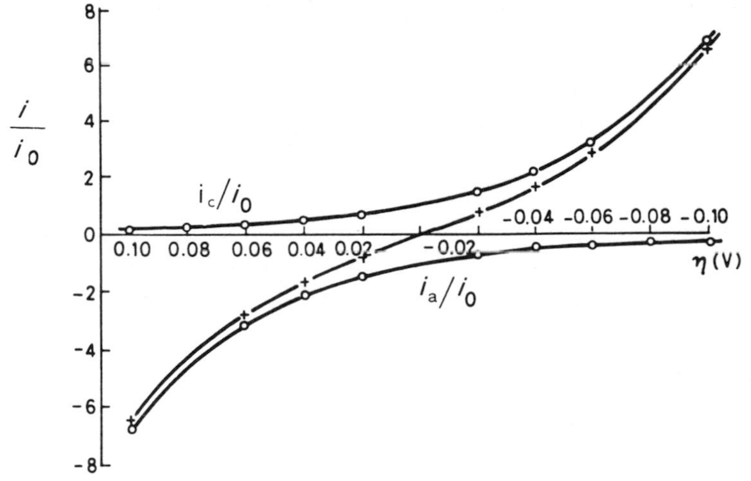

Figure 7.6 The ratio i/i_0 (calculated from equation (7.18)) plotted as a function of overvoltage η. i_c/i_0, i_a/i_0 are the partial cathodic and anodic ratios. $\alpha = 0.5$.

For higher overvoltages the full expression (7.18) must be used as has been demonstrated with the calculations given. In Figure 7.6 are shown calculated variations of $i/i_0, i_c/i_0$ and i_a/i_0 with overvoltage for $\alpha = 0.5$.

The asymmetrical variations of the same current ratios for $\alpha = 0.25$ and 0.75 are presented in Figure 7.7.

For a large overvoltage to the cathodic reaction, only the first exponential term in equation (7.18) is significant, the second being very small by comparison. The dependence of cathodic current density on overvoltage may then be given by

$$\ln i = \ln i_0 - \frac{\alpha n F \eta}{RT} \tag{7.21}$$

Explicitly, η is then given by

$$\eta = \frac{2.3RT}{\alpha nF} \log i_0 - \frac{2.3RT}{\alpha nF} \log i \tag{7.22}$$

Conversely, for a large overvoltage to the anodic reaction, only the second term of equation (7.18) is of significance and

$$\eta = -\frac{2.3RT}{(1-\alpha)nF} \log i_0 + \frac{2.3RT}{(1-\alpha)nF} \log i \tag{7.23}$$

Equations (7.22) and (7.23) are of identical form to an empirical equation proposed by Tafel, viz.

$$\eta = a + b \log i \tag{7.24}$$

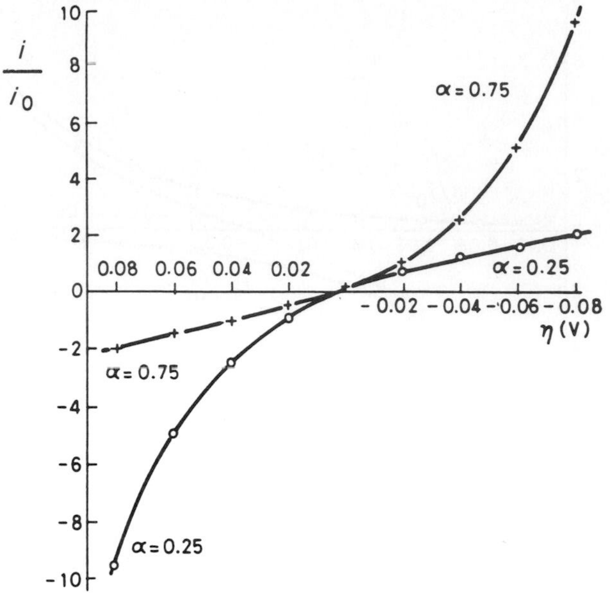

Figure 7.7 The ratio i/i_0 (calculated from equation (7.18)) plotted as a function of overvoltage η for $\alpha = 0.25$ and for $\alpha = 0.75$.

Graphs of η versus $\log i$ are known as Tafel plots, and examples are shown in Figure 7.8. The linear portions of the asymptotes correspond to the Tafel equation and have slopes (Tafel slopes) of magnitude b; $2.3RT/\alpha nF$ for the cathodic and $2.3RT/(1-\alpha)nF$ for the anodic process. An experimental line does not continue and cut the $\log i$ axis, since i refers to a *net* current density. This will approach zero as η approaches zero. The extrapolated asymptotes intersect on the line $\eta = 0$ at $\log i_0$.

7.4 The modern approach to the interpretation of electrode reactions

Strong hydration interaction between ions and water dipoles has been ignored in the above treatment of electrode reactions. However, it is clear that whether considering reactions such as

$$M^{3+} + e \rightleftharpoons M^{2+}$$

at electrodes or

$$M^{3+} + N^{2+} \rightleftharpoons M^{2+} + N^{3+}$$

in solution, changes in hydration shells must be an important part of any fuller theoretical model devised in attempting to explain the nature of electron-transfer reactions. The many attempts to cope with this problem

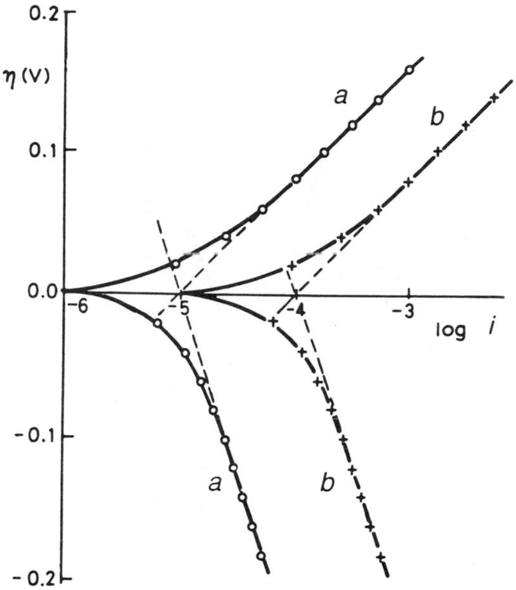

Figure 7.8 Tafel plots constructed using data from Figure 7.7. (a) $\alpha = 0.25$; $I = 10^{-5}$ A. (b) $\alpha = 0.25$; $I = 10^{-4}$ A.

have found some sort of culmination in the work of Marcus. This has been particularly successful in accounting for observed relationships between parameters characteristic of reactions occurring in a homogeneous manner in solution and those, essentially heterogeneous, occurring at electrodes.

In terms of the M^{3+} and M^{2+} energy parabolas shown in Figure 7.9, solvent redistribution may be represented by the function ΔG_s. ΔG_s, the reorganization energy, is a free-energy term representing the difference in hydration of reactants and products. The activation energy, $\Delta G^{\ddagger}$, is related to ΔG and ΔG_s by the following expression

$$\Delta G^{\ddagger} = \frac{(\Delta G + \Delta G_s)^2}{4\Delta G_s} \tag{7.25}$$

The contributing terms in the Butler–Volmer equation may be expressed in terms of rate constants as well as rates or currents. Thus, for a cathodic process

$$k_c = k_0 \exp\left[-\frac{\alpha n F \eta}{RT}\right] \tag{7.26}$$

which emphasizes the direct relationship between the transfer coefficient and

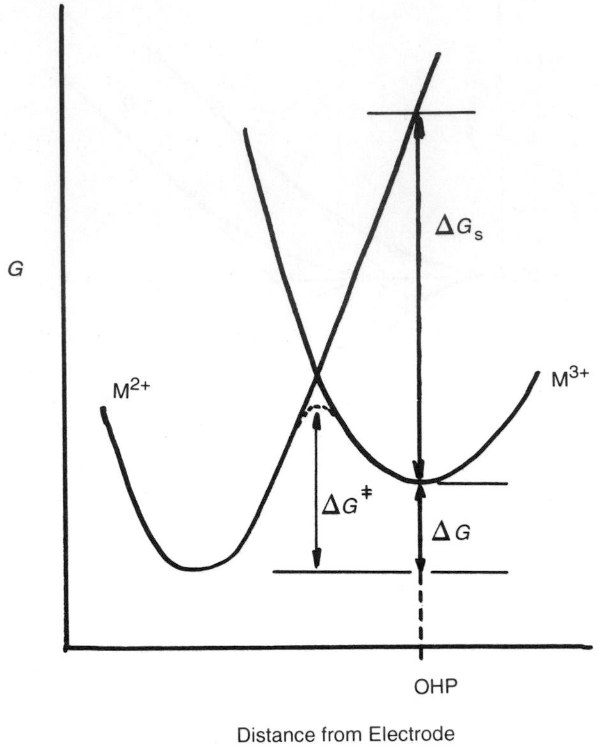

Figure 7.9 Intersecting free energy parabolas for oxidized and reduced species showing relationship between ΔG, $\Delta G^{\ddagger}$ and ΔG_S, the reorganization energy. OHP = outer Helmholtz plane.

the rate constant of the process to which it applies. Equation (7.26) may alternatively be written as

$$\ln k_c = \text{constant} - \frac{\alpha nFE}{RT}$$

Therefore

$$\alpha = -\left(\frac{\partial \ln k_c}{\partial E}\right) = \frac{\partial \Delta G^{\ddagger}}{\partial E} \tag{7.27}$$

which, in terms of equation (7.25) gives

$$\alpha = \frac{1}{2}\left(\frac{\Delta G}{\Delta G_s} + 1\right) \tag{7.28}$$

It is instructive to consider some of the implications of equation (7.28): its use makes predictions consistent with the kinetic treatment given so far and introduces further insights into the nature of the energy profiles of electrode processes.

Firstly, if $\Delta G = 0$, corresponding to the reversible case where $\eta = 0, \alpha$ is seen to have the value 1/2. If $\Delta G_s \sim \Delta G, \alpha \sim 1$; this means that almost all the overvoltage η is being used in generation of the transition state (see Figure 7.5). The implication of this condition being established so close to the electrode surface is that the transition state has a structure much closer to the product than to the reactant. By contrast, if $\Delta G_s \sim -\Delta G, \alpha = 0$; the transition state is now formed near to the OHP and will resemble M^{n+}. These cases all imply a fairly *small* value for ΔG_s. If at the other extreme ΔG_s is large, and very much larger than ΔG, it is seen that α again has the value 1/2 but here the electrode process is *slow*. The wording of this last statement should be noted carefully: electron transfer as such occurs rapidly, in accordance with the Frank–Condon principle. Marcus has elaborated on the structural and environmental changes which are necessary to allow such adiabatic electron transfer. An important outcome of the Marcus model is that it predicts relationships which are experimentally verifiable. In particular, for a redox process which occurs both homogeneously in solution and at an electrode, the following relationship has been established between the two rate constants.

$$\frac{k_{\text{elec}}}{(k_{\text{hom}})^{1/2}} \sim 10^{-2}\,\text{m}^{1/2}\,\text{mol}^{1/2}\,\text{s}^{-1/2} \qquad (7.29)$$

Thus for the process $V^{3+} \rightarrow V^{2+}$ for which $k_{\text{elec}} = 4 \times 10^{-5}\,\text{m}^2\,\text{s}^{-1}$ and $k_{\text{hom}} = 1.6 \times 10^{-5}\,\text{m}^3\,\text{mol}^{-1}\,\text{s}^{-1}$ equation (7.29) is seen to hold exactly. Many reactions show the ratio to have a value of the order given above but there are exceptions.

Electrode processes for which $n > 1$ take place in one-electron steps of which one is rate determining. For such cases the interpretation of an experimentally determined value of αn is not always clear: for what value of n should be used to calculate α if the contributing one-electron stages occur with very different values? For example, a known overall two-electron reduction may show an experimental value of $\alpha n = 1.4$ which might suggest $\alpha = 0.7$. But if the first transfer is fast with $\Delta G_s \sim \Delta G, \alpha_1 = 1$ so $\alpha_2 = 0.4$ and while 0.4 now reflects quite logically the position of the transition state on the distance axis, 0.7 does not.

7.5 Electrolysis and overvoltage

It is necessary now to consider a number of types of overvoltage, their source and control and the way in which they influence the course of an electrochemical reaction.

7.5.1 *Activation overvoltage* (η_A)

A slow electron transfer, such as that considered in previous sections, has a high activation energy. If such a reaction is to proceed at a reasonablc rate and produce an efficient quantity of product, a significant increase of applied potential over the equilibrium value is necessary. This excess potential is known as activation overvoltage (η_A) This description emphasizes that the slow, rate-determining step in the process is the electron transfer due to the high activation-energy barrier which it must cross.

Two further types of overvoltage are of importance and may occur simultaneously with activation overvoltage.

7.5.2 *Resistance overvoltage* (η_R)

The most common form of resistance overvoltage arises from the passage of electric current through an electrolyte solution surrounding the electrode. Such a solution is not of infinite conductivity and shows resistance to the current flow, with the result that an ohmic (IR) drop in potential occurs between the working electrodes. This effect may be offset by insulating the solution of the reference electrode from the working solution by enclosing the former in a fine glass capillary, the open end of which is brought as close to the surface of the electrode under investigation as is compatible with uniform field force over the surface of this electrode (Figure 7.10)

The optimum position of this *Luggin capillary*, as it is called, is usually a matter of experiment.

A less common form of ohmic overvoltage is caused by the formation, on the surface of the electrode, of an adherent layer of reaction product which is a relatively poor conductor of electricity. Surface oxide films show such behaviour, their resistance being such that overvoltages of several hundred volts may be produced.

7.5.3 *Concentration overvoltage* (η_C)

This is a small, but important effect (particularly for some electroanalytical techniques) which arises due to concentration changes induced in the vicinity of electrodes by electrochemical reactions occurring there.

Consider the simplest of all cells in which two identical electrodes of metal M are placed in a solution of M^{n+} ions and let the electrode equilibria be established rapidly, i.e. if no potential is applied, but the cell is simply short-circuited, no current will flow since the potential of both electrodes is the same.

If even a small potential difference is applied between the electrodes, the balance is destroyed, one electrode becoming a cathode the other an anode. At the former M^{n+} ions are discharged at a faster rate than they dissolve

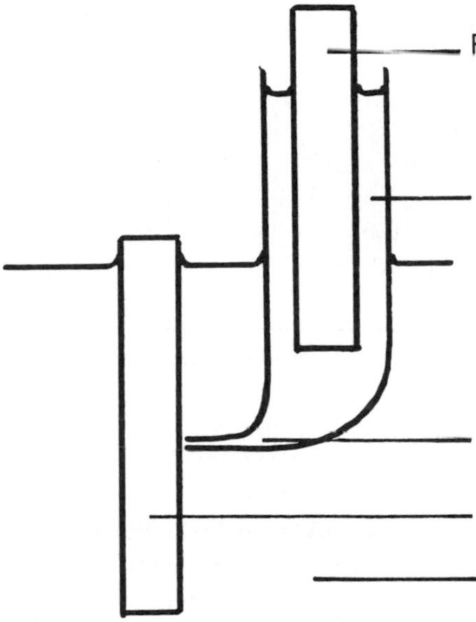

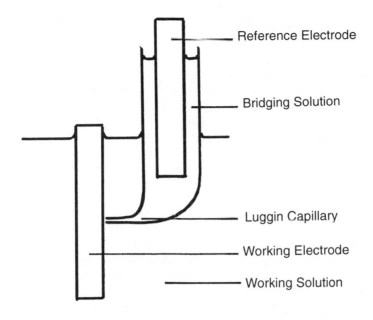

Figure 7.10 The principle of the Luggin capillary.

and at the anode M passes into solution more rapidly than M^{n+} ions are discharged.

The total amount of material discharged at the cathode or dissolved from the anode may be calculated from Faraday's laws, from a knowledge of the quantity of electricity passed. However, by reference to the Hittorf mechanism of electrolysis, we see that, at the cathode, only a fraction t_+ of the material deposited there has reached there by electrical migration. The remaining fraction must be made good from the layer of solution in the immediate vicinity of the electrode surface. From the moment that electrolysis starts, therefore, the solution close to the cathode surface shows a concentration decrease. For the same current to flow, and therefore the same rate of deposition of M^{n+}, a more negative potential will be required. Similarly, a more positive potential will be required at the anode since here only a fraction t_+ of the metal ions formed by dissolution are removed by migration so that the concentration of the anode solution increases. The effect is to produce a back emf, so that to maintain the current flow, the applied emf must be increased by this amount. An electrode whose potential deviates from its equilibrium value due to these causes is said to be *concentration polarized.* Stirring of the electrolyte solution or rotation or vibration of the electrodes can serve to reduce the extent of concentration polarization but does not eliminate it entirely.

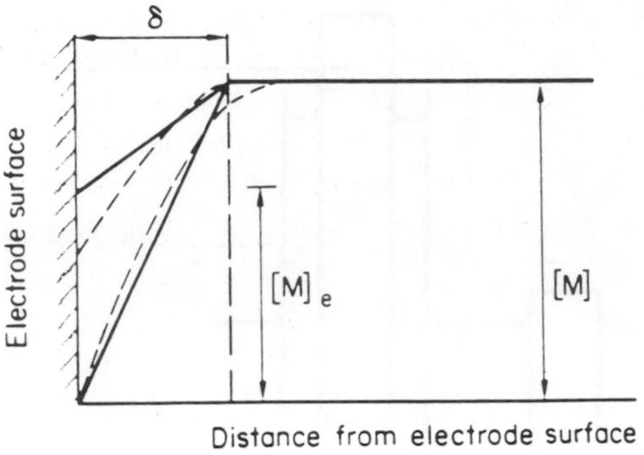

Figure 7.11 Representation of the concentration gradient in the vicinity of an electrode at which reduction is taking place.

At an electrode surface there is a diffusion layer across which there is a concentration variation from the surface to the edge of the layer. To a good approximation this variation may be regarded as uniformly linear as shown in Figure 7.11. The thickness of the layer, δ, has a magnitude of the order of 0.05 cm under static conditions, which becomes reduced to the region of 0.001 cm with rapid stirring.

For a steady current through the cell, the concentration of the surface layer of solution at the cathode falls and is made good by the diffusion of material from the bulk. A steady state is rapidly reached where M^{n+} ions removed by deposition are replaced by those arriving by both electrical migration and diffusion. The rate of arrival of M^{n+} ions by diffusion is given by

$$\frac{D([M^{n+}] - [M^{n+}]_e)}{\delta} \tag{7.30}$$

D is the diffusion coefficient of the species M^{n+}, and is defined as the amount of M^{n+} transported per unit area across unit diffusion layer thickness under unit concentration gradient in unit time.

The rate of arrival at the cathode of M^+ ions by migration is, for a current density i

$$\frac{t_+ i}{nF} \tag{7.31}$$

Since the total amount deposited is given by i/nF, it follows that, under steady-state conditions.

$$\frac{i}{nF} = \frac{t_+ i}{nF} + \frac{D([M^{n+}] - [M^{n+}]_e)}{\delta} \tag{7.32}$$

As the potential impressed between the electrodes is increased, the current density will increase so long as the concentration gradient across the diffusion layer can increase to maintain the supply of M^{n+}. A point will, however, be reached where M^{n+} becomes zero. Since the concentration gradient has now reached it maximum value, the rate of supply of $[M^{n+}]$ by diffusion has reached its maximum value and the electrode process can proceed no faster. Equation (7.32) now becomes

$$\frac{i_{\text{lim}}}{nF} = \frac{t_+ i_{\text{lim}}}{nF} + \frac{D[M^{n+}]}{\delta} \tag{7.33}$$

Comparing equations (7.32) and (7.33) it is seen that

$$\frac{i(1-t_+)}{i_{\text{lim}}(1-t_+)} = \frac{([M^{n+}] - [M^{n+}]_e)}{[M^{n+}]} \tag{7.34}$$

or

$$\frac{[M^{n+}]_e}{[M^{n+}]} = \frac{i_{\text{lim}}}{i_{\text{lim}} - i} \tag{7.35}$$

so that the concentration overvoltage, η_c, may be written

$$\eta_c = \frac{RT}{nF} \ln \frac{[M^{n+}]_e}{[M^{n+}]} = \frac{RT}{nF} \ln \frac{i_{\text{lim}}}{i_{\text{lim}} - i} \tag{7.36}$$

7.5.4 Summary of overvoltage phenomena and their distinguishing features

The overvoltage of an individual electrode may be expressed as the sum of contributions from activation, concentration and resistive film overvoltages:

$$\eta = \eta_A + \eta_C + \eta_R$$

the use of the Luggin capillary virtually eliminating the *IR* contribution. Hence, η_R is often absent.

There are a number of distinguishing features of the above three forms of overvoltage which allow the effects to be identified experimentally.

1. η_R unlike η_A and η_C, appears and disappears instantaneously when the polarizing circuit is made or broken.
2. η_A increases rapidly and exponentially after a polarizing current is caused to flow and decreases in a complementary way when the current flow is stopped. The exponential growth and decay are in accordance with the concept of η_A as a function of the activation energy of an electrode process. The magnitude of η_A is strongly affected by the physical and chemical nature of the electrode material.

3. η_C grows and decays slowly on application or interruption of the current flow at a rate characteristic of the diffusion coefficients of the species involved. η_C is unique in being the only form of overvoltage affected by stirring and is unaffected by the nature of the surface of the electrode material.

7.6 Hydrogen and oxygen overvoltage

The evolution of hydrogen and oxygen are well-known phenomena during electrolysis of dilute aqueous solutions of acids and bases between inert metal electrodes. If bright platinum electrodes are used for the electrolysis, it is found that in most cases the minimum potential difference which must be applied between them before gas bubbles appear is close to 1.7 V. That a similar value should usually be observed is hardly surprising since the same overall chemical process is occurring, viz. the decomposition of water, although the electrode reactions differ depending upon whether the solutions are acidic or alkaline.

In acid

$$\text{at the cathode} \quad 2H^+ + 2e \rightarrow H_2$$
$$\text{at the anode} \quad H_2O \rightarrow \tfrac{1}{2}O_2 + 2H^+ + 2e$$

In alkali

$$\text{at the cathode} \quad 2H_2O + 2e \rightarrow H_2 + 2OH^-$$
$$\text{at the anode} \quad 2OH^- \rightarrow \tfrac{1}{2}O_2 + H_2O + 2e$$

7.6.1 Decomposition potentials and overvoltage

A graph of current, or current density, versus potential gives a decomposition curve and allows decomposition potentials to be determined (Figure 7.12). Decomposition potentials are never well-defined and can only be obtained approximately by extrapolation of the rising part of the curve to the potential axis. Nevertheless the value of approximately 1.7 V for the electrolysis of water is sufficiently different from the theoretical value for it to be apparent that there is a large cell overvoltage even when resistance and concentration effects are taken into account. The theoretical decomposition potential may be calculated as follows: the reaction occurring in the electrolysis cell is

$$H_2O(l) \rightarrow H_2(g) + \tfrac{1}{2}O_2(g)$$

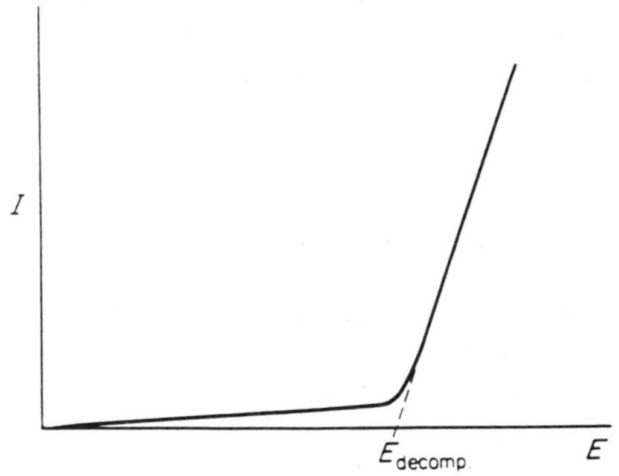

Figure 7.12 Decomposition curve. $E_{decomp.}$ is the minimum potential difference which must be applied between a pair of electrodes before decomposition occurs and a current flows. $E_{decomp.}$ has little theoretical significance since it is made up of two individual electrode potentials and their associated overvoltages.

for which $\Delta G^{\ominus} \sim 238\,140$ joules. Now $\Delta G^{\ominus} = -nFE^{\ominus}$ where, for the present case $E^{\ominus}$ is the standard emf of the hydrogen–oxygen cell; therefore,

$$E^{\ominus} = -\frac{238\,140}{2 \times 96\,500} = -1.23\,\text{V}$$

The cell overvoltage observed when using bright platinum electrodes is thus of the order of 0.5 V. An experimental decomposition voltage has no theoretical significance of its own, since a moment's consideration will show that it consists of two individual electrode potentials and the *IR* drop between them. These individual potentials will be made up of the thermodynamic values plus the overvoltages; the latter comprise contributions from activation, concentration and film resistance overvoltage.

7.6.2 *Individual electrode overvoltages*

Individual electrode overvoltages may be determined experimentally by means of the circuit in Figure 7.13. Here a constant electrolysis current density is maintained by a high-tension battery/series resistance combination to polarize the electrodes. Each electrode in turn is then combined with a reference electrode and the emf values of the two cells successively formed in this way measured via the potentiometer. Since the reference electrode potential is known, the potential of the anode and cathode may be determined at the current density imposed. The Luggin capillary, brought as close to the electrode surfaces as possible, largely removes the *IR* contribution to the

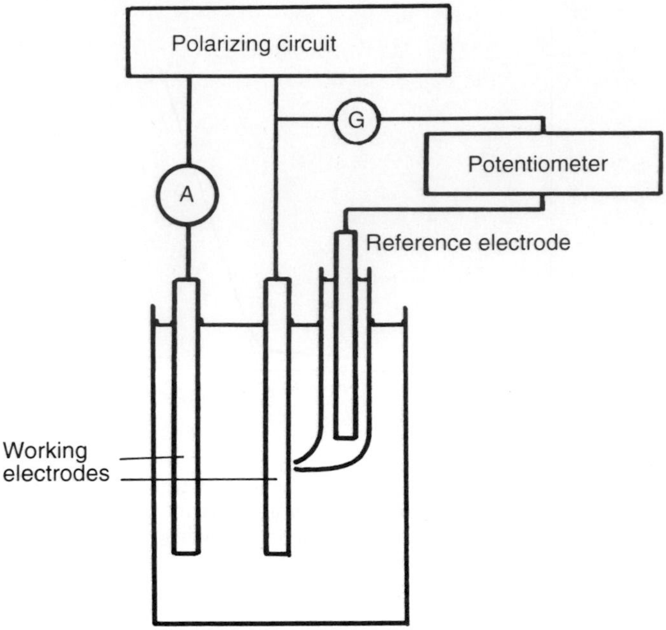

Figure 7.13 Determination of individual electrode overvoltages and elimination of the *IR* correction.

measured overpotential. By these means the cell overvoltage of about 0.5 V for the above case may be shown to comprise an overvoltage of about 0.1 V at the cathode and about 0.4 V at the anode. Table 7.1 shows approximate hydrogen overvoltages for various electrode materials in dilute sulphuric acid.

Table 7.1 Selected hydrogen overvoltages for various electrode materials in dilute sulphuric acid

Metal	η(V)
Hg	0.78
Zn	0.70
Sn	0.53
Cu	0.23
Ni	0.21
Pt (bright)	0.10
Pt (platinized)	0.005

7.7 Theories of hydrogen overvoltage

The essential stages in the overall process of hydrogen discharge and gas evolution at a cathode may be assumed to be as follows:

1. H_3O^+ ions diffuse from the bulk solution to the edge of the double layer.
2. H_3O^+ ions are transferred across the layer.
3. Dehydration of H_3O^+.
4. H^+ receives an electron from the electrode.

Stages (2), (3) and (4) constitute the discharge reaction which may be expressed as

$$M + H_3O^+ + e \rightarrow M\text{–}H + H_2O \tag{7.37}$$

(H atom adsorbed onto electrode surface)

5. Formation of hydrogen molecules from hydrogen atoms. This can occur in one of two possible ways, viz.

   (i) $M\text{–}H + M\text{–}H \rightarrow 2M + H_2$ (7.38)

   or

   (ii) $M\text{–}H + H_3O^+ + e \rightarrow M + H_2 + H_2O$ (7.39)

   (i.e. a new H_3O^+ ion interacts on the electrode surface—this is often called the electrochemical reaction)
6. Desorption of hydrogen molecules.
7. Formation of bubbles and evolution of gas.

The problem is to decide which stages are rate determining. If stage (1) were to be rate determining, then the process overall would be diffusion limited due to concentration polarization. Usually, one of the stages (2) to (5) is the slowest. Which one depends upon many factors such as the operating conditions and the nature of the cathode.

Tafel proposed that step 5(i), equation (7.38), of the above scheme is the slow stage. He was able, on the basis of this assumption, to derive a form of equation (7.24) in which the constant b takes the form $2.3RT/2F$, i.e. approximately 0.03 at 298 K. Such a value is rarely encountered in practice although it is observed with low overvoltage metals and with platinized platinum.

For many metals showing higher overvoltages, stages (2) to (4), producing the discharge reaction (7.37) would seem to contain the slow step. Again it is possible to derive an expression of the form of equation (7.24) but for which $b \sim 0.118$ at 298 K. This value is indeed observed for many of the higher overvoltage metals.

A form of equation (7.24), identfying constant b, but also predicting a dependency of overvoltage on the pH of the solution may be derived assuming that the slow step is the electrochemical reaction (7.39).

It is clear that no one theory can account for all classes of behaviour under all conditions. For low overvoltage metals, it seems probable that the catalytic process is rate determining, while for metals showing higher overvoltages, slow ion-discharge theories are more successful in explaining observed behaviour. In a smaller number of instances, the electrochemical theory is more successful, for example in explaining the behaviour of silver and nickel in alkaline solution where η is observed to be a function of the pH of the solution.

For electrode materials with high catalytic activity, it would appear that the slow stage in the hydrogen evolution proccss can be the removal of molecular hydrogen from the electrode surface by diffusion.

Problems

7.1 If the Tafel constants, a and b, have the values 1.54 V and 0.119 V respectively for the reduction of hydrogen ions at a lead cathode, calculate the values of the transfer coefficient α and the exchange current density.

7.2 When dilute sodium hydroxide was electrolysed using a nickel cathode, the overpotential was found to be 0.394 V to maintain a current density of 0.01 A cm^{-2} and 0.148 V to maintain a current density of 0.0001 A cm^{-2}. Calculate the transfer coefficient and exchange current density for the hydrogen/hydrogen ion equilibrium at a nickel cathode in the given medium.

7.3 At cathode overvoltages indicated, the following values of cathodic current were obtained at a platinum electrode of area 1.5 cm^2 immersed in a solution of Fe^{2+} and Fe^{3+} ions at 298 K. Use these data to calculate the transfer coefficient and the exchange current density.

η(V)	0.02	0.05	0.07	0.10	0.12	0.15	0.20
I(mA)	3.20	9.95	17.03	35.18	55.89	110.78	343.62

7.4 Given that the Tafel constants for the deposition of zinc on platinum are a = 0.280 V and b = 0.059 V, respectively, show why it is not possible to plate zinc on to platinum by electrolysis of a zinc salt at unit activity in neutral solution at a current density of 1 mA cm^{-2}.

7.5 Given the values of α and i_0 calculated in question 7.3, for a $Pt/Fe^{3+}/Fe^{2+}$ electrode, estimate the current density for an anodic overvoltage of 0.07 V.

7.6 For the discharge of hydrogen ions from dilute sulphuric acid at a platinum electrode at 298 K, the following current densities were observed for the range of cathodic overvoltages indicated

η(mV)	20	50	70	100	120	150	200	250
i(m A cm^{-2})	0.57	1.40	2.05	3.36	4.56	7.16	15.05	31.55

From the appropriate Tafel plot calculate the transfer coefficient, α, and the exchange current density, i_0.

7.7 For a $Pt/Ce^{4+}/Ce^{3+}$ electrode the exchange current density, i_0 has the value 4.0×10^{-5} A cm^{-2} while the transfer coefficient is 0.75. Taking the standard potential of the electrode as +1.61 V and assuming unit activity for both ion species, calculate the currents flowing through an electrode of area 1 cm^2 at applied potentials of 1.30, 1.40, 1.50, 1.61, 1.70, 1.80 and 1.90 V.

7.8 If for the electrode equilibrium $Cu^{2+} + 2e \rightleftharpoons Cu$, the transfer coefficient is 0.5 and the exchange current density is 2.5×10^{-5} A cm^{-2}, calculate the Tafel constants at 298 K and estimate the overpotential to deposit copper from a solution of unit activity at this temperature at a current density of 5×10^{-3} A cm^{-2}.

PART II APPLICATIONS

8 Determination and investigation of physical parameters

8.1 Applications of the Debye–Hückel equation

The various forms of the equations resulting from the Debye-Hückel theory find practical application in the determination of activity coefficients and make possible the determination of thermodynamic data. Two important cases will be considered here.

8.1.1 Determination of thermodynamic equilibrium constants

Let us consider as an example the dissociation of a 1:1 weak electrolyte

$$AB \rightleftharpoons A^+ + B^-$$

The thermodynamic dissociation constant K_T is given by

$$K_T = \frac{[A^+][B^-]}{[AB]} \frac{\gamma_{A^+}\gamma_{B^-}}{\gamma_{AB}} = K\frac{\gamma_\pm^2}{\gamma_{AB}}$$

Therefore

$$K_T \sim K\gamma_\pm^2 \tag{8.1}$$

where K is the concentration, or conditional dissociation constant. For a weak electrolyte in dilute solution γ_{AB} for the undissociated, and therefore non-ionic, species is very nearly unity. Taking logarithms of equation (8.1) and substituting for $\gamma_\pm$ from the limiting law expression, we obtain

$$\log K = \log K_T + 2A\sqrt{I} \tag{8.2}$$

K_T may therefore be determined from measured values of K over a range of ionic strength values and extrapolating the K versus $\sqrt{I}$ plot to $\sqrt{I} = 0$. This is a general technique for the determination of all types of thermodynamic equilibrium constants, e.g. solubility, stability and acid dissociation constants.

8.1.2 Dependence of reaction rates on ionic strength

In the treatment of ionic reactions by Brønsted and Bjerrum an equilibrium is considered to exist between reactant ions and a 'critical complex', the latter bearing close resemblance to the activated complex of the theory of absolute reaction rates. Thus, for the reaction scheme

$$A^{z_A} + B^{z_B} \rightleftharpoons \left(x^{(z_A+z_B)}\right)^{\ddagger} \longrightarrow \text{products} \tag{8.3}$$

we may write, for the pre-equilibrium

$$K = \frac{[x^{\ddagger}]}{[A][B]}\frac{\gamma_{\ddagger}}{\gamma_A\gamma_B} \tag{8.4}$$

(omitting charges for clarity) so that the rate, v with which A and B react may be expressed by

$$v = k[A][B] = k_0[A][B]\frac{\gamma_A\gamma_B}{\gamma_{\ddagger}} \tag{8.5}$$

where

$$k = k_0\frac{\gamma_A\gamma_B}{\gamma_{\ddagger}} \tag{8.6}$$

k_0, k being the rate constants at infinite and finite dilution, respectively. In logarithmic form equation (8.6) becomes

$$\log k = \log k_0 + \log\gamma_A + \log\gamma_B - \log\gamma_{\ddagger} \tag{8.7}$$

in which activity coefficients may be expressed by equation (2.15) thus,

$$\log k = \log k_0 - \frac{A\sqrt{I}}{1 + Ba\sqrt{I}}\,[z_A^2 + z_B^2 - (z_A + z_B)^2]$$

Therefore

$$\log k = \log k_0 + \frac{2Az_Az_B\sqrt{I}}{1 + Ba\sqrt{I}} \tag{8.8}$$

or,

$$\log k \sim \log k_0 + 2Az_Az_B\sqrt{I} \tag{8.9}$$

in very dilute solution, or,

$$\log k \sim \log k_0 + 1.02z_Az_B\sqrt{I} \tag{8.10}$$

for water as solvent at 298 K.

These last equations take account of the salt effect observed for reactions between ions, the slopes of graphs of log k/k_0 versus $\sqrt{I}$ being very close to those predicted by equation (8.10) at low concentrations (Figure 8.1). At higher concentrations, deviations from linearity occur and these are particularly noticeable for reactions having $z_Az_B = 0$, e.g. for a reaction

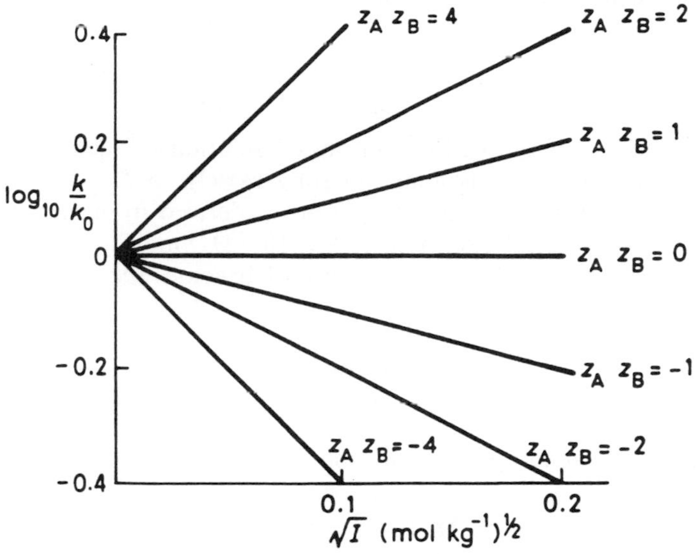

Figure 8.1 Theoretical variations of rate with $\sqrt{I}$ for reactions showing different values of the product $z_A z_B$.

between an ion and a neutral molecule. According to equation (8.10) such reactions should show no variation of rate with ionic strength and this is indeed the case up until about $I = 0.1$. Above this point, increasing ionic strength does cause the rate to vary. The reason for this is the bI term of the Hückel equation (2.19). In this case it may be readily shown that

$$\log k = \log k_0 + (b_A + b_B - b_{\ddagger})I \qquad (8.11)$$

so that $\log k/k_0$ in this case becomes a linear function of I rather than of $\sqrt{I}$. This has been experimentally verified.

8.2 Determination of equilibrium constants by conductivity measurements

8.2.1 Solubilities of sparingly soluble salts

The conductivity of a saturated solution of a sparingly soluble salt may be found by subtracting from the observed value in water the conductivity of water itself. For such low concentrations of electrolyte the value for water, however pure, will now be significant, the small number of ions produced by the electrolyte giving a conductance of the same order of magnitude as water; for example, κ for an aqueous AgCl solution saturated at 298 K was found to be $3.42 \times 10^{-4}\,\Omega^{-1}\,m^{-1}$. κ for the water used was $1.60 \times 10^{-4}\,\Omega^{-1}\,m^{-1}$.

Therefore
$$\kappa(AgCl) = (3.42 - 1.60) \times 10^{-4}\,\Omega^{-1}\,m^{-1}$$
$$= 1.82 \times 10^{-4}\,\Omega^{-1}\,m^{-1}$$

The molar conductivity, Λ_s, for a saturated solution of such an insoluble electrolyte may be taken as approximating closely to Λ_0. This quantity may be calculated, from values of λ_0^+ and λ_0^- derived from measurements on more soluble electrolytes, to be $1.38 \times 10^{-2}\,\Omega^{-1}\,mol^{-1}$. Thus equation (4.4a) may be used to calculate C which now becomes the solubility:

$$C \sim \frac{\kappa}{\Lambda_0} = \frac{1.82 \times 10^{-4}}{1.38 \times 10^{-2}}$$
$$= 1.32 \times 10^{-5}\,mol\,dm^{-3}$$

From this the solubility product, K_s, may be obtained as

$$K_s \sim [Ag^+][Cl^-] = 1.74 \times 10^{-10}$$

8.2.2 *The ionic product of self-ionizing solvents*

For water, which we have seen ionizes according to $H_2O + H_2O \rightleftharpoons H_3O^+ + OH^-$,

$$K_w = [H_3O^+][OH^-]$$

It is necessary to determine κ for very pure water. Then, knowing the ion conductivities for water at infinite dilution

$$C = [H_3O^+] = [OH^-] = \frac{\kappa}{(\lambda_{H^+})^0 + (\lambda_{OH^-})^0} \tag{8.12}$$

Hence C may be found and $K_w = C^2$.

8.2.3 *Dissociation constants of weak electrolytes, e.g. weak acids*

Dissociation constants may be determined from conductance data by use of equation (4.8) which, on rearrangement, gives

$$C\Lambda_C = K_a \frac{\Lambda_0^2}{\Lambda_C}\left(1 - \frac{\Lambda_C}{\Lambda_0}\right)$$
$$= K_a\left(\frac{\Lambda_0^2}{\Lambda_C} - \Lambda_0\right)$$

A plot of $C\Lambda_C$ versus $1/\Lambda_c$ gives a straight line of slope $K_a\Lambda_0^2$ and intercept $-K_a\Lambda_0$ from both of which K_a is determinable.

In order to determine the thermodynamic constant K_a^T it is necessary to determine K_a at different low concentrations and to obtain corresponding values of the degree of dissociation, α. From equation (8.1)

$$\log K_a^T = \log K_a - 2A\sqrt{I}$$

A being the Debye–Hückel constant for water as solvent, and for a weak acid, HA

$$I = \tfrac{1}{2}([H^+] + [A^-]) \quad \text{and} \quad [H^+] = [A^-] = \alpha C$$

therefore

$$I = \alpha C$$

therefore,

$$\log K_a^T = \log K_a - 2A\sqrt{(\alpha C)} \tag{8.13}$$

where K_a^T is the true thermodynamic dissociation constant of the acid, a plot of $\log K_a$ versus $\sqrt{(\alpha C)}$ giving a straight line of intercept $\log K_a^T$.

8.3 Thermodynamics of cell reactions

Measurement of the emf of a reversibly operating cell as well as its temperature coefficient enable ΔG,ΔH and ΔS for the cell reaction to be determined. ΔG at a given temperature follows directly from the cell emf E, by application of the equation

$$\Delta G = -nFE \tag{8.14}$$

ΔH may be expressed in terms of ΔG in the form of the Gibbs–Helmholtz equation

$$\Delta H = \Delta G - T\left[\frac{\partial(\Delta G)}{\partial T}\right]_P \tag{8.15}$$

which, in terms of the cell emf, takes the form

$$\Delta H = -nFE - T\left[\frac{\partial(-nFE)}{\partial T}\right]_P \tag{8.16}$$

or,

$$\Delta H = -nFE + nFT\left(\frac{\partial E}{\partial T}\right)_P \tag{8.17}$$

$(\partial E/\partial T)_P$ being the temperature coefficient of the cell emf at constant pressure. In practice this has to be determined very carefully by determining E for the cell over a wide range of temperature so that the tangent to the E–T plot at a given temperature may be drawn and its slope measured.

Once ΔG and ΔH have been found, ΔS may be calculated from

$$\Delta G = \Delta H - T\Delta S \tag{8.18}$$

or

$$\Delta S = \frac{(\Delta H - \Delta G)}{T}$$

therefore,

$$\Delta S = nF\left(\frac{\partial E}{\partial T}\right)_P \tag{8.19}$$

Since cell emf values are measured in volts, the units of ΔH and ΔG are joules and those of ΔS, joules deg^{-1}. Positive, negative and almost zero temperature coefficients have been observed, although negative coefficients are the most usual, signifying that for most cells the electrical energy obtainable from them is less than ΔH because some heat is produced as the cell operates. For positive coefficients the energy obtainable is greater than ΔH and in this case heat must be absorbed from the surroundings when the cell operates and unless a supply of heat is maintained the temperature of such a cell will fall. For a cell with a near zero temperature coefficient, the electrical energy is almost identical with the enthalpy change and this is found, for example, with the Daniell cell. It was somewhat unfortunate that early work with a Daniell cell appeared to confirm the erroneous belief that the electrical energy of a reversible cell was always equal to $-\Delta H$ of the cell reaction.

In principle, electrochemical measurements provide a most sophisticated means of determining such data particularly in the light of the precision with which potentials may be measured when due experimental precautions are taken. It is essential, however, that the reaction under study does actually occur in the cell to the exclusion of all others. It is also vital, of course, that both the electrodes used in the cell behave reversibly.

8.4 Determination of standard potentials and mean ion activity coefficients

Of particular use in such determinations are cells of the Harned type of which the simplest is represented by

$$\underset{}{H_2(1\,\text{atm}),\ Pt}|\underset{m}{HCl}|AgCl(s), Ag$$

m being the molality of the hydrochloric acid. Since the overall cell reaction is

$$\tfrac{1}{2}H_2 + AgCl(s) \longrightarrow Ag(s) + H^+ + Cl^- \tag{8.20}$$

$$-\Delta G = nFE \underset{(n=1)}{=} RT\ln K - RT\ln\frac{a_{Ag}a_{H^+}a_{Cl^-}}{a_{H_2}^{1/2}a_{AgCl}} \tag{8.21}$$

or

$$E = E^{\ominus} - \frac{RT}{F}\ln a_{H^+}a_{Cl^-} = E^{\ominus} - \frac{2RT}{F}\ln a_{\pm} \tag{8.22}$$

In terms of the molality of the hydrocholoric acid equation (8.22) becomes

$$E = E^{\ominus} - \frac{2RT}{F}\ln m - \frac{2RT}{F}\ln\gamma_{\pm} \tag{8.23}$$

$\gamma_{\pm}$ being the mean ion activity coefficient of the acid. Equation (8.23) may more usefully be expressed in the form

$$\left(E + \frac{2RT}{F}\ln m\right) = E^{\ominus} - \frac{2RT}{F}\ln\gamma_{\pm} \tag{8.24}$$

It is seen that this equation provides a valuable route to the determination of $\gamma_{\pm}$ but that in order to do this it is necessary to know $E^{\ominus}$. If a number of measurements of E are made for a range of values of m extending into the region where the Debye–Hückel limiting law holds, i.e. where $\ln\gamma_{\pm} \propto m^{\frac{1}{2}}$, $E^{\ominus}$ may be obtained by plotting the left hand side of equation (8.24) versus $m^{\frac{1}{2}}$. As $m \to 0, \gamma_{\pm} \to 1$ so that extrapolation of the line to $m^{\frac{1}{2}} = 0$ gives $E^{\ominus}$ as intercept.

It is normally better, however, to use forms of the Hückel equation, e.g.

$$-\ln\gamma_{\pm} = A\left(\frac{\sqrt{m}}{1+\sqrt{m}} - Bm\right) \tag{8.25}$$

so that

$$\left[E + \frac{2RT}{F}\ln m - \frac{A\sqrt{m}}{1+\sqrt{m}}\right] = E^{\ominus} - ABm \tag{8.26}$$

A graph of the left-hand side of equation (8.26) versus m may be extrapolated to the condition $m = 0$ to give an intercept of $E^{\ominus}$.

Once $E^{\ominus}$ is known, equation (8.26) may be used to calculate $\gamma_{\pm}$ for any molality of acid. The general principle of the technique may be extended to other electrolytes provided that the cell can be devised so that each electrode is reversible with respect to one of the ions.

Another possibility is to combine two cells of the above type back-to-back in the form of a cell without transference. The emf of such a cell is given by

$$E = \frac{2RT}{F}\ln\frac{m_2(\gamma_{\pm})_2}{m_1(\gamma_{\pm})_1} \tag{8.27}$$

If $(\gamma_{\pm})_1$ is known at molality m_1, $(\gamma_{\pm})_2$ at molality m_2 (or any other molality) may be determined by using equation (8.27) in the form

$$\frac{EF}{2\times 2.3RT} = \log\frac{m_2}{m_1} + \log\frac{(\gamma_{\pm})_2}{(\gamma_{\pm})_1} \tag{8.28}$$

or,

$$\log(\gamma_\pm)_2 = \frac{EF}{4.6RT} + \log(\gamma_\pm)_1 - \log\frac{m_2}{m_1} \tag{8.29}$$

$(\gamma_\pm)$ may be determined by the Debye–Hückel relation in dilute solution, i.e.

$$-\log\gamma_\pm = \frac{A\sqrt{m}}{1 + Ba\sqrt{m}} \tag{8.30}$$

(for a 1:1 electrolyte). Rearrangement and expansion of equation (8.30) gives

$$\begin{aligned} -\log\gamma_\pm &= A\sqrt{m}(1 + Ba\sqrt{m})^{-1} \\ &= A\sqrt{m}(1 - Ba\sqrt{m} + \cdots) \\ &= A\sqrt{m} - ABam \end{aligned} \tag{8.31}$$

It was shown by Hitchcock that $\log\gamma_\pm$ may also be given by

$$\log\gamma_\pm = B' - \log\frac{(\gamma_\pm)_1}{(\gamma_\pm)_2} \tag{8.32}$$

where B' is a further constant. Combining equation (8.32) with equation (8.31) gives

$$\log\frac{(\gamma_\pm)_1}{(\gamma_\pm)_2} - A\sqrt{m} = B'(ABa)m \tag{8.33}$$

If the left-hand side of equation (8.33) is plotted against m, a linear plot is obtained of slope (ABa) and intercept B'. When B' is known, the value of $\log\gamma_\pm$ at any molality may be calculated by means of equation (8.32)

If transport numbers for electrolyte ions are known reliably over the concentration range used in a cell, it is possible to determine $\gamma_\pm$ values from cells with transference. This, however, is a less usual practice.

Amalgam electrodes have proved useful in the determination of values for alkali chlorides and hydroxides using cells such as

$$\mathrm{M(Hg)|MCl|AgCl(s), Ag}$$

or

$$\mathrm{H_2, Pt|MOH|M(Hg)}$$

8.5 The determination of transport numbers

The concept of transport number was introduced in Chapter 4 where it was seen that the complementary movement of cations and anions under the influence of an applied electric field could be used to identify the behaviour of individual ion species.

There are essentially three approaches to the determination of transport numbers: the method due to Hittorf (of which there are several modifications), the moving boundary method (of which there are several forms) and methods based on measurements of cell emf.

8.5.1 Determination by the Hittorf method

In this method the cathode, anode and central portions of an electrolysis cell are made physically distinct and solutions contained within them (catholyte, anolyte and central) are analysed after controlled electrolysis for a known period of time.

Figure 8.2 shows the Hittorf mechanism of electrolysis when using electrodes made of the metal whose cations are those of the electrolyte in solution and when Q coulombs are passed. It is seen that the concentration of the electrolyte in the central portion of the cell remains unchanged at the end of the electrolysis.

Consider an electrolyte, represented by MX, whose constituent ions are M^{z+} and X^{z-} where $|z_+| = |z_-|$ (an example would be copper sulphate giving Cu^{2+} and SO_4^{2-}).

At the cathode M^{z+} ions are discharged according to

$$M^{z+} + ze \longrightarrow M$$

while at the anode M dissolves,

$$M \longrightarrow M^{z-} + ze$$

As a consequence of Faraday's laws (Chapter 1), after the passage of Q coulombs, $Q/|z_+|F$ moles of M have been deposited on the cathode surface. Of these $Q/|z_+|F$ moles only an amount $t_+Q|z_+|F$ is provided by the migration process. The remainder must be provided by electrolyte close to the cathode surface. Most of the species M^{z+} will thus be stripped out of the immediate vicinity of the cathode soon after electrolysis is started.

A concentration gradient will now be set up between the solution at the surface of the cathode and regions of the solution further away. Across this gradient M^{z+} and X^{z-} ions will diffuse and by the time Q coulombs have passed, $t_-Q/|z_+|F$ moles of M^{z+} will have been provided by diffusion, and discharged as M, to make up the total of $Q/|z_+|F$ moles of M deposited.

When a steady current flows, the concentration gradient at the electrode surface automatically adjusts itself to maintain just the correct rate of diffusion. Diffusion of both ion species of the electrolyte ensures that there is always sufficient of the anion being provided in this region so that the rate

Anode region	Q coulombs passed	Cathode region
$\frac{Q}{\lvert z_+\rvert F}$ moles of M^{z+} gained by dissolution		$\frac{Q}{\lvert z_+\rvert F}$ moles of M^{z+} lost by discharge
$\frac{t_+Q}{\lvert z_+\rvert F}$ moles of M^{z+} lost by migration out	→	$\frac{t_+Q}{\lvert z_+\rvert F}$ moles of M^{z+} gained by migration in
$\frac{t_-Q}{\lvert z_-\rvert F}$ moles of X^{z-} gained by migration in	←	$\frac{t_-Q}{\lvert z_-\rvert F}$ moles of X^{z-} lost by migration out
Net gain of M^{z+} $=\frac{(1-t_+)Q}{\lvert z_+\rvert F}=\frac{t_-Q}{\lvert z_+\rvert F}$ moles		Net loss of M^{z+} $=\frac{(1-t_+)Q}{\lvert z_+\rvert F}=\frac{t_-Q}{\lvert z_+\rvert F}$ moles
Gain of electrolyte MX $=\frac{t_-Q}{\lvert z_+\rvert F}(M^{z+})+\frac{t_-Q}{\lvert z_-\rvert F}(X^{z-})$		Loss of electrolyte MX $=\frac{t_-Q}{\lvert z_+\rvert F}(M^{z+})+\frac{t_-Q}{\lvert z_-\rvert F}(X^{z-})$
$=\frac{t_-Q}{\lvert z_+\rvert F}$ moles MX	No net gain or loss of electrolyte	$=\frac{t_-Q}{\lvert z_+\rvert F}$ moles MX

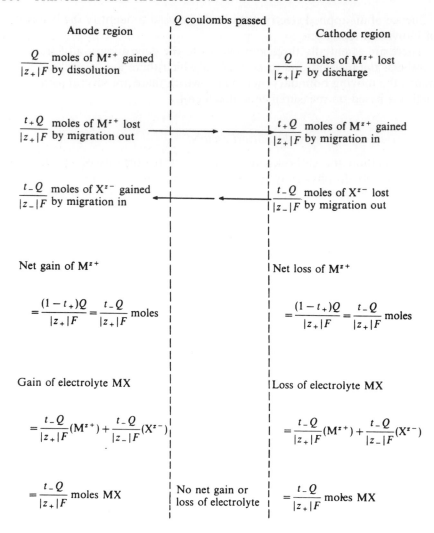

Figure 8.2 Hittorf electrolysis mechanism for the case where the electrodes are of the metal whose cation is that of the electrolyte in solution.

of migration of the anion away from the cathode is maintained. It is seen from Figure 8.2 that a net loss of $t_-Q/|z_+|F$ moles of electrolyte occurs in the catholyte solution.

Similar arguments apply to the anode region. After the passage of Q coulombs, $Q/|z_+|F$ moles of M^{z+} have dissolved from the anode of which $t_+Q/|z_+|F$ migrate out, leaving a net gain of $t_-Q/|z_+|F$ moles of M^{z+} in the anolyte. This is complemented by the migration into the anolyte of $t_-Q/|z_-|F$ moles of X^{z-}. Thus a net gain of $t_-Q/|z_+|F$ moles of electrolyte occurs in the anolyte solution.

As to the central portion, whatever it has gained from the anode region it has donated to the cathode region and vice versa. It thus experiences neither a net gain nor a net loss of electrolyte during the course of the electrolysis unless this is so prolonged that the diffusion processes, referred to above, extend so far out from the electrode surfaces that they become significant here and invalidate the calculations.

It is important to note that for the above case, in which it is the cations which react at the electrodes, the losses and gains in both electrode regions are functions of the transport number of the anion. It is evident that the transport number of the cation cannot be determined independently but must be calculated from the expression

$$t_+ + t_- = 1$$

For the case where the electrolyte is of a more general type whose ions are dispersed in solution according to

$$\mathrm{M}_{\nu+}\mathrm{X}_{\nu-} \longrightarrow \nu_+\mathrm{M}^{z+} + \nu_-\mathrm{X}^{z-}$$

then the respective gain and loss of electrolyte in anolyte and catholyte is again given by,

$$\frac{t_-Q}{|z_+|F}(\mathrm{M}^{z+}) + \frac{t_-Q}{|z_-|F}(\mathrm{X}^{z-})$$

Here, however,

$$z_+ \neq z_-$$

but rather

$$\nu_+z_+ = \nu_-z_-$$

i.e. the gain and loss become

$$\frac{t_-Q}{|z_+|F}(\mathrm{M}^{z+}) + \left(\frac{\nu_-}{\nu_+}\right)\left(\frac{t_-Q}{|z_+|F}\right) = \frac{t_-Q}{|z_+|F} \quad \text{moles of } \mathrm{M}_{\nu+}\mathrm{X}_{\nu-}$$

For example, in the case of a copper(II) halide, CuX_2, for which $z_+ = 2$ and $z_- = 1, \nu_+ = 1$ and $\nu_- = 2$, the gain and loss are

$$\frac{t_-Q}{2F}(\mathrm{Cu}^{2+}) + \frac{2t_-Q}{2F}(\mathrm{X}^-) = \frac{t_-Q}{2F}(\mathrm{CuX_2})$$

Electrolysis processes may not always be as simple as this and caution must be exercised in calculating transport numbers in that it is necessary to have firm prior knowledge of precisely what electrode reactions occur. For instance, for the electrolysis of metal salts with inert electrodes, such as platinum, hydrogen and not the metal cation may be discharged; similarly, oxygen may be discharged in preference to the anion. Even if the electrolyte

Q coulombs passed

Anode region	Cathode region
$\frac{Q}{\lvert z_-\rvert F}$ moles of X^{z-} lost by discharge	$\frac{Q}{\lvert z_+\rvert F}$ moles of M^{z+} lost by discharge
$\frac{t_-Q}{\lvert z_-\rvert F}$ moles of X^{z-} gained by migration in	$\frac{t_-Q}{\lvert z_-\rvert F}$ moles of X^{z-} lost by migration out
$\frac{t_+Q}{\lvert z_+\rvert F}$ moles of M^{z+} lost by migration out	$\frac{t_+Q}{\lvert z_+\rvert F}$ moles of M^{z+} gained by migration in
Net loss of X^{z-}	Net loss of M^{z+}
$=(1-t_-)\frac{Q}{\lvert z_-\rvert F}$ moles	$=(1-t_+)\frac{Q}{\lvert z_+\rvert F}$ moles
$=\frac{t_+Q}{\lvert z_-\rvert F}$ moles	$=\frac{t_-Q}{\lvert z_+\rvert F}$ moles
Loss of electrolyte MX	Loss of electrolyte MX
$=\frac{t_+Q}{\lvert z_+\rvert F}+\frac{t_+Q}{\lvert z_-\rvert F}$	$=\frac{t_-Q}{\lvert z_+\rvert F}+\frac{t_-Q}{\lvert z_-\rvert F}$
$=\frac{t_+Q}{\lvert z_+\rvert F}$ moles MX	$=\frac{t_-Q}{\lvert z_+\rvert F}$ moles MX

No net gain or loss of electrolyte

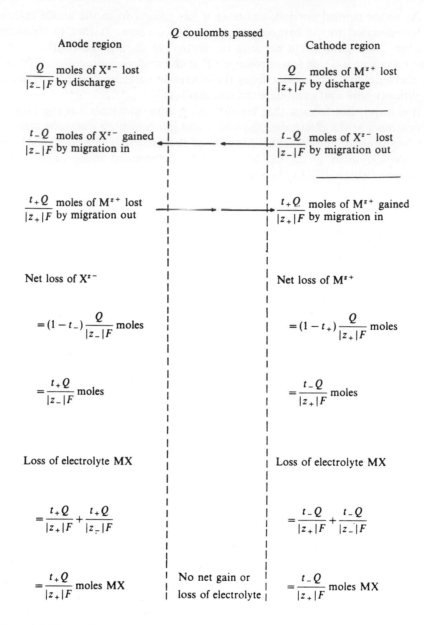

Figure 8.3 Hittorf electrolysis mechanism for the case in which anion and cation are discharged at inert electrodes.

anion and cation are discharged, the net result using inert electrodes will be quite different from the case considered above. The mechanism in this case is shown in Figure 8.3. Here it is seen that there is a net loss of electrolyte

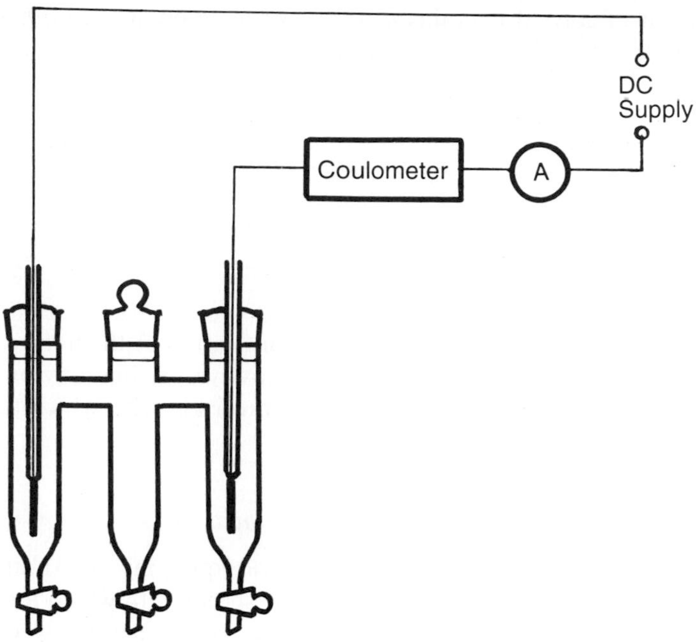

Figure 8.4 Experimental circuit for Hittorf determination of transport numbers.

from both anode and cathode regions. Again, however, no gain or loss occurs in the central region.

For this case at the cathode M^{z+} cations are discharged,

$$M^{z+} + ze \longrightarrow M$$

while at the anode X^{z-} anions are discharged

$$X^{z-} \longrightarrow X + ze$$

The essential experimental circuit for Hittorf's method is shown in Figure 8.4.

A current of 10–20 mA is passed for 1–2 hours. Smaller currents passed for longer times encourage the undesirable diffusion into the central compartment. After electrolysis, samples of solution from anode and/or cathode compartment are withdrawn and analysed. It is also advisable to analyse a sample of solution from the central compartment to check that the composition in this region has, in fact, remained unchanged.

It is most important that concentrations of the compartments be referred to a given weight of solvent, since concentration changes in the electrolyte solution are associated with volume changes. This is due to the fact that ionic species are hydrated and carry their hydration shells with them during their movements through a solution during electrolysis. Suppose that a given weight of solvent contains n_0 moles of a cation species initially and n after electrolysis. If n_e is the number of moles deposited cathodically (determined by the series coulometer), t_+n_e is the number added to the cathode region by migration. Thus,

$$n - n_0 = n_e - t_+n_e$$

and therefore,

$$t_+ = \frac{n_e + n_0 - n}{n_e} \tag{8.34}$$

Hittorf's method provides an excellent demonstration of the nature of electrolysis processes. On the practical level, however, it has obvious drawbacks, not least of which is the usually low precision with which the small concentration changes may be determined.

8.5.2 Determination by moving boundary methods

Such methods represent direct applications of equations (4.54) and (4.55) whereby transport numbers are related to the speeds with which ions move. Moving boundary techniques are based upon the observed rate of movement, under the influence of an applied emf, of a sharp boundary between solutions of two different electrolytes having an anion or cation in common. Measurement of the rate of movement of a sharp boundary presents few problems, since, even if the solutions do not differ in colour, the difference in their refractive indices makes the boundary between them easily distinguishable. A schematic diagram of the relation between two such solutions is shown in Figure 8.5 for the determination of the transport number of a cation. For anion transport numbers, two electrolytes with a common cation are used. If the transport number of M_1^+ is required, $M_2^+X^-$ is referred to as the indicator electrolyte, M_1^+ being sometimes referred to as the leading ion. It is necessary for this latter ion to have higher conductivity than M_2^+.

Let us suppose that the boundary moves from AB to CD when a quantity It coulombs of electricity is passed; let the volume swept out by the boundary be V. Thus, in the volume bounded by AB and CD, the $M_1^+X^-$ initially present is completely replaced by $M_2^+X^-$ after the passage of It coulombs. Or, C_1VF charges pass through a section of the tube in time t. This in turn must be equal to the fraction t_1^+It of the quantity of electricity passed.

Therefore

$$t_1^+ = \frac{C_1VF}{It} \tag{8.35}$$

Apart from the density and conductivity conditions indicated in Figure 8.5 there is a particular value of the ratio C_1/C_2 necessary for the boundary to

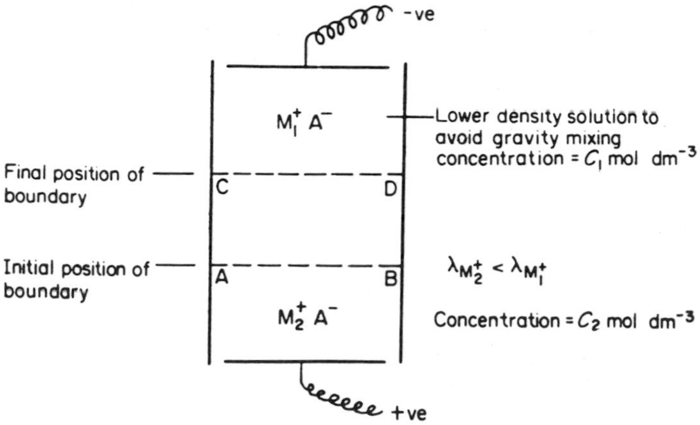

Figure 8.5 Principle of the moving boundary method.

remain sharp. This may be shown as follows: suppose that the boundary moves from AB to CD on the passage of 1 Faraday, so that the quantity of M_1^+ transported across the boundary is t_1^+ moles. Since this was originally contained in a volume V, it follows that

$$t_1^+ = C_1 V \tag{8.36}$$

Similarly, for the indicator electrolyte,

$$t_2^+ = C_2 V \tag{8.37}$$

Or, combining equations (8.36) and (8.37)

$$\frac{C_2}{C_1} = \frac{t_2^+}{t_1^+} \tag{8.38}$$

It appears from this last equation that only by accurate fore-knowledge of t_2^+/t_1^+ could the correct value of C_2 be decided. In practice it is only found necessary for equation (8.38) to hold within about 10%. Since the indicator solution is chosen to have a lower conductivity than that of the electrolyte under study, the field gradients must differ in the two solutions if the ions M_1^+ and M_2^+ are to move at the same speed and maintain a sharp boundary.

Since $\lambda_{M_2^+} < \lambda_{M_1^+}$, the potential gradient (for a given current) is steeper in the indicator solution than in the leading solution. If M_1^+ ions tended to lag behind the boundary into the indicator solution, they would immediately be accelerated by the steeper potential gradient towards the boundary. Similarly, if M_2^+ ions tended to move in front of the boundary, the smaller potential gradient would serve to slow them down (Figure 8.6)

The simplest use of the moving boundary method is that in which an autogenic boundary is formed, i.e. one formed spontaneously at the start of an electrolysis.

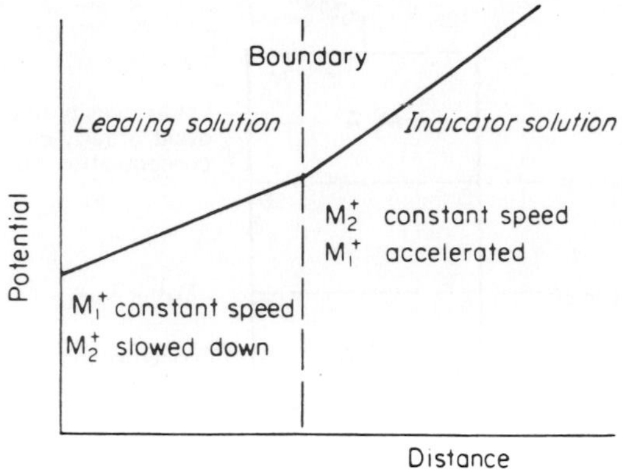

Figure 8.6 Conditions for maintenance of a sharp boundary.

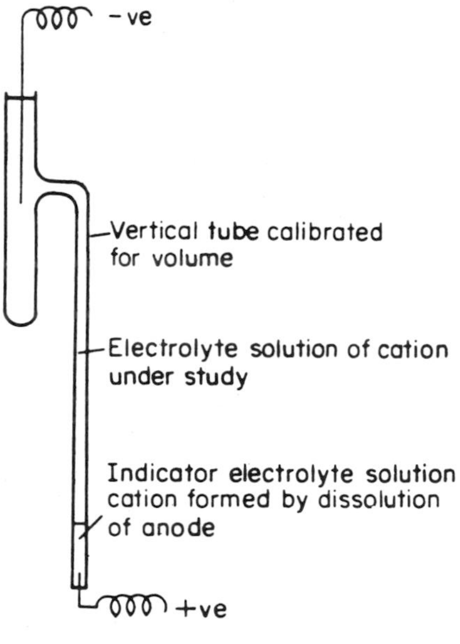

Figure 8.7 The formation of an autogenic boundary.

The principles are shown in Figure 8.7. The cathode is chosen as appropriate to the electrolyte used: if the transport number of metal ion M^+ is to be determined, an electrolyte MX is used, using a cathode of metal M. The anode might be copper or cadmium.

On application of a potential the anode dissolves and a solution of CuX_2 or CdX_2 is formed near to its surface. This is the indicator solution; the boundary between MX and CuX_2 or CdX_2 is followed in order to obtain a value of the transport number of M^+ by application of equation (8.35).

8.5.3 Determination using cell emf

Comparison of the expressions for the emf values of cells with and without transport, for example for the cells

$$\underset{(a_\pm)_1}{Zn|ZnSO_4} \vdots \underset{(a_\pm)_2}{ZnSO_4|Zn} \qquad \text{emf } E_1$$

and

$$\underset{(a_\pm)_1}{Zn|ZnSO_4,\ Hg_2SO_4}|Hg|\underset{(a_\pm)_2}{Hg_2SO_4,\ ZnSO_4|Zn} \qquad \text{emf } E_2$$

shows that the first involves the transport number, t_-, of the sulphate anion, whilst the second does not: thus

$$E_1 = t_- \frac{RT}{F} \ln \frac{(a_\pm)_2}{(a_\pm)_1} \tag{8.39}$$

and

$$E_2 = \frac{RT}{F} \ln \frac{(a_\pm)_2}{(a_\pm)_1} \tag{8.40}$$

and

$$\frac{E_1}{E_2} = t_- \tag{8.41}$$

It must be borne in mind that t_- represents the average value of the transport number for the two solutions of mean ion activities $(a_\pm)_1$ and $(a_\pm)_2$. The values of these latter quantities should therefore in practice be as close to one another as is compatible with sufficiently large emf values to be precisely measured.

8.5.4 Interpretation and application of transport numbers

In section 8.5.1 it was pointed out how necessary it is to know and understand the electrode process which occurs at the anode and cathode during transport number measurements so that one is quite sure which ion it is for which a certain value is determined. One must also pay strict attention to the way in which an electrolyte may ionize and the forms in which the ions are present in solution under given conditions. In this way experimental results which appear unrealistic may be explained rationally. The cadmium ion shows a transport number of about 0.4 in very dilute solutions of cadmium iodide. With progressively higher electrolyte concentrations, however, the value drops sharply below 0.4, passes through zero and finally attains negative values. At the same time the transport number of the iodide ion apparently increases beyond 0.6, passes through, and eventually exceeds, unity.

The above observations are easily explained when the nature of the ions present in more concentrated solutions of cadmium iodide is understood. In such solutions this electrolyte exists largely as Cd^{2+}, and CdI_4^{2-}, i.e. as the simple cadmium ion with a double charge and as a doubly charged anionic complex ion. This latter will obviously migrate in a direction opposite to that of the simple ion. The fact that measured cadmium transport numbers approach negative values indicates that the conductivity of the complexed form is greater than that of the uncomplexed (aquo) form so that a net loss of cadmium occurs at the cathode.

A more fundamental use of transport numbers is in the determination of ion conductivities from measurements of conductivities of electrolytes; an ion conductivity being a fraction t_+ of the observed value for the electrolyte. In order to obtain infinite dilution values of ion conductivities, molar conductivities and transport numbers also have to be obtained by extrapolation to these conditions. In some cases this process may be unreliable. Once, however, the conductivity of a given ion is reliably obtained it is possible to obtain those of any oppositely charged partner in a variety of electrolytes by subtraction.

8.6 Determination of equilibrium constants by measurements of potential

8.6.1 Dissociation constants of weak acids

Such constants may be determined approximately by application of the Henderson–Hasselbalch equation (see equation (3.50)), viz.

$$\mathrm{pH} \sim \mathrm{p}K_\mathrm{a} + \log \frac{b}{a-b} \tag{8.42}$$

where a is the number of moles of a weak monobasic acid in solution and b the number of moles of a strong base added to it so that some of the acid is converted to the salt of the base. It is clear that if $b = a/2$ then $\mathrm{pH} = \mathrm{p}K_\mathrm{a}$, so that, if half the equivalent amount of strong base is added to the weak acid, the pH of the resulting solution is approximately the $\mathrm{p}K_\mathrm{a}$ value.

Figure 8.8 shows the experimental curve for the titration of 25 cm^3 of 0.1 $\mathrm{mol\,dm^{-3}}$ ethanoic acid with 0.1 $\mathrm{mol\,dm^{-3}}$ sodium hydroxide at 298 K. At the half-equivalence point, pH = 4.74, i.e.

$$\mathrm{p}K_\mathrm{a} = 4.74 \text{ and } K_\mathrm{a} = 1.82 \times 10^{-5}$$

At the equivalence point equation (3.41) predicts that the pH should be given by

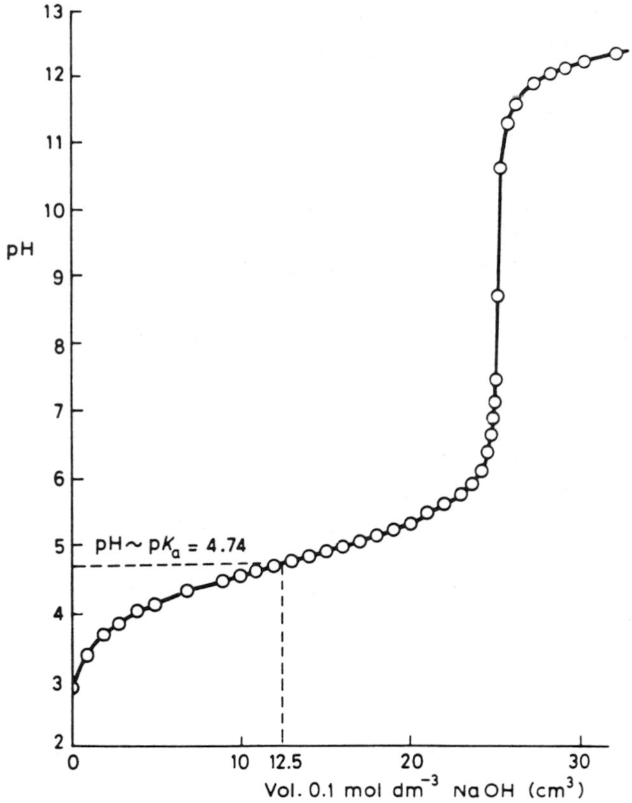

Figure 8.8 Variation of pH with volume of added 0.1 mol dm^{-3} sodium hydroxide during titration of 25 cm^3 0.1 mol dm^{-3} ethanoic acid.

$$\begin{aligned} pH &= \tfrac{1}{2}(pK_w + pK_a + \log C) \\ &= \tfrac{1}{2}(14 + 4.74 + (-1.30)) \\ &= 8.72 \end{aligned}$$

This value agrees closely with the value in Figure 8.8.

More accurate determinations are based on emf measurements of Harned cells in which the central compartment is occupied by a solution of the weak acid, one of its salts formed with a strong base, and a strong electrolyte with its cation derived from the strong base and its anion common with that of the electrode anion, e.g.

$$Pt, H_2(1\,atm)|HA(m_1), NaA(m_2), NaCl(m_3), AgCl(s)|Ag$$

The emf of this cell if given by

$$E_{cell} = E_{AgCl} - E_{H_2} \tag{8.43}$$

therefore,

$$E_{cell} = E^{\ominus}_{AgCl} - \frac{RT}{F} \ln a_{Cl^-} - \frac{RT}{F} \ln a_{H^+} \tag{8.44}$$

or,

$$E_{cell} = E^{\ominus}_{AgCl} - \frac{RT}{F} \ln a_{H^+} a_{Cl^-} \tag{8.45}$$

therefore,

$$\frac{F(E_{cell} - E^{\ominus}_{AgCl})}{RT} = -\ln m_{H^+} m_{Cl^-} - \ln \gamma_{H^+} \gamma_{Cl^-} \tag{8.46}$$

Now

$$K_a = \left(\frac{m_{H^+} m_{A^-}}{m_{HA}}\right) \left(\frac{\gamma_{H^+} \gamma_{A^-}}{\gamma_{HA}}\right)$$

Therefore,

$$\ln K_a = \ln \frac{m_{H^+} m_{A^-}}{m_{HA}} \ln \frac{\gamma_{H^+} \gamma_{A^-}}{\gamma_{HA}} \tag{8.47}$$

If the right-hand side of equation (8.47) is added to the right-hand side of equation (8.46) while the left-hand side of equation (8.47) is subtracted from it, the equality in equation (8.46) is in no way affected, thus

$$\frac{F(E_{cell} - E^{\ominus}_{AgCl})}{RT} = -\ln \frac{m_{HA} m_{Cl^-}}{m_{A^-}} - \ln \frac{\gamma_{HA} \gamma_{Cl^-}}{\gamma_{A^-}} - \ln K_a$$

or

$$\left[\frac{F(E_{cell} - E^{\ominus}_{AgCl})}{2.3RT} + \log \frac{m_{HA} m_{Cl^-}}{m_{A^-}}\right] = -\log \frac{\gamma_{HA} \gamma_{Cl^-}}{\gamma_{A^-}} - \log K_a \tag{8.48}$$

The emf of the cell is measured at various values of m_1, m_2 and m_3 and the left-hand side of equation (8.48) plotted as a function of the ionic strength of the solution. At the condition $I \to 0$, each $\gamma \to 1$ and the right-hand side of the equation approaches $-\log K_a$.

Since the sodium chloride is completely dissociated, $m_{Cl^-} = m_3$. HA is partly dissociated into H_3O^+ and A^- so that $m_{HA} = m_1 - m_{H_3O^+} \sim m_1$ if $m_{H_3O^+}$ is small. Also, the A^-ions originate partly from NaA and partly from HA so that $m_{A^-} = m_2 + m_{H_3O^+} \sim m_2$ if $m_{H_3O^+}$ is small. Thus equation (8.48) may be reasonably approximated to

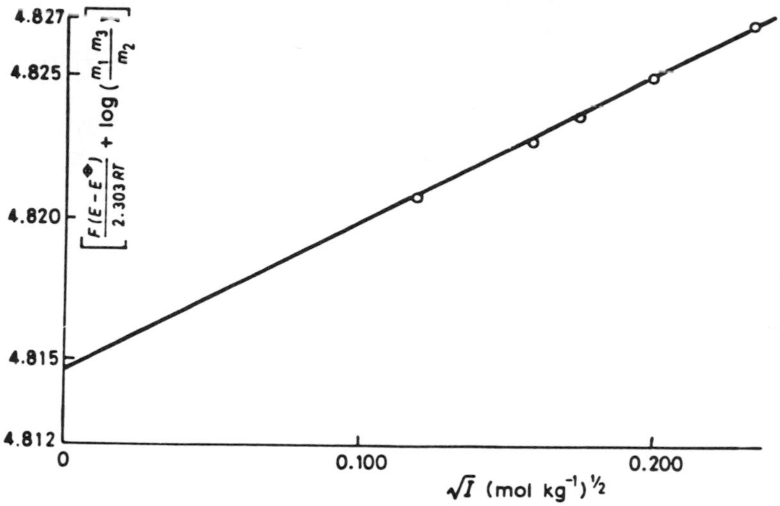

Figure 8.9 Determination of the dissociation constant of n-butyric acid in aqueous solution according to equation (8.48). $-\log K_a = 4.8147$; $K_a = 1.52 \times 10^{-5}$.

$$\left[\frac{F(E_{\text{cell}} - E^{\ominus}_{\text{AgCl}})}{2.3RT} + \log\left(\frac{m_1 m_3}{m_2}\right)\right] = -\log\frac{\gamma_{\text{HA}}\gamma_{\text{Cl}^-}}{\gamma_{\text{A}^-}} - \log K_a \qquad (8.49)$$

Figure 8.9 shows the appropriate plot for the system

$$\text{Pt, H}_2|\text{butyric acid } (m_1), \text{ sodium butyrate } (m_2), \text{NaCl}(m_3), \text{ AgCl(s)}|\text{Ag}$$

Use of equation (8.49) is restricted to acids with pK_a values in the region of 4–5; $m_{H_3O^+}$ must be calculated in the case of stronger acids and the effects of hydrolysis must be taken into account for weaker acids. The important feature of the method is that it is independent of pH measurements. pK_a values determined in this way may be used to standardize the practical pH scale which will be discussed in section 8.7.

A further method involves forming a half-cell by immersing a hydrogen electrode in the solution of the acid whose dissociation constant is required. A complete cell is made by coupling this half-cell with a suitable reference electrode via a salt bridge. The acid is titrated with a strong base, the pH of the solution being measured potentiometrically after each addition of base.

It is necessary to calibrate the reference electrode and salt bridge with standard buffers if agreement is to be obtained between values of dissociation constants obtained by this method and those resulting from Harned cell measurements, which are independent of pH.

We require a formal expression for $a_{H_3O^+}$ which is provided by the equation for the thermodynamic dissociation constant of the acid, viz.

$$a_{H_3O^+} = \frac{K_a a_{HA}}{a_{A^-}} = K_a \left(\frac{m_{HA}}{m_{A^-}}\right)\left(\frac{\gamma_{HA}}{\gamma_{A^-}}\right) \qquad (8.50)$$

Let the overall concentration of the acid be m_a; during the course of the titration m_{HA} will decrease as A^- is formed and m_{A^-} increases. At all stages, however,

$$m_a = m_{HA} + m_{A^-} \qquad (8.51)$$

If m_b represents the varying concentration of base added during the titration, it is seen that

$$m_b + m_{H^+} = m_{A^-} + m_{OH^-} \qquad (8.52)$$

i.e.

$$m_b = m_{A^-} + m_{OH^-} - m_{H^+}$$

Substitution for m_{HA} from equation (8.51) and for m_{A^-} from equation (8.52) into equation (8.50) gives

$$\begin{aligned} a_{H_3O^+} &= K_a \left[\frac{m_a - m_{A^-}}{m_b - m_{OH^-} + m_{H^+}}\right]\left(\frac{\gamma_{HA}}{\gamma_{A^-}}\right) \\ &= K_a \left[\frac{m_a - m_b + m_{OH^-} - m_{H^+}}{m_b - m_{OH^-} + m_{H^+}}\right]\left(\frac{\gamma_{HA}}{\gamma_{A^-}}\right) \\ &= K_a \left(\frac{m_a - B}{B}\right)\left(\frac{\gamma_{HA}}{\gamma_{A^-}}\right) \qquad (8.53) \end{aligned}$$

where $B = m_b - m_{OH^-} + m_{H^+}$ or,

$$\mathrm{pH} = \mathrm{p}K_a + \log\frac{B}{m_a - B} + \log\frac{\gamma_{A^-}}{\gamma_{HA}} \qquad (8.54)$$

The last term on the right-hand side of equation (8.54) may be expressed in terms of the Debye–Hückel theory i.e.,

$$\log \frac{\gamma_{A^-}}{\gamma_{HA}} = -A\sqrt{I} + bI \tag{8.55}$$

so that equation (8.54) becomes

$$A\sqrt{I} - \log \frac{B}{m_a - B} + pH = pK_a + bI \tag{8.56}$$

pK_a may now be determined by plotting the left-hand side of equation (8.56) for each point of the titration curve, against ionic strength. The value is obtained, by extrapolation to zero ionic strength, as the intercept on the ordinate axis. Calculation of B is facilitated by the application of similar approximations to those used in the Harned method.

In order for the values of pK_a to be in agreement with those obtained by the Harned method, it is vital that the potential of the reference electrode/salt bridge combination is given a value such that the pH values determined during the titration agree with those obtained by calculation from true dissociation constants given by the Harned method.

8.6.2 The ionization constant of water

Again, a cell without a liquid junction is used, e.g.

$$\text{Pt, H}_2|\text{MOH}(m_1),\ \text{MCl}(m_2)|\text{AgCl(s), Ag}$$

for which

$$E = E^{\ominus} - \frac{RT}{F} \ln m_{H^+} m_{Cl^-} \gamma_{H^+} \gamma_{Cl^-} \tag{8.57}$$

Now

$$\begin{aligned} K_w &= \frac{a_{H^+} a_{OH^-}}{a_{H_2O}} \\ &= \frac{\gamma_{H^+} \gamma_{OH^-} m_{H^+} m_{OH^-}}{a_{H_2O}} \end{aligned} \tag{8.58}$$

Combining equations (8.57) and (8.58) and noting that $a_{H_2O} = 1$, we obtain

$$\begin{aligned} E &= E^{\ominus} - \frac{RT}{F} \ln \frac{m_{Cl^-} \gamma_{Cl^-} K_w}{m_{OH^-} \gamma_{OH^-}} \\ &= E^{\ominus} - \frac{RT}{F} \ln K_w - \frac{RT}{F} \ln \frac{m_{Cl^-}}{m_{OH^-}} - \frac{RT}{F} \ln \frac{\gamma_{Cl^-}}{\gamma_{OH^-}} \end{aligned} \tag{8.59}$$

or,

$$\left[E - E^{\ominus} + \frac{RT}{F}\ln\frac{m_2}{m_1}\right] = -\frac{RT}{F}\ln K_w - \frac{RT}{F}\ln\frac{\gamma_{Cl^-}}{\gamma_{OH^-}} \tag{8.60}$$

As before, the left-hand side of equation (8.60) is plotted as a function of ionic strength (or $\surd$ (ionic strength), which gives a rather better plot) and $-\ln K_w$ is determined as the intercept at $I = 0$ when $\gamma_{Cl^-}/\gamma_{OH^-} = 1$.

8.6.3 Solubility products

The problem with such determinations is to devise suitable electrodes and cells. One method is to use the sparingly soluble material as part of an electrode, e.g. to determine K_{AgCl} one could employ a silver–silver chloride electrode. K_{AgCl} may then be measured approximately by coupling the electrode with a reference electrode and determining its potential.

The potential of the silver electrode is given by

$$E = E^{\ominus}_{Ag} + \frac{RT}{F}\ln a_{Ag^+} \tag{8.61}$$

therefore,

$$E = E^{\ominus}_{Ag} + \frac{RT}{F}\ln K_{AgCl} - \frac{RT}{F}\ln a_{Cl^-} \tag{8.62}$$

since

$$\left(a_{Ag^+} = \frac{K_{AgCl}}{a_{Cl^-}}\right)$$

and

$$\ln K_{AgCl} = (E - E^{\ominus}_{Ag})\frac{F}{RT} + \ln a_{Cl^-} \tag{8.63}$$

In order for equation (8.63) to be used it is required to know $E^{\ominus}_{Ag}$ and a_{Cl^-} the former may be determined, the latter assumed approximately equal to $a_{\pm}$ of KCl. Since suitable electrode systems can usually be devised the method is used fairly widely even through it is not particularly accurate.

An alternative method can present serious problems in that is not always easy to devise a suitable cell. For the determination of K_{AgCl} the following cell could be used

$$\mathrm{Ag, AgCl(s), HCl}|\underset{1\,\mathrm{atm}}{\mathrm{Cl_2}}\mathrm{, Pt}$$

This cell has also the further disadvantages that chlorine electrodes are difficult to use since chlorine attacks platinum while formation of HCl and

HClO can alter the composition of the solution. The emf of the cell is given by

$$E = E_{\mathrm{Cl_2}} - E_{\mathrm{Ag}}$$

therefore

$$E = E^{\ominus}_{\mathrm{Cl_2}} - \frac{RT}{F}\ln a_{\mathrm{Cl^-}} - E^{\ominus}_{\mathrm{Ag}} - \frac{RT}{F}\ln K_{\mathrm{AgCl}} + \frac{RT}{F}\ln a_{\mathrm{Cl^-}} \qquad (8.64)$$

therefore

$$E = \left(E^{\ominus}_{\mathrm{Cl_2}} - E^{\ominus}_{\mathrm{Ag}}\right) - \frac{RT}{F}\ln K_{\mathrm{AgCl}} \qquad (8.65)$$

from which K_{AgCl} may be found when the standard potential of the silver and of the chlorine electrode are known.

8.6.4 *Equilibrium constants of redox reactions*

Standard redox potentials may be used to determine whether reactions proceed quantitatively and whether, therefore, they may be usefully employed analytically. It is well known, for example, that Ce^{4+} ions oxidize Fe^{2+} ions and that this reaction is used for the titrimetric determination of Fe(II). The usefulness of such reactions may be assessed by considering the redox potentials for the individual redox systems. Thus, we have

$$E_{\mathrm{Fe^{3+}|Fe^{2+}}} = E^{\ominus}_{\mathrm{Fe^{3+}|Fe^{2+}}} + \frac{RT}{F}\ln\frac{a_{\mathrm{Fe_{3+}}}}{a_{\mathrm{Fe^{2+}}}}$$

$$(= +0.76\,\mathbf{V})$$

and

$$E_{\mathrm{Ce^{4+}|Ce^{3+}}} = E^{\ominus}_{\mathrm{Ce^{4+}|Ce^{3+}}} + \frac{RT}{F}\ln\frac{a_{\mathrm{Ce^{4+}}}}{a_{\mathrm{Ce^{3+}}}}$$

The reaction for the cell obtained by combining the two electrodes is

$$\mathrm{Ce^{4+} + Fe^{2+} \longrightarrow Fe^{3+} + Ce^{3+}} \qquad (8.66)$$

If two such half-cells are coupled, the activity of Fe^{3+} ions increases while that of Fe^{2+} ions decreases, causing an increase of the ratio $a_{\mathrm{Fe^{3+}}}/a_{\mathrm{Fe^{2+}}}$. At the same time, a decrease in $a_{\mathrm{Ce^{4+}}}$ and an increase in $a_{\mathrm{Ce^{3+}}}$ leads to a decrease in the ratio $a_{\mathrm{Ce^{4+}}}/a_{\mathrm{Ce^{3+}}}$. Thus, it is seen that the potentials approach each other and meet at the condition of equilibrium when we may write

$$E^{\ominus}_{\mathrm{Fe^{3+}/Fe^{2+}}} + \frac{RT}{F}\ln\left(\frac{a_{\mathrm{Fe^{3+}}}}{a_{\mathrm{Fe^{2+}}}}\right)_{\mathrm{eq}} = E^{\ominus}_{\mathrm{Ce^{4+}/Ce^{3+}}} + \frac{RT}{F}\ln\left(\frac{a_{\mathrm{Ce^{4+}}}}{a_{\mathrm{Ce^{3+}}}}\right)_{\mathrm{eq}} \qquad (8.67)$$

therefore,

$$E^{\ominus}_{Ce^{4+}/Ce^{3+}} - E^{\ominus}_{Fe^{3+}/Fe^{2+}} = \frac{RT}{F}\ln\left(\frac{a_{Fe^{3+}}}{a_{Fe^{2+}}}\right)_{eq} - \frac{RT}{F}\ln\left(\frac{a_{Ce^{4+}}}{a_{Ce^{3+}}}\right)_{eq}$$

$$= \frac{RT}{F}\ln\left(\frac{a_{Fe^{3+}}a_{Ce^{3+}}}{a_{Fe^{2+}}a_{Ce^{4+}}}\right)_{eq} \tag{8.68}$$

therefore,

$$1.61 - 0.76 = \frac{RT}{F}\ln K$$

K being the equilibrium constant for the redox process. For the case K is seen to be large ($\sim 2.39\times10^{14}$ at 298 K) so that the reaction is quantitatively useful since the equilibrium is also established *rapidly*.

8.6.5 Formation (stability) constants of metal complexes

In a cell of the type

$$\underset{1}{M|MX}||\underset{2}{MX|M}$$

considered in section 6.9.2, the emf will be given approximately by a form of equation (6.45) written in terms of concentrations, viz.

$$E \sim \frac{RT}{F}\ln\frac{[M^+]_2}{[M^+]_1}$$

If a complexing agent is added to the left-hand solution, $[M^+]_1$, is reduced and the emf is altered.

For example, if a concentration cell based on silver electrodes placed in different concentrations of silver nitrate is used, the value of E will be different in the absence and presence of a measured concentration of ammonia added to the left-hand half-cell solution. The reason for this is the formation of the complex $Ag(NH_3)_2^+$ by the reaction

$$Ag^+ + 2NH_3 \rightleftharpoons Ag(NH_3)_2^+$$

characterized by the formation constant

$$\beta_2 = \frac{[Ag(NH_3)_2]^+}{[Ag^+][NH_3]^2} \tag{8.69}$$

From the observed emf it is possible to calculate the concentration of free uncomplexed Ag^+ in the solution containing the complex. This in turn leads to the concentration of the complex and of unbound ammonia so that substitution of these values in equation (8.69) yields a value of β_2.

A frequently used method to investigate such systems is based on pH titrations. The shape of the titration curve of a complexing agent, with acidic or basic properties with an appropriate base or acid respectively will be modified in the presence of a metal ion. Analysis of the difference can yield formation constant data for both single and consecutively formed complex species.

Polarographic half-wave potentials (Chapter 9) and molar ion conductivities or diffusion coefficients have been effectively used for the determination of consecutive overall and stepwise formation constants.

8.7 The experimental determination of pH

8.7.1 The hydrogen electrode

From a consideration of the formal definition of pH, viz.

$$\text{pH} = -\log a_{\text{H}^+}$$

it is clear that an electrode in equilibrium with a solution containing hydrogen ions will adopt a potential which is a function of the concentration of hydrogen ions and therefore of the pH of the solution. Since the equilibrium at the electrode is

$$\text{H}^+ + \text{e} \rightleftharpoons \tfrac{1}{2}\text{H}_2$$

the potential adopted is given by

$$E = E^{\ominus} + \frac{RT}{F} \ln \frac{a_{\text{H}^+}}{(a_{\text{H}_2})^{\frac{1}{2}}}$$

or

$$E = \frac{RT}{F} \ln a_{\text{H}^+} - \frac{RT}{2F} \ln P_{\text{H}_2}$$

since $E^{\ominus} = 0$ by definition for this electrode as the primary standard. Further, if the partial pressure of hydrogen is 1 atmosphere.

$$E = \frac{RT}{F} \ln a_{\text{H}^+} \tag{8.70}$$

Exact potentiometric determination of pH using the hydrogen electrode is not as easy as it might at first appear. In principle, it could be coupled with a suitable reference electrode to form the cell

$$\text{reference electrode}||H_3O^+|H_2, Pt$$

for which the emf may be written

$$E = E_{H_2|H_3O^+} - E_{ref} = \frac{2.3RT}{F}\log a_{H_3O^+} - E_{ref} \tag{8.71}$$

assuming that the liquid junction potential is eliminated. Or, in terms of pH,

$$pH = -\frac{(E + E_{ref})F}{2.3RT} \tag{8.72}$$

However, even if the liquid junction potential is eliminated it is required to know E_{ref}, since the value of pH obtained will be strongly influenced by its value. The value of E_{ref} is calculated assuming that ionic activity coefficients are determined only by the total ionic strength and not by the individual chemical properties of the species.

All that one can hope to do by such a method is to determine pH values which are consistent with those calculated from thermodynamic constants using equation (8.56). Thus, it is necessary to reassess the values of potential adopted by reference electrodes used for this purpose and also to take appropriate account of liquid junction potentials.

If the pH value determined by using equation (8.72) is to identify with that used in equation (8.56) to give the thermodynamic dissociation constant of an acid, then the following equality is necessary

$$-\frac{(E + E_{ref})F}{2.3RT} = pK_a + \log\frac{B}{m_a - B} - A\sqrt{I} + bI \tag{8.73}$$

or,

$$E + \frac{2.3RT}{F}\left(pK_a + \log\frac{B}{m_a - B} - A\sqrt{I}\right) = -E_{ref} - \frac{2.3RT}{F}bI \tag{8.74}$$

Experimentally, the hydrogen electrode is immersed in a series of solutions containing weak acids and their salts, the true dissociation constants of the acids being known. Connection of the chosen reference electrode to this solution is achieved by a means of a saturated potassium chloride salt bridge. When the left-hand side of equation (8.74) is plotted as a function of ionic strength, a reassessed value of E_{ref} is obtained as intercept on the I axis. No account has been taken of the liquid junction potential which exists between the salt bridge and the electrolyte solution and whose value varies with both the nature of the buffer used and its concentration. Such variations only produce uncertainities of the order of tenths of millivolts in the value of E_{ref} obtained and the scale of pH values thus obtained constitutes the conventional pH scale.

In practice, the use of the hydrogen electrode for pH determinations is severely limited in that is may not be used in solutions containing reducible materials and is easily poisoned by catalytic poisons.

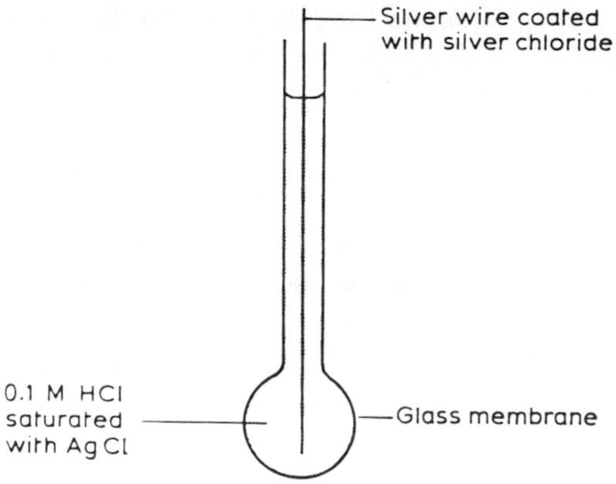

Figure 8.10 Components of the glass electrode.

8.7.2 The glass electrode

The glass electrode is the most widely used indicator electrode for pH determinations used in the laboratory. It operates on the principle that the potential difference between the surface of a glass membrane and a solution is a linear function of pH. A standard solution of know pH must be in contact with the other side of the membrane and in this is immersed a reference electrode (silver/silver chloride or calomel). The construction is shown schematically in Figure 8.10.

The arrangement may be represented as

$$\mathrm{Ag, AgCl(s)|0.1\ M\ HCl|glass|solution}$$

When used in practice it must be coupled with a further reference electrode also dipping into the working solution, e.g.

$$\mathrm{Ag, AgCl(s)|0.1\ M\ HCl|glass|solution||KCl\ (saturated),\ Hg_2Cl_2(s), Hg}$$

Since the potential of the silver/silver chloride electrode is constant and the potential difference between the inner surface of the glass membrane and the hydrochloric acid solution is constant, the only potential difference which can vary is that between the outer surface of the membrane and the working solution. The overall potential of the system is thus a function of the pH of the working solution only.

The above cell forms the basis for the practical pH scale which may be defined by

$$\mathrm{pH_X} = \mathrm{pH_S} - \frac{(E_\mathrm{X} - E_\mathrm{S})F}{2.3RT} \tag{8.75}$$

Here, pH_S, represents the value for a standard buffer solution while E_S is the emf of the cell when this buffer is present. pH_X and E_X represent the pH value and cell emf respectively when the buffer solution is replaced by the solution undcr study. pH_S values are obtained initially by the methods previously described whereby reference electrodes are recalibrated by the use of equation (8.74).

The mode of action of the glass electrode is very complex and, of all the theories put forward, no single one can account for all the observed properties. It is very likely, however, that an important stage involves the absorption of hydrogen ions into the lattice of the glass membrane. The potential of the glass electrode/calomel cell may be expressed as

$$E = K + \frac{RT}{F} \ln a_{H^+} \tag{8.76}$$

Here K is not a true constant but varies on a day-to-day basis for any electrode. It is for this reason that a glass electrode must always be standardized at regular intervals with buffer solutions of know pH. At least two such solutions should be used covering a range of pH values to ensure the constancy of K over this selected range. The variation of K is a function of the asymmetry potential of the glass electrode which is determined by the differing responses to pH of the inner and outer surfaces of the membrane. This difference may well originate in the different conditions of strain in the two surfaces.

The glass electrode works reliably in the range pH 1–9 and it is unaffected by poisoning and oxidizing and reducing agents. In alkaline solutions, at pH > 9, particularly when sodium ions are present, the pH values recorded tend to be lower than the true values and this is due to the infiltration of sodium ions into the glass lattice. Such 'alkaline errors' are minimized by the use of special glasses; lithium glasses, for instance, extend the range reliably to pH 12. At the acid extreme of the scale, reliability is again questionable for pH < 1, where it appears that interference from anion surface adsorption occurs.

Problems

8.1 The dissociation constant of 2,4-dinitrophenol at 298 K has been observed to have the following values in varying concentrations of sodium chloride

$K \times 10^5$	8.344	8.424	8.541	8.630	8.776	8.923	9.162
m (NaCl)	0.0001	0.0002	0.0004	0.0006	0.0010	0.0015	0.0025

Use these data to determine the thermodynamic dissociation constant of 2,4-dinitrophenol at 298 K. (It may be assumed that the ionization of 2,4-dinitrophenol itself makes negligible contribution to the ionic strength of the solutions.)

8.2 The effect of ionic strength upon the rate constant for the reaction between the persulphate ion and the iodide ion has been reported by C.V. King and M.B. Jacobs (1931) *J. Am.*

Chem. Soc., **53**, 1704. The following table collects selected rate constant data, one set in which the ionic strength is maintained by the addition of KCl, the other in which KI is used for this purpose.

$-\log k$	$\sqrt{I}$ (KCl used)	$-\log k$	$\sqrt{I}$ (KI used)
0.987	0.0452	0.996	0.0336
0.979	0.0495	0.983	0.0524
0.951	0.0604	0.943	0.0675
0.935	0.0667	0.917	0.0797
0.928	0.0803	0.893	0.0903
0.900	0.0919	0.889	0.0998
0.876	0.1022	0.860	0.1138
0.854	0.1116	0.836	0.1239
0.801	0.1358	0.801	0.1376
0.772	0.1564	0.770	0.1580

What information may be obtained from these data?

8.3 A saturated solution of silver chloride at 298 K was found to have a conductivity of $2.733 \times 10^{-6}\,\Omega^{-1}\,\mathrm{cm}^{-1}$. The conductivity of the water used for preparing the solution was $0.880 \times 10^{-6}\,\Omega^{-1}\,\mathrm{cm}^{-1}$. The limiting molar conductivities at infinite dilution at 298 K for silver nitrate, hydrochloric acid and nitric acid are 133.36, 426.15 and $421.26\,\Omega^{-1}\,\mathrm{cm}^2\,\mathrm{mol}^{-1}$ respectively. Calculate the solubility of silver chloride at 298 K and its thermodynamic solubility product.

8.4 The ion conductivities for H_3O^+ and OH^- ions at infinite dilution at 298 K are $3.4981\,\Omega^{-1}\,\mathrm{m}^2\,\mathrm{mol}^{-1}$ and $1.9830\,\Omega^{-1}\,\mathrm{m}^2\,\mathrm{mol}^{-1}$ respectively. The conductivity of very pure water at 298 K is $5.498 \times 10^{-6}\,\Omega^{-1}\,\mathrm{m}^{-1}$. Calculate a value for the ionic product of water K_W at 298 K.

8.5 The following values of the molar conductivity of ethanoic acid were obtained at 298 K. By an appropriate graphical method obtain an approximate value for the dissociation constant of the acid. Why must values obtained by such means be regarded as approximate?

$C\,(\mathrm{mol\,dm^{-3}})$	0	0.0005	0.001	0.002	0.005	0.01	0.02
$\Lambda\,(\Omega^{-1}\,\mathrm{cm^2\,mol^{-1}})$	390.7	66.5	48.5	35.5	23.8	16.2	10.7

8.6 The Clark cell,

$$\mathrm{Zn(10\%\ amalgam\ in\ Hg}|\mathrm{ZnSO_4.7H_2O(s),\ Hg_2SO_4(s)}|\mathrm{Hg}$$
(satd. aq. soln.)

has the following values of emf for the temperatures shown

T (K)	293	298	308	318	328
E (V)	1.4267	1.4202	1.4062	1.3908	1.3740

Estimate the Gibbs free energy change, the entropy change and the enthalpy change for the cell reaction at 308 K.

8.7 The standard emf of the cell Ag|AgBr(s)|AgBr(aq)|Ag is 0.726 V at 298 K. Use this information to calculate the solubility product and solubility of silver bromide in water.

8.8 The cell H_2(1 atm) Pt|HCl aq. (m)|AgCl|Ag was found, at 298 K, to have the following values of emf for various molalities of hydrochloric acid.

$m\,(\mathrm{mol\,kg^{-1}})$	0.0004	0.0036	0.0100	0.0400	0.0900
E (V)	0.6266	0.5160	0.4656	0.3974	0.3577

Determine the standard emf of the cell and calculate the activity coefficient of hydrochloric acid in the most concentrated of the five solutions.

8.9 In a Hittorf experiment the anode compartment of the apparatus contained 0.0758 g of silver nitrate plus 10 g water. After electrolysis, the solution had the composition of 0.2701 g silver nitrate and 28.755 g water. In the time of the electrolysis 0.01857 g of copper were deposited in a series copper coulometer. Estimate the transport numbers of the silver and nitrate ions.

8.10 The boundary formed between an aqueous solution of hydrochloric acid (0.01 mol dm^{-3}) and sodium chloride contained in a vertically mounted tube of internal diameter 5 mm, moved a distance of 7.2 mm in 5 min with a constant current flow of 5.5 mA. Estimate the transport number of the hydrogen ion.

8.11 The emf of the cell Pt, H_2(1 atm)|HCl(aq. 0.025 mol kg^{-1}, AgCl(s)|Ag is 0.4196 V at 298 K. Calculate the pH of the hydrochloric acid in the cell if the standard potential of the silver/silver chloride reference electrode is 0.2225 V at 298 K. Compare the value obtained using the Debye–Hückel limiting law.

8.12 The emf of the cell

Pt, H_2(1 atm)|ethanoic acid (aq. 0.0082 mol kg^{-1}),
sodium ethanoate, (aq. 0.0075 mol kg^{-1}),
sodium chloride, (aq.0.0079 mol kg^{-1}), AgCl(s)| Ag

is 0.6253 V at 298 K. The standard potential of the silver/silver chloride electrode is 0.2225 V. Calculate an approximate value for the dissociation constant of ethanoic acid.

8.13 The variation of the ionic product of water (K_W) with temperature was reported by H.S. Harned and W.J. Hamer (1933) *J. Am. Chem. Soc.*, **55**, 2194. Some of their results are tabulated below; use them to estimate the heat of ionization of water at 288, 298 and 308 K. Hence, calculate the heat of neutralization of a strong monobasic acid by a strong base in dilute aqueous solution at 298 K.

T(K)	278	288	298	308	318	328
K_W ($\times 10^{14}$)	0.186	0.452	1.008	2.088	4.016	7.297

9 Electroanalytical techniques

9.1 What constitutes electroanalysis?

Much of this chapter is concerned with the analytical applications of current–voltage relationships which form the basis of the wide range of voltammetric methods. These have their foundation in the understanding of electrode processes considered in Chapter 7 and to some extent in the electrochemical thermodynamics of Chapter 6. This link with mechanistic aspects of electrochemistry has led to the inclusion here of a small amount of material on the elucidation of mechanism. Such analysis of *behaviour* rather than of *quantity* may offend a purist view of analysis, but contextually seems appropriate for the present coverage.

A further link with material presented earlier derives from the distinctive range of electrometric titrations. All concentration-dependent electrochemical parameters may, in principle, be made the basis of a titration method. This is no less true of the use of diffusion currents in voltammetry to produce the range of amperometric techniques, than it is of the similar exploitation of conductivities and electrode potentials in the respective techniques of conductimetric and potentiometric titration. Quantitative electrochemical generation of a titrant—a direct application of Faraday's laws—makes possible the design of coulometric titrations.

9.2 Conductimetric titrations

Variations of conductivity may be used to follow the courses of acid–base and precipitation reactions. A drawback of the latter is the possible contamination of the electrodes by the precipitate formed. A grave disadvantage of any conductivity-based titration is its non-applicability in the presence of high concentrations of electrolyte species other than those required to be determined. This is in contrast to many other electroanalytical techniques where such electrolytes not only do not interfere, but offer distinct advantages.

Conductimetric titration curves for acid–base reactions depend upon the relative strengths of the acids and bases used. In order to maintain straight-line variations of conductivity, it is best to use a titrant concentration considerably greater than that of titrand.

1. *Titration of strong acid by strong base.* The titration graph will have

the form shown in Figure 9.1(a). Rounding of the graph near the equivalence point is due to water dissociation. This latter is immaterial since it is only necessary to take a number of points well to either side of the equivalence point and to extrapolate the two linear segments accordingly. The explanation of the shape of the graph is very simple: over the region AB the fast-moving hydrogen ions of the acid are replaced by the more slowly moving base cations with a consequent fall in conductivity. After all the hydrogen ions are removed, the conductivity rises between B and C as an excess of hydroxyl ions is added to the solution.

2. *Titration of weak acid by strong base.* The type of titration graph obtained in this case is shown in Figure 9.1(b). The conductivity initially rises from A to B as the salt, e.g. sodium ethanoate is formed. Any contribution to the overall conductivity by that of hydrogen ions is largely suppressed by the buffering action of the ethanoate ion. Beyond the equivalence point, the conductivity increases from B to C due to the increasing concentration of hydroxyl ions.

3. *Titration of strong acid by weak base.* The titration plot will take the form shown in Figure 9.1(c). An initial rapid decline over the region AB is due to the replacement of the mobile hydrogen ions by cations of the weak base. From B to C the weak base is added in excess to a solution of its salt so that its ionization is suppressed. Consequently, the conductivity of the excess hydroxyl ions is neglible.

4. *Titration of weak acid by weak base.* In such cases, titration curves of the type shown in Figure 9.1(d) are obtained. Over the region AB, corresponding to the initial addition of weak base, the ionization of the weak acid is suppressed by the buffer action so that the conductivity falls. As the salt is progressively formed, the number of ions in solution rises with consequent increase in conductivity over BC. In the region CD, addition of the weakly ionized base to a solution of its salt causes the conductivity to almost level off.

9.3 Potentiometric titrations

Potentiometric techniques use measurement of electrode potentials and their variation with the changing chemical environment induced by the progressive interaction of a titrant and a titrand. Such methods should be distinguished from those where an electrode potential is used for direct measurements of concentration of an analyte as is the case with a potentiometric sensor.

9.3.1 Zero current potentiometry

For this technique it is simply required to have access to an indicator electrode whose potential is a function of the activity of the species to be

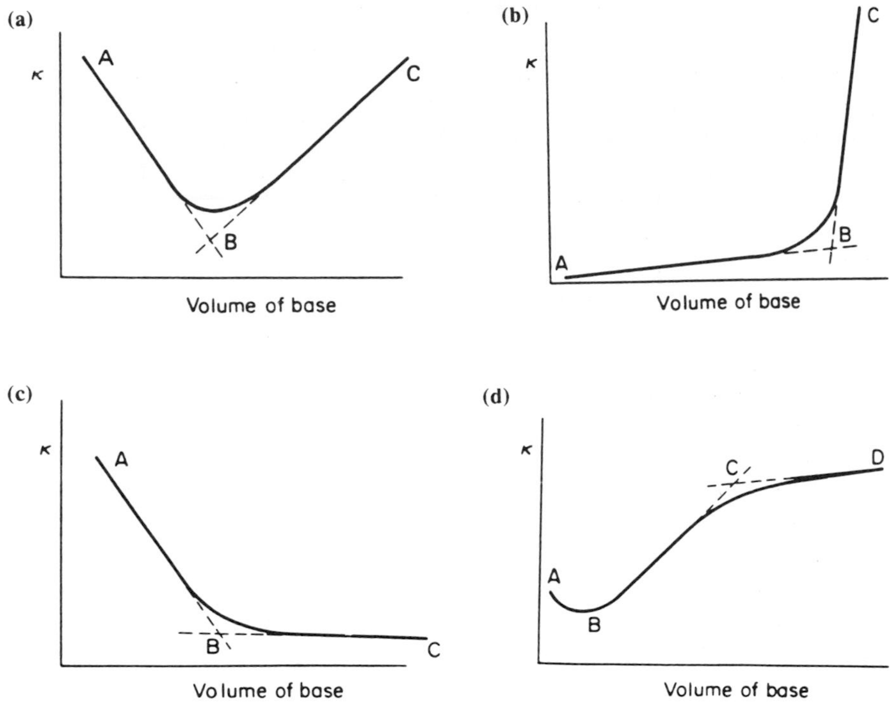

Figure 9.1 Shapes of conductimetric titration graphs for titration of: (a) strong acid by strong base; (b) weak acid by strong base; (c) strong acid by weak base; (d) weak acid by weak base (for this case the equivalence point is insufficiently clear to be exploited).

titrated. A cell is made by placing this component with a suitable reference electrode in the solution to be titrated. The cell is connected within the circuit shown schematically in Figure 9.2, its emf being measured after each addition of titrant: zero current is maintained through the indicating galvanometer by balancing the cell emf against the potentiometer voltage.

The equivalence point of the titration may be determined either from the inflexion point of the graph of indicator electrode potential versus titrant volume (Figure 9.3(a)) or, with appropriate instrumentation, from the derivative plot of dE/dv versus v (Figure 9.3(b)).

Although the potential–volume plot is always of the same essential shape, its definition is dependent upon the equilibrium constant of the titration reaction and its stoichiometry. The equivalence point may be identified with the inflexion point for a 1:1 reaction with a large equilibrium constant and may be satisfactorily determined for a large number of acid–base, precipitation, redox and, in particular, complexometric titrations.

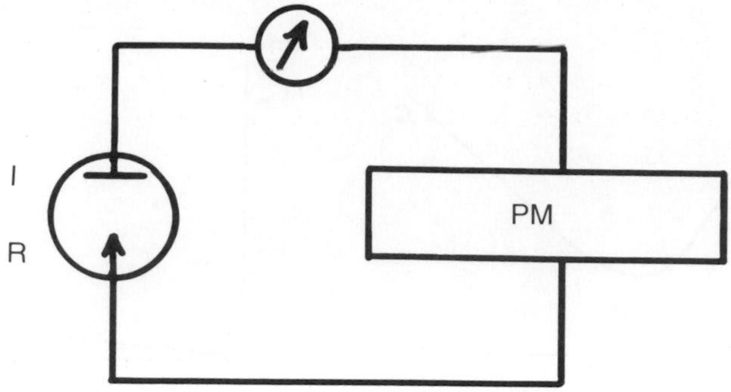

Figure 9.2 Basic circuit for classical potentiometry. I = indicator electrode; R = reference electrode; PM = potentiometer device.

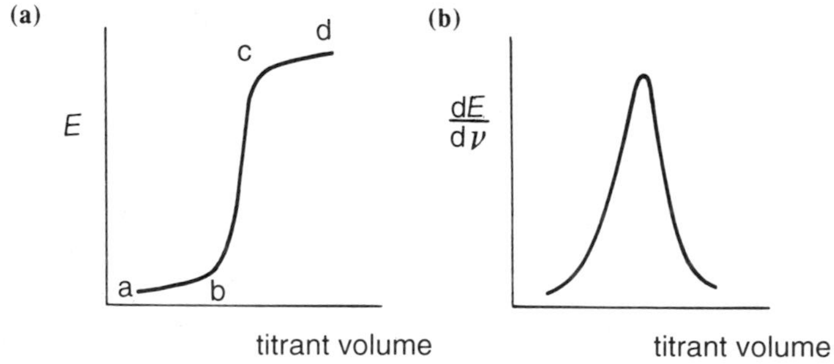

Figure 9.3 Potentiometric titration curves. See text for explanation.

Electrode reactions need not proceed reversibly for a satisfactory potentiometric end-point to to be determined. What is essential is that there shall be a large well-defined potential change in the region of the end-point.

Consider the oxidation of Fe^{2+} ions by Ce^{4+} ions, a reaction already considered in section 8.6.4 where it was seen that the redox potential of the Fe^{3+}/Fe^{2+} couple (+0.76 V) is significantly different from that of the Ce^{4+}/Ce^{3+} couple (+1.61 V). With a solution of Fe^{2+} ions placed in the cell fitted with a platinum indicator and a calomel reference electrode and a burette delivering a solution of Ce^{4+} ions, a titration curve of the form shown in Figure 9.3(a) may be obtained. As soon as a small amount of Ce^{4+} solution is added to the solution of Fe^{2+} ions, some oxidation takes place, i.e.

$$Fe^{2+} + Ce^{4+} \rightarrow Fe^{3+} + Ce^{3+}$$

The indicator and reference electrodes are now dipping into a solution containing mainly Fe^{2+} ions but a small amount of Fe^{3+} ions as well. The potential of the platinum electrode will now adopt a characteristic value which is a function of the activity ratio of Fe^{3+} to Fe^{2+} ions as expressed in the Nernst equation,

$$E_{Fe^{3+}/Fe^{2+}} = E^{\ominus}_{Fe^{3+}/Fe^{2+}} + \frac{RT}{F} \ln \frac{a_{Fe^{3+}}}{a_{Fe^{2+}}}$$

As more Ce^{4+} ions are added the ratio of Fe^{3+} to Fe^{2+} ions increases. The potential adopted by the indicator electrode does not, however, vary greatly. Its value increases only slowly over the range a–b in accordance with the Nernst equation. A point will be reached where all the Fe^{2+} ions have been removed by oxidation and a small excess of Ce^{4+} ions has been added. The potential of the indicator electrode will now be a function of the ratio of activities of Ce^{4+} and Ce^{3+} ions according to

$$E_{Ce^{4+}/Ce^{3+}} = E^{\ominus}_{Ce^{4+}/Ce^{3+}} + \frac{RT}{F} \ln \frac{a_{Ce^{4+}}}{a_{Ce^{3+}}}$$

The region c–d shows similar slight variation of potential with titrant volume as is seen over the region a–b. The regions a–b and c–d are separated widely because of the large difference in standard potentials of the two redox systems. It is seen that in the region of the equivalence point, the potential of the two couples is the same—a fact already used in section 8.6.4 to calculate the equilibrium constant of the redox reaction. Between the two, virtually plateau, regions the potential shows a sharp increase.

9.3.2 *Constant current potentiometry*

Constant current techniques often prove to be useful for redox titrations involving couples of which at least one behaves in an irreversible manner, i.e. equilibrium is not established instantaneously at the electrode surface. As a result of this, the potential jump in the region of the end-point is too small to be useful. The potentials of electrodes which behave irreversibly show considerable variation in value when they pass a current. The constant current required for such titrations has to be determined in a separate experiment prior to the titration itself. In Figure 9.4 are shown current–voltage curves for two redox couples one of which behaves reversibly, the other irreversibly. It is seen that at zero current the potentials are very close whereas, with increasing current, they begin to diverge considerably. In some cases the divergence can be so large that, when the titration is subsequently performed, there is no need to plot the titration graph since

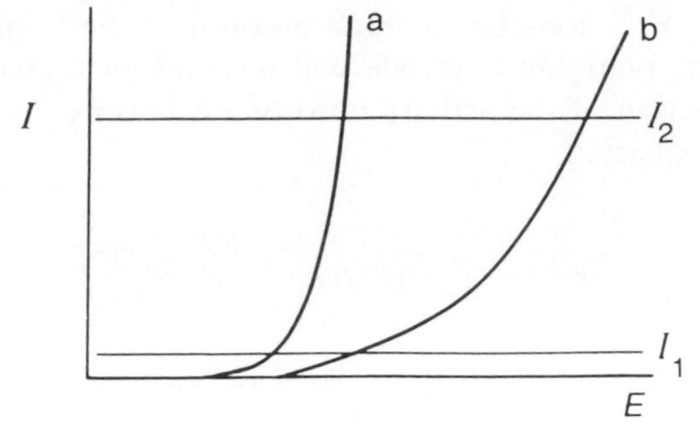

Figure 9.4 Current potential curves for titrant, (a) (reversible) and titrand, (b) (irreversible): constant current I_2 gives better separation of potentials than does I_1.

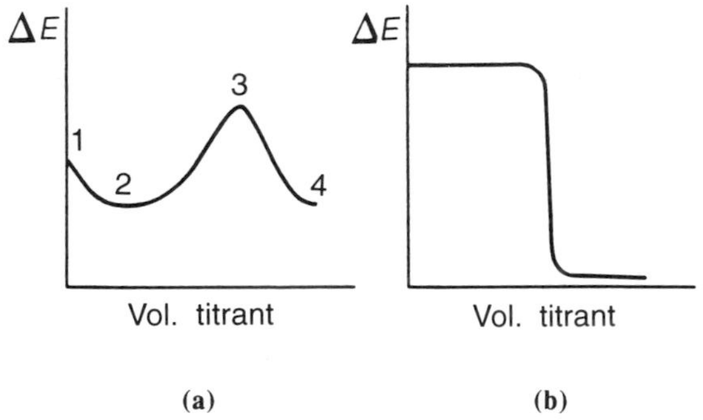

Figure 9.5 Potentiometric titration curves using two indicator electrodes. (a) Both redox couples reversible. (b) One couple reversible, the other irreversible. See text for explanation.

the end-point is directly determinable by a large change in electrode potential.

9.3.3 Potentiometry with two indicator electrodes

When two indicator electrodes are used the potential difference between them is plotted as a function of titrant concentration. The shapes of the individual titration curves vary with the degree of irreversibility of the couples as shown in Figure 9.5. Titration of Fe(II) by Ce(IV) gives a curve of the type shown in Figure 9.5(a) and the reactions occurring at various points on the curve will be considered briefly. The most important stages of the titration occur at the points on the curve labelled 1–4.

1. The only electroactive species present are Fe^{2+} and H_3O^+ so that the only reactions possible are

$$H_3O^+ + e \rightarrow \tfrac{1}{2}H_2 + H_2O \quad \text{at the cathode}$$

and

$$Fe^{2+} \rightarrow Fe^{3+} + e \qquad \text{at the anode}$$

Since these two reactions occur at well-separated potentials, ΔE is large.

2. At this point the titration reaction is half completed, so that $[Fe^{2+}] = [Fe^{3+}] = [Ce^{3+}]$. Possible reactions are now

$$Fe^{3+} + e \rightarrow Fe^{2+} \qquad \text{at the cathode}$$

and

$$\left.\begin{array}{l} Fe^{2+} \rightarrow Fe^{3+} + e \\ \text{or} \\ Ce^{3+} \rightarrow Ce^{4+} + e \end{array}\right\} \text{at the anode}$$

Since the second suggested anode reaction only occurs at a very positive potential it is an unlikely contribution. ΔE is now small since the cathode and anode reactions occur at almost the same potential (the Fe^{3+}/Fe^{2+} couple is fairly reversible)

3. Here the end-point has been reached; all the Fe^{2+} has been oxidized to the Fe^{3+} state and all the Ce^{4+} has been reduced to Ce^{3+}. The electrode reactions are thus

$$Fe^{3+} + e \rightarrow Fe^{2+} \quad \text{at the cathode}$$

and

$$Ce^{3+} \rightarrow Ce^{4+} + e \quad \text{at the anode}$$

ΔE is again very large, since the latter reactions occur at widely differing potentials.

4. As more and more Ce^{4+} ions enter the solution beyond the end-point, the reactions are likely to be

$$Ce^{4+} + e \rightarrow Ce^{3+} \quad \text{at the cathode}$$

and

$$Ce^{3+} \rightarrow Ce^{4+} + e \quad \text{at the anode}$$

ΔE drops rapidly from the end-point to a very small value consistent with the high degree of reversibility of the Ce^{4+}/Ce^{3+} system.

In the case of titration of an irreversible titrand by a reversible titrant, the change of ΔE in the region of the equivalence point is shown in Figure 9.5(b).

9.4 Classical voltammetric techniques

Voltammetry is the generic term for techniques which make use of current–voltage curves under conditions of concentration polarization of an indicator electrode.

The total rate of deposition of a metal ion, M, in terms of the rates of the mass-transfer processes migration and diffusion has been expressed by

$$\frac{i}{nF} = \frac{t_+ i}{nF} + \frac{D([M^{n+}] - [M^{n+}]_e)}{\delta} \qquad \text{(see equation (7.32))}$$

or

$$\frac{i_{lim}}{nF} = \frac{t_+ i_{lim}}{nF} + \frac{D}{\delta}[M^{n+}] \qquad \text{(see equation (7.33))}$$

The last expression holds under conditions of maximum diffusion rate when the surface concentration of the cation is zero.

Equation (7.33) relates the limiting current density to the concentration of the species M^{n+}. Unfortunately it also involves its transport number: since this varies with conditions and concentration, the equation as it stands does not hold very promising analytical possibilities. However, there is a very simple way of almost completely eliminating the electrical migration effect.

It has been seen earlier that t_+ for a given ion species is diminished in the presence of other ions: it then serves to carry a *smaller fraction* of the total current. Addition of a sufficient excess of another electrolyte, showing no electrochemical reactions which interfere with M^{n+}, causes t_+ to become vanishingly small. The rate of arrival of M^{n+} at the cathode is then controlled solely by diffusion. The added species is known as an indifferent, base, or supporting electrolyte: potassium chloride is useful in this role and is commonly used. Thus equation (7.33) becomes

$$\frac{i_{lim}}{nF} = \frac{D}{\delta}[M^{n+}] = \frac{I_{lim}}{AnF}$$

A being the surface area of the electrode and I_{lim} the limiting diffusion-controlled current, explicitly given by

$$I_{lim} = \frac{nFAD[M^{n+}]}{\delta} \tag{9.1}$$

It has not always been easy to exploit equation (9.1) for quantitative analysis: this is because the prime requirement for quantitative analysis—reproducibility—is not always easily obtained with solid electrodes due to contamination by the products of electrolysis. However, partly due to improved methods of pretreatment and partly because of the speed of voltage scans with modern techniques, solid electrodes are now quite as important as the classical voltammetric electrode material, viz. mercury.

9.4.1 Polarography

This is the particular form of voltammetry based on the use of a dropping mercury (micro-)electrode (DME). This consists of mercury in the form of small drops issuing from the end of a fine-bore capillary.

Mercury in contact with a solution of an electrolyte such as potassium chloride approaches very closely in its behaviour to that required of an ideal polarized electrode (see Chapter 5). Within a certain fairly wide range of potentials, no ions are discharged at its surface, nor are they dissolved from it. This means that the potential applied to it may be varied at will without the establishment of electrochemical equilibria. In practice, *a near-ideal polarized electrode* may be recognized by the fact that, over a range of applied potential, no current flows. A *polarized* electrode shows *no current change* over a range of applied potential.

Despite the obvious practical complication of working with a dynamic rather than a static electrode, the DME shows a number of advantages.

1. Drops are reproducibly formed so that currents, although varying with drop growth and detachment from the capillary, are also reproducible.
2. A fresh surface of electrode is continuously presented to the electrolyte solution, eliminating chemical contamination and an ‘electrode history’.
3. Since the DME is a micro-electrode, solutions may be partially electrolysed very many times without measurable reduction in concentration.
4. The hydrogen overvoltage for mercury as an electrode material is, as discussed in Chapter 7, extremely high. In fact, the cathodic polarization limit is caused by discharge of electrolyte cations rather than of hydrogen ions. The current–voltage characteristics of the DME in potassium chloride solution are shown in Figure 9.6.
5. Since so many electro-active species undergo electron exchange processes on mercury in the near-ideal polarized range of 0 to -1.8 V, they show analytically exploitable signals superimposed on the charging current baseline shown in Figure 9.6.

Advantages of the DME are tempered by a number of disadvantages:

1. The cathodic versatility of mercury is not matched by its anodic behaviour. Dissolution sets in at about $+0.4$ V vs. the saturated calomel electrode potential. So far as most metal ions are concerned, this is of little importance since most of them are reduced at potentials considerably more negative than this although copper and silver are exceptions.
2. Oxygen is reduced in a two-stage process as follows:

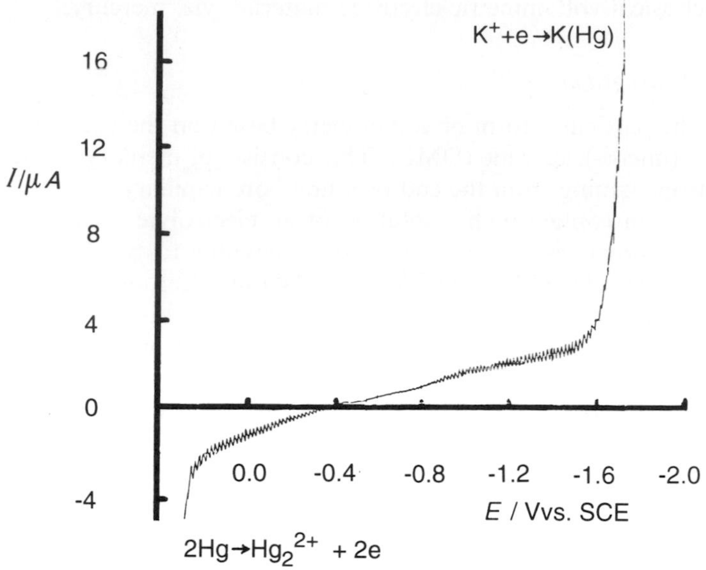

Figure 9.6 Current–voltage curve for 0.1 mol dm^{-3} potassium chloride at a dropping mercury electrode.

In acidic media.

Stage 1.	$O_2 + 2H^+ + 2e \rightarrow H_2O_2$
Stage 2.	$H_2O_2 + 2H^+ + 2e \rightarrow 2H_2O$
Overall	$O_2 + 4H^+ + 4e \rightarrow 2H_2O$

In neutral and alkaline media.

Stage 1.	$O_2 + 2H_2O + 2e \rightarrow H_2O_2 + 2OH^-$
Stage 2.	$H_2O_2 + 2e \rightarrow 2OH^-$
Overall	$O_2 + 2H_2O + 4e \rightarrow 4OH^-$

These reactions are seen to identify with the reverse of the anodic processes occurring in the electrolysis of water considered in section 7.6.

The current–voltage curve for oxygen extends over most of the cathodic working range. Dissolved oxygen must therefore be removed from a working solution by flushing with some inert gas, such as nitrogen, before electrolysis is attempted.

3. Mercury is toxic, but sensible precautions make its use safe.

Schematic circuits for voltammetry generally and for classical d.c. polarography in particular are shown in Figure 9.7. The essential features of a

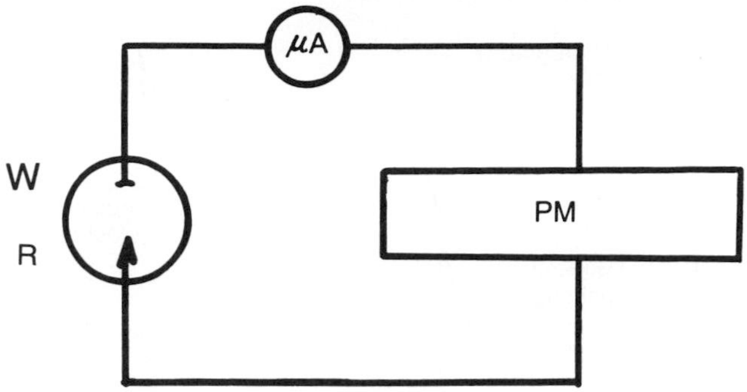

(a)

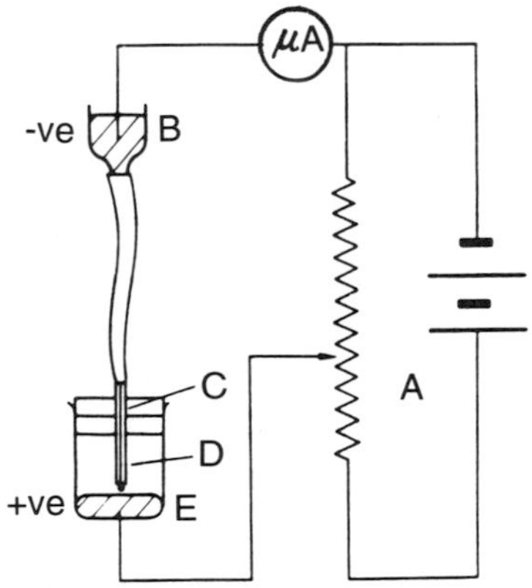

(b)

Figure 9.7 (a) Schematic circuit for classical (two-electrode) voltammetry. The circuit is similar to that for potentiometry except that a *controlled* microscopic electrolytic process is induced and the current measured by a microammeter. W = working microelectrode; R = reference electrode; PM = potentiometer device. (b) Basic polarographic circuit. A = potentiometer for variation of applied potential between C and E; B = mercury reservoir; C = dropping mercury electrode; D = working solution containing depolarizer(s) and supporting electrolyte; E = mercury pool anode (in precise measurements of potential this is replaced by a true reference electrode such as a saturated calomel electrode).

polarographic current–voltage curve are shown in Figure 9.8. Such curves are known as polarograms or 'polarographic waves'. They show development of a *faradaic* current signal superimposed on the residual *non-faradaic* (charging) current. It is seen that this residual current cuts the voltage axis, usually at about −0.4 or −0.5 V vs. SCE, and corresponds to the potential of zero charge (Chapter 5).

The current plateau corresponds to the condition that metal ions are reduced as fast as they reach the electrode by natural diffusion. In the range of applied potential where this constant diffusion current flows, the

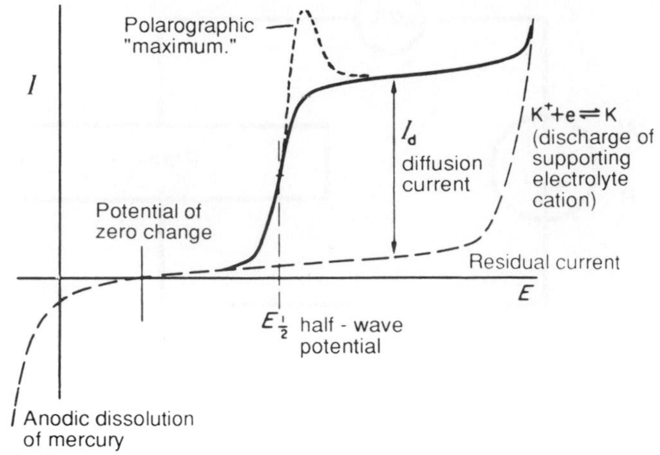

Figure 9.8 Essential features of a polarographic current–voltage curve or wave (polarogram). The polarogram corresponding to reduction of a depolarizer is superimposed on the supporting electrolyte curve in the region of the 'residual current'.

DME is clearly polarized—it is in fact described as *concentration polarized.* Reducible and oxidizable species which show such characteristics are often described as *depolarizers*: reference to the polarogram in Figure 9.8 makes this terminology clear. The 'polarographic maximum' indicated on the main wave profile derives from unusual (tangential rather than radial) mass transfer. Such behaviour seems to originate in a charge distribution between the capillary and the solution end of a mercury drop. The effect may be eliminated by the addition to the solution of a 'maximum suppressor' such as gelatin or (better) Triton-X-100. Such species function in this context through their surface activity: they are added at very low levels ($< 1\%$) since too much can distort the profile and lower the limiting currents of polarographic waves.

9.4.2 Characteristics of diffusion-controlled polarographic waves

Mean diffusion currents are expressed in the Ilkovic equation

$$\bar{I}_{\mathrm{d}} = 607nD^{\frac{1}{2}}m^{\frac{2}{3}}t^{\frac{1}{6}}C \tag{9.2}$$

Here n is the number of electrons transferred in the electrode reaction, D is the diffusion coefficient of the depolarizer ($\mathrm{cm^2\,s^{-1}}$), m is the rate of flow of mercury ($\mathrm{mg\,s^{-1}}$), t is the drop time (s) and C is the depolarizer concentration ($\mathrm{mmol\,dm^{-3}}$). With these units and the numerical constant given in equation (9.2) $\bar{I}_{\mathrm{d}}$ is given in microamps. The potential at the midpoint of the wave, where $\bar{I} = \bar{I}_d/2$, known as the *half-wave potential* ($E_{\frac{1}{2}}$), is

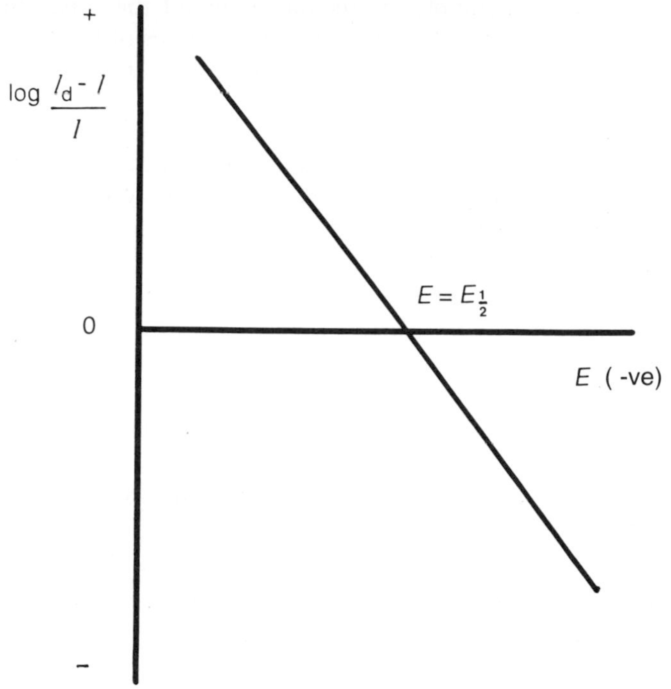

Figure 9.9 Logarithmic analysis of a reversible polarographic wave. Slope = $nF/2.3RT$.

characteristic of a given depolarizer for fixed solution conditions. As such, half-wave potentials may be used to identify qualitatively components in a mixture of depolarizers. Values are, however, extremely sensitive to the presence of different complexing species, including supporting electrolyte anions, and $E_{\frac{1}{2}}$ values should be used for 'fingerprinting' only with extreme caution. Provided that the reduction process occurs reversibly the currents and corresponding potentials on the rising portions of waves are related through the Heyrovsky–Ilkovic equation

$$E = E_{\frac{1}{2}} + \frac{RT}{nF} \ln (\bar{I}_d - \bar{I})/\bar{I} \tag{9.3a}$$

Equation (9.3a) is seen to be very similar to the Nernst equation and may be regarded as its polarographic equivalent. A plot of $\ln[(\bar{I}_d - \bar{I})/\bar{I}]$ versus E may be used to identify both n and $E_{\frac{1}{2}}$ (Figure 9.9).

The sensitivity of $E_{\frac{1}{2}}$ values to complexation has two significant applications. The separation of otherwise overlapping and interfering polarographic signals may be effected by the addition of an appropriate ligand to a working solution. Modern techniques with improved signal resolution have obviated the necessity for such procedures to some extent, but multiple ion analysis (a distinct advantage of polarography) is usually effected by working in several complexing media to spread the waves in differing patterns. A separation of 0.2 V between the $E_{\frac{1}{2}}$ values of following waves is regarded as the minimum for satisfactory analysis. Additionally, accurate measurement of the shift in $E_{\frac{1}{2}}$ with ligand concentration has formed the basis of a range of methods for determining formation (or stability) constants. It is usually necessary for the reduction of both aqua and otherwise complexed ions to take place reversibly. Under these conditions it may be shown that for a complexing reaction

$$\mathrm{M} + N\mathrm{X} \rightleftharpoons \mathrm{MX}$$

the shift in half-wave potential, $\Delta E_{\frac{1}{2}}$, is given by

$$\frac{0.4343nF}{RT} \cdot \Delta E_{\frac{1}{2}} \sim \log \beta_N + N \log [\mathrm{X}] \tag{9.4}$$

where β_N is the formation constant of MX_N and [X] is the free ligand concentration.

For irreversible reductions a wave has less than the theoretical slope for the number of electrons transferred and equation (9.3a) must be modified to

$$E = E_{\frac{1}{2}} + \frac{RT}{\alpha nF} \ln \frac{\bar{I}_{\mathrm{d}} - \bar{I}}{\bar{I}} \tag{9.3b}$$

α being the transfer coefficient. It is instructive to compare the expected shapes for two reductions having the same value for n, one involving a rapid electron transfer (reversible), the other a slow one (irreversible) as shown in Figure 9.10. For wave (a), diffusion is rate determining over the entire wave profile, the mass-transfer process being always slower than the electron-exchange rate. For wave (b), the electron transfer process is rate determining over lower parts of the wave, even these small currents requiring large (activation) overvoltage to sustain them. It is not until the later part of the wave, where even more overvoltage is applied, that the electron-transfer rate becomes of such a magnitude that this process gives way to the diffusion process in determining the rate. Thus, it is important to realize that, even for slow processes, the limiting currents are subject to diffusion control and as such are given by the Ilkovic equation. It follows that both reversible and irreversible diffusion controlled waves may be exploited analytically. For analysis it is possible to draw a calibration graph of diffusion current versus concentration and to read 'unknown' concentrations from this. Sometimes it may be more convenient to employ standard addition techniques, whereby the increase in a current signal caused by the addition of a known concentration of the depolarizer to be determined, enables the concentration producing the original signal to be calculated by simple proportion.

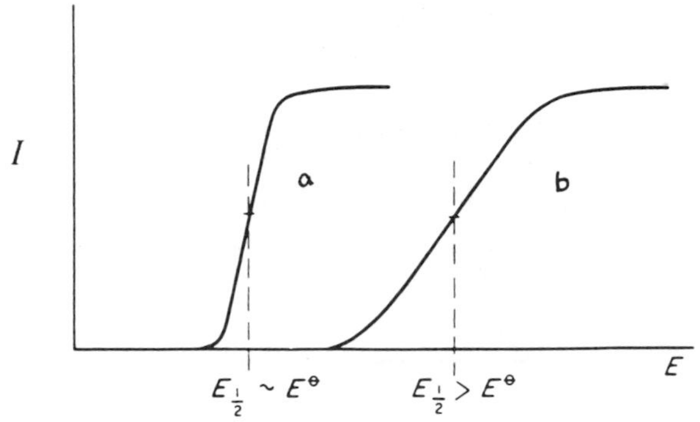

Figure 9.10 Comparison of polarograms for (a) a reversible and (b) an irreversible reduction. The limiting currents in both cases being diffusion controlled and the number of electrons transferred equal.

In order to base analytical methods upon the application of the Ilkovic equation, it is essential to establish that limiting currents produced by depolarizers are diffusion controlled. This is by no means always the case. Rearrangement of equation (9.2) in the form

$$\bar{I}_d = (607nD^{1/2}C)(m^{2/3}t^{1/6})$$

emphasizes the presence of two types of variable in the Ilkovic equation—those concerned with the solution and those concerned with the electrode. When electrode factors are maintained constant, a linear relationship between $\bar{I}_d$ and concentration is maintained. If solution factors are kept constant, $\bar{I}_d \propto m^{2/3}t^{1/6}$. By Poiseuille's equation, the rate of flow of a liquid (v) through a capillary under a head of liquid is directly proportional to the height of the column, h. Therefore $v \propto m \propto h$, and since also $v \propto 1/t, \bar{I}_d \propto h^{2/3}h^{-1/6} = h^{1/2}$. Thus, a diffusion- controlled wave shows a linear relationship between $\bar{I}_d$ and the square root of the height of the mercury reservoir.

The simplest experimental arrangement may be used for concentration determinations in the range 10^{-5} mol dm^{-3} to 10^{-2} mol dm^{-3}. Below about 10^{-4} mol dm^{-3}, however, the ratio of faradaic current to the concentration-independent non-faradaic current becomes progressively smaller until the latter predominates. All modern instrumental refinements of the polarographic method are aimed at improving this ratio: these are considered in section 9.6.

Solid electrodes offer some advantage in this regard: if suitable pretreatment can be devised and implemented to obtain reproducible results, the problem of residual currents does not arise. Additionally, they allow probing of the anodic potential range which mercury does not.

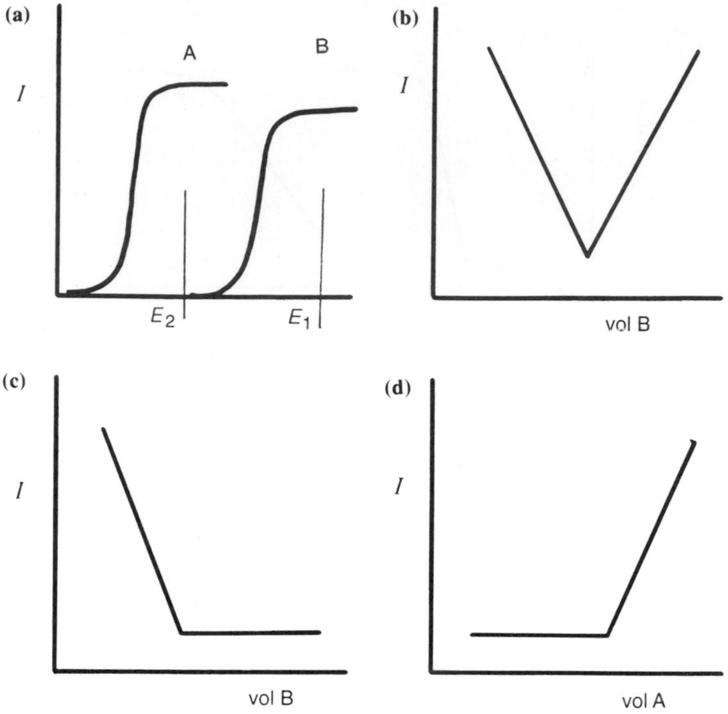

Figure 9.11 (a) Selected polarographic characteristics of titrant A and titrand B for amperometry; (b) titration of A by B at applied potential E_1; (c) titration of A by B at applied potential E_2; (d) titration of B by A at potential E_2.

9.4.3 Amperometric titrations

Because of the direct proportionality of diffusion current and bulk depolarizer concentration, titration of such species by following the diffusion current which they generate at a fixed potential can be particularly convenient. It is not necessary for both titrant and titrand to be electro-active although the more conventional form of titration curve is obtained when this situation obtains. A number of commonly encountered circumstances are summarized in Figure 9.11.

Titration of species B as titrand by A as titrant (or indeed of A by B) with an applied potential E_1 (Figure 9.11(a)) yields the titration curve shown in Figure 9.11(b).

Titration of A by B at an applied potential E_2 where only A is electro-active produces the curve of Figure 9.11(c), while titration of B by A at this same potential yields Figure 9.11(d). Curves of the type shown in Figure 9.11(c) and 9.11(d) are also obtained when either titrant or titrand are electro-inactive and are particularly common with analysis based on quantitative complexation. The DME may be used successfully for amperometry, the electrode even functioning satisfactorily in stirred solutions which are heavily loaded with the precipitated products of a titration reaction. Solid electrodes are, however, particularly useful in this context even though these may have their current responses affected by an electrochemical history. Identification of a titration end-point is not affected by what would be a distinct disadvantage in direct volumetric analysis. This allows full advantage to be taken of controlled forced convection in the form of rotated electrodes to enhance measured current signals. A singular advantage of the techniques of amperometry is the extreme simplicity of the apparatus which may be used for accurate analysis. The modification whereby a constant small voltage is applied between a pair of identical indicator electrodes and the resultant current measured directly during the course of the titration complements the similar variation used in potentiometry and allows the use of 'dead-stop' techniques.

9.4.4 Wave characteristics and the mechanism of electrochemical processes

The profile of a current–voltage curve can reflect a more or less complex interplay of sequential electron-transfer processes further complicated by the involvement of associated chemical reactions. Analysis of polarographic and other voltammetric signals has led to much insight into the nature of electrode processes and has contributed to the development of industrial processes and to the identification of the controlled parameters required for their efficient exploitation.

Consider first a few possibilities arising from sequential electron transfer: these may be understood in terms of the scheme

$$M^{2+} \xrightarrow{e} M^{+}$$

$$M^{+} \xrightarrow{e} M^{0}$$

When both electrons are transferred rapidly the reaction may be expressed as

$$M^{2+} \xrightarrow{2e} M^{0}$$

and the polarogram will have the profile characteristic of a reversible two-electron reduction (Figure 9.12(a)). For example Cu^{2+} in non-complexing supporting electrolytes shows such a wave: in reality it is somewhat obscured by the fact that the half-wave potential is close to 0.0 V vs. SCE so that the foot of the polarogram is overtaken by the anodic curve corresponding to the dissolution of mercury. However, in complexing media, such as ammonia, the lower oxidation state is sufficiently stabilized relative to the higher one that it shows separate reduction. Two one-electron reversible waves are now observed (Figure 9.12(b)).

If the first step is slower than the second, i.e.

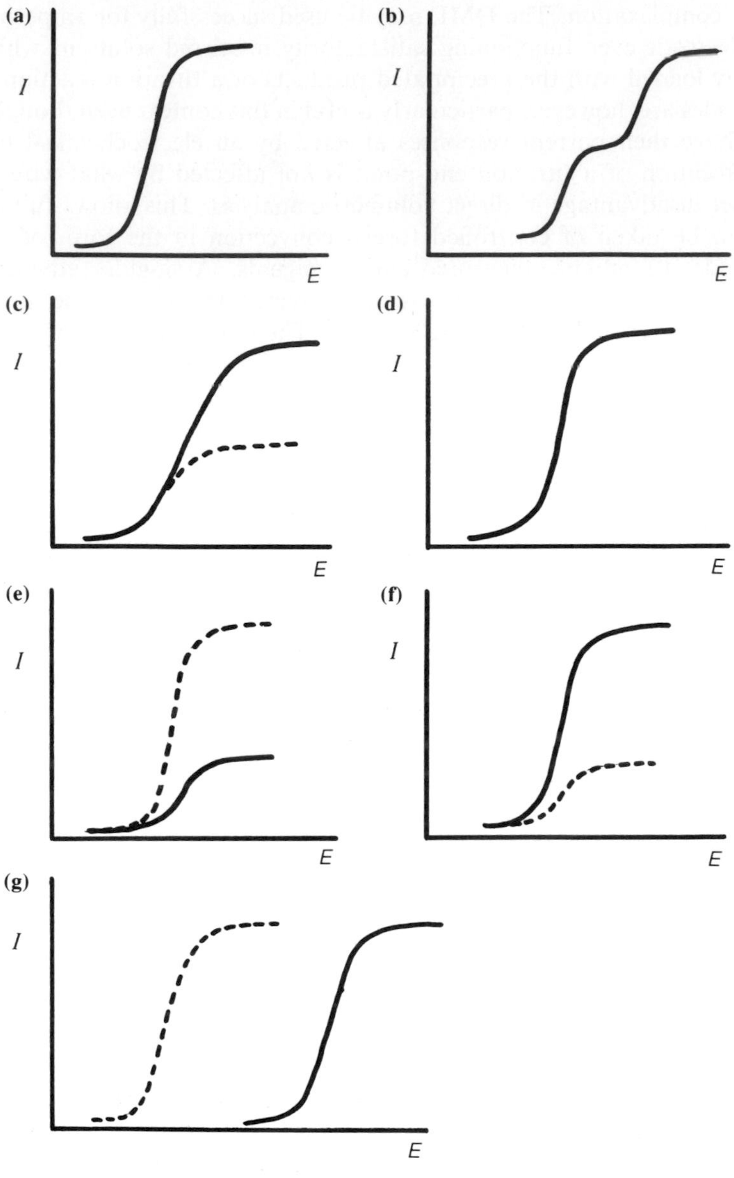

Figure 9.12 (a) Reversible 2-electron reduction wave, both electrons transferred rapidly. (b) Separation into equal one-electron stages through complexation producing a relatively more stable lower oxidation state intermediate. (c) Slope of overall two-electron wave controlled by slow first electron transfer. (d) Slope of overall two-electron wave controlled by fast first electron transfer. (e) Formation of 'kinetic' wave smaller than the hypothetical profile based on a diffusion-controlled current. The CE mechanism. (f) Formation of a 'catalytic' wave by regeneration of depolarizer at electrode surface. The EC' mechanism. (g) Shift of wave profile to more negative potentials where reduction product undergoes subsequent chemical reaction. The EC mechanism.

$$M^{2+} \xrightarrow{e(\text{slow})} M^{+}$$

$$M^{+} \xrightarrow{e(\text{fast})} M^{0}$$

the shape and slope of the wave will be dominated by this slow rate-determining step whose wave profile cannot be seen in isolation but is indicated by the dashed curve in Figure 9.12(c). The overall two-electron wave will have the slope of this first stage but an overall height the same as in Figure 9.12(a).

If the second step is slower than the first, i.e.

$$M^{2+} \xrightarrow{e(\text{fast})} M^{+}$$

$$M^{+} \xrightarrow{e(\text{slow})} M^{0}$$

the ready supply of M^+ via the first stage means that, although the overall height of the wave remains the same as in the previous case, the overall profile is much steeper (Figure 9.12(d)).

One of the earliest types of polarographic behaviour recognized as different to the diffusion-controlled situation characterized by the Ilkovic equation is that in which a *kinetic* wave is formed. In the more modern classification of coupled electrode and homogeneous reactions these are said to proceed by the CE mechanism, i.e. a chemical followed by an electrochemical reaction. The general scheme may be expressed as

$$A \underset{k_{-1}}{\overset{k_1}{\rightleftharpoons}} B \xrightarrow{\pm ne} C$$

and is exemplified by the behaviour of formaldehyde viz.

$$(HO)_2CH_2 \underset{k_{-1}}{\overset{k_1}{\rightleftharpoons}} H_2O + CH_2O \xrightarrow[+2e]{2H^+} CH_3OH \qquad (9.5)$$

Formaldehyde exists in aqueous solution largely as the electro-inactive hydrate. The reducible anhydrous molecule is formed from the hydrate only slowly and a wave is produced which is smaller than would be expected for diffusion control (Figure 9.12(e)). A further distinguishing feature is that the height of the wave is not influenced by the mercury reservoir height.

A further type of behaviour is in polarographic terminology known as a *catalytic* wave: it is often significantly higher than the expected diffusion-controlled wave (Figure 9.12(f)) and originates in the regeneration of the depolarizer at the electrode surface by chemical reaction with some solute species. Such behaviour is frequently represented by the scheme

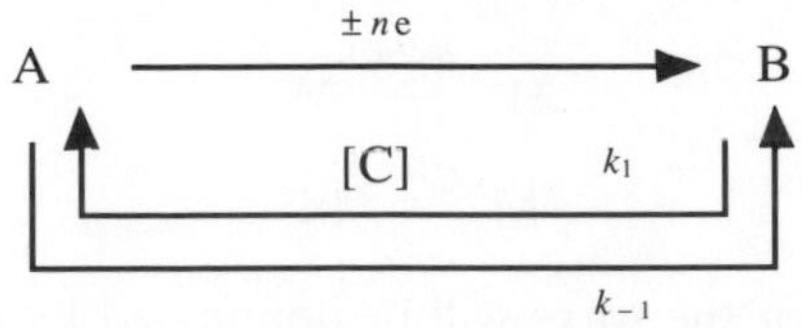

and is shown by the Fe^{3+}/Fe^{2+} system in the presence of hydrogen peroxide to oxidize the lower valency–state product. In general terminology this is known as an *EC′* mechanism. In a modification termed an EC mechanism, viz.

$$A \xrightarrow{\pm ne} B \underset{k_{-1}}{\overset{k_1}{\rightleftharpoons}} C$$

the product of an electrochemical reaction has its concentration reduced by a subsequent chemical reaction which produces a secondary product. The effect is to shift the position of the reduction or oxidation wave—to more negative potentials for the former and to more positive potentials for the latter case—although the height remains unaltered (Figure 9.12(g)). The oxidation of *p*-aminophenol on a platinum electrode shows such behaviour, viz.

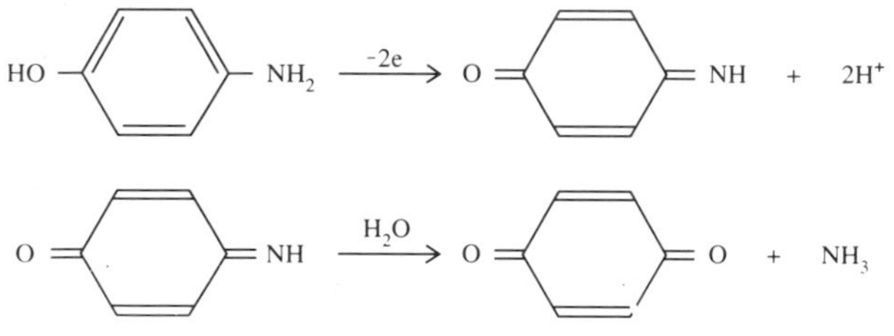

(9.6)

The sequence of behaviour represented in Figure 9.12 more or less exhausts the capabilities of electrochemical techniques themselves to *imply* mechanisms. Analysis of electrochemical behaviour can only *indirectly* lead to the elucidation of mechanisms: this is particularly the case in respect of intermediates formed at an electrode surface.

Predictions of expected experimental characteristics can be made to some extent on the basis of theoretical models. Matching the predictions of a model with what is actually seen has proved to be a powerful approach but may lead to only qualified support for a particular mechanism. Differences in electrochemical parameters reflecting different mechanisms may be very small: in some instances different proposed routes lead to identical experimental behaviour. Thus, while it is possible from current–voltage curves to *infer* an ECE mechanism with the scheme

$$A \xrightarrow{\pm ne} B$$

$$B \xrightarrow{k_1} C$$

$$C \xrightarrow{\pm n'e} D$$

the experimental behaviour could, in fact, equally well be accounted for by the following disproportionation (DISP) sequence

$$A \xrightarrow{\pm ne} B$$

$$B \xrightarrow{k_1} C$$

$$B + C \longrightarrow A + D$$

Such problems have led to the harnessing of various spectroscopic techniques to voltammetric investigations to allow direct identification of short-lived intermediates. These approaches are complemented by more modern electrochemical methods based on cyclic voltammetry (section 9.7) and the ring–disc electrode (section 9.6)

9.5 Modern polarographic methods

Modern techniques use a three-electrode rather than a two-electrode system of the sort shown in Figure 9.2. Figure 9.13 shows the essential features of the circuit which is aimed at overcoming the disadvantage of the *IR* drop which can distort classical polarograms. In order to minimize such effects before the introduction of potentiostats it was necessary to use very high concentrations of supporting electrolytes. These often needed to be higher even than those required for the suppression of the migration effect and introduced the further complication of requiring very high purity electrolytes for trace analysis of depolarizers.

With the three-electrode arrangement current only flows between indicator and counter electrodes: the potentiostat probes the potential difference between indicator and reference electrodes. If the *IR* drop causes a *difference* between the potential difference probed and that required, the potentiostat responds by increasing the potential applied to the indicator electrode.

9.5.1 *Variation of current during the life of mercury drops*

At applied potentials corresponding to electrochemical processes taking place, the faradaic current varies in the manner shown in Figure 9.14.

In the short period just before detachment, the current remains virtually constant. If this may be instrumentally sampled, a *net* current, free

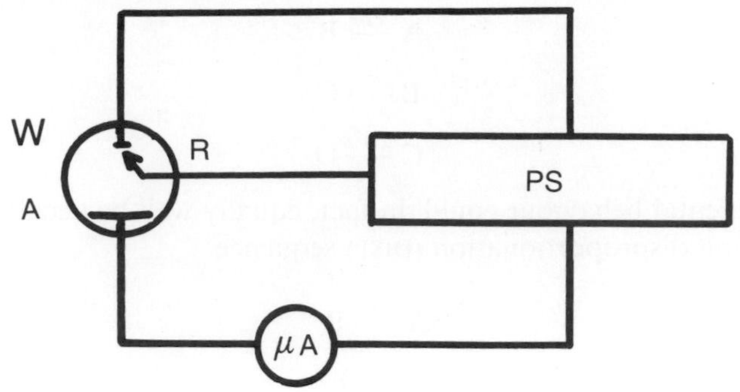

Figure 9.13 Circuit for three-electrode voltammetry. The microammeter responds to current flowing between the working and auxiliary/counter electrodes. W = working electrode; R = reference electrode; A = auxiliary or counterelectrode; PS = potentiostat (this probes the potential difference between W and R).

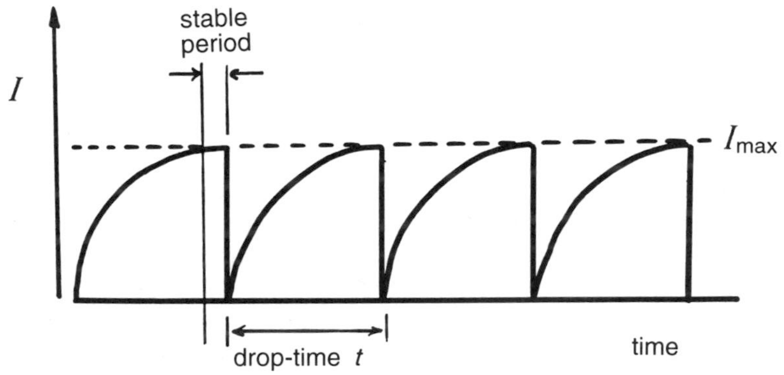

Figure 9.14 Variation of faradaic current in a sequence of drops during an electrode process sensed by the DME.

of oscillations, may be recorded during a synchronized potential scan and some improvement in sensitivity gained from the resulting (stepped) polarogram. Although such sampled (or Tast) d.c. polarography now has limited analytical application it occupies an important place in developments leading towards modern pulse techniques. These evolved via methods based upon the superimposition of square-wave voltage profiles on a conventional d.c. voltage scan.

At the end of each square-wave voltage half-cycle the a.c. component of the cell current contains an improved faradaic to non-faradaic ratio compared to its value at the start. While both components will have decayed during the period in question they do so to different extents (Figure 9.15).

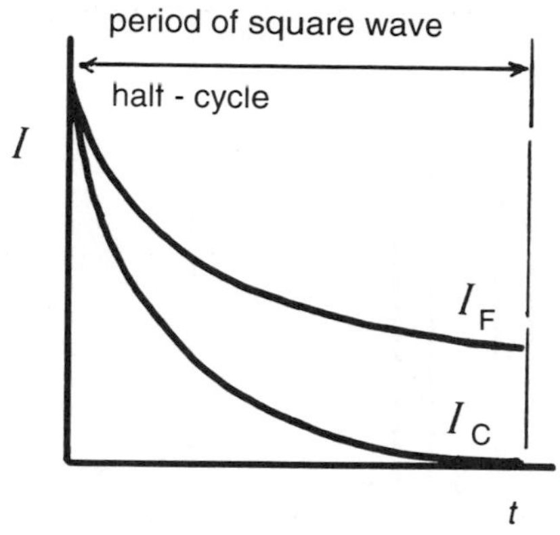

Figure 9.15 Decay of faradaic (I_F) and capacitance (I_C) a.c. components during a square/rectangular wave voltage half-cycle.

9.5.2 *Pulse polarography*

Pulse techniques combine the use of rectangular potential waveforms with synchronization of their application during the stable condition close to the end of the life of a mercury drop. It is usual to control the drop-life to 1/2, 1 or 2 s by electromechanical detachment of drops from a DME system with a long (12–15 s) natural period. More recently, the static mercury dropping electrode has been developed which eliminates the gravitational reservoir generation of electrode material. In this device a solenoid-operated plunger system supplies mercury drops from a low-level reservoir. This system is safer than the traditional one, allows rapid formation of stable drops and uses wider-bore capillaries less susceptible to blockage. Voltage pulses of increasing magnitude are applied in synchronization near to the end of each drop-life of a succession of drops. Resultant net faradaic current components are recorded instrumentally and displayed as a function of applied potential. The pulse profile and consequent polarogram are shown in Figure 9.16.

The pulse polarogram is seen to have the essential shape of a conventional polarographic wave but generated in a sequence of steps rather than continuous output oscillations. Such changes improve the detection limits of the classical technique by approximately an order of magnitude, i.e. to about 10^{-6} mol dm^{-3}.

9.5.3 *Differential pulse polarography*

This technique is based on the effects of superimposing a series of equal square-wave impulses (5–100 mV) on a linearly increasing d.c. voltage.

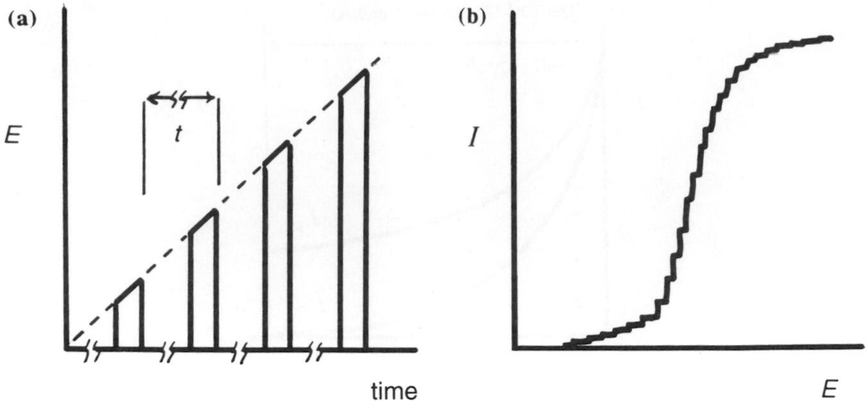

Figure 9.16 (a) Potential pulse-time profile for pulse polarography; (b) stepped (sampled) pulse polarogram.

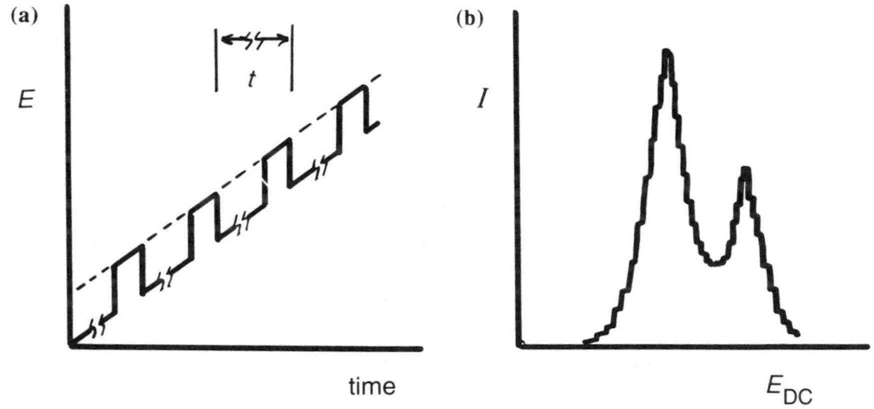

Figure 9.17 (a) Potential pulse-time profile superimposed on linear potential ramp for differential pulse polarography; (b) Sequential differential pulse polarogram with peaked form giving improved resolution between following signals.

Synchronization is effected such that a pulse is applied once during the last stages of the life of each drop. The current signal recorded in this case is the difference between the current flowing just before the imposition of a pulse and that flowing during its last few milliseconds. The combined potential profile and a resultant polarogram are shown in Figure 9.17. Sharply peaked, stepped signals allow analysis at depolarizer concentrations of the order of 10^{-8} mol dm^{-3} and significantly improve resolution between adjacent polarograms.

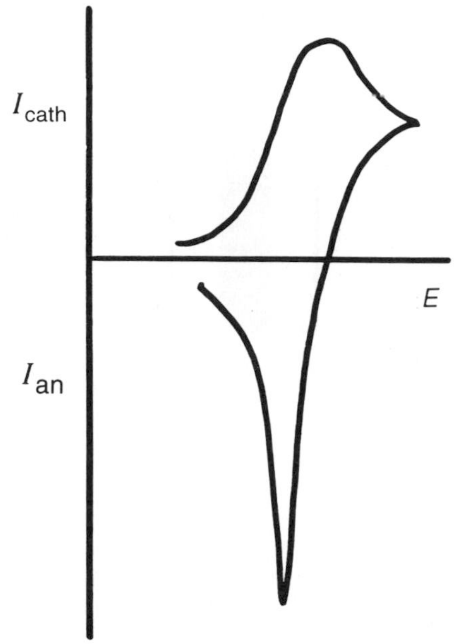

Figure 9.18 Current–voltage pattern for slow cathodic deposition followed by anodic stripping.

9.5.4 Stripping voltammetry

If a solution of a silver ions is subject to voltammetric analysis at a carbon electrode with a slowly varying potential whose direction is reversed, after the diffusion limit is reached, and restored to its original value, a current–voltage curve of the type shown in Figure 9.18 is obtained.

The enhanced anodic signal is caused by the accumulation of silver on the electrode during the slow forward sweep. This principle may be used in anodic stripping analysis which amounts to a convenient preconcentration technique. A stationary hanging mercury drop may be used when the metal forms an amalgam.

9.6 Voltammetry based on forced controlled convection

Controlled hydrodynamic conditions referred to when considering amperometry have also been used in direct analysis.

Any technique which is based on controlled convection, whether this is by controlled stirring of the working solution, by rotation or vibration of an electrode within such a solution, or by the flow of solution past stationary electrodes (as used in some high-performance liquid chromatography detector systems) may be classed under the heading of *hydrodynamic voltammetry*. Two cases will be considered briefly here.

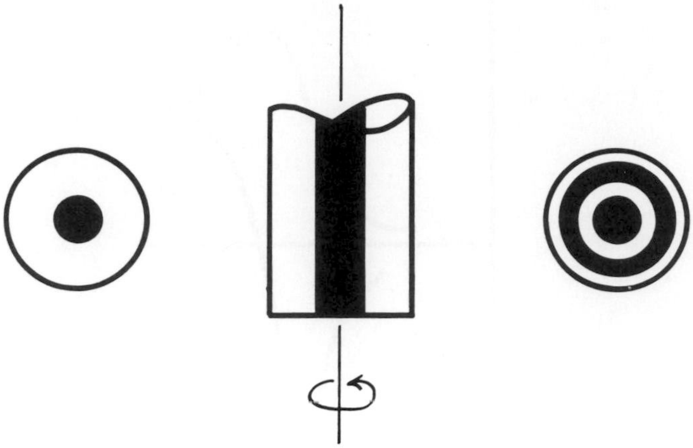

(a) (b) (c)

Figure 9.19 (a) Section through rotating disc electrode: conducting electrode material is surrounded co-axially by insulating support. (b) Orientation of electrode during rotation when exposed disc is presented to a solution. (c) Section through ring–disc electrode. The materials of ring and disc may be the same or different and are separated coaxially by the insulating support.

9.6.1 Rotating disc voltammetry

A disc electrode, typically rotated at a rate of 20 rotations per second, which is precisely controllable, is capable of generating equally controllable currents. The structure of such a probe is shown in Figure 9.19(a), (b). The magnitude of the current signals generated is expressible in terms of an equation similar in form to equation (9.1) except that the diffusion layer thickness, δ, is now expressed in terms of hydrodynamic parameters, viz.

$$\delta = \frac{1.62 D^{1/3} \nu^{1/6}}{\omega^{1/2}} \tag{9.7}$$

Here ω is the angular velocity of rotation and ν is the kinetic viscosity of the solution. Substitution of equation (9.7) into equation (9.1) yields the Levich equation,

$$I_{\text{lim}} = \frac{nFAD^{2/3}\omega^{1/2}[\mathrm{M}^{n+}]}{1.62\nu^{1/6}} \tag{9.8}$$

9.6.2 The ring–disc electrode

In this device an additional ring surrounds the central disc and is separated from it by a narrow ring of insulating material (Figure 9.19(c)). In some modifications the disc and ring are made of different materials.

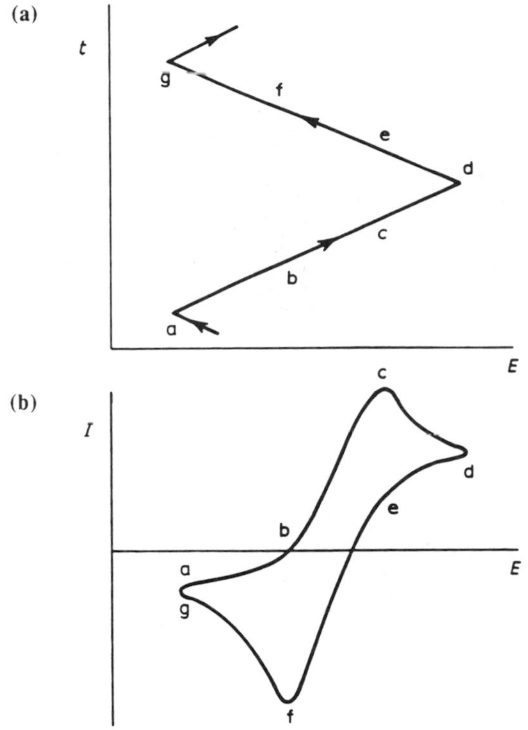

Figure 9.20 (a) Symmetrical saw-tooth potential–time signal used in cyclic voltammetry. (b) Corresponding cyclic voltammogram for a near-reversible system.

The system has found particular application in mechanistic/kinetic studies since it allows detection of unstable products formed by electrode reactions at the disc region. This may be achieved by subjecting the latter to a conventional linear potential scan while maintaining a constant potential applied to the ring at a value characteristic of the limiting current region of the species generated at the disc.

A modification using impingement of electrolyte solution on a small-scale ring–disc system has allowed realization of the wall-jet ring–disc electrode considered in Chapter 10.

9.7 Cyclic voltammetry

In this technique the applied potential is varied with time in a symmetrical saw-tooth wave form as shown in Figure 9.20(a), while the resulting current is detected and recorded over the entire cycle of forward and reverse sweeps (Figure 9.20(b)).

It is usual for solid microelectrodes to be used and glassy or pyrolytic carbon or carbon-paste probes have proved to be convenient. Species which are reduced and deposited in the forward scan of each cycle are re-oxidized in the reverse scan. The technique is useful in the elucidation of mechanism particularly in respect of the identification of intermediates. With the very high sweep rates which may be generated by modern electronic techniques it has proved possible to identify short-lived intermediates which decompose within the time scale of conventional voltage scan frequencies. In Figure 9.20(b) is shown the form of a cyclic voltammogram expected for a near-reversible system. The greater the separation between the peaks for forward and reverse scans, the more irreversible is the electrode process.

9.8 Ultramicroelectrodes

A very small plane electrode surface with a diameter less than about 50 μm offers particular features in respect of the establishment of steady-state diffusion-controlled currents. Such currents, because of such a small electrode area, will be extremely small; the nanoamp scale now replaces that of the microamp.

Although *currents* are so small, *current densities* are very high and have a quite different distribution at the very small plane surface to that at a larger one. The comparative distributions are shown in Figure 9.21. The predominating influence of edge effects shown in Figure 9.21(b) produces an essentially hemispherical charge density distribution. The electrode thus behaves as though it were hemispherical (Figure 9.21(c)): this offers some theoretical advantage in that it is significantly easier to solve current equations for such a profile than it is for a plane. Three practical advantages are important:

1. A supporting electrolyte need not be used since steady-state currents are established in their absence.
2. A three-electrode, potentiostatic polarizing system is not required. Currents are so small that reference electrodes can now sustain them without associated *IR* problems.
3. Such electrodes offer some promise as *in vivo* analytical probes, although there can be disadvantages arising from the insulating effects of antibodies resulting from rejection of electrode material.

The steady-state diffusion-limited currents may be expressed by equations of elegant simplicity, viz.

$$I_d = knFDrC \tag{9.9}$$

where k is a numerical constant of value 4 or 2π, according as to whether

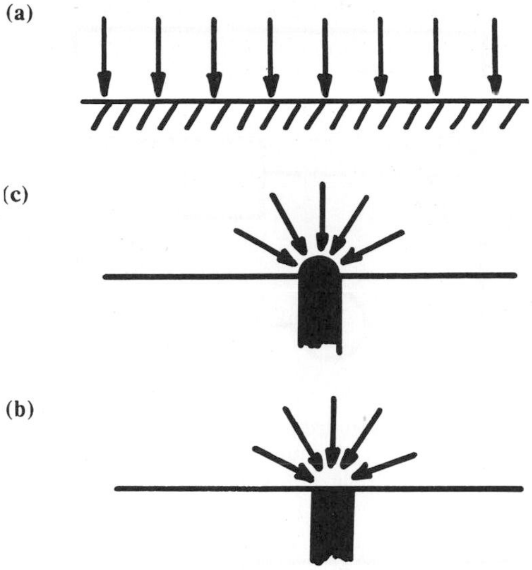

Figure 9.21 (a) Current density distribution at a large planar electrode where edge effects are insignificant. (b) Near-hemispherical current density distribution at a very small planar electrode where edge effects are significant: (c) Current density distribution at a hemispherical ultramicroelectrode.

the electrode is planar (Figure 9.21(b)) or hemispherical (Figure 9.21(c)) relative to its insulating support.

9.9 Electrogravimetry

As the name implies, methods under this heading combine electrodeposition and weighing, i.e. total electrolytic removal of a species from solution followed by measurement of the change in weight of a suitable cathode.

A restricting feature is the obvious requirement that only *one* species must react with the working electrode: in the earliest constant-current methods, a progressive increase in applied voltage is required to sustain that current during the electrolysis. This increases the possibility of interference from other metal depositions or even hydrogen evolution. Controlled potential gravimetry (Figure 9.22) ensures deposition well removed from other processes: however, this means that a long time may be required for the electrolysis to reach completion recognized by a condition of zero current.

9.10 Coulometric methods

The circuit required for controlled potential coulometry is shown in Figure 9.23. Here the electronic integrator provides direct measurement of the

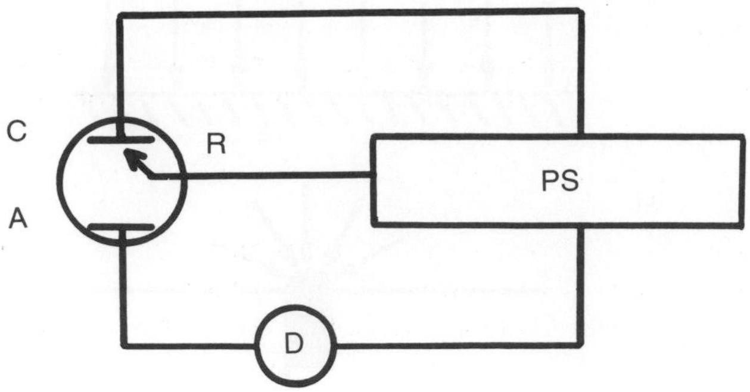

Figure 9.22 Schematic circuit for controlled potential electrogravimetry. PS = potentiostat; C = cathode (large-area mercury pool or platinum mesh); R = reference electrode; A = auxiliary electrode; D = current detector.

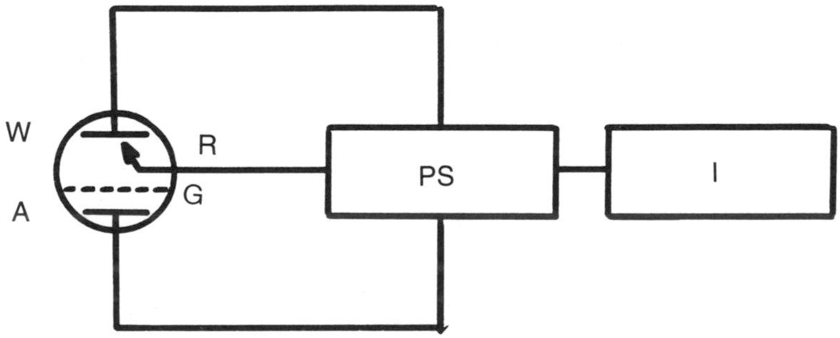

Figure 9.23 Schematic circuit for controlled potential coulometry. W = working electrode; A = auxiliary electrode; R = reference electrode; PS = potentiostat; I = electronic integrator to measure directly the quantity of electricity passed during electrolysis; G = separator between working and auxiliary electrode compartments.

quantity of electricity passed during electrolysis of a required analyte. A contributing feature to the high precision and accuracy achievable is the physical separation of working and counterelectrode compartments. The range of potentials available for such analysis is defined by the most frequently used electrode materials, namely mercury and platinum.

Coulometric methods require large area electrodes (mercury pool or platinum mesh) combined with rapid stirring to encourage rapid completion of electrolysis. This is in sharp contrast to voltammetry where microelectrodes are used in either quiet solutions or with carefully controlled convection. Coulometry is essentially destructive while voltammetry involves such a small amount of actual decomposition that in practical terms its techniques are non-destructive.

The most efficient analyses will be those based on reversible electrode reactions since applied potentials near to a realizable $E^{\ominus}$ will ensure efficient electrolysis. Irreversible processes are more problematic in that applied potentials often well removed from $E^{\ominus}$, to ensure a practicable electrolysis rate, introduce difficulties associated with interference from other electroactive species, secondary reactions and low current efficiencies.

Coulometric *titrations* are based on the generation of a titrant by electrolytic means: conventional standardization of titrant is thus obviated although it is, of course, imperative that the generation occurs with 100% efficiency. A particular advantage is that it is sometimes possible to use unstable titrants which would be unusable by conventional volumetric means. This makes available some quantitatively useful analytical reactions which find no place with other titration methods.

Problems

9.1 Calculate the limiting current density for an electrode at which silver is deposited from an aqueous solution containing silver ions at a concentration of 0.1 mol dm^{-3}, in the presence of excess indifferent electrolyte, if the diffusion layer thickness is 0.05 cm. Estimate the effect on the magnitude of the limiting current if the solution is stirred rapidly.

9.2 The following mean currents, for the applied potentials indicated, were obtained for a reversible, diffusion-controlled reduction of a metal ion M^{n+} to metal M at a dropping mercury electrode at 298 K (all mean current values have been corrected for the appropriate residual current in the presence of excess KCl at 298 K).

E (V) vs. SCE	0.97	0.98	0.99	1.01	1.02	1.03	1.04	1.05
$\bar{I}$ (μA)	2.134	4.255	7.718	17.100	20.644	22.831	25.000	25.000

By means of an appropriate graph, determine the half-wave potential ($E_{1/2}$) and the number of electrons (n) transferred per metal ion during the reduction process.

Given that the concentration of M^{n+} was 2.98×10^{-3} mol dm^{-3}, that the rate of flow of mercury (m) was 3.299 mg s^{-1} and that the drop-time (t) was 2.47 s, estimate the value of the diffusion coefficient (D) of the hydrated M^{n+} cation.

9.3 For a metal ion M^{n+} at a concentration of 3.06×10^{-3} mol dm^{-3} and in the presence of excess inert electrolyte, the following data were obtained using the same dropping mercury electrode system as used in Question (9.2).

E (V) vs. SCE	1.00	1.02	1.03	1.05	1.07	1.09	1.10	1.15
$\bar{I}$(μA)	3.809	7.025	9.207	13.733	17.881	20.678	23.800	23.800

(Current values have again been corrected for the residual, capacitance effect.)

It was found that the plot of $\bar{I}_d$ versus square root of mercury reservoir height was linear and passed through the origin of the graph. Determine the half-wave potential ($E_{1/2}$) and assess the nature of the electrode process.

9.4 The method of standard addition has been used to determine traces of nickel impurity in cobalt salts by various polarographic techniques.

Reagent grade $CoSO_4.7H_2O$(300 g) was dissolved in about 50 cm^3 water in a 100 cm^3 volumetric flask; after addition of 2 cm^3 conc. HCl and 5 cm^3 pure pyridine plus 5 cm^3 of 0.2% gelatin, the solution was diluted to the mark.

Exactly 75 cm^3 of this solution were transferred to a polarographic cell and a d.c. polarogram with I_d = 1.97 μA was obtained for the conditions used. After adding 4 cm^3 of a 9.24×10^{-3} mol dm^{-3} solution of AnalaR $NiCl_2$, a polarogram with I_d = 3.95 μA was obtained.

Explain the principles behind the observations and estimate the percentage of nickel in the original sample as both Ni and $NiSO_4.7H_2O$. Derive any equation required for the estimation.

9.5 Two metal ion species M^{2+} and N^{2+} in non-complexing supporting electrolytes show almost coincident polarographic waves so that analysis for M^{2+} in the presence of N^{2+} is not possible.

In the search for some ligand species, X, which complexes with N^{2+} rather than with M^{2+}, estimate the approximate stability constant required, at a ligand concentration of 0.2 mol dm^{-3}, to induce the required wave separation of about 0.2 V for d.c. polarography.

Make clear any assumptions underlying the calculation.

10 Electrochemical sensors

10.1 Ion-selective electrodes

Although a number of devices, which offer analysis via a direct relationship between an electrochemical parameter and an analyte concentration, may be broadly described as sensors, the term tends to be used in relation to a number of features of experimental convenience. These concern the size, exclusivity of response to particular species, robustness and the ease and speed of practical application.

The glass electrode, considered in Chapter 8 in the context of pH measurement, is the best known example of an ion-selective electrode (ISE). Its operation depends on the modification of a half-cell potential by the response to hydrogen ions of a glass membrane separating the half-cell from a solution whose pH is required. Many ion-selective membrane materials have been developed, the response probed by all of them being a function of ion-exchange reactions at their surface and ion-conduction inside them. They fall into three categories, viz. glass, solid state and liquid.

10.1.1 Glass membrane electrodes

It is well known that glass electrodes require soaking before use to allow certain components of the glass and the solution to exchange and produce an 'ion-exchanging layer' on the membrane surface. The condition of the electrode after this operation is completed is summarized in Figure 10.1. Water molecules become associated with part of the membrane and play a part in the ion-exchange process. Glass is unique in that, while it is an ionic conductor for small ions such as sodium, it is hydrogen ions which are involved most exclusively in the ion-exchange process even at high pH values and when the activity of ions such as Na^+ is high. In the latter condition, the membrane shows response to these ions as evidenced by the 'alkaline error'.

Electrodes specifically designed for the determination of Na^+ or K^+ have had the composition of the glass membrane so devised to show an enhanced alkaline error while at the same time having the preference for hydrogen ions in the exchange process suppressed.

By analogy with the expression representing the formal definition of pH it is possible to define the quantity pX by

$$pX = -\log a_X \qquad (10.1)$$

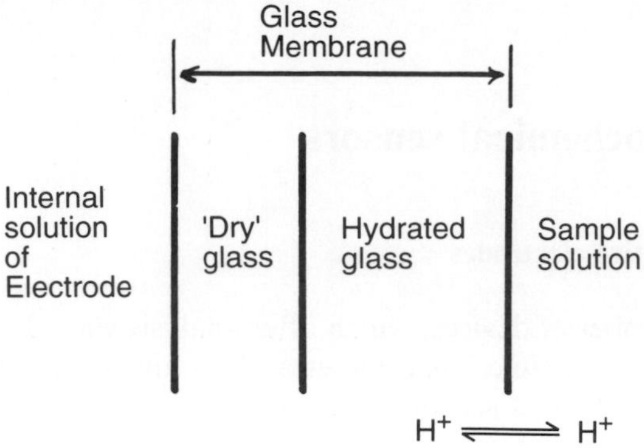

Figure 10.1 Arrangement of a glass membrane electrode. Compare with the glass electrode structure shown in Figure 8.10.

for an electrode responsive to ion X, whose activity a_X is related to the potential adopted by the electrode by

$$E = K + \frac{2.3RT}{nF} \log a_X \tag{10.2}$$

so that

$$E = K - \frac{2.3RT}{nF} \mathrm{pX} \tag{10.3}$$

K representing the proportion of sensed potential due to constants of the system.

10.1.2 Solid-state electrodes

Here the membrane is either a single crystal or a solid ion-exchanging material. For instance, for halide ion-sensitive electrodes the membrane may be a solid silver salt, the potential of the electrode being given by

$$E = E^{\ominus} + \frac{2.3RT}{F} \log a_{Ag^+} \tag{10.4}$$

At the surface of the membrane there is a certain activity of silver ions which is a function of the halide activity of the solution into which the electrode dips (Figure 10.2).

The activity of silver ions may be expressed in terms of the solubility product of the silver halide by

$$a_{Ag^+} = \frac{K_{AgX}}{a_{X^-}} \tag{10.5}$$

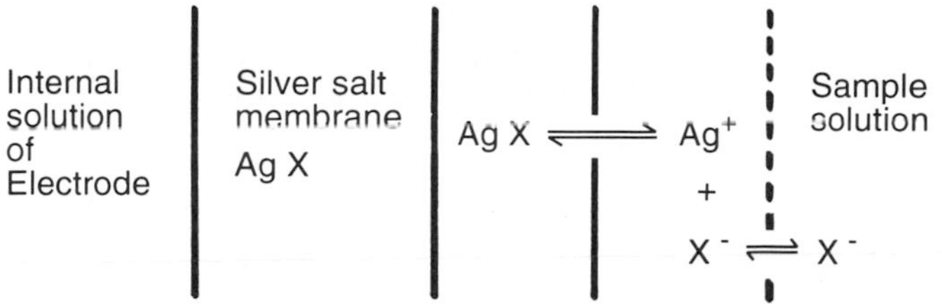

Figure 10.2 Arrangement of components and interactions in a solid-state halide ion-sensitive electrode.

therefore,

$$E = E^{\ominus} + \frac{2.3RT}{F}(\log K_{\mathrm{AgX}} - \log a_{\mathrm{X}^-})$$

$$E = K - \frac{2.3RT}{F}\log a_{\mathrm{X}^-} \qquad (10.6)$$

10.1.3 Liquid membrane electrodes

A liquid membrane comprises a thin, porous inert support impregnated with an ion-exchanging material in the liquid phase. This material is often an organic species dissolved in some organic solvent, in which case use in non-aqueous solutions may be restricted. There are two types of liquid membranes: (a) charged liquid membranes which usually contain the ion for which the electrode is sensitive (the operation here is very similar to the solid state case), and (b) neutral liquid membranes which do not contain the ion to be detected but a polymeric species whose geometry offers selectivity to particular ion species by allowing encapsulation within a cavity of appropriate size. The antibiotic valinomycin, incorporated into a polyvinyl chloride matrix or dissolved in diphenylether operates very effectively in this way with K^+ ion-selective electrodes. The situation across the membrane is shown in Figure 10.3.

The principle has been widely and effectively used in electrodes based on exploitation of the tailor-made cavities available within cryptands and crown ethers.

10.2 Problems with ion-selective electrodes

The most important practical difficulty with ion-sensitive electrodes is the interference from ion species in solution other than the one required. Such electrodes must therefore be devised so that their *selectivity* for the ion in

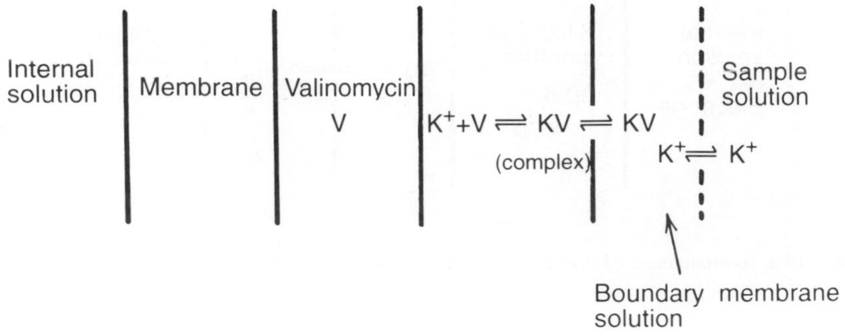

Figure 10.3 Arrangement of components and interactions in the K^+-sensitive valinomycin electrode.

question is as high as possible. Glass electrodes for pH determination are the most selective of all ion-sensitive electrodes but it has been seen that, even here, serious interferences may occur. When other ions interfere the equation for the potential adopted by the electrode must be modified to

$$E = K + \frac{2.3RT}{nF}\log\left(a_{\mathrm{i}} + K_{\mathrm{ij}}a_{\mathrm{j}} + K_{\mathrm{ik}}a_{\mathrm{k}} + K_{\mathrm{il}}a_{\mathrm{l}}\ldots\right) \tag{10.7}$$

where a_{i} = activity of ion to be determined; a_{j}, a_{k}, $a_{\mathrm{l}}, \ldots$ = activities of interfering ions; and K_{ij}, K_{ik}, $K_{\mathrm{il}}\ldots$ = 'selectivity constants' of the electrode towards particular ions. These selectivity constants are defined as follows

$$K_{\mathrm{ij}} = \frac{u_{\mathrm{j}}}{u_{\mathrm{i}}}({}_{\mathrm{e}}K_{\mathrm{ij}}) \tag{10.8}$$

where $u_{\mathrm{i}}, u_{\mathrm{j}}$ are ion mobilities and ${}_{\mathrm{e}}K_{\mathrm{ij}}$ is the equilibrium constant for the reaction:

$$\mathrm{j_{solution} + i_{membrane} \rightleftharpoons j_{membrane} + i_{solution}}.$$

In cases where neutral membranes are used, the mobilities used are those of the complexes of the membrane material formed by the determined and interfering ions.

The Nernstian basis of all such potentiometric sensor devices carries with it a more fundamental disadvantage. The signal–activity relationship is a *logarithmic* one, small errors in measurement of potential generating large errors in analysis unless care is taken. A matter sometimes overlooked is that such probes respond to *activity* rather than concentration.

Other disadvantages can be the comparatively short lives of some devices and interference from electronic noise: matters addressed by the significant industry which has developed around this major feature of modern analytical chemistry.

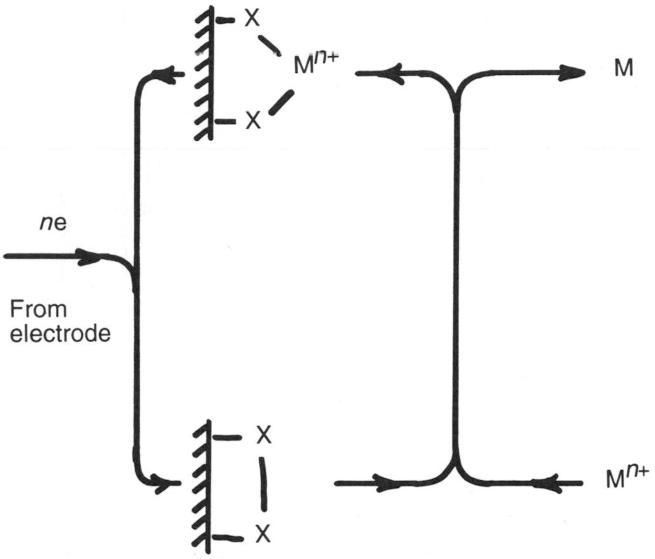

Figure 10.4 Scheme for chemically modified electrode based on interaction of metal ion M^{n+} with adsorbed analyte-specific ligand species X_2.

10.3 Chemically modified electrodes

A chemically modified electrode (CME) is one whose surface is purposely contaminated with material which reacts selectively and reversibly with a chosen analyte. Such a material might be an adsorbed ligand with which a metal ion binds (Figure 10.4).

Although a simple idea, there are significant difficulties to be overcome in the construction of such probes. There must be effective adsorption of the analyte-specific species (represented by X_2 in Figure 10.4) onto the otherwise inert conductor. Interaction of this adsorbed material with the studied analyte must be rapid and strong in comparison with that of competing species: it must not be so strong, however, that the interaction process cannot be reversed and the probe used repeatedly with X_2 in position for further analyses. It is also necessary for there to be an electron transfer between the complexed analyte and conductor.

Operation of such *amperometric* sensors depends upon measurement of the steady-state current flowing in consequence of a constant potential maintained between the conductor and a suitable reference electrode. The latter would invariably be built into the—often significantly miniaturized—probe. The easier and more accurate measurement of currents compared

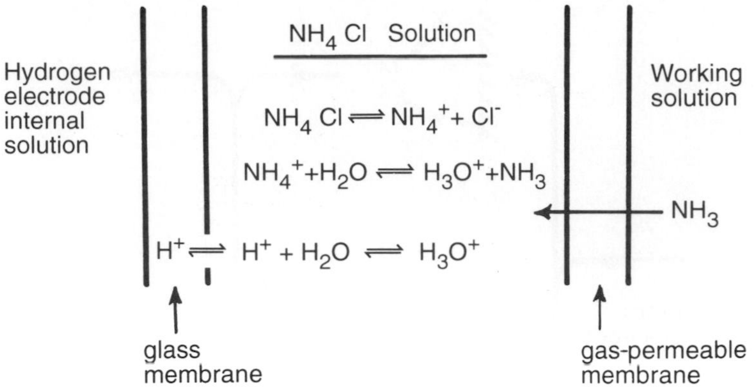

Figure 10.5 Components and interactions in an ammonia-sensing electrode.

with potentials, combined with the direct proportionality of signal and concentration, has attracted considerable commercial investment in the development of amperometric sensors.

10.4 Gas-sensing electrodes

These may be of potentiometric or amperometric type. The former are based on a conventional potentiometric sensor whose response to its specific analyte is modified by quantitative interaction of dissolved gas with this species. An example is an ammonia-sensing electrode based on a hydrogen electrode operating in a solution of ammonium chloride (Figure 10.5). This is combined with a thin layer at the glass interface enclosed by a gas-permeable polymer membrane. Diffusion of NH_3 molecules from the working solution disturbs the electrolyte equilibrium of the internal solution and thus its pH as sensed by the glass electrode. The difference in potential caused is a function of the logarithm of NH_3 concentration.

Amperometric gas-sensing electrodes have a similar construction but they function by responding to the steady-state diffusion-controlled electrochemical reactions of dissolved gases. This may be understood by reference to the schematic arrangement of an oxygen sensor shown in Figure 10.6.

The steady-state diffusion of O_2 molecules through both the separating membrane and the electrolyte solution of the sensing region results in a steady current signal (I) if the potential of the indicator electrode (E) is suitably poised with respect to the reference (R).

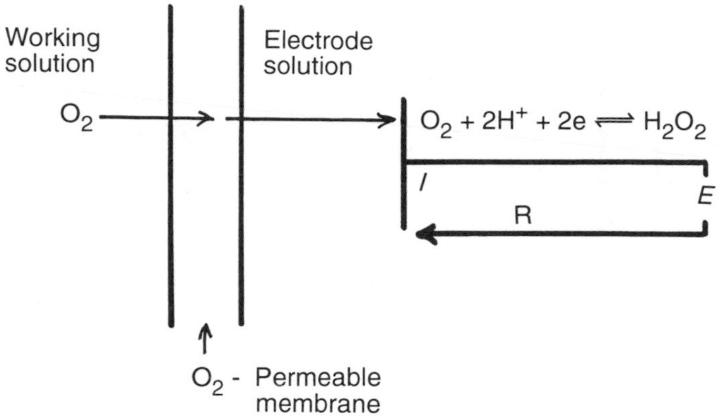

Figure 10.6 Scheme for an amperometric oxygen-sensing electrode.

10.5 Enzyme electrodes

The chemical specificity of enzymes has been made the basis of both potentiometric and amperometric devices. As with chemically modified electrodes, the principle is simple but the problems of immobilizing an enzyme onto an appropriate probe, controlling containment of solutions by means of membranes and ensuring reproducibility are formidable ones. Nevertheless there have been some notable successes and a great deal of effort and funding is currently directed towards the development of such sensors. Urease has been successfully used in a urea probe, shown schematically in Figure 10.7, a device based on a modified ammonium ion selective electrode.

Difficulties have sometimes been encountered in establishing equilibria and steady potentials for such systems and it is more usual for enzyme electrodes to be of the amperometric type.

The enzyme glucose oxidase (GOD) has been used in a variety of electrode systems to detect glucose. Hydrogen peroxide, produced by reoxidation of the reduced enzyme ($GODH_2$) may be detected via its anodic reaction.

$$\underset{(G)}{\text{glucose}} + \underset{(GOD)}{\text{glucose oxidase}} \longrightarrow \underset{(GODH_2)}{\text{reduced enzyme}} + \underset{(GL)}{\text{gluconolactone}}$$

$$GODH_2 + O_2 \longrightarrow H_2O_2 + GOD$$
$$H_2O_2 - 2e \rightleftharpoons O_2 + 2H^+$$

The amperometric signal is directly proportional to the concentration of glucose in the medium to be analysed and may be measured by connec-

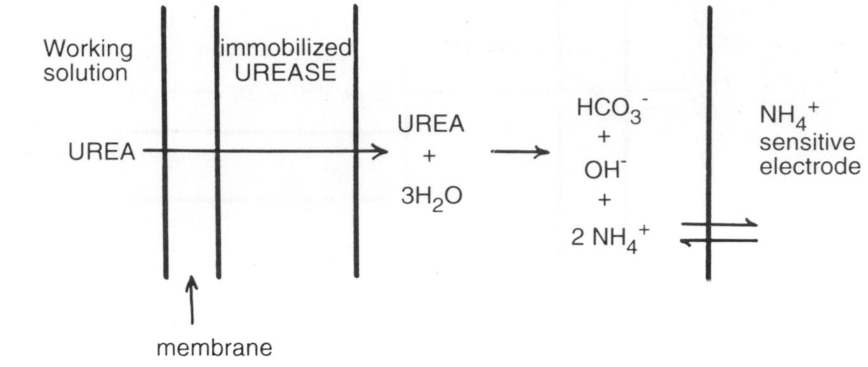

Figure 10.7 Scheme for a potentiometric enzyme electrode.

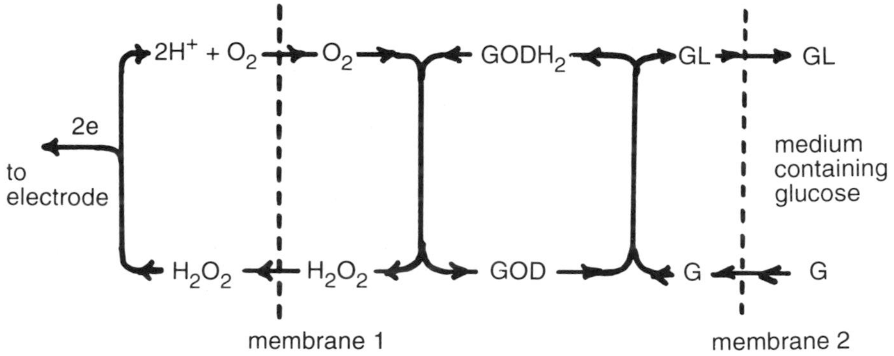

Figure 10.8 Arrangement for an amperometric glucose sensor based on glucose oxidase with hydrogen peroxide as electron mediator.

tion with a reference electrode. The sequence of reactant–product cycles is shown in relation to the schematic probe structure in Figure 10.8.

A major disadvantage of this system is the necessity for two membranes to contain the two chemical cycles and to separate them from the electrochemical step. Further, the performance of the probe is somewhat affected by changes in the prevailing concentration of oxygen.

It has been found possible to use single membrane systems in conjunction with electron mediators based upon ferrocene derivatives. The sequence of chemical–electrochemical cycles now becomes simplified to that shown in Figure 10.9. Re-oxidation of reduced enzyme is effected by, for instance, the ferrocenium ion, $FcCO_2H^+$, formed by oxidation of ferrocene

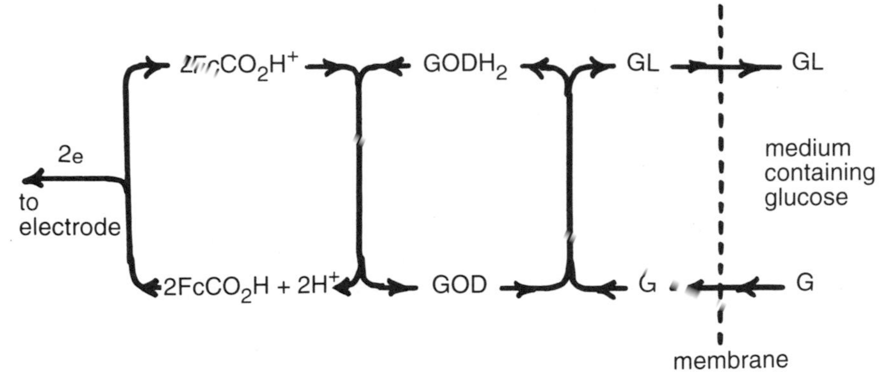

Figure 10.9 Scheme of chemical–electrochemical cycles for a ferrocene-mediated enzyme electrode.

carboxylic acid, $FcCO_2H$

$$GODH_2 + 2FcCO_2H^+ \longrightarrow GOD + 2FcCO_2H + 2H^+$$

followed by the amperometric signalling reaction

$$2FcCO_2H - 2e \rightleftharpoons 2FcCO_2H^+$$

A glucose concentration-dependent current is generated when the potential of the working electrode is made positive to the correct extent with respect to the reference (usually SCE). This requires potentiostatic control to a value which is dependent upon the ferrocene derivative used.

An even simpler system, at least superficially, is based upon the initially reduced form of the enzyme becoming re-oxidized directly at the electrode surface due to the modification of the latter. This has usually involved incorporation of a conducting organic salt which both allows strong adsorption of both forms of the enzyme and aids electron transfer. The arrangement now becomes simplified to that in Figure 10.10.

10.6 Sensors based on modified metal oxide field effect transistors (MOSFETs)

Introduction of such transistor devices brings a quite new dimension to sensor systems. Their effectiveness, coupled with the capability of miniaturization and stacking into multiple probes leads on to the possibility of making systems capable of sensing several analytes. The essential features

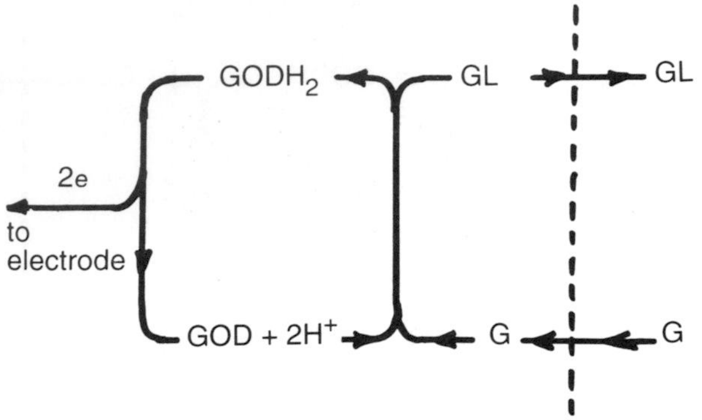

Figure 10.10 Unmediated enzyme electrode reactions.

of a MOSFET are shown in Figure 10.11, the significant property for exploitation in analysis being the modulation of the current, I_{SD}, by the gate voltage, E_G. In its application as a sensor the device is modified to what is known as a ChemFET—a Chemically Sensitive Field Effect Transistor. This involves replacement of the gate electrode by some kind of analyte sensor together with a reference electrode and solution of species to be determined. In essence the device operates on both potentiometric and amperometric characteristics. The primary analytical interaction of analyte with sensor produces a response in E_G. This, in turn, affects the value of I_{SD} which is the quantity measured as a concentration indicator.

10.7 The wall-jet ring–disc electrode (WJRDE)

A modification of the ring–disc electrode (Chapter 9) is directed at analytical application. The ring–disc system is held stationary while a jet of analyte-containing solution is made to impinge on the disc component. This can be effected on a very small scale and the usual arrangement is shown in Figure 10.12.

The jet can be so controlled that its scattering at the disc (Figure 10.12, inset) produces a reproducible pattern leading to a controllable mass-transfer profile not unlike that considered briefly in section 9.6.

The device has been successfully used in the determination of protein concentrations by microtitration with electrolytically generated bromine. The potential difference between ring and disc may be poised such that Br_2 is generated at the disc and reduced at the ring. Under such circumstances a plot of ring current versus disc current is linear as shown in Figure 10.13, curve (a).

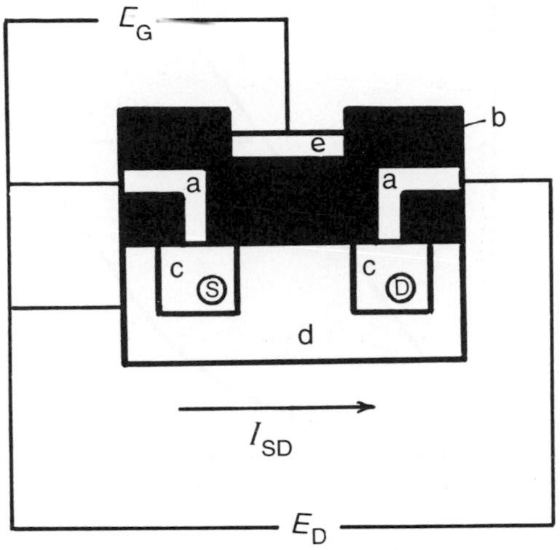

Figure 10.11 The features of a MOSFET. In conversion to an analytical sensor as a ChemFET, the gate electrode is replaced by an analyte sensor plus reference electrode and incorporates a cell to contain the solution to be analysed. (a) conductor, (b) insulator, (c) n-type semiconductors (S = source, D = drain), (d) p-type semiconductor, (e) gate electrode. E_G = gate voltage, I = current. E_D = drain voltage.

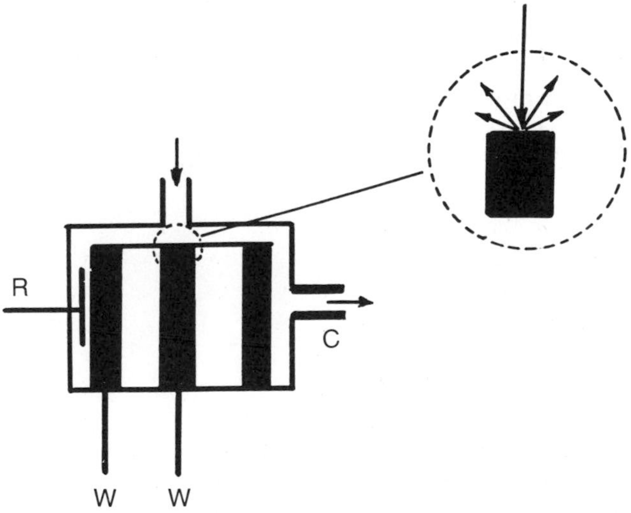

Figure 10.12 Principle of the wall-jet ring–disc electrode. W = working disc and ring electrodes. R = reference electrode. C = auxiliary or counter electrode incorporated into the outlet.

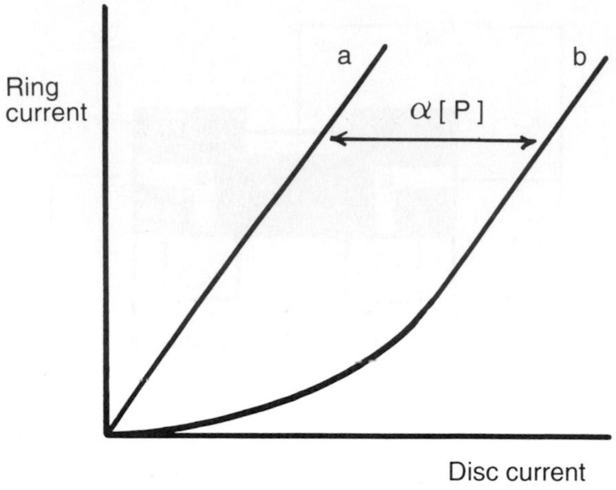

Figure 10.13 Titration of protein with electrolytically-generated bromine:

$$\textit{At the disc}: 2Br^- - 2e \longrightarrow Br_2$$
$$\textit{At the ring}: Br_2 + 2e \longrightarrow 2Br^-$$

The reaction $P + nBr_2 \longrightarrow PBr_{2n}$ reduces the ring current.
For explanation of curves (a) and (b) see text.

In the presence of protein, however, which will typically react rapidly with bromine in a molecular ratio of one to several hundreds, the amount of unreacted bromine available to react at the ring electrode is greatly reduced. The graph of ring current to disc current is now shifted, relative to its position in the absence of protein, by an amount proportional to the concentration of protein (Figure 10.13, curve (b)).

11 The exploitation of electrode processes

11.1 Mixed potentials and double electrodes

The potential adopted by a metal in aqueous solution is not always governed by the M^{n+}/M equilibrium which might be expected for simple cases. Other reactions may occur when the metal is thermodynamically unstable in aqueous solution. In particular, the reduction of hydrogen ions may interfere with the electrode equilibrium. Consider a metal such as zinc, with a fairly negative electrode potential, in an acidic solution containing its cations. If more positive potentials were to be imposed upon the zinc electrode the current density would vary according to the Tafel line as the metal ionized (assuming only activation polarization) a situation shown in Figure 11.1.

The standard potential of hydrogen is considerably more positive than that of zinc, but at potentials more negative than $E^{\ominus}_{H}$, the rate of hydrogen discharge increases according to its characteristic Tafel line. At the point where the two Tafel lines meet, both reactions must occur at the same current density, so that the potential corresponding to this intersection point is a steady one which is known as a mixed or corrosion potential. The situation may alternatively be represent by the current–potential relationships for the two reactions (Figure 11.2).

The actual potential adopted by the system for different currents will follow the resultant line AB which at $E = E_{corr.}$ gives equal currents flowing in opposite directions. The species co-reducing need not be hydrogen but may be any species reducing at a more positive than M^{n+}.

11.1.1 Pourbaix diagrams

Graphs of reversible metal electrode potentials versus pH of solutions into which they dip at fixed temperature and pressure, provide important information regarding the thermodynamic stability of various phases. Such Pourbaix diagrams provide a thermodynamic basis for the explanation of corrosion reactions. It must, however, be emphasized that the construction of such diagrams takes no account of the kinetics of reactions which occur under the conditions represented by various areas appearing on them. This means that they should be used with some caution when attempting to predict corrosion behaviour. A much simplified Pourbaix diagram for iron in aqueous solution is shown in Figure 11.3. Here the dashed lines labelled (a)

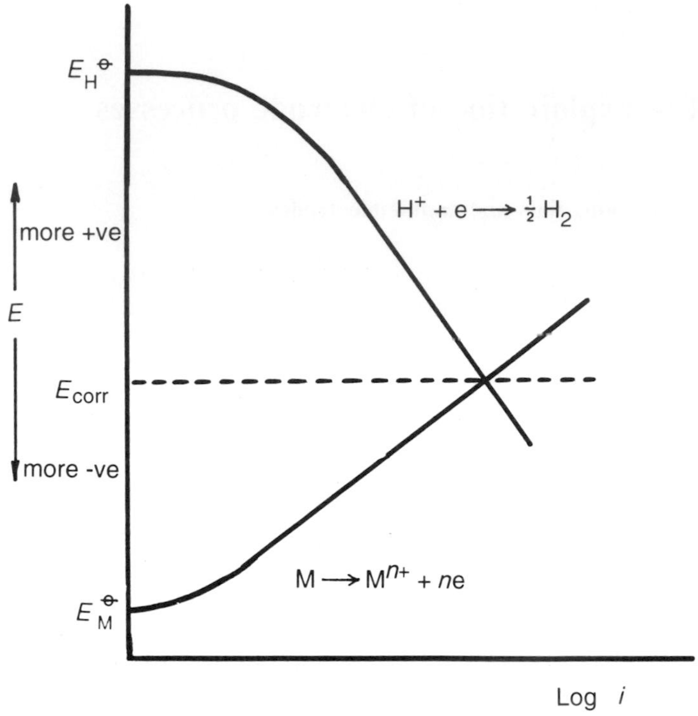

Figure 11.1 Mixed or corrosion potential in acid solution for a metal (e.g. zinc) with an electrode potential which is negative relative to hydrogen. Graphs showing such intersecting Tafel lines are often known as Evans diagrams.

and (b) give the pH dependence of the equilibrium potentials of hydrogen and oxygen electrodes. The area labelled 'passivation' corresponds to the formation of solid compound on the metal surface which protects the metal from attack.

High hydrogen overvoltages prevent high-purity metals from corroding at anything but very slow rates. In the presence of more noble metal impurities, however, corrosion rates may become greatly accelerated, the more noble metal regions acting as 'local cathodes' at which hydrogen evolution may take place.

In neutral and alkaline solutions, the mixed potential attained is usually insufficient to cause hydrogen evolution, at least at atmosphere pressure. In such cases oxygen, either adsorbed at the surface or in solution, may itself become a depolarizer undergoing reduction by the following cathodic processes.

$$O_2 + 2H_2O + 4e \longrightarrow 4OH^-$$

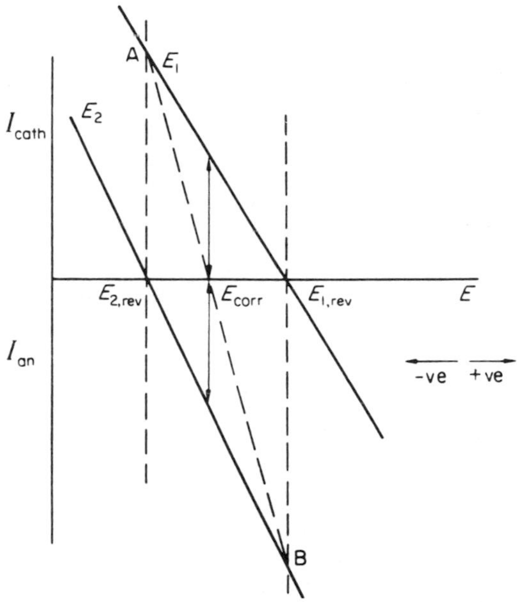

Figure 11.2 Resultant current–potential plot A–B for a mixed electrode: at the corrosion potential equal currents flow in opposite directions.

and

$$O_2 + 4H_3O^+ + 4e \longrightarrow 6H_2O$$

Such partial cathodic processes may well occur via adsorbed oxygen or oxides. Areas where oxide or oxygen are not adsorbed then correspond to anodes, and in this way local cells may be set up over the whole of the metal surface.

Corrosion rates involving oxygen consumption vary in a quite different manner to those involving hydrogen evolution. While the latter type starts very slowly, even in the presence of noble metal impurities, increasing dissolution exposes increasing areas of noble metal, to form an increasing number of local cathodes. The rate then accelerates rapidly. In corrosion involving oxygen, initial rates are high and drop to low steady values very rapidly, the rapid decline corresponding to the removal of adsorbed oxygen, after which the corrosion rate becomes largely dependent on the rate of diffusion of oxygen to the metal surface. In the rusting of iron, for instance, the initial reactions appear to be dissolution of iron to give Fe^{2+} ions at local anodes.

These Fe^{2+} ions then combine with hydroxyl ions, formed at local cathodes, to produce Fe(II) hydroxide. If sufficient oxygen is available, the

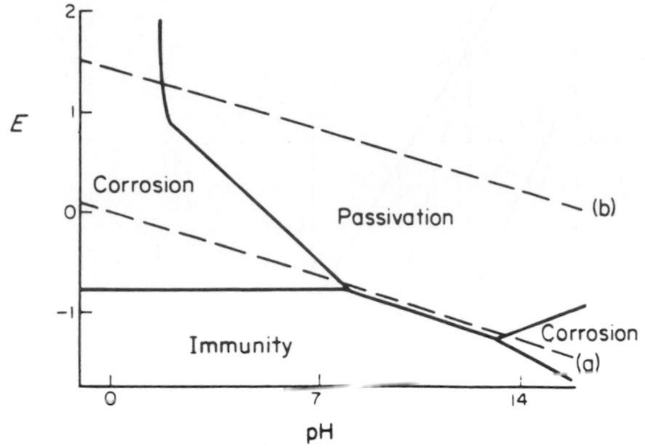

Figure 11.3 Simplified Pourbaix diagram for the iron–water system: the dashed lines (a) and (b) are the pH dependencies of E for hydrogen and oxygen electrodes, respectively, i.e.

$$\text{(a)}\ 2H + 2e \rightleftharpoons H_2 \quad (pH < 7)$$

$$2H_2O + 2e \rightleftharpoons H_2 + 2OH^- \quad (pH > 7)$$

$$\text{(b)}\ \tfrac{1}{2}O_2 + 2H^+ + 2e \rightleftharpoons H_2O \quad (pH < 7)$$

$$\tfrac{1}{2}O_2 + H_2O + 2e \rightleftharpoons 2OH^- \quad (pH > 7)$$

latter may be oxidized to rust, $Fe_2O_3 \cdot H_2O$. In practice the products of intermediate oxidation stages (Fe_3O_4 and hydrates) also occur due to lack of oxygen. Corroded surfaces show formation of these products in preferential layers, the true rust deposit being outermost.

The above reactions are not observed with metals which form an oxide surface layer in air, except in the neighbourhood of cracks or large pores in the layer. At such points anodic dissolution may occur. In cases where the anodically formed ions produce oxide after reacting with hydroxyl ions formed at a local cathode, the fault in the protective film becomes sealed against further corrosion. Should different products be formed, the development of further local cells is encouraged and severe local corrosion may result. These effects are summarized in Figure 11.4.

11.1.2 Corrosion prevention

Corrosion reactions may be minimized by two means—by covering surfaces with protective films or by using inhibition processes. Steel, for example may be protected to various degrees by surface layers of chromium, nickel, zinc or tin.

Cracks in a surface film of plated metal with a more positive electrode

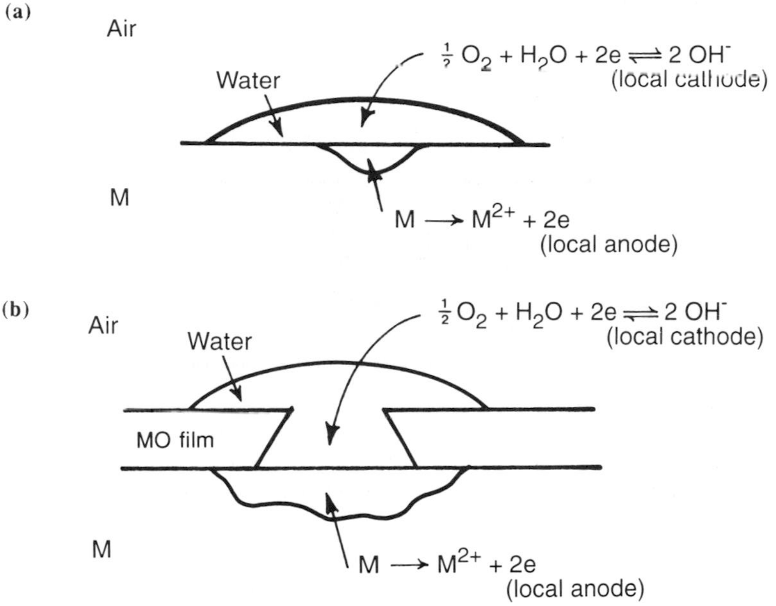

Figure 11.4 Corrosion of metal M in vicinity of oxygen-containing water droplets. (a) Dissolution of M occurs at local anode. (b) Dissolution of M occurs at local anode formed at site of a pore in protective oxide layer. Reaction with cathodically generated OH ions re-seals the oxide layer by: $M^{2+} + 2OH^- \rightarrow MO + H_2O$.

potential than ion, give rise to local cells in which the exposed, more base, metal becomes the anode and the erstwhile protective layer the cathode. This is a problem which can develop with tin plate as used in the canning industry; development of holes in the tin layer leads to the formation of a corrosion cell *between* the tin layer and the exposed iron at the bottom of the hole, a situation shown in Figure 11.5.

If the protective layer is of a less noble metal, such as is the arrangement in galvanized objects and other situations where zinc prevents ion from corroding, zinc becomes the anode and iron the cathode so that the status of the iron is preserved (Figure 11.6). This encouragement of an alternative corrosion process to the one causing harm is known as 'cathodic protection' and is employed extensively in the protection of metals in marine use and in pipelines.

Oxide layers formed by some metals in the atmosphere are most useful if they are physically tough and damage resistant. In any case, if damage should occur, the protective layer rapidly reforms. Thicker coatings may be obtained by anodizing, i.e. by anodically polarizing the material by electrolysis with some cathode in a suitable electrolyte.

Inhibiting reactions form an important part of corrosion prevention.

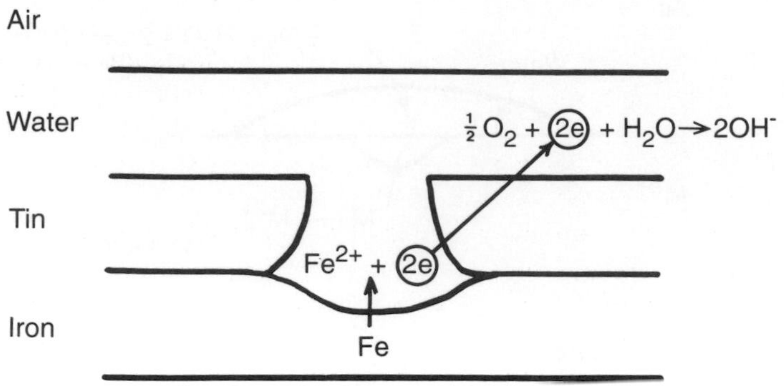

Figure 11.5 Corrosion of iron through hole in tinplate: a corrosion cell forms between the tin layer and the iron exposed at the bottom of the hole.

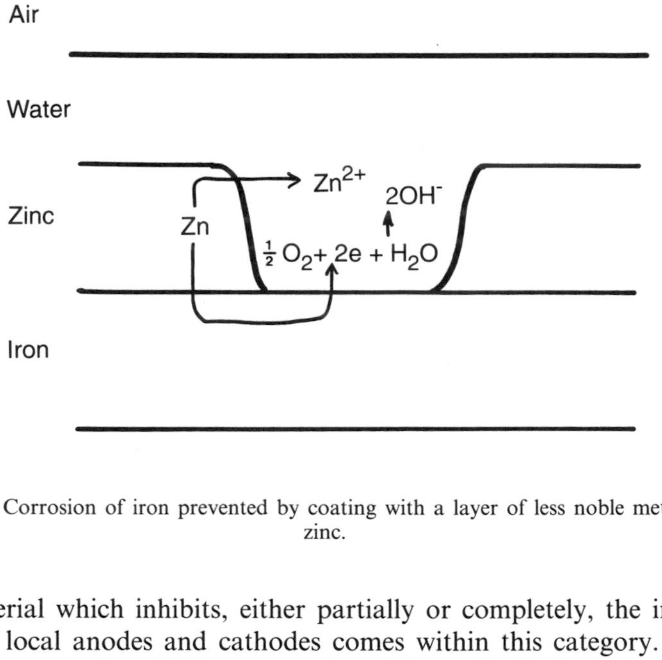

Figure 11.6 Corrosion of iron prevented by coating with a layer of less noble metal such as zinc.

Any material which inhibits, either partially or completely, the initial reaction at local anodes and cathodes comes within this category. Anodic inhibitors are species which can complex with anodically formed metal ions to produce protective layers. It is, however, vital that such species should be present in sufficient concentration to give protection at all local anodes, otherwise corrosion at a relatively few unprotected sites will cause extreme damage. Cathodic inhibitors increase the discharge overvoltage of hydrogen ions or prevent the formation of molecular hydrogen from atoms.

11.2 Electrochemical processes as sources of energy

Electrical energy may be produced through the operation of a chemical reaction taking place in a galvanic cell. The earliest, most rudimentary form of galvanic cell consisted of alternate sheets of copper and zinc separated by wet cloth. This arrangement subsequently gave rise, in a rather different form, to the Daniell cell.

Since we cannot normally expect to obtain reactions giving a free energy change of more than about 200 kJ per Faraday, application of $\Delta G = -nEF$ leads to the conclusion that about 2 V is the maximum emf obtainable from a simple cell. Higher voltages may, of course, be produced through appropriate banking of large numbers of cells.

There are three types of cell

1. *Primary cells*
   These are based upon reactions which are not reversible so that recharging is out of the question and once the cell reaction has proceeded to completion, the cell is discarded.
2. *Secondary cells*
   Such cells are based upon almost reversible electrode processes. All processes occurring during their discharge while used as a source of emf may to a large extent be reversed in the recharging process. The overall efficiency of the recharging reaction may be significantly reduced by side reactions.
3. *Fuel cells*
   In fuel cells an attempt is made to exploit to the fullest extent the *free energy* change of selected reactions to produce electrical energy. The functioning of fuel cells is fundamentally different to that of batteries: while the latter *store* electrical energy, fuel cells convert the energy of chemical processes *directly* into electricity.

11.2.1 Primary cells

As an example the Leclanché cell may be considered which may be represented as follows

$$Zn|NH_4Cl, \underset{\uparrow}{ZnCl_2}|MnO_2, C$$

(forming a gel with starch)

Its construction is shown in Figure 11.7.

Such 'dry cells', with the electrolyte medium thickened by the use of suitable additives, may be used in any position without spillage. The reactions occurring in the cell are complex but the behaviour of the system may be largely explained in terms of the following:

At the anode

$$Zn \longrightarrow Zn^{2+} + 2c\ (E^{\ominus} = -0.76\,V)$$

At the cathode

$$MnO_2 + H_3O^+ + e \longrightarrow MnOOH + H_2O$$

followed by

$$2MnOOH \longrightarrow Mn_2O_3 + H_2O$$

Evolution of hydrogen gas at the cathode would be most undesirable and, in any case, would cause serious losses of energy. It is to prevent such an occurrence that the cathode is surrounded by manganese dioxide (the depolarizer), which discourages hydrogen formation by undergoing other reactions preferentially. The manganese dioxide proves to be more efficient in this respect when it contains lattice defects which may be artifically induced.

The Leclanché cell is irreversible, and therefore incapable of recharging, because of the occurrence of side reactions such as

$$OH^- + NH_4^+ \longrightarrow H_2O + NH_3$$

$$2NH_3 + Zn^{2+} + 2Cl^- \longrightarrow Zn(NH_3)_2Cl_2$$

(a sparingly soluble complex which forms a crystalline deposit)

$$Zn^{2+} + 2OH^- \longrightarrow ZnO + H_2O$$

The cell provides a cheap source of electrical energy with an emf of about 1.6 V, but since the cathode potential is a function of pH, this value falls rapidly on continuous discharge.

A more constant voltage is produced by the Ruben–Mallory cell in which a large excess of hydroxyl ions renders its operation less sensitive to pH change. This cell usually takes the form

$$Hg|HgO, KOH, Zn(OH)_2|Zn$$

Air or oxygen cells are modifications of the Leclanché cell in that the cathode is activated carbon in contact with atmospheric oxygen. Two forms are

$$Zn|NH_4Cl|C(O_2)$$

or

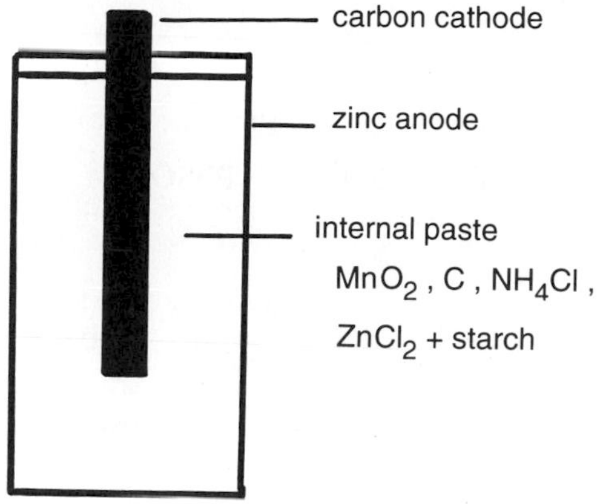

Figure 11.7 Components and their orientation within a Leclanché cell.

$$Zn|NaOH|C(O_2)$$

Such cells give maximum emf of about 1.5 V and have satisfactory voltage–time characteristics. Their major disadvantage is that they cannot be highly loaded due to the slow rate of oxygen polarization so that they operate most satisfactorily for very low currents or in intermittent use.

11.2.2 Secondary cells

The lead–acid battery is an example of a galvanic cell in which electrode processes are almost reversible. The action is based upon the pre-electrolysis of a solution of sulphuric acid saturated with lead sulphate between lead electrodes. Lead is deposited at the cathode, while Pb^{2+} ions are oxidized to the Pb^{4+} state at the anode. The Pb^{4+} ions are subsequently hydrolysed and deposited as PbO_2. In practice the electrodes are usually in the form of grids of a lead/antimony alloy (for mechanical strength) filled with a paste of red lead and litharge in sulphuric acid. As initially constructed, therefore, the system corresponds to the situation existing in the fully discharged cell.

During preliminary charging the electrode used as anode forms porous PbO_2; that used as the cathode forms spongy lead. During the subsequent discharge process, where the polarity is reversed and in which the cell acts spontaneously, the following reactions occur:

At the Pb anode

$$Pb + SO_4^{2-} \longrightarrow PbSO_4 + 2e$$

At the PbO_2 *cathode*

$$\text{(i) } PbO_2 + 4H^+ \longrightarrow Pb^{4+} + 2H_2O$$
$$\text{(ii) } Pb^{4+} + 2e \longrightarrow Pb^{2+}$$
$$\text{(iii) } Pb^{2+} + SO_4^{2-} \longrightarrow PbSO_4$$

Overall

$$Pb + PbO_2 + 2SO_4^{2-} + 4H^+ \underset{\text{charge}}{\overset{\text{discharge}}{\rightleftharpoons}} 2PbSO_4 + 2H_2O$$

The potential of the PbO_2 electrode, from a consideration of the electrochemical step (ii), is given by

$$E_{PbO_2} = E^{\ominus}_{Pb^{4+}/Pb^{2+}} + \frac{RT}{2F} \ln \frac{a_{Pb^{4+}}}{a_{Pb^{2+}}} \tag{11.1}$$

but, since $a_{Pb^{4+}} \sim a_{Pb^{2+}}$, and both are very small since the solution is saturated with PbO_2 and $PbSO_4$,

$$E \sim E^{\ominus}_{Pb^{4+}/Pb^{2+}} \sim +1.70\,V$$

The potential of the lead electrode is given by

$$E_{Pb} = E^{\ominus}_{Pb^{2+}/Pb} + \frac{RT}{2F} \ln a_{Pb^{2+}} \sim -0.28\,V \tag{11.2}$$

Therefore, the cell emf $\sim 1.70 - (-0.28) = +1.98\,V$.

Charging and discharging curves for a lead accumulator are shown in Figure 11.8 from which it is evident that the processes are not completely reversible. Mixed potentials occurring at the electrodes cause them to corrode and give rise to 'spontaneous discharge', so that, even when no current is being drawn from the cell, the following (irreversible) reactions occur.

At the Pb electrode

$$Pb + 2H^+ \longrightarrow Pb^{2+} + H_2$$

Consideration of this reaction stresses the importance of eliminating any metal which has a lower hydrogen overvoltage than lead. Even traces of such metals will 'poison' the battery beyond repair.

At the PbO_2 *electrode*

$$PbO_2 + Pb + 2H_2SO_4 \longrightarrow 2PbSO_4 + 2H_2O$$

This constitutes attack of the lead supporting the effective electrode material. The sulphate deposit from both this spontaneous discharge and from the normal discharge, coagulates with time and retards further electrode

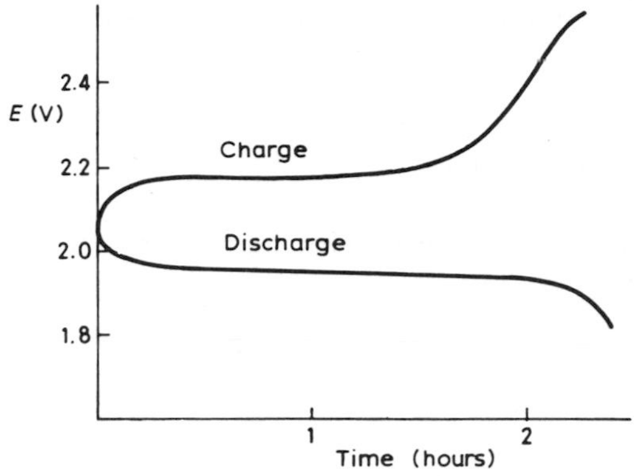

Figure 11.8 Constant current charging/discharging curves for the lead–acid battery.

processes—an effect known as 'sulphation'. Regular charging can reduce such effects to a minimum.

The efficiency of such cells may be expressed in two ways, current efficiency and energy efficiency.

Current efficiency (ε_I). This may be defined by

$$\varepsilon_I = \frac{\text{charge produced while discharging}}{\text{charge taken up during charging}}$$

The product of I (amps) and time (hours) at any point on the charging or discharging curve of Figure 11.8 gives the charge taken up or given out respectively in ampere hours. Thus,

$$\text{Current efficiency } \varepsilon_I = \frac{(It)_{\text{discharge}}}{(It)_{\text{charge}}} \tag{11.3}$$

which is about 95% for the lead acid cell.

Energy efficiency ε_U. This may be defined by

$$\varepsilon_U = \frac{\text{energy given out during discharge}}{\text{energy taken up while charging}}$$

$$= \frac{E_{\text{discharge}} \times (It)_{\text{discharge}}}{E_{\text{charge}} \times (It)_{\text{charge}}} \tag{11.4}$$

For the lead–acid battery this ratio is about 0.8. The voltage obtainable during discharge will be less than the reversible value, E_{rev}, by overvoltage and IR corrections, thus

$$E_{\text{discharge}} = E_{\text{rev}} - \eta - IR \tag{11.5}$$

η collecting both activation and concentration overvoltage effects. Conversely the voltage required for charging is in excess of E_{rev} according to

$$E_{\mathrm{charge}} = E_{\mathrm{rev}} - \eta' - IR \tag{11.6}$$

η' again collecting overvoltages.

It is seen therefore that the current and energy efficiencies are related through the expression

$$\varepsilon_{\mathrm{U}} = \left(\frac{E_{\mathrm{dis}} - \eta - IR}{E_{\mathrm{ch}} - \eta' - IR} \right) \varepsilon_I \tag{11.7}$$

Activation contributions to the overvoltage terms are small for this system, by far the larger part of both η and η' arising from concentration polarization effects. The two energies may be obtained from the areas under the respective curves, the difference between them being the energy loss.

Although the lead–acid cell has high current and energy efficiencies, it leaves much to be desired when considered in the more important practical terms of its energy output to weight ratio. While improved output-to-weight ratios may be obtained by using larger surface area grid plates, these are less strong mechanically and current surges are liable to break up the delicate PbO_2 and $PbSO_4$ deposits to cause sludging. It is often more economical to use more durable electrode components at the expense of some efficiency.

In the Edison alkaline battery, a 20% solution of potassium hydroxide is electrolysed between a cathode of iron/iron(II) hydroxide and an anode of nickel hydroxide. These electrode materials are pressed into perforated steel containers along with mercury and finely divided nickel at cathode and anode respectively to raise the conductivity (which is low for hydroxides). During the charging process the following reactions occur.

At the $Fe(OH)_2$ *electrode* (*cathode*)

$$Fe(OH)_2 + 2e \longrightarrow Fe + 2OH^-$$

At the $Ni(OH)_2$ *electrode* (*anode*)

$$Ni(OH)_2 + OH^- \longrightarrow NiOOH + H_2O + e$$

The overall charging/discharging reaction may therefore be written

$$2NiOOH + Fe + 2H_2O \underset{\text{charge}}{\overset{\text{discharge}}{\rightleftharpoons}} 2Ni(OH)_2 + Fe(OH)_2$$

Side reactions such as

$$2NiOOH + Ni(OH)_2 \longrightarrow Ni_3O_2(OH)_4$$

cause the process to be less reversible than those in the lead–acid cell, the current efficiency being about 80% and the energy efficiency about 60%. The emf produced initially is close to the reversible value of 1.4 V, but drops rapidly to a fairly steady 1.3 V.

The lead–acid cell has proved to have the best characteristics for the regular and routine use required in motor cars. With modern sealed units the hazards associated with the domestic handling of such cells have been largely eliminated. This was not always so and people of a certain age with a rural upbringing will remember the weekly recharge of the 'accumulator' for powering the 'wireless' in homes without an electricity supply! The more easily handled Leclanché cell is unfortunately not rechargeable but the nickel–cadmium battery offers both advantages and operates on the following processes:

At the anode

$$Cd + 2H_2O \longrightarrow Cd(OH)_2 + 2H^+ + 2e$$

At the cathode

$$2NiOOH + 2H^+ + 2e \longrightarrow 2Ni(OH)_2$$

The overall process therefore is

$$Cd + 2NiOOH + 2H_2O \underset{\text{charge}}{\overset{\text{discharge}}{\rightleftharpoons}} Cd(OH)_2 + 2Ni(OH)_2$$

The apparent attractiveness of systems based upon more electropositive metals, such as those of the alkali group, coupled with the more electronegative non-metallic elements is rarely a straightforward matter to exploit. Quite apart from the reactivity of many of the materials, the complications of some systems are almost overwhelming; even so, some remarkably difficult ones have been made workable. The sodium–sulphur cell, although electrochemically simple and rechargeable works on the electrode reactions

$$2Na \longrightarrow 2Na^+ + 2e$$
$$S + 2e \longrightarrow S^{2-}$$

Overall

$$2Na + S \rightleftharpoons Na_2S$$

The electrolyte medium is solid alumina, and since an inert atmosphere of argon has to be used the cell requires careful sealing. Added to these factors the cell will only operate effectively at temperatures of the order of 300–400°C and the highly corrosive environment requires a mild steel container lined with chromium. It is hardly a suitable system for domestic purposes but has seen some pilot application for commercial traction purposes.

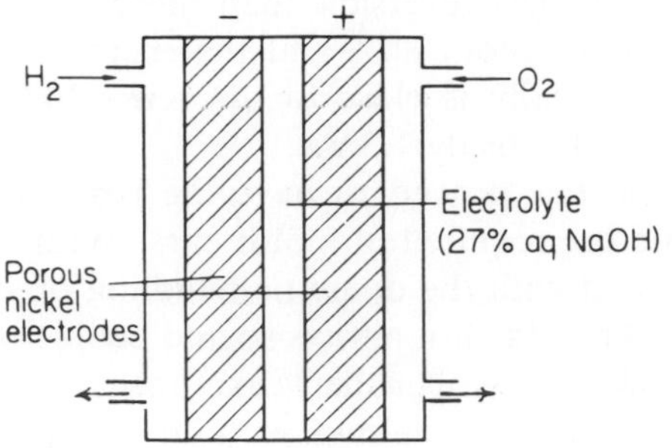

Figure 11.9 The Bacon hydrogen/oxygen fuel cell.

There is considerable current interest in aluminium as an electrode material and a rather unique combination of aluminium anode with air cathode has resulted in a cell which is rechargeable by mechanical rather than by electrochemical means. Electrode and overall reactions are

$$Al \longrightarrow Al^{3+} + 3e$$

$$\tfrac{3}{4}O_2 + \tfrac{3}{2}H_2O + 3e \longrightarrow 3OH^-$$

Overall

$$Al + \tfrac{3}{4}O_2 + \tfrac{3}{2}H_2O \rightleftharpoons Al(OH)_3$$

The electrolyte medium may be either sodium chloride ('saline') or sodium hydroxide ('alkaline').

Since the gelatinous aluminium hydroxide tends to adhere to the electrodes and cell walls and inhibit the reactions, it is necessary to agitate the solution to encourage precipitation. Removal of water from the electrolyte implied from the cell reaction means that this requires replenishing periodically along with the aluminium anode which dissolves. The recharging is thus effectively replacement of both the anode material and water; the aluminium lost in the discharge process may, however, be regenerated from the precipitated hydroxide.

11.2.3 Fuel cells

A fuel cell is based on the reversal of cause and effect of an electrolytic process and *generates a voltage* by supplying the products to appropriately constructed electrodes. The arrangement of components in the Bacon hydrogen/oxygen fuel cell is shown in Figure 11.9. Operation of the cell depends upon the following processes

$$2H_2 \longrightarrow 4H^+ + 4e;\ O_2 + 4H^+ + 4e \longrightarrow 2H_2O\ \text{(in acid)}$$

or

$$2H_2 + 4OH^- \longrightarrow 4H_2O + 4e; \; O_2 + H_2O + 4e \longrightarrow 4OH^- \text{ (in alkali)}$$

The overall reaction is the production of water for both cases

$$2H_2 + O_2 \longrightarrow 2H_2O$$

For this reaction as written, tabulated thermodynamic data suggest the following values for free energy and enthalpy functions

$$\Delta G^{\ominus} \sim 476\,\text{kJ}$$

and

$$\Delta H^{\ominus} \sim 484\,\text{kJ}$$

From the first value it is clear that the theoretical voltage which could be developed follows from $\Delta G^{\ominus} = -nE_{eq}F$ as

$$E_{eq} = \frac{-476\,000}{-4 \times 96\,500} = 1.23\,\text{V}$$

In practice, voltages of the order of 1.0 to 1.1 V are obtained, which implies an efficiency of about 80%.

Another, more fundamental way of viewing the efficiency (ε) is in terms of the thermodynamics of the process as

$$\varepsilon = \frac{\Delta G}{\Delta H} = \frac{476}{484} \times 100 \sim 98\%$$

which is impressively high.

Indeed, when written in terms of the temperature coefficient of the cell voltage (see equation (8.17))

$$\varepsilon = \frac{E}{E - T\left(\dfrac{\partial E}{\partial T}\right)_P}$$

which implies that for a cell with a *positive* temperature coefficient, $\varepsilon >$ 100%!

Such a state of affairs cannot be realized because of the unavoidable overvoltage effects. Even so at a maximum these are likely to reduce the efficiencies by no more than 30%. Thus, for the case concerned, at least 68% efficiency should be realizable.

Combustion processes, and therefore all mechanical devices dependent for their operation upon them, are *intrinsically* of much lower efficiency. This is because the two types of process are *fundamentally* different: while the voltage of a fuel cell is directly dependent on ΔG of the reaction upon which it functions, the output of a heat engine depends on the *heat evolved* when the chosen fuel is *burnt*, i.e. $-\Delta H$, the enthalpy change for the combustion process. Now the efficiency of even an ideal heat engine is given in terms of the heat (Q_2) taken in at a higher temperature (T_2) and that given out (Q_1) at a lower temperature (T_1) as

$$\frac{Q_2 - Q_1}{Q_2} \tag{11.8}$$

and, since

$$\frac{Q_2}{T_2} = \frac{Q_1}{T_1} \qquad \text{(from the definition of entropy)}$$

it follows that

$$\varepsilon = \frac{T_2 - T_1}{T_2} \tag{11.9}$$

The value of this quantity might typically be 30%, which implies that 70% is lost through the thermodynamic nature of the proces—the so-called Carnot losses. Add to this the additional likely 15% loss due to the frictional effects of the engine and the final resultant efficiency could well be as low as 15%. This simple balance sheet approach brings home the likely advantages to be gained by developing fuel cell systems.

For an efficient cell, all processes must occur rapidly. Reactants must be able to reach the electrodes easily so that porous electrodes with large internal surface areas, saturated with electrolyte, are used. The pore sizes are often graded from large on the gas side of an electrode to small on the electrolyte side. Good catalyst materials ensure rapid electrochemical reactions and, in combination with a higher temperature, help to suppress the cathodic formation of perhydroxyl ions by the reaction.

$$O_2 + 4H_2O + 6e \longrightarrow 4OH^- + 2OH_2^-$$

With such precautions the hydrogen/oxygen cell can be made to show efficiencies of up to 75%. A Bacon-type cell has been successfully employed in space projects where the water produced (at the rate of about a pint per kilowatt-hour) is used to supplement the water supply.

Many other fuel cell systems have been devised with varying degrees of success. Provided that certain inherent difficulties can be overcome, a wide variety of designs and modes of operation could become available for specific purposes. Their use for traction purposes to replace engines with high pollution risk is a major attraction. At the other extreme artificial hearts, powered by fuel cells consuming food fuels, have been suggested as a further possibility.

At present, a great drawback is that while the attractive prospect of the use of cheap fuels such as hydrocarbons presents itself, difficulties are encountered in practice by the poisoning of catalytic surfaces by intermediates.

11.3 Electrocatalysis and electrosynthesis

Inert electrodes which enable electron transfer between them and species in solution to take place at rates which are dependent upon the magnitudes of imposed electric fields, come within the classical definition of a catalyst. Rates of electron-transfer reactions are a function of both the electrode material and the applied potential but the electrode remains unchanged—a phenomenon known as *electrocatalysis.*

While in a broad sense, all such processes may be classed as heterogeneous, the term in this context is generally reserved for the more overtly heterogeneous character of reactions undergone by species during their adsorption on an electrode surface. This is a particularly common effect with organic compounds. The interplay of the adsorption and electron-transfer processes is a subtle one, the former being largely a function of the nature of the electrode surface, the latter of the potential that it has imposed upon it. Voltammetric/polarographic curves allow the study, albeit on a very small scale, of adsorption effects and indicate the properties and conditions required for scale-up synthetic applications.

The Arrhenius equation for the rate (current) of a heterogeneous electrochemical reaction whose overall activation energy comprises contributions of a chemical ($\Delta G^{\ddagger}_{\text{chem}}$), an adsorption ($\Delta G^{\ddagger}_{\text{ads}}$) and an overpotential ($\alpha\eta F$) kind may be written as

$$I = (\text{constant}) \left[-\frac{\Delta G^{\ddagger}_{\text{chem}} + \Delta G^{\ddagger}_{\text{ads}} - \alpha\eta F}{RT} \right] \tag{11.10}$$

Equation (11.10) emphasizes the significance of the magnitude of the $\Delta G^{\ddagger}_{\text{ads}}$ term on the rate; even more significant is the powerful control that η exerts on the reaction for it if becomes of such a magnitude that

$$\eta = \left[\frac{\Delta G^{\ddagger}_{\text{chem}} + \Delta G^{\ddagger}_{\text{ads}}}{\alpha F} \right] \tag{11.11}$$

the reaction takes place with effectively zero activation energy.

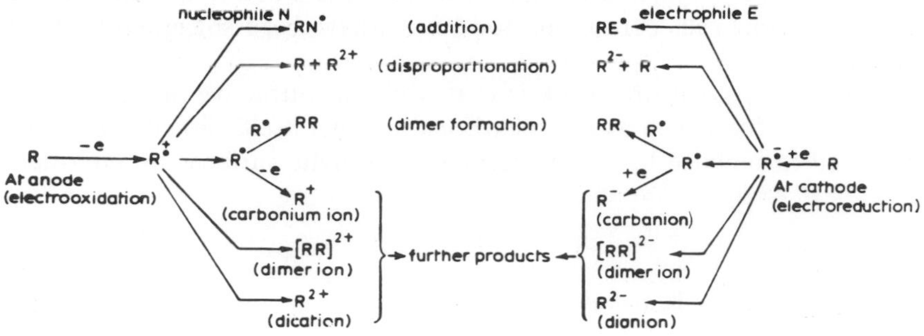

Figure 11.10 The main electro-organic reaction routes.

Electrochemical properties of different surface structures, even of the same material, can be quite distinct; indeed the nature of electrode surfaces may change during the course of electrolysis even to the extent of losing their effectiveness and showing passivity. Insulating oxide layers are not an uncommon feature at anodes and such (usually undesirable) effects may be reversed by a variety of mechanical and chemical methods.

Platinum has proved to be a versatile electrocatalyst; this versatility owes as much to its chemical and electrochemical stability as to its catalytic properties. It is possible, for instance, to oxidize ethene to carbon dioxide in a multistage process for which the rate-determining step is the reaction between adsorbed C_2H_4 and OH^-. The affinity of platinum for species of this sort is such that, while the bonding is of sufficient strength and duration to provide adequate opportunity for adsorbed species to react, it is not so overpowering as to prevent products removing themselves. Other noble metals such as osmium and iridium are often inferior to platinum for this reason.

Almost every kind of organic reaction may, in principle, be initiated electrochemically and, although this has been known for a very long while, it is only comparatively recently that electro-organic synthesis has emerged as a more general discipline from a somewhat empirical background. A major problem is that there are so many variables which, without fine tuning, give rise to alternative products. Variations of the electrode materials, solvent, electrolyte, pH and applied voltage can significantly alter the nature and/or proportions of products. While voltammetric methods may be used on a small scale to identify appropriate conditions, a major problem is to scale up laboratory experiments to arrangements of pilot and industrial size. Solvents themselves can cause something of a problem since those which offer ready solubility to species which must be used are often of low conductivity and mixed aqueous/non-aqueous media and the addition of supporting electrolytes may sometimes be necessary. Some idea of the range of electrochemically promoted organic reactions, when R is an appropriate starting material, may be gained by scrutiny of Figure 11.10.

One example each of anodically and cathodically initiated processes will be considered briefly.

11.3.1 Anodically initiated process

The Kolbe reaction, based upon carboxylic acids, may be used to prepare a variety of products via both radical and carbonium ion mechanisms. Both the mechanism which operates and the products which result are dependent on the electrode material, the concentration of acid and the pH of the dispersive medium. A platinum anode favours a two-electron formation of radicals followed by dimerization—the classical Kolbe hydrocarbon chain extension, i.e.

$$2RCOO^- \xrightarrow{-2e} 2CO_2 + 2R^\bullet$$
$$2R^\bullet \longrightarrow RR$$

With carbon anodes, carbonium ions are formed which can, in the presence of other species, lead to a wide variety of products including alcohols, alkenes, esters, ethers and ketones.

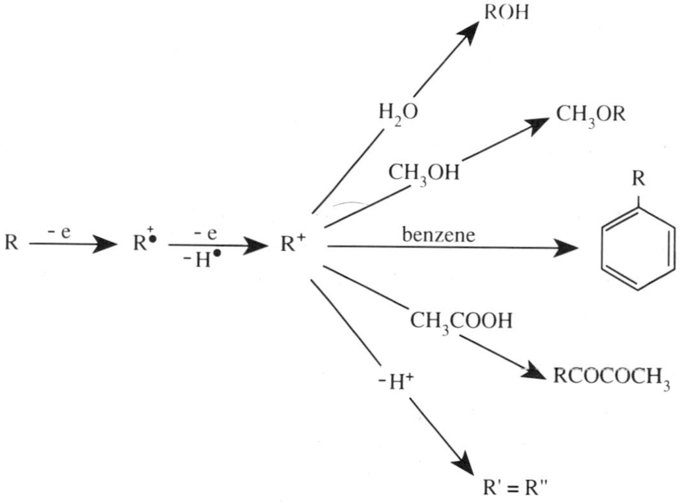

11.3.2 Cathodically initiated process

At present the largest electro-organic synthesis in the world is that of adiponitrile by the hydrodimerization of acrylonitrile which overall may be expressed by

$$2CH_2{=}CHCN + 2e + 2H^+ \longrightarrow NCCH_2CH_2CH_2CH_2CN$$

The low solubility of acrylonitrile in water alone necessitates the use of 30% tetraethylammonium *p*-toluene sulphonate in water; this not only improves the solubility of the reactant but provides the all-important high conductivity. If the pH is not strictly controlled to the range 7.5 to 9.5, side reactions become significant. In any case, the best yield of adiponitrile is some 90% the remainder being propionitrile. The balance of products may be completely altered if the quaternary ammonium salt is replaced by an alkali metal salt, the major product then being propionitrile.

11.4 Electrochemistry on an industrial scale

Electrochemical principles are used so widely on an industrial scale that for the present purposes it is necessary to be very selective in deciding on representative examples.

Where industry has a choice between electrochemical or other means to a commercial end, cheapness of the electricity supply will do much to influence the decision made. This is seen particularly in the range of methods used for the extraction of metals from their ores, although there are a number of instances such as those of aluminium and magnesium where there is little choice but to use electrochemistry.

A long-established application of the processes of electrodeposition is to be found in the electroplating industry. The nature and quality of an electrodeposit is strongly dependent upon the choice of conditions and the addition of certain materials (often strongly surface active organic species) known as 'brighteners'. Even nowadays the production of high-quality plated finishes is something of an art rather than the application of an understood science, the choice of many working conditions and materials being largely empirical. With the development of conducting paints which can key satisfactorily to plastic surfaces, it is now possible to electroplate metal finishes on to these materials.

The deposition process may be logically extended to the application of electroforming in which often intricate metal components may be made via deposition upon a cathode of appropriate shape and material. It is even possible to co-deposit more than one metal at a time to produce electrochemically formed alloys. In another context, careful control of the potentials of electrodes in large scale electrolyses forms the basis of electrorefining and separation processes.

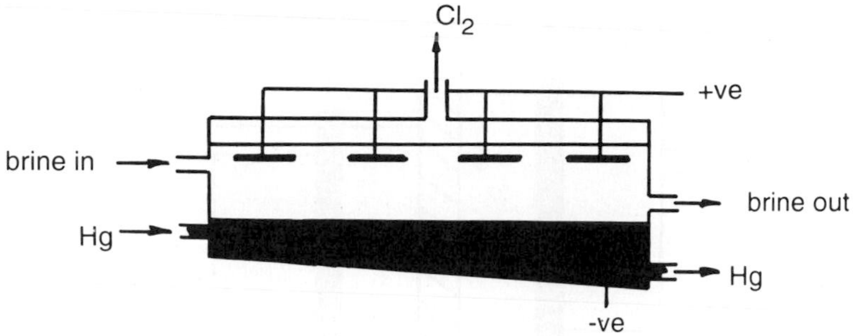

Figure 11.11 Flow system for electrolysis of brine between carbon anodes and mercury cathode. The foundation of the 'chlor-alkali' industry.

The reverse dissolution reactions at anodically polarized electrodes have been developed into the technologies of electropolishing and electromachining. Anodic dissolution of a thin layer of metal from the bulk often leaves a finish comparable in appearance to one which has been mechanically polished—but considerably more easily and cheaply and often of longer durability. Electrochemical 'drilling' of holes in metal components by such means, using a precision-operated controllable cathode, can be as rapid as any conventional power tool and may be performed with considerable accuracy.

To complement the earlier consideration of the world's largest electro-organic synthesis, it seems appropriate to briefly review the development of the world's largest scale electrolytic process. This is the electrolysis of brine to give chlorine and caustic soda and constitutes the 'chlor-alkali' industry.

Traditionally mercury was used as the cathode, making use of the high hydrogen overvoltage so that sodium was discharged in preference to hydrogen. With a carbon or graphite anode the two electrode reactions were

At the anode

$$2Cl^- \longrightarrow Cl_2 + 2e$$

At the cathode

$$2Na^+ + 2e + 2Hg \longrightarrow 2NaHg\,(\text{as amalgam})$$

Based on a flow system (Figure 11.11), the amalgamated mercury on treatment with water produced caustic soda after which the regenerated mercury was recycled. This system raised considerable technical and environmental problems. On the one hand, the continuous destruction of the anodes required continuous adjustment to retain their position relative to the liquid

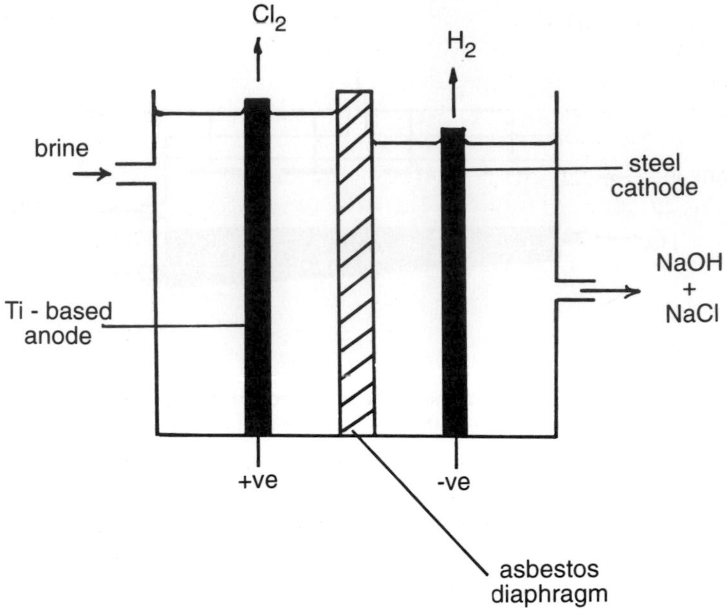

Figure 11.12 Principle of the diaphragm cell: generation of hydrogen at the cathode by $2H_2O + 2e \rightleftharpoons H_2 + 2OH^-$ produces dilute NaOH directly.

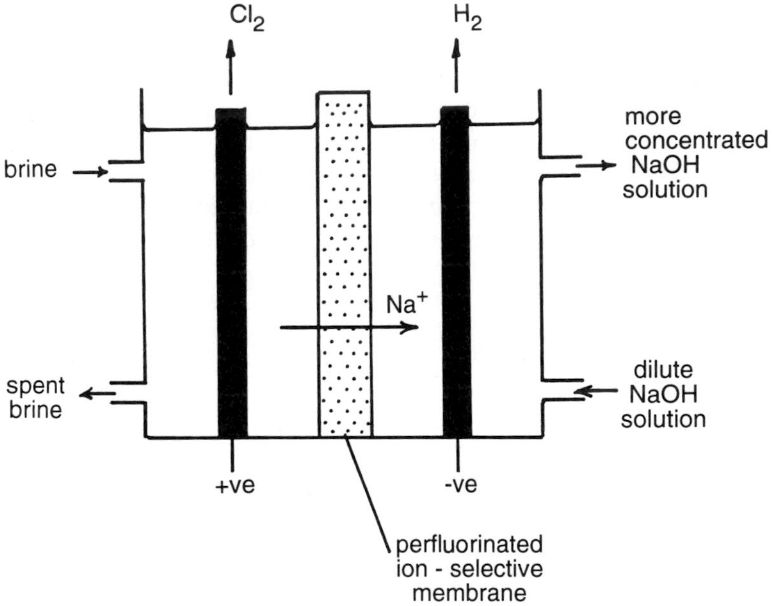

Figure 11.13 Principle of modern membrane cell for the electrolysis of brine.

cathode; on the other hand, the toxicity of mercury led to the development of arrangements involving alternative cathode materials.

Electrolysis between a steel cathode and a titanium-based anode generates chlorine and hydrogen. Caustic soda is effectively produced directly in the cathode region: this caustic solution must be prevented from mixing with that from the anode region where interaction of NaOH and Cl_2 would produce hypocholorite and chlorate. This may be achieved by separating the two regions by an asbestos diaphragm. The schematic arrangement of the *diaphragm cell* is shown in Figure 11.12.

There are a number of advantages offered by the diaphragm cell: the use of mercury is avoided, the applied voltage required is about 3.5 V (about 1V less than that needed for the mercury cell) and the anodes are 'dimensionally stable'. These are offset by the fact that the solution of sodium hydroxide produced is rather dilute (10–12%) and requires concentration. The additional cost of the associated evaporation process makes the two methods economically comparable.

The more modern *membrane cell* has the purely physical barrier between anode and cathode compartments replaced by an ion-selective membrane. Na^+ ions may cross the perfluorinated membrane but there is no flow of solution *between* the compartments (Figure 11.13). A significantly stronger solution of caustic soda is produced—of the order of 40%.

Problems

11.1 The zinc chloride battery, operating on the reaction $Zn + Cl_2 \rightarrow ZnCl_2$, has possibilities for exploitation as a source of power for driving vehicles in that a bank of 118 such cells, connected in series, has a capacity of approximately 90 kWh. Calculate (i) the standard emf of the cell, (ii) the mass of solid chlorine hydrate ($Cl_{12}.8\ H_2O$) which must be stored in the bank in order to sustain power capacity.

11.2 The following reactions may be made to operate in fuel cells at 298 K

$$CH_4(g) + 2O_2(g) = CO_2(g) + 2\ H_2O(l)$$
$$\Delta H^{\ominus}_{298} = -890.4\,\text{kJ mol}^{-1}; \qquad \Delta G^{\ominus}_{298} = -818.0\,\text{kJ mol}^{-1}$$
$$C_2H_6(g) + \tfrac{7}{2}O_2(g) = 2CO_2(g) + 3\ H_2O(l)$$
$$\Delta H^{\ominus}_{298} = -1560\,\text{kJ mol}^{-1}; \qquad \Delta G^{\ominus}_{298} = -1467.5\,\text{kJ mol}^{-1}$$
$$C_3H_8(g) + 5O_2(g) = 3CO_2(g) + 5\ H_2O(l)$$
$$\Delta H^{\ominus}_{298} = -2220\,\text{kJ mol}^{-1}; \qquad \Delta G^{\ominus}_{298} = -2108.0\,\text{kJ mol}^{-1}$$
$$CH_3OH(l) + \tfrac{3}{2}O_2(g) = CO_2(g) + 2\ H_2O(l)$$
$$\Delta H^{\ominus}_{298} = -764.0\,\text{kJ mol}^{-1}; \qquad \Delta G^{\ominus}_{298} = -706.9\,\text{kJ mol}^{-1}$$

For each case calculate (i) the number of electrons transferred overall in the cell reaction, (ii) the reversible emf of the cell at 298 K, (iii) the maximum efficiency of the cell.

Further reading

Chapter 1 (and general coverage)

Bockris, J.O'M. and Drazic, D.M. (1972) *Electrochemical Science*, Taylor and Francis, London.
Bockris, J.O'M. and Khan, S.U.M. (1979) *Quantum Electrochemistry*, Plenum Press, New York and London.
Bockris, J.O'M. and Reddy, A.K.N. (1973) *Modern Electrochemistry*, Plenum Press, New York.
Brett, C.M.A. and Brett, A.M.O.B. (1993) *Electrochemistry: Principles, Methods and Applications*, Oxford University Press, Oxford.
Goodisman, J. (1987) *Electrochemistry: Theoretical Foundations*, Wiley-Interscience, New York.
Hibbert, D.B. (1993) *Introduction to Electrochemistry*, Macmillan, Basingstoke and London.
Koryta, J. (1987) *Principles of Electrochemistry*, Wiley, Chichester.
Koryta, J., Dvorak, J. and Kavan, L. (1993) *Principles of Electrochemistry*, 2nd edn., Wiley, New York.

Chapter 2

Davies, C.W. (1962) *Ion Association*, Butterworths, London.
Robinson, R.A. and Stokes, R.H. (1959) *Electrolyte Solutions*, 2nd edn., Butterworths, London.

Chapter 3

Gurney, R.W. (1967) *Ionic Processes in Solution*, McGraw-Hill, London.
Prue, J.E. (1966) *Ionic Equilibria*, Pergamon Press, Oxford.
Rossotti, H. (1978) *The Study of Ionic Equilibria*, Longmans, London.

Chapter 4

Robbins, J. (1972) *Ions in Solution 2*. Clarendon Press, Oxford.
Selley, N.J. (1977) *Experimental Approach to Electrochemistry*, Edward Arnold, London.

Chapter 5

Delahay, P. (1965) *Double Layer and Electrode Kinetics*, Wiley-Interscience, New York.
Parsons, R. (1961) Structure of the electrical double layer and its influence on the rates of electrode reactions. In *Advances in Electrochemistry and Electrochemical Engineering*, Vol. 1 (Eds H. Gerischer and C.W. Tobias), Interscience, New York.
Sparnay, M.J. (1972) *The Electrical Double Layer*, Pergamon Press, Oxford.

Chapter 6

Ives, D.J.G. and Janz, G.I. (1961) *Reference Electrodes*, Academic Press, New York.
Koryta, J. (1991) *Ions, Electrodes and Membranes*, Wiley, Chichester.
Newman, J. (1973) *Electrochemical Systems*, Prentice-Hall, Englewood Cliffs, New Jersey.

Chapter 7

Albery, W.J. (1975) *Electrode Kinetics*, Clarendon Press, Oxford
Bauer, H.H. (1972) *Electrodics*, Thieme, Stuttgart.
Brenet, J.P. and Traone, K. (1971) *Transfer Coefficients in Electrochemical Kinetics*, Academic Press, London.
Erdey-Gruz, T. (1972) *Kinetics of Electrode Processes*, Adam Hilger, London.
Fried, I. (1973) *The Chemistry of Electrode Processes*, Academic Press, London.
Hush, N.S. (1971) *Reactions of Molecules at Electrodes*, Wiley-Interscience, New York.
Pletcher, D. (1991) *A First Course in Electrode Processes*, The Electrochemical Consultancy, Romsey.
Thirsk, H.R. and Harrison, J.A. (1972) *A Guide to the Study of Electrode Kinetics*, Academic Press, London.

Chapter 8

Sawyer, D.T. and Roberts, J.L. (1974) *Experimental Electrochemistry for Chemists*, Wiley, New York.
Yaeger, E. and Salkind, A.J. (Eds) *Techniques of Electrochemistry*, Vol. I (1972), Vol. II (1973), Wiley-Interscience, New York.

Chapter 9

Adams, R.N. (1969) *Electrochemistry at Solid Electrodes*, Dekker, New York.
Albery, W.J. and Hitchman, M.L. (1971) *Ring–Disc Electrodes*, Oxford University Press, Oxford.
Bard, A.J. and Faulkner, L.R. (1980) *Electrochemical Methods, Fundamentals and Applications*, Wiley, New York.
Crow, D.R. (1969) *Polarography of Metal Complexes*, Academic Press, London.
Crow, D.R. (1986) Voltammetry. In *Metals Handbook, Materials Characterization*, Vol. 10, 9th edn. ASM International, Materials Park, Ohio.
Franklin Smyth, W. (1992) *Voltammetric Determination of Molecules of Biological Significance*, Wiley, Chichester.
Galus, Z. (1976) *Fundamentals of Electrochemical Analysis*, Ellis Horwood, Chichester.
Riley, T. and Tomlinson, C. (1987) *Principles of Electroanalytical Methods*, ACOL/Wiley, Chichester.
Southampton Electrochemistry Group (1985) *New Instrumental Methods in Electrochemistry*, Ellis Horwood, Chichester.

Chapter 10

Ammann, D. (1986) *Ion-Selective Microelectrodes*, Springer-Verlag, Berlin.
Covington, A.K. (1979) *Ion-Selective Electrode Methodology*, Vols I and II, CRC Press, Boca Raton, Florida.
Edmonds, T.E. (1988) *Chemical Sensors*, Blackie, London.
Evans, A. (1987) *Potentiometry and Ion-Selective Electrodes*, Wiley, New York.

Chapter 11

Baizer, M.M. and Lund, H. (1991) *Organic Electrochemistry*, Dekker, New York.
Fry, A.J. (1989) *Synthetic Organic Electrochemistry*, Wiley, New York.
Hine, F. (1985) *Electrode Processes and Electrochemical Engineering*, Plenum Press, New York.
Kyriacou, D.K. (1981) *Basics of Electroorganic Synthesis*, Wiley, New York.
McDougall, A. (1976) *Fuel Cells*, Macmillan, London.

Pletcher, D. and Walker, F.C. (1990) *Industrial Electrochemistry*, Chapman and Hall, London.
Shono, T. (1990) *Electroorganic Synthesis*, Academic Press, New York.
Wrangler, G. (1985) *Introduction to Corrosion and Protection of Metals*, Chapman and Hall, London.

Solutions to problems

Chapter 2

2.1 According to equation (2.13)

$$\kappa = \left(\frac{2 \times 10^3 e^2 N}{\varepsilon_0 \varepsilon k T}\right)^{1/2} \sqrt{I} = 0.503 \times 10^{12} \sqrt{\frac{I}{\varepsilon T}}$$

therefore radius $= 1/\kappa = (1.988 \times 10^{-12})\sqrt{\varepsilon T/I}$ for $1/\kappa$ in m.
(i) for 1:1 electrolyte in water at 298 K the following values are obtained:

I (mol kg^{-1})	0.1	0.01	0.001	0.0001
$1/\kappa$ (nm)	0.96	3.04	9.62	30.41

(ii) for 1:2 electrolyte in water at 298 K the following values arise:

I (mol kg^{-1})	0.3	0.03	0.003	0.0003
$1/\kappa$ (nm)	0.55	1.76	5.55	17.58

Corresponding values in DMF are:

(i)	0.66	2.08	6.57	20.79
(ii)	0.38	1.20	3.80	12.00

2.2 By equation (2.14)

$$I = \tfrac{1}{2}\sum m_i z_i^2 = \tfrac{1}{2}[(1 \times 0.0015 \times 2^2) + (2 \times 0.0015 \times 1^2)]$$
$$= 0.0045 \text{ mol kg}^{-1}.$$

(i) $\gamma_{Mg^{2+}} = 0.730$; $\gamma_{Cl^-} = 0.924$;
(ii) $\gamma_\pm = 0.854$. This may be obtained either from the individual activity coefficients using equation (2.7) or by the limiting law expression for $\log \gamma_\pm$ given in equation (2.17).

2.3 (i) $I = \frac{1}{2}[(1 \times 0.001 \times 3^2) + (3 \times 0.001 \times 1^2)] = 0.006$ mol kg^{-1}
(ii) $I = \frac{1}{2}[(1 \times 0.001 \times 3^2) + (3 \times 0.001 \times 1^2) + (1 \times 0.002 \times 1^2) + (1 \times 0.002 \times 1^2)] = 0.008$ mol kg^{-1}

2.4 (i) $\gamma_\pm = 0.761$ for $I = 0.006$ mol kg^{-1}
$\gamma_\pm = 0.730$ for $I = 0.008$ mol kg^{-1}
(ii) $\gamma_\pm = 0.776$ for $I = 0.006$ mol kg^{-1}
$\gamma_\pm = 0.749$ for $I = 0.008$ mol kg^{-1}

2.5 $B = \left(\dfrac{2 \times 10^3 N^2 \epsilon^2}{\varepsilon_0 \varepsilon RT}\right)^{1/2}$

$$= \left[\frac{2 \times 10^3 \times (6.023 \times 10^{23})^2 \times (1.6021 \times 10^{-19})^2}{8.8542 \times 10^{-12} \times 78.54 \times 8.314 \times 298}\right]^{1/2}$$

$$= 3.288 \times 10^9 \mathrm{m^{-1} mol^{-1/2} kg^{1/2}}$$

$$A = \left[\frac{\epsilon^2 N}{2.303 RT \times 8\pi\varepsilon_0\varepsilon}\right] B$$

$$= \left[\frac{(1.6021 \times 10^{-19})^2 \times 6.023 \times 10^{23}}{2.303 \times 8.314 \times 298 \times 8 \times 3.142 \times 8.8542 \times 10^{-12} \times 78.54}\right] \times 3.288 \times 10^9$$

$$= 0.509 \mathrm{mol^{-1/2} kg^{1/2}}$$

2.6 Use the Debye-Hückel equation in the form of equation (2.18), viz.

$$-\frac{A\,|z_+ z_-|\,\sqrt{I}}{\log \gamma_\pm} = 1 + Ba\sqrt{I}$$

and plot values of LHS versus $\sqrt{I}$ to give slope = Ba and intercept = 1. Slope = 1.40 mol kg^{-1} which, with $B = 3.29 \times 10^9$ m^{-1}mol$^{-1/2}$ kg$^{1/2}$, gives a = 0.43nm.

2.7 Application of equation (2.20), $q = z_i z_j \epsilon^2 / 8\pi\varepsilon_0\varepsilon kT$ yields q = 0.357nm, 1.428 nm and 3.213 nm for the respective electrolyte types.

Chapter 3

3.1 (i) 3.00, (ii) 2.70, (iii) 11.30, (iv) 11.48, (v) 3.38, (vi) 10.78. Cases (i)–(iv) are strong acids or bases and allow direct assessment of [H_3O^+] or [OH^-]. For case (v) consideration of weak acid dissociation equilibrium yields pH = $\frac{1}{2}$(pK_a − log C), while for case (vi) consideration of weak base dissociation equilibrium yields pOH = $\frac{1}{2}$(pK_b − log C).

3.2 From pH = $\frac{1}{2}$(pK_a − log C), pK_a = 4.44, whence pH of 0.15 mol dm^{-3} solution = 2.63.

3.3 From the Henderson–Hasselbalch equation (equation (3.50)) the required [salt] to [acid] ratio is 5.534. Therefore, since original number of moles of acid = 0.04.

$$\frac{\text{salt}}{\text{acid}} = 5.534 = \frac{\text{vol. of NaOH}}{\text{vol. of acid} - \text{vol. of NaOH}}$$

if vol. of NaOH = x, then $x/200 - x = 5.534$ giving x = 169.4 cm^3.

3.4 (i) pH = $\frac{1}{2}$[pK_w − pK_b − log C] (equation (3.43)) yields pH = 4.97;
(ii) pH = $\frac{1}{2}$[pK_w + pK_a + log C] (equation (3.41)) yields pH = 8.88;
(iii) pH = $\frac{1}{2}$[pK_w + pK_a − pK_b] (equation (3.48)) yields pH = 7.01.

3.5 (i) pH = 4.94 (via equation (3.50)); (ii) pH = 9.55 (via equation (3.52)).

3.6 25 cm^3 of 0.15 mol dm^{-3} isopropylamine contain 0.00375 mol. 10 cm^3 of 0.12 mol dm^{-3} HCl contain 0.00120 mol, therefore amount of salt = 0.00120 mol and amount of base remaining = 0.00255 mol. Application of equation (3.52) gives pH = 10.30.

3.7 (i) 9.60, (ii) 10.19, (iii) 2.67, (iv) 2.09. pH values for (i) and (ii) are situated on either side of pK_{a2} and are given by equation (3.54) viz.

$$\mathrm{pH} \sim \mathrm{p}K_{a2} + \log \frac{[\mathrm{G}^-]}{[\mathrm{G}^\pm]}$$

pH values for (iii) and (iv) are situated on either side of pK_{a1} and are given by

$$\mathrm{pH} \sim \mathrm{p}K_{a1} + \log \frac{[\mathrm{G}^\pm]}{[\mathrm{G}^+]} \qquad \text{(see equation (3.53))}$$

3.8 Solution is 0.05 mol dm^{-3} in each component, with ionic strength given by

$$\begin{aligned} I &= \tfrac{1}{2}\{(C_{\mathrm{Na}^+} \times 1^2) + (C_{\mathrm{H_2PO_4^-}} \times 1^2) + (C_{\mathrm{HPO_4^{2-}}} \times 2^2)\} \\ &= \tfrac{1}{2}\{(0.15 \times 1^2) + (0.05 \times 1^2) + (0.05 \times 2^2)\} \\ &= 0.2\ \mathrm{mol\,dm^{-3}}. \end{aligned}$$

Second dissociation refers to $H_2PO_4^- \rightleftharpoons H^+ + HPO_4^{2-}$ therefore,

$$K_{a2} = \frac{a_{\mathrm{H}^+} a_{\mathrm{HPO_4^{2-}}}}{a_{\mathrm{H_2PO_4^-}}} = \frac{a_{\mathrm{H}^+} C_{\mathrm{HPO_4^{2-}}} \gamma_{\mathrm{HPO_4^{2-}}}}{C_{\mathrm{H_2PO_4^-}} \gamma_{\mathrm{H_2PO_4^-}}}$$

For equal concentrations of the two forms

$$K_{a2} = a_{\mathrm{H}^+} \frac{\gamma_{\mathrm{HPO_4^{2-}}}}{\gamma_{\mathrm{H_2PO_4^-}}}$$

γ values may be calculated from the extended Debye–Hückel expression to be $\gamma_{\mathrm{HPO_4^{2-}}} =$ 0.2349 and $\gamma_{\mathrm{HPO_4^-}} = 0.6968$ giving $a_{\mathrm{H}^+} = 1.881 \times 10^{-7}$ and pH = 6.73.

3.9 (i) pH $\cong \frac{1}{2}(\mathrm{p}K_a + 1) = 2.43$; (ii) pH$\cong 3.86 + \log(2.5)/(7.5) = 3.38$ (see equation (3.50)); (iii) pH $\approx \mathrm{p}K_a = 3.86$; (iv) pH$\cong 3.86 + \log(7.5)/(2.5) = 4.16$ (see equation (3.50)); (v) pH$\cong \frac{1}{2}(14.00+3.86-1.301) = 8.28$ (see equation (3.41)); (vi) $[\mathrm{OH}^-] = 0.001$ mol 30 cm^{-3}. i.e. 0.0033 mol dm^{-3}; pH = 12.52.

3.10 From pH $\cong \frac{1}{2}[(\mathrm{p}K_w)_C + \mathrm{p}K_a + \log C]$ (see equation (3.41))

$$\begin{aligned} (K_w)_C &= 1.452 \times 10^{-14} = [\mathrm{H_3O^+}][\mathrm{OH^-}] \\ (K_w)_T &= [\mathrm{H_3O^+}][\mathrm{OH^-}]\gamma_\pm^2;\ \gamma_\pm = 0.822 \\ (K_w)_T &= 1.452 \times 10^{-14}(0.676) = 0.982 \times 10^{-14}. \end{aligned}$$

3.11 $\mathrm{p}K_b = 3.911$; $\gamma_\pm = 0.813$. From equation (3.43)

$$\begin{aligned} \mathrm{pH} &\cong \tfrac{1}{2}[(\mathrm{p}K_w)_C - \mathrm{p}K_b - \log C] \\ (K_w)_C &= 1.698 \times 10^{-15};\ (K_w)_T = 1.122 \times 10^{-15} \end{aligned}$$

The values stress the fairly strong temperature dependence of K_w.

Chapter 4

4.1 From data for KCl, cell constant $s = \kappa/G = 1.58\,\text{cm}^{-1}$.
From data for NaCl, $\kappa = 6.032 \times 10^{-4}\,\Omega^{-1}\,\text{cm}^{-1}$.
Therefore by equation (4.4c)

$$\Lambda = \frac{1000\kappa}{C} = \frac{0.6032}{0.005} = 120.6\,\Omega^{-1}\,\text{cm}^2\,\text{mol}^{-1}$$
$$= 1.206 \times 10^{-2}\,\Omega^{-1}\,\text{m}^2\,\text{mol}^{-1}.$$

4.2 From the Kohlrausch Law expressed by equation (4.9),

$$\Lambda_0 \text{ for propionic acid} = (0.8590 + 4.2126 - 1.2156) \times 10^{-2}\,\Omega^{-1}\,\text{m}^2\,\text{mol}^{-1}$$
$$= 3.856 \times 10^{-2}\,\Omega^{-1}\,\text{m}^2\,\text{mol}^{-1}.$$

4.3 (i) Use the Einstein equation, $D = \dfrac{uRT}{zF}$ (see equation (4.36))

thus $$D = \frac{6.4 \times 10^{-8}}{38.95} = 1.64 \times 10^{-9}\,\text{m}^2\,\text{s}^{-1}$$

(ii) Use the Nernst–Einstein equation, $\lambda = \dfrac{z^2F^2D}{RT}$ (see equation (4.24))

thus, $= 38.95 \times 96500 \times 1.64 \times 10^{-9} = 0.616 \times 10^{-2}\,\Omega^{-1}\,\text{m}^2\,\text{mol}^{-1}$

(iii) Use the Stokes–Einstein equation, $D = \dfrac{RT}{6\pi r\eta N}$ (see equation (4.38))

whence $r = 0.149$ nm.

This is larger than the crystallographic value and is a partial reflection of the effects of solvation.

Chapter 5

5.1 $I = 0.003\,\text{mol}\,\text{dm}^{-3}$; $1/\kappa = (1.988 \times 10^{-12})\sqrt{\varepsilon T/I}$ (see question 2.1).
Therefore here $1/\kappa = \delta = 5.55$nm.

5.2 Use the Smoluchowski equation in the form

$$\zeta = \frac{\eta v}{\varepsilon_0\varepsilon V} = \frac{\eta u_0}{\varepsilon_0\varepsilon} \quad \text{(see equation (5.14))}$$
$$= \frac{8.94 \times 10^{-4} \times 3.5 \times 10^{-8}}{78.5 \times 8.8542 \times 10^{-12}}$$

Therefore

$$\zeta = 44.8\,\text{mV}$$

Since $I = 0.04\,\text{mol}\,\text{dm}^{-3}$;

$$\frac{1}{\kappa} = 1.988 \times 10^{-12}\sqrt{\frac{78.5 \times 298}{0.04}}$$
$$= 1.520 \times 10^{-9}\,\text{m}$$

$$\therefore \kappa a = 0.658 \times 10^9 \times 2.5 \times 10^{-7} = 165$$

When such values of $\kappa a \gg 1$ operate, the Smoluchowski equation is applicable; it does not apply, however, to small particles such as were encountered in Chapter 2 with a characteristically of the order of 0.4 nm (and $\kappa a \sim 0.26$ for $I = 0.04\,\text{mol dm}^{-3}$).

5.3 Use equation (5.18) directly to yield a value for Φ of $5.89 \times 10^{-5}\,\text{cm}^3\,\text{s}^{-1}$.

Chapter 6

6.1 (i) Left hand
$Zn \longrightarrow Zn^{2+} + 2e$
Right hand
$Cu^{2+} + 2e \longrightarrow Cu$
Overall, $Zn + Cu^{2+} \longrightarrow Zn^{2+} + Cu$

(ii) Left hand
$Pb \longrightarrow Pb^{2+} + 2e$
Right hand
$Hg_2Cl_2 + 2e \longrightarrow 2Hg + 2Cl^-$
Overall, $Pb + Hg_2Cl_2 \longrightarrow 2Hg + Pb^{2+} + 2Cl^-$

(iii) Left hand
$Zn \longrightarrow Zn^{2+} + 2e$
Right hand
$2Fe^{3+} + 2e \longrightarrow 2Fe^{2+}$
Overall, $Zn + 2Fe^{3+} \longrightarrow Zn^{2+} + 2Fe^{3+}$

$\gamma_{\pm(ZnSO_4)} = 0.458$ (ionic strength $= 0.04\,\text{mol kg}^{-1}$),
$\gamma_{\pm(CuSO_4)} = 0.757$ (ionic strength $= 0.004\,\text{mol kg}^{-1}$),
$\gamma_{\pm(Pb(NO_3)_2)} = 0.708$ (ionic strength $= 0.03\,\text{mol kg}^{-1}$).

(i) Applying the Nernst equation (equation (6.23)) for both half-cells,

$$E_{Zn^{2+}/Zn} = -0.760 + \frac{0.0257}{2}\ln(0.01 \times 0.458) = -0.829\text{ V}$$

$$E_{Cu^{2+}/Cu} = +0.340 + \frac{0.0257}{2}\ln(0.001 \times 0.757) = +0.248\text{ V}$$

whence $E_{cell} = +0.248 - (-0.829) = 1.077$ V.

(ii) $E_{cell} = +0.242 - (-0.194) = 0.436$V
(iii) For the redox half-cell

$$E_{Fe^{3+}/Fe^{2+}} = E^{\ominus} + \frac{RT}{F}\ln\frac{a_{Fe^{3+}}}{a_{Fe^{2+}}}$$
$$= +0.760 + 0.0257\ln\frac{[Fe^{3+}]\,\gamma_{Fe^{3+}}}{[Fe^{2+}]\,\gamma_{Fe^{2+}}}$$

Since $-\log\gamma_{M^{z+}} = z^2A\sqrt{I}/(1+\sqrt{I})$ and here the ionic strength is $0.022\,\text{mol kg}^{-1}$, it may be shown that $\gamma_{Fe^{3+}} = 0.256$ and $\gamma_{Fe^{2+}} = 0.546$. Therefore, $E_{Fe^{3+}/Fe^{2+}} = +0.760 + 0.0257\ln(0.256/0.546) = +0.741$V.
Thus $E_{cell} = +0.741 - (-0.829) = 1.570$V.

6.2 (i) $Pt|FeSO_4(aq.), Fe_2(SO_4)_3(aq.), H^+||Ce_2(SO_4)_3(aq.), Ce(SO_4)_2(aq.), H^+|Pt$

$$E^{\ominus} = 0.850\text{ V}$$

(ii) $Zn|ZnSO_4(aq.)||Cl^-(aq.), AgCl, Ag$
$E^{\ominus} = 0.982$ V.

6.3 $E^{\ominus}_{AgCl} = E^{\ominus}_{cell} = 0.222$ V (since $E^{\ominus}_{H_2} = 0.00$ V)

6.4 $E_{right} - E_{left} = \left[0.222 - \frac{RT}{F}\ln m\gamma_{Cl^-}\right] - 0.242 = 0.051$ V, giving $\gamma_{Cl^-} = 0.792$.
This compares with the values of 0.718 obtained from the limiting law and 0.772 obtained from the extended law.

6.5 $E^{\ominus}_{\text{cell}} = 0.014\,\text{V}; \Delta G^{\ominus} = -nE^{\ominus}F = -2.7\text{kJ mol}^{-1}$

6.6 Liquid junction potential $= 0.004\,\text{V} = (2t_- - 1)RT/F \ln(0.0735/0.00898)$, yields $t_- = 0.537$ and $t_+ = 0.463$.

6.7 The situation, both initially and after equilibrium is reached, may be summarized, for the general case involving protein Na_yR, as

NaCl, initial concn. $= C_1$	Na_yCl, initial concn. $= C_2$
At equilibrium have:	At equilibrium have:
Na^+ Cl^-	yNa^+ R^{y-} Cl^-
$(C_1 - x)$ $(C_1 - x)$	$(yC_2 + x)$ C_2 x

Therefore, $(C_1 - x)(C_1 - x) = C_2x$ (for $y = 1$) and $C_2x + 2C_1x - C_1^2 = 0$ from which the required relationship follows.

6.8 (i) $C_1 = 0.1\,\text{mol dm}^{-3}: yC_2 = 10 \times 0.01 = 0.1\text{mol dm}^{-3}$
$x/C_1 = 0.1/0.1 + 0.2 = 0.333$ (using expression derived in question 6.7)
(ii) 0.286
(iii) 0.250

6.9 (i) Using equation (6.70) and the scheme from which it derives,

$$x = (K_w C)^{1/3} = [H^+]_{\text{inside}} = [OH^-]_{\text{outside}}$$

$$\therefore pH_{\text{inside}} = -\log[H^+] = -1/3[\log K_w + \log C] = -1/3[-14 - 1] = 5$$

$$pOH_{\text{outside}} = 5, \text{therefore } pH_{\text{outside}} = 9.$$

(ii)
$$\Delta\phi = \frac{RT}{F} \ln \frac{[H^+]_{\text{inside}}}{[H^+]_{\text{outside}}} = \frac{2.3RT}{F}[pH_{\text{outside}} - pH_{\text{inside}}]$$
$$= 0.236\,\text{V}$$

6.10 Schematically, the situation at equilibrium is:

R^- H^+ Cl^-	H^+ Cl^-
(x) $(x + y - z)$ $(y - z)$	z z

where x, y are the initial concentrations of RH and HCl, z is the equilibrium concentration of HCl in the right hand compartment. At equilibrium $[H^+]_L[Cl^-]_L = [H^+]_R[Cl^-]_R$

$$(x + y - z)(y - z) = z^2$$

$pH_L = 2.72 = -\log(x + y - z)$; $pH_R = 3.37 = -\log z$,
and since $\log(x + y - z) + \log(y - z) = 2\log z$.
$\log(y - z) = 2.72 - 6.74 = -4.02$. Therefore $(y - z) = 9.55 \times 10^{-5}\,\text{mol dm}^{-3}$
but since $\log(x + y - z) = -2.72$, $(x + y - z) = 1.905 \times 10^{-3}\,\text{mol dm}^{-3}$.
Thus $x = 1.810 \times 10^{-3}\,\text{mol dm}^{-3}$, and relative molecular mass of the acid $= (15 \times 10^3)/(1.81) = 8290$.

6.11 ${}^{\text{I}}m_+{}^{\text{I}}m_- = {}^{\text{II}}m_+{}^{\text{II}}m_- = {}^{\text{II}}m^2 : {}^{\text{I}}m_+ = z^{\text{I}}m_{R^{z-}} + {}^{\text{I}}m_-$

Therefore $^{II}m^2 = (z^{I}m_{R^{z-}} + {}^{I}m_-){}^{I}m_-$

$$\approx \left({}^{I}m_- + \frac{z^{I}m_{R^{z-}}}{2}\right)^2 \text{ if square terms in } {}^{I}m_{R^{z-}} \text{ are neglected.}$$

Therefore ${}^{I}m_- \cong {}^{II}m - \dfrac{z^{I}m_{R^{z-}}}{2}$

Therefore $\Delta\phi \cong \dfrac{RT}{F}\ln\dfrac{{}^{I}m_-}{{}^{II}m_-} = \dfrac{RT}{F}\ln\left(\dfrac{{}^{II}m_- - \frac{z^{I}m_{R^{z-}}}{2}}{{}^{II}m}\right) \approx \dfrac{RT}{2F}\dfrac{z^{I}m_{R^{z-}}}{{}^{II}m}$

6.12 Graph of $\Delta\phi$ versus $m_{Hb^{z-}}$ has slope equal to $(RT/2F)(z/{}^{II}m)$ and has the value -0.240 V kg mol^{-1} so that $z = -8.5$.

Chapter 7

7.1 $\eta = a + b\log i$

$$a = -\frac{0.0591}{\alpha}\log i_0;\ b = \frac{0.0591}{\alpha},\ \text{therefore } \alpha = \frac{0.0591}{0.119} = 0.497$$

$$1.54 = -\frac{0.0591}{0.497}\log i_0,\ \text{therefore } \log i_0 = -12.951 \text{ and } i_0 = 1.12 \times 10^{-13}\,\text{A cm}^{-2}$$

7.2 For current densities i_1, i_2, overvoltages η_1, η_2 may be expressed as

$$\eta_1 = a + b\log i_1 \text{ and } \eta_2 = a + b\log i_2$$

Therefore

$$(\eta_2 - \eta_1) = b(\log i_2 - \log i_1)$$

$$0.394 - 0.148 = b(-2.000 - (-4.000)),\ \text{therefore } b = 0.123$$

$$0.394 = a + (0.123 \times -2.00),\ a = 0.640\text{ V.}$$

$$b = 0.123 = \frac{0.0591}{\alpha},\ \text{therefore } \alpha = 0.480$$

$$a = -\frac{0.0591}{0.48}\log i_0,\ \text{therefore } \log i_0 = -5.198;$$

$$i_0 = 6.34 \times 10^{-6}\,\text{A cm}^{-2}$$

7.3 Draw up a table of η and $\ln i$ as follows

η (V)	0.02	0.05	0.07	0.10	0.12	0.15	0.20
i (mA cm^{-2})	2.13	6.63	11.35	23.45	37.26	73.85	229.08
$\ln i$ (mA cm^{-2})	0.756	1.892	2.429	3.155	3.618	4.302	5.434

Plot η versus $\ln i$. Slope $= -22.78 = -\alpha F/RT$, therefore $\alpha = 0.58$.
Intercept for linear portion $= 0.90 = \ln i_0$, therefore $i_0 = 2.45$ mA cm^{-2}

7.4 $E = E^{\ominus} = -0.760$ V, for i = 1 mA cm^{-2}
$\eta = 0.280 + 0.0591\log(0.001) = 0.280 - 0.177 = 0.101$ V
Therefore plating occurs at $-0.760 - 0.101 = -0.861$ V
In neutral solution pH = 7, therefore $[H^+] \sim 10^{-7}$ mol dm^{-3}

$$E = E^{\ominus} = \frac{RT}{F}\ln a_{H^+} \sim 0.0591\log(10^{-7}) = -0.414\text{ V}$$

Thus zinc cannot be plated at −0.861 V because hydrogen is liberated at voltages 0.45 V more positive than this.

7.5
$$\frac{i}{i_0} = \exp\left[\frac{-\alpha F\eta}{RT}\right] - \exp\left[\frac{(1-\alpha)F\eta}{RT}\right] \quad \text{(see equation (7.18))}$$

$$\text{for } \eta = 0.07; i/i_0 = 0.206 - 3.143 = -2.937$$

$$i = 2.937 \times 2.5 = 7.34\,\text{mA cm}^{-2}$$

7.6 Draw up the following table

η (m V)	20	50	70	100	120	150	200	250
$\ln i$ (mA cm^{-2})	−0.562	0.336	0.718	1.212	1.517	1.969	2.711	3.452

Intercept of linear portion = −0.25 = $\ln i_0$

$$i_0 = 0.779\,\text{mA cm}^{-2}$$

$$\frac{\ln i}{\eta} = \frac{\alpha F}{RT} = 38.95\alpha = 14.808\,\text{V}^{-1} \text{ (from graph)}$$

Therefore $\alpha = \dfrac{14.808}{38.95} = 0.38$

7.7 Using equation (7.18) the following table may be drawn up for the anodic and cathodic overvoltages corresponding to the values of applied potential to be considered.

E (V)	1.30	1.40	1.50	1.61	1.70	1.80	1.90
η (V)	0.31	0.21	0.11	0.00	−0.09	−0.19	−0.29
i/i_0	−20.46	−7.73	−2.88	0.00	13.45	257.18	4777.21
I (mA)	−0.82	−0.30	−0.12	0.00	0.54	10.29	191.1

(Taking anodic currents as negative)

7.8
$$\eta = \frac{-2.303RT}{\alpha nF}\ln i_0 + \frac{2.303RT}{\alpha nF}\ln i$$

Therefore

$$\eta = 0.278 + 0.059\log i, \quad a = 0.278, \quad b = 0.059$$

$$\eta = 0.278 + 0.059\log(5 \times 10^{-3}) = 0.142\,\text{V}.$$

Chapter 8

8.1 Plot $\log K$ versus $\sqrt{I}$ according to equation (8.2).
Intercept = −4.0912 = $\log K_T$; $K_T = 8.11 \times 10^{-5}$

8.2 Graphs of $\log k$ versus $\sqrt{I}$ are linear and lie on the same line for data obtained with both electrolytes. Effect is therefore confirmed as due to the ionic medium. Slope $\sim$ 1.83, i.e. $z_A z_B \sim 2$, according to equation (8.10), consistent with the rate-determining step being reaction between $S_2O_8^{2-}$ and I^-.

8.3 $\kappa_{AgCl} = 1.853 \times 10^{-6}\,\Omega^{-1}\,\text{cm}^{-1}$; $\Lambda^{\infty}_{AgCl} = 138.25\,\Omega^{-1}\,\text{cm}^2\,\text{mol}^{-1}$

$$\text{Solubility} = \frac{1.853 \times 10^{-3}}{138.25} = 1.340 \times 10^{-5}\,\text{mol dm}^{-3}$$

Thus concentration solubility product = $1.796 \times 10^{-10}\,\text{mol}^2\text{dm}^{-6}$

$$I = 1.34 \times 10^{-5} \text{mol dm}^{-3} \text{ and therefore } \gamma_{\pm} = 0.996$$

Thermodynamic $K_s = (1.796 \times 10^{-10})(0.996)^2 = 1.78 \times 10^{-10}$.

8.4 $C = \dfrac{\kappa}{\Lambda} = 1.003 \times 10^{-4} \text{ mol m}^{-3} = 1.003 \times 10^{-7} \text{ mol dm}^{-3}, K_w = 1.006 \times 10^{-14}$.

8.5 $\alpha \cong \dfrac{\Lambda}{\Lambda_0}$; plot of $\left(\dfrac{\alpha^2}{1-\alpha}\right)$ versus $1/C$ (equations (4.6), (4.8))
yields $K_a = 1.74 \times 10^{-5} \text{ mol dm}^{-3}$ as slope.
The expression for α is approximate and the value of K_a is a concentration constant

8.6 Plot of E versus T (graph shows clear curvature) gives

$$\left(\frac{\partial E}{\partial T}\right)_P = -0.00148 \text{V K}^{-1} \text{ at } 308 \text{ K}.$$

$$\Delta G_{308 \text{ K}} = -nEF = -271.4 \text{ k J mol}^{-1}$$

$$\Delta H_{308 \text{ K}} = -nEF + TnF\left(\frac{\partial E}{\partial T}\right)_P = -359.4 \text{ kJ mol}^{-1}$$

$$\Delta S_{308 \text{ K}} = nF\left(\frac{\partial E}{\partial T}\right)_P = -285.6 \text{ J K}^{-1} \text{ mol}^{-1}$$

8.7

$$E_{\text{AgBr}} = E^{\ominus}_{Ag} + \frac{RT}{F} \ln a_{\text{Ag}^+} \quad \text{(see equation (8.61))}$$

$$= E^{\ominus}_{\text{Ag}} + \frac{RT}{F} \ln K_{\text{AgBr}} - \frac{RT}{F} \ln a_{\text{Br}^-} \quad \text{(see equation (8.62))}$$

$$= E^{\ominus}_{\text{AgBr}} - \frac{RT}{F} \ln a_{\text{Br}^-}$$

$$\text{Or, } E^{\ominus}_{\text{AgBr}} - E^{\ominus}_{\text{Ag}} = \frac{RT}{F} \ln K_{\text{AgBr}} = -0.726 \text{ V}$$

Therefore

$$\ln K_{\text{AgBr}} = -\frac{0.726}{0.0257} = -28.249$$

$$K_{\text{AgBr}} = 5.390 \times 10^{-13} \sim [\text{Ag}^+][\text{Br}^-]$$

$$\text{Solubility} = 7.34 \times 10^{-7} \text{mol kg}^{-1}$$

8.8 A plot of $[E_{\text{cell}} + (2RT/F) \ln m]$ versus $m^{1/2}$ gives an intercept of $E^{\ominus}$. At $m^{1/2} = 0$ (cquation (8.24)) $E^{\ominus} = 0.2225$ V. Substitution into equation (8.24) gives $\gamma_{\pm} = 0.798$.

8.9 At the start for every 10 g water there is $0.0758/170 = 0.0004459$ mole of silver; at end for every 28.755 g water there is $0.2701/170 = 0.0015888$ mole of silver (n). If the solution had remained unchanged in composition, 28.755 g of water would have been associated with $(0.0004459) \times (28.755)/10 = 0.0012822$ mole of silver (n_0). Copper deposited in coulometer $= 0.01857/31.75 = 0.0005849$ mole (n_e).
By equation (8.34),

$$t_+ = \frac{n_e + n_0 - n}{n_e}$$

whence

$$t_{Ag^+} = 0.476; t_{NO_3^-} = 0.524$$

8.10 $C = 10 \text{mol m}^{-3}$; $V = 1.4139 \times 10^{-7} \text{m}^3$; $It = 1.65$ coulombs. By equation (8.35) $t_{H^+} = 0.827$.

8.11

$$\begin{aligned} E_{cell} &= E_{AgCl} - EH_2 \\ &= E^{\ominus}_{AgCl} - \frac{RT}{F} \ln a_{Cl^-} - \frac{RT}{F} \ln a_{H^+} \quad \text{(see equation (8.22))} \\ &\cong E^{\ominus}_{AgCl} - \frac{2RT}{F} \ln a_{H^+} (\text{assuming } \gamma_{H^+} \approx \gamma_{Cl^-}) \end{aligned}$$

Therefore
$0.4196 - 0.2225 = -0.1182 \log a_{H^+}$, giving pH = 1.668.
By the Debye–Hückel limiting law

$$\log \gamma_{H^+} \approx -0.509 \times 1^2 \times \sqrt{0.025} = -0.0805; \gamma_{H^+} \approx 0.8308$$

Therefore

$$a_{H^+} \approx 0.02077, \text{ giving pH} \approx 1.683$$

8.12 Data may be substituted into equation (8.49) assuming that the value of the activity quotient is approximately unity. Thus

$$\frac{0.6253 - 0.2225}{2.303RT/F} + \log \left[\frac{0.0079 \times 0.0082}{0.0075} \right] \approx -\log K_a$$

whence $K_a = 1.77 \times 10^{-5}$.

8.13 Isochore equation $\mathrm{d} \ln K_w / \mathrm{d}T = \Delta H_i / RT^2$ or $\ln K_w = -\Delta H_i / RT + \text{const}$. A graph of K_w versus $1/T$ yields a slight curve of negative slope from which ΔH_i may be estimated at the three temperatures as:
58.1 k J mol^{-1}(288 K), 56.4 k J mol^{-1}(298 K), and 54.6 k J mol^{-1}(308 K)
Heat of neutralization at 298 K = -56.4 k J mol^{-1}.

Chapter 9

9.1

$$\begin{aligned} i_{lim} &= \frac{DnF[Ag^+]}{\delta} \quad \text{(see equation (9.1))} \\ &= \frac{1.64 \times 10^{-5} \times 96500 \times 10^{-4}}{0.05} = 3.17 \text{ mA cm}^{-2} \end{aligned}$$

For rapid stirring, $\delta \sim 0.001$ cm, so i_{lim} is increased 50-fold to about 158.5 mA cm^{-2}.

9.2 A plot of $\log\{(\bar{I}_d - \bar{I})/\bar{I}\}$ versus E according to the Heyrovsky–Ilkovic equation (see equation (9.3)) gives $E_{1/2} = -1.00$ V vs. SCE. Slope = 0.0293 = $0.0591/n$, therefore $n = 2$.

$$\begin{aligned} D &= \left[\frac{\bar{I}_d}{607nm^{2/3}t^{1/6}C} \right]^2 \quad \text{from equation (9.2)} \\ &= 7.19 \times 10^{-6} \text{ cm}^2 \text{ s}^{-1} \end{aligned}$$

9.3 The data are characteristic of an irreversible reduction. The same plot as for the previous question yields a value for the slope of $0.056 = 0.0591/\alpha n$. Since the limiting current is clearly diffusion controlled, its magnitude indicates that $n = 2$, so that $\alpha n = 1.06$.

9.4 Let the unknown concentration be C_1 and that of the standard solution be C_s: then by the Ilkovic equation

$$C_1 = \frac{I_1}{k}; \quad \Delta C = \frac{\Delta I}{k}$$

After addition of volume v of standard solution of concentration C_s to a volume V of the working solution of concentration C_1, the new concentration is

$$\left[C_s\left(\frac{v}{V+v}\right) + C_1\left(\frac{V}{V+v}\right)\right]$$

$$\therefore \Delta C = \left[C_s\left(\frac{v}{V+v}\right) + C_1\left(\frac{V}{V+v}\right)\right] - C_1$$

$$= (C_s - C_1)\left(\frac{v}{V+v}\right)$$

$$\therefore k = \frac{\Delta I}{\Delta C} = \frac{\Delta I(V+v)}{v(C_s - C_1)}$$

$$\therefore C_1 = \frac{I_1}{k} = \frac{I_1 v(C_s - C_1)}{\Delta I(V+v)}$$

which leads to

$$C_1 = \frac{I_1 v C_s}{\Delta I(V+v) + I_1 v}$$

Substitution of the experimental data into this last equation gives

$$C_{Ni^{2+}} = \frac{1.97 \times 4 \times 9.24 \times 10^{-3}}{(1.98 \times 79) + (1.97 \times 4)}$$

$$= 4.43 \times 10^{-4}\,\text{mol dm}^{-3}$$

$$= 0.026\,\text{g dm}^{-3}\,\text{Ni}^{2+}$$

$$= 0.0026\,\text{g Ni}^{2+}\text{ per }100\,\text{cm}^3$$

$$\therefore \%\text{Ni}^{2+} = \frac{0.0026 \times 100}{3} = 0.087\%$$

9.5 From equation (9.4) the shift of half-wave potential $\Delta E_{1/2}$ induced by ligand concentration [X] to form a complex MX_N is given by:

$$\frac{0.4343nF}{RT} \cdot \Delta E_{1/2} \sim \log \beta_N + N \log[X]$$

To apply the equation it must be assumed that both aqua and complexed ion are reduced reversibly. Then if complexation extends to formation of MX, substitution of data yields $\beta_1 = 2.9 \times 10^7$: if MX_2 is formed, it is found that $\beta_2 = 1.5 \times 10^8$.

Chapter 11

11.1 (i) $E^{\ominus}_{Zn^{2+}/Zn} = -0.76\,\text{V}$; $E^{\ominus}_{Cl_2/Cl^-} = +1.36\,\text{V}$ (see Table 6.1)
$E^{\ominus}_{cell} = 2.12\,\text{V}$.

(ii) Energy stored = 90 kWh. = 90 × 1000 × 60 × 60 J
Voltage of 118 cells = 250 V.

$$\text{Quantity of electricity stored} = \frac{90 \times 1000 \times 60 \times 60}{250}$$

$$= 1.296 \times 10^6 \text{C}$$

$$\text{Amount of } Cl_2 \text{ required} = \frac{1.296 \times 10^6}{2 \times 96500} = 6.72 \text{ mol } Cl_2 = 477.12 \text{ g } Cl_2$$

Corresponding amount of hydrate required = 638.4g.

11.2 (i) The overall reaction involving methane derives from the two half-cell reactions

$$CH_4 + 2H_2O = CO_2 + 8H^+ + 8e$$

$$2O_2 + 8H^+ + 8e = 4H_2O$$

$$\overline{CH_4 + 2O_2 = CO_2 + 2H_2O}$$

Similar treatment of all the cases cited yields the values 8, 14, 20 and 6 for the number of electrons transferred in each cell reaction.

(ii) Using $\Delta G^\ominus = -nE^\ominus F$ (see equation (6.19)), the respective $E^\ominus$ values are: 1.060 V, 1.086 V, 1.092 V, 1.221 V.

(iii) Since efficiency = $\Delta G/\Delta H$, the respective maximum efficiencies are: 91.9%, 94.0%, 95.0%, 92.5%.

Appendix I The electrical potential in the vicinity of an ion

By definition, the electrical potential, ϕ, at some point is the work done in bringing a unit positive charge from infinity (where $\phi = 0$) to that point (Figure I.i).

The concentration of positive and negative ions (N_+, N_-) at the point where the potential is ϕ may be found from the Boltzmann distribution law, thus

$$N_+ = N_+^0 \, \mathrm{e}^{-(z_+ \epsilon \phi / kT)}$$

and

$$N_- = N_-^0 \, \mathrm{e}^{+(z_- \epsilon \phi / kT)} \qquad \text{(I.i)}$$

where ϵ is the unit electronic charge, k the Boltzmann constant, z_+, z_- the number of charges carried by positive and negative ions respectively, and N_+^0, N_-^0 the number of ions of each type per unit volume in the bulk.

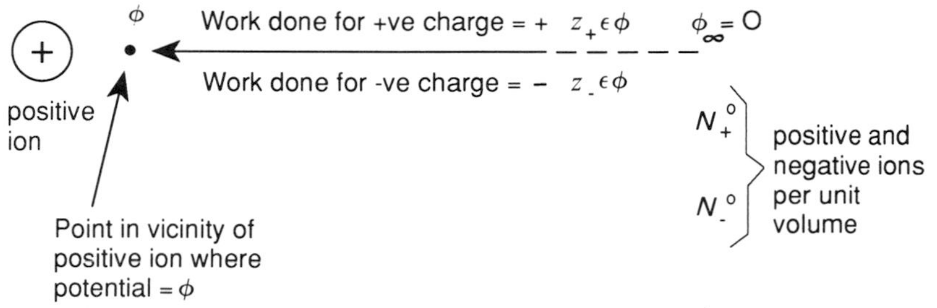

Figure I.i The work of bringing charges from infinity ($\phi_\infty = 0$) to a point near a selected ion where the potential = ϕ.

It is seen that these equations are consistent with the expected fact that there are, on average, more negative ions than positive ions in the vicinity of a given positive ion and vice versa.

The electrical density (ρ) at the point where the potential is ϕ is the excess positive or negative electricity per unit volume at that point. It is easily seen that for the present case this must be

$$\begin{aligned} \rho &= N_+ z_+ \epsilon - N_- z_- \epsilon \\ &= N_+^0 z_+ \epsilon \, \mathrm{e}^{-(z_+ \epsilon \phi / kT)} - N_-^0 z_- \epsilon \, \mathrm{e}^{+(z_- \epsilon \phi / kT)} \end{aligned} \qquad \text{(I.ii)}$$

For the simplest case of a 1:1 electrolyte

$$z_+ = z_- = 1$$

and

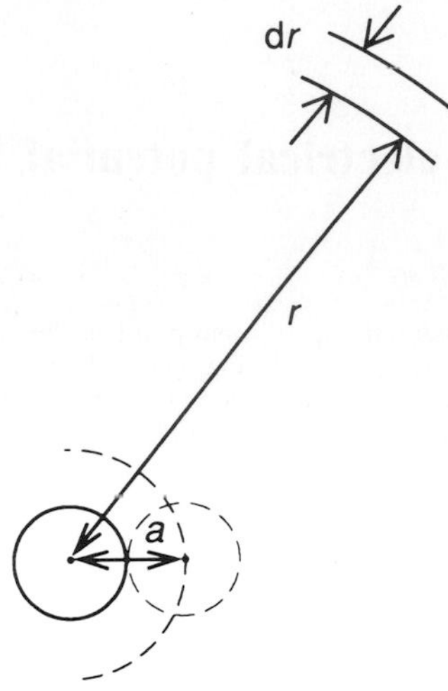

Figure I.ii Model for calculation of charge of an ion atmosphere about a central ion.

$$N_+^0 = N_-^0 = N_i$$

i.e.

$$\rho = N_i \epsilon \left[e^{-\epsilon\phi/kT} - e^{\epsilon\phi/kT} \right] \tag{I.iii}$$

If it is assumed that $\epsilon\phi/kT \ll 1$, the two exponential terms may be expanded and all but the first terms in the expansions neglected so that equation (I.iii) becomes

$$\rho \sim -2N_i \left[\frac{\epsilon^2 \phi}{kT} \right] \tag{I.iv}$$

For the more general case where $z_+, z_- \neq 1$, equation (I.iv) becomes modified to

$$\rho \sim -\Sigma N_i z_i^2 \left[\frac{\epsilon^2 \phi}{kT} \right] \tag{I.v}$$

The electrostatic potential and charge density are also related in the Poisson equation, viz.

$$\frac{\partial^2 \phi}{\partial x^2} + \frac{\partial^2 \phi}{\partial y^2} + \frac{\partial^2 \phi}{\partial z^2} = -\frac{\rho}{\varepsilon_0 \varepsilon} \tag{I.vi}$$

where ε_0 is the permittivity of a vacuum ($8.854 \times 10^{-12}\,C^2\,N^{-1}\,m^{-2}$) and ε is the relative permittivity, or dielectric constant, of the solvent. x, y, z are the rectangular coordinates of the point at which the potential is ϕ. In terms of polar coordinates equation (I.vi) becomes

$$\frac{1}{r^2} \frac{\partial}{\partial r} \left(r^2 \frac{\partial \phi}{\partial r} \right) = -\frac{\rho}{\varepsilon_0 \varepsilon} \tag{I.vii}$$

Substituting for ρ from equation (I.v) we have that

$$\frac{1}{r^2} \frac{\partial}{\partial r} \left(r^2 \frac{\partial \phi}{\partial r} \right) = \frac{\epsilon^2}{\varepsilon_0 \varepsilon k T} \Sigma N_i z_i^2 \tag{I.viii}$$

which we will express as $\kappa^2 \phi$, where

$$\kappa = \left[\frac{\epsilon^2 \Sigma N_i z_i^2}{\varepsilon_0 \varepsilon k T}\right]^{1/2} \quad \text{(I.ix)}$$

A general solution of equation (I.viii) takes the form

$$\phi = \frac{A\,\mathrm{e}^{-\kappa r}}{r} + \frac{A'\mathrm{e}^{\kappa r}}{r} \quad \text{(I.x)}$$

in which A, A' are integration constants. The second term may, in fact, be ignored, since as $r \to \infty, \phi \to 0$, thus A' must be zero, i.e. ϕ must be finite even for very large values of r. Thus,

$$\phi = \frac{A\,\mathrm{e}^{-\kappa r}}{r} \quad \text{(I.xi)}$$

and since

$$\kappa^2 \phi = -\frac{\rho}{\varepsilon_0 \varepsilon}$$

substitution of ϕ from equation (I.xi) into the expression

$$\rho = -\kappa^2 \phi \varepsilon_0 \varepsilon$$

yields

$$\rho = -A\left(\frac{\kappa^2 \varepsilon_0 \varepsilon}{r}\right)\mathrm{e}^{-\kappa r} \quad \text{(I.xii)}$$

For electroneutrality, the total negative charge of the atmosphere about a given positively charged central ion is $-z_i\epsilon$. The total charge of the atmosphere is determined by considering the charge carried by a spherical shell of thickness dr and distance r from the central ion and integrating from the closest distance that atmosphere and central ions may approach out to infinity (Figure I.ii). Thus

$$\int_a^\infty 4\pi r^2 \rho\,\mathrm{d}r = -z_i\epsilon \quad \text{(I.xiii)}$$

Therefore,

$$A\kappa^2 \varepsilon_0 \varepsilon \int_a^\infty 4\pi r\,\mathrm{e}^{-\kappa r}\,\mathrm{d}r = z_i\epsilon \quad \text{(I.xiv)}$$

Integration by parts gives A as

$$A = \left(\frac{z_i\epsilon}{4\pi\varepsilon_0\varepsilon}\right)\left(\frac{\mathrm{e}^{\kappa a}}{1+\kappa a}\right) \quad \text{(I.xv)}$$

So that the potential ϕ may now be expressed by

$$\phi = \left(\frac{z_i\epsilon}{4\pi\varepsilon_0\varepsilon}\right)\left(\frac{\mathrm{e}^{\kappa a}}{1+\kappa a}\right)\left(\frac{\mathrm{e}^{-\kappa r}}{r}\right) \quad \text{(I.xvi)}$$

When r approaches a, the distance of closest approach, equation (I.xvi) becomes

$$\phi = \left(\frac{z_i\epsilon}{4\pi\varepsilon_0\varepsilon a}\right)\left(\frac{1}{1+\kappa a}\right) = \left(\frac{z_i\epsilon}{4\pi\varepsilon_0\varepsilon a}\right) - \left(\frac{z_i\epsilon}{4\pi\varepsilon_0\varepsilon}\right)\left(\frac{\kappa}{1+\kappa a}\right) \quad \text{(I.xvii)}$$

or, in the most general terms

$$\phi = \pm\left(\frac{z_i\epsilon}{4\pi\varepsilon_0\varepsilon a}\right) \mp \left(\frac{z_i\epsilon}{4\pi\varepsilon_0\varepsilon}\right)\left(\frac{\kappa}{1+\kappa a}\right) \quad \text{(I.xviii)}$$

or,

$$\phi = \phi_0 + \phi_i \quad \text{(I.xvix)}$$

where ϕ_0 is the contribution of the ion itself to ϕ while ϕ_i is the contribution of its atmosphere.

Appendix II Significance of the constant κ in the Debye–Hückel equation

In Appendix I, κ was expressed as

$$\kappa = \left[\frac{\epsilon^2}{\varepsilon_0 \varepsilon k T} \Sigma N_i z_i^2\right]^{1/2} \qquad \text{(see I.ix)}$$

in which $N_i = NC_i$, where N is the Avogadro constant and C_i is the ion concentration in $\mathrm{mol\,m^{-3}}$. Thus,

$$\kappa = \left(\frac{\epsilon^2 N}{\varepsilon_0 \varepsilon k T} \Sigma C_i z_i^2\right)^{1/2}$$

or

$$\kappa = \left(\left(\frac{2\epsilon^2 N}{\varepsilon_0 \varepsilon k T}\right) \tfrac{1}{2}\Sigma C_i z_i^2\right)^{1/2} \qquad \text{(II.i)}$$

It is seen that equation (II.i) contains the expression $\frac{1}{2}\Sigma C_i z_i^2$. This is very similar in form to the expression defining the ionic strength, I, of the solution, viz.

$$I = \tfrac{1}{2}\Sigma m_i z_i^2 \qquad \text{(II.ii)}$$

where m_i represents the concentration of each ion of the electrolyte in the units $\mathrm{mol\,kg^{-1}}$.

Now if C_i in $\mathrm{mol\,m^{-3}}$, as used above, is replaced by c_i in $\mathrm{mol\,dm^{-3}}$, then $N_i = 10^3\,Nc_i$, and if the solution is of such dilution that $1\mathrm{dm^3}$ corresponds closely to $1\mathrm{dm^3}$ of pure solvent, i.e. 1 kg for the case of water, we may write

$$N_i = 10^3\,Nm_i$$

Thus equation (II.i) becomes

$$\kappa = \left(\left(\frac{2 \times 10^3 \epsilon^2 N}{\varepsilon_0 \varepsilon k T}\right) \tfrac{1}{2}\Sigma m_i z_i^2\right)^{1/2}$$

or

$$\kappa = \left(\frac{2 \times 10^3 \epsilon^2 N}{\varepsilon_0 \varepsilon k T}\right)^{1/2} \sqrt{I} \qquad \text{(II.iii)}$$

Appendix III Derivation of the Lippmann equation

The expression $d\gamma = \Sigma\Gamma_i d\tilde{\mu}_i = 0$ (see equation (5.3)) may be applied to the four distinct phases of the following cell

$$\begin{array}{cccc} Pt(H_2) \mid & HCl \mid & Hg \mid & Pt \\ 1 & 2 & 3 & 4 \end{array}$$

in which the hydrogen electrode may be regarded as non-polarizable and the mercury electrode as ideally polarizable.

Application of equation (5.3) to the interface between phase 2 and phase 3, gives

$$-d\gamma = [\Gamma_{Hg^+} d\tilde{\mu}_{Hg^+} + \Gamma_{e^-} d\tilde{\mu}_{e^-}] + [\Gamma_{H_3O^+} d\tilde{\mu}_{H_3O^+} \Gamma_{Cl^-} d\tilde{\mu}_{Cl^-} + \Gamma_{H_2O} d\mu_{H_2O}] \qquad \text{(III.i)}$$

Now, for an electrolyte, the chemical potential is the sum of the potentials of its component ions, e.g.

$$\mu = \nu_+\mu_+ + \nu_-\mu_- \qquad \text{(III.ii)}$$

therefore,

and

$$\left.\begin{array}{ll} \mu_{Hg} = \tilde{\mu}_{Hg^+} + \tilde{\mu}_{e^-} & \text{in phase 3} \\ \mu_{HCl} = \tilde{\mu}_{H_3O^+} + \tilde{\mu}_{Cl^-} & \text{in phase 2} \end{array}\right\} \qquad \text{(III.iii)}$$

therefore,

$$-d\gamma = [\Gamma_{Hg^+} d\mu_{Hg} - (\Gamma_{Hg^+} - \Gamma_{e^-}) d\tilde{\mu}_{e^-}] + [\Gamma_{Cl^-} d\mu_{HCl} + (\Gamma_{H_3O^+} - \Gamma_{Cl^-}) d\tilde{\mu}_{H_3O^+} + \Gamma_{H_2O}\, d\mu_{H_2O}] \qquad \text{(III.iv)}$$

Now

$$\Gamma_i = \frac{n_i}{A} \qquad \text{(III.v)}$$

i.e. the number of particles of species i per unit area. It is possible to express the total number of charges per unit area, adsorbed by the interface from phase 2 and phase 3 as

and

$$\left.\begin{array}{r} (\Gamma_{Hg^+} - \Gamma_{e^-})F = \sigma_3 \\ (\Gamma_{H_3O^+} - \Gamma_{Cl^-})F = \sigma_2 \end{array}\right\} \qquad \text{(III.vi)}$$

For electroneutrality at the interface

$$\sigma_2 + \sigma_3 = 0 \qquad \text{(III.vii)}$$

Also, for equilibrium to be maintained across the interface between mercury and platinum (phase 3 and phase 4)

$$(d\tilde{\mu}_{e^-})_3 = (d\tilde{\mu}_{e^-})_4 \qquad \text{(III.viii)}$$

Similarly, for equilibrium across the interface between phases 1 and 2,

$$(d\tilde{\mu}_{H_3O^+})_2 = -(d\tilde{\mu}_{e^-})_1 \qquad \text{(III.ix)}$$

Substituting equation (III.vi)–(III.ix) into equation (III.iv) yields

$$-\mathrm{d}\gamma = \Gamma_{\mathrm{Hg}^+}\mathrm{d}\mu_{\mathrm{Hg}} - \frac{\sigma_3}{F}(\mathrm{d}\tilde{\mu}_{\mathrm{e}^-})_4 + \Gamma_{\mathrm{Cl}^-}\mathrm{d}\mu_{\mathrm{HCl}} + \frac{\sigma_2}{F}(\mathrm{d}\tilde{\mu}_{\mathrm{e}^-})_1 + \Gamma_{\mathrm{H_2O}}\mathrm{d}\mu_{\mathrm{H_2O}} \qquad \text{(III.x)}$$

therefore

$$-\mathrm{d}\gamma = \Gamma_{\mathrm{Hg}^+}\mathrm{d}\mu_{\mathrm{Hg}} - \frac{\sigma_3}{F}[(\mathrm{d}\tilde{\mu}_{\mathrm{e}^-})_4 - (\mathrm{d}\tilde{\mu}_{\mathrm{e}^-})_1] + \Gamma_{\mathrm{Cl}^-}\mathrm{d}\mu_{\mathrm{HCl}} + \Gamma_{\mathrm{H_2O}}\mathrm{d}\mu_{\mathrm{H_2O}} \qquad \text{(III.xi)}$$

We may assume the equality of σ_2 and σ_3 since, as the mercury electrode is completely polarizable, no charge may be transferred across the interface; for the same reason the compositions of the phases must remain constant so that $\mathrm{d}\mu$ terms = 0.

Now, at a given temperature and pressure, the potential of the hydrogen electrode is affected only by the activity of HCl and is not affected by an applied external voltage, E. Any variations, $\mathrm{d}E$ in E, may therefore be regarded as changes $\mathrm{d}(\Delta\phi)$ at the Hg/HCl interface. Therefore,

$$(\mathrm{d}\tilde{\mu}_{\mathrm{e}^-})_1 - (\mathrm{d}\tilde{\mu}_{\mathrm{e}^-})_4 = F(\phi_1 - \phi_4) = F\,\mathrm{d}E \qquad \text{(III.xii)}$$

therfore, equation (III.xi) becomes

$$-\mathrm{d}\gamma = \Gamma_{\mathrm{Hg}^+}\mathrm{d}\mu_{\mathrm{Hg}} + \Gamma_{\mathrm{Cl}^-}\mathrm{d}\mu_{\mathrm{HCl}} + \Gamma_{\mathrm{H_2O}}\,\mathrm{d}\mu_{\mathrm{H_2O}} + \sigma_3\,\mathrm{d}E \qquad \text{(III.xiii)}$$

or,

$$\left(\frac{\partial\gamma}{\partial E}\right)_{\mathrm{P,T},\mu} = -\sigma_3 \qquad \text{(III.xiv)}$$

Equation (III.xiv) is known as the Lippmann equation.

Appendix IV Potentials in the diffuse double layer

For a single (x) direction it is possible, by analogy with equation (I.vi) to write

$$\frac{\partial^2 \phi}{\partial x^2} = \frac{\epsilon^2 \phi}{\varepsilon_0 \varepsilon k T} \Sigma N_i z_i^2 = \kappa^2 \phi = \frac{-\rho}{\varepsilon_0 \varepsilon} \tag{IV.i}$$

whereas $1/\kappa$ in the Debye–Hückel theory is regarded as the effective radius of the ion atmosphere about an ion, here it is to be identified with δ the thickness of the diffuse double layer. For equation (IV.i) the general solution follows by analogy with that for (I.vi), viz.

$$= A\,\mathrm{e}^{-\kappa x} + B\,\mathrm{e}^{\kappa x} \tag{IV.ii}$$

and since $\phi \rightarrow 0$ as $x \rightarrow \infty$, $B = 0$. Now,

$$\frac{-\rho}{\varepsilon_0 \varepsilon} = \kappa^2 \phi$$

where ρ = charge/unit volume of electrolyte solution. Therefore,

$$\rho = -\varepsilon_0 \varepsilon \kappa^2 \phi \tag{IV.iii}$$

$$= -A \varepsilon_0 \varepsilon \kappa^2\, \mathrm{e}^{-\kappa x} \tag{IV.iv}$$

(by combining equations (IV.ii) and (IV.iii)).

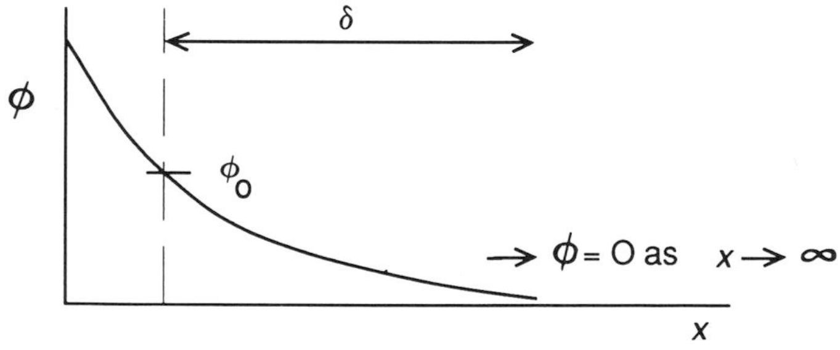

Figure IV.i The diffuse double layer in the region bounded by the conditions $\phi = \phi_0$ and $\phi = 0$.

Let the charge density at the electrode surface (i.e. at $x = 0$) be σ/unit area. This then is equal in magnitude, but of opposite sign, to the total volume charge in solution, i.e.

$$\sigma = -\int_a^{\infty} \rho\, \mathrm{d}x \tag{IV.v}$$

a being the distance of closest approach of ions to the surface. Therefore,

$$\sigma = -A\varepsilon_0\varepsilon\kappa^2 \int_a^{\infty} e^{-\kappa x}\, dx \qquad \text{(IV.vi)}$$

$$= A\varepsilon_0\varepsilon\kappa e^{-\kappa a} \qquad \text{(IV.vii)}$$

therefore

$$A = \frac{\sigma}{\varepsilon_0\varepsilon\kappa} e^{\kappa a} \qquad \text{(IV.viii)}$$

When the last expression for A is substituted into that for ϕ (equation (IV.ii)), we obtain

$$\phi = \frac{\sigma}{\varepsilon_0\varepsilon\kappa} e^{\kappa(a-x)} \qquad \text{(IV.ix)}$$

Index

American Drama of the Twentieth Century

Longman Literature in English Series

General Editors: David Carroll and Michael Wheeler
Lancaster University

For a complete list of titles see pages viii and ix

American Drama of the Twentieth Century

Gerald M. Berkowitz

Longman
London and New York

Addison Wesley Longman Limited
Edinburgh Gate,
Harlow, Essex CM20 2JE, England
and Associated Companies throughout the world.

Published in the United States of America
by Addison Wesley Longman Inc., New York

First published 1992
Third impression 1997

ISBN 0 582 01602 9 CSD
ISBN 0 582 01601 0 PPR

British Library Cataloguing-in-Publication Data
A catalogue record for this book is
available from the British Library

Library of Congress Cataloging-in-Publication Data
Berkowtiz, Gerald M.
American drama of the twentieth century / Gerald M. Berkowitz.
p. cm. – (Longman literature in English Series)
Includes bibliographical references (p.) and index.
ISBN 0–582–01602–9 (csd) – ISBN 0–582–01601–0 (ppr)
1. American drama – 20th century – History and criticism.
I. Title. II. Series.
PS350.B47 1992
812'.509–dc20 92-9396 CIP

Transferred to digital print on demand 2001
Printed and bound by Antony Rowe Ltd, Eastbourne

Contents

Editors' Preface

The multi-volume Longman Literature in English Series provides students of literature with a critical introduction to the major genres in their historical and cultural context. Each volume gives a coherent account of a clearly defined area, and the series, when complete, will offer a practical and comprehensive guide to literature written in English from Anglo-Saxon times to the present. The aim of the series as a whole is to show that the most valuable and stimulating approach to the study of literature is that based upon an awareness of the relations between literary forms and their historical contexts. Thus the areas covered by most of the separate volumes are defined by period and genre. Each volume offers new and informed ways of reading literary works, and provides guidance for further reading in an extensive reference section.

In recent years, the nature of English studies has been questioned in a number of increasingly radical ways. The very terms employed to define a series of this kind – period, genre, history, context, canon – have become the focus of extensive critical debate, which has necessarily influenced in varying degrees the successive volumes published since 1985. But however fierce the debate, it rages around the traditional terms and concepts.

As well as studies on all periods of English and American literature, the series includes books on criticism and literary theory, and on the intellectual and cultural context. A comprehensive series of this kind must of course include other literatures written in English, and therefore a group of volumes deals with Irish and Scottish literature, and the literatures of India, Africa, the Caribbean, Australia and Canada. The forty-seven volumes of the series cover the following areas: Pre-Renaissance English Literature, English Poetry, English Drama, English Fiction, English Prose, Criticism and Literary Theory, Intellectual and Cultural Context, American Literature, Other Literatures in English.

David Carroll
Michael Wheeler

Longman Literature in English Series

General Editors: David Carroll and Michael Wheeler
Lancaster University

Pre-Renaissance English Literature

* English Literature before Chaucer *Michael Swanton*
English Literature in the Age of Chaucer
* Engish Medieval Romance *W. R. J. Barron*

English Poetry

* English Poetry of the Sixteenth Century *Gary Waller (Second Edition)*
* English Poetry of the Seventeenth Century *George Parfitt (Second Edition)*
English Poetry of the Eighteenth Century, 1700–1789
* English Poetry of the Romantic Period, 1789–1830 *J. R. Watson (Second Edition)*
* English Poetry of the Victorian Period, 1830–1890 *Bernard Richards*
English Poetry of the Early Modern Period, 1890–1940
* English Poetry since 1940 *Neil Corcoran*

English Drama

English Drama before Shakespeare
* English Drama: Shakespeare to the Restoration, 1590–1660 *Alexander Leggatt*
* English Drama: Restoration and Eighteenth Century, 1660–1789 *Richard W. Bevis*
English Drama: Romantic and Victorian, 1789–1890
English Drama of the Early Modern Period, 1890–1940
English Drama since 1940

English Fiction

* English Fiction of the Eighteenth Century, 1700–1789 *Clive T. Probyn*
* English Fiction of the Romantic Period, 1789–1830 *Gary Kelly*
* English Fiction of the Victorian Period, 1830–1890 *Michael Wheeler (Second Edition)*

* English Fiction of the Early Modern Period, 1890–1940 *Douglas Hewitt*
English Fiction since 1940

English Prose

* English Prose of the Seventeenth Century, 1590–1700 *Roger Pooley*
English Prose of the Eighteenth Century
English Prose of the Nineteenth Century

Criticsim and Literary Theory

Criticism and Literary Theory from Sidney to Johnson
Criticism and Literary Theory from Wordsworth to Arnold
Criticism and Literary Theory from 1890 to the Present

The Intellectual and Cultural Context

The Sixteenth Century
* The Seventeenth Century, 1603–1700 *Graham Parry*
* The Eighteenth Century, 1700–1789 *James Sambrook (Second Edition)*
The Romantic Period, 1789–1830
* The Victorian Period, 1830–1890 *Robin Gilmour*
The Twentieth Century: 1890 to the Present

American Literature

American Literature before 1880
* American Poetry of the Twentieth Century *Richard Gray*
* American Drama of the Twentieth Century *Gerald M. Berkowitz*
* American Fiction 1865–1940 *Brian Lee*
* American Fiction since 1940 *Tony Hilfer*
* Twentieth-Century America *Douglas Tallack*

Other Literatures

Irish Literature since 1800
Scottish Literature since 1700
Australian Literature
* Indian Literature in English *William Walsh*
African Literature in English: East and West
Southern African Literatures
Caribbean Literature in English
* Canadian Literature in English *W. J. Keith*

* *Already published*

Acknowledgements

We are indebted to the author's agent on behalf of the Estate of Thornton Wilder for permission to reproduce extracts from the play *Our Town* by Thornton Wilder (Coward-McCann, Inc., 1938), copyright 1938, © 1957 by Thornton Wilder, an extract from the play 'Pullman Car Hiawatha' from *The Long Christmas Dinner and Other Plays in One Act* by Thornton Wilder (Coward-McCann, Inc., 1931), copyright 1931 by Yale University Press & Coward-McCann, Inc. Copyright © 1959 by Thornton Wilder, and extracts from the play *The Skin of Our Teeth* by Thornton Wilder (Harper & Bros, 1942), copyright 1942, © 1970 by Thornton Wilder.

Author's Preface

I want to thank the General Editors of this series, David Carroll and Michael Wheeler, first for offering me the opportunity to write this book, and, second, for wisely knowing when to give me my head and when to crack the whip.

The Department of English of Northern Illinois University granted me a partial relief from teaching at a point in my research when extra time was particularly valuable and appreciated.

Alfred Weiss, who taught me most of this in the first place, generously read the manuscript to help make sure I got it right; my debts to him continue to mount. Having grown up in what I refer to later in this book as the Broadway era, when virtually the only venue on the continent for the best in American theatre was New York City, I was doubly fortunate in living near New York and in having parents who introduced me to the theatre and encouraged my interest in it. For that, among the innumerable gifts they have given me, this book is offered as an inadequate tribute.

Gerald M. Berkowitz
October 1991

To My Parents

Chapter 1
Introduction

The American drama is, for all practical purposes, the twentieth-century American drama. There were plays written and performed on the American continent well before there was a United States, and during the nineteenth century the American theatre was widespread and active. But, as was also true in much of Europe, it was, with rare exceptions, not the home of a particularly rich or ambitious literature. The theatre was a broadly popular light entertainment form, much like television today; it is possible to do artistically ambitious work on American commercial television, but television is not likely to be the first medium to come to the mind of a serious writer. This is not to say that the playwrights of the nineteenth century were without talent, but that, like television writers, they were more likely to be artisans skilled at producing the entertaining effects that audiences wanted, than artists looking to illuminate the human condition or challenge received values.

Yet in America, as in Europe, a change in the kind of literature being written for the theatre began to become apparent in the last years of the nineteenth century. As with many historical and artistic developments in American culture, this was much less a matter of an organized 'movement' than of trial and error and accidents of personality; an individual writer might not be consciously innovating, but something in his work might attract audiences or inspire other writers, so that the art form lurched forward a step. There is, for example, little evidence that James A. Herne considered *Margaret Fleming* (1890) revolutionary in any way, but with hindsight we can see that his version of the mildly sensational melodrama typical of the period raises moral questions that its contemporaries do not, and that those questions give the play a distinctly twentieth-century feel. The process was slow and unsteady, with false starts and relapses, but by the second decade of the twentieth century artistically ambitious writers were venturing into drama and finding it able to carry a weight of psychological insight and philosophical import it had not been asked to carry before.

This was very much a rebirth of an art form; with little in the recent history of the genre to build on, the first generations of twentieth-century American dramatists had to discover for themselves what shape the twentieth-century American drama would take. It is not surprising, then, that the years from, say, 1900 to 1930 saw a great variety of dramatic styles and vocabularies, as playwrights experimented with epic, symbolism, expressionism, verse tragedy and the like, finding out as they went along what a play could and could not do. Foremost among the experimenters was Eugene O'Neill, of whom it is only a slight exaggeration to say that during the 1920s he never wrote two plays in the same style. O'Neill, whose father had been a star of the nineteenth-century theatre, and who thus had a sharp awareness of its literary limitations, was consciously experimenting, trying to shape and stretch the medium so it could do what he wanted it to – express his profoundly thought-out insights and philosophies. But even the less determinedly innovative writers of the period found themselves making up the rules as they went along.

Inevitably, some experiments failed. One reason O'Neill kept changing styles was that many of them disappointed him, while some of the '–isms' that were briefly successful in Europe proved unamenable to American topics and tastes, or simply uncommercial. Through trial and error, however, one particular dramatic mode came to the fore. Theatrically effective, easy for audiences to relate and respond to, remarkably flexible in its adaptability to the demands of different authors, the natural voice of American drama was revealed by the 1930s to be in realistic contemporary middle-class domestic melodrama and comedy.

Realistic contemporary middle-class domestic melodrama – each of those words is worth examining and defining. 'Realism' does not mean the uncensored photographic and phonographic record of external reality. Dramatic realism is an artifice as much as any other mode, violating reality in order to give the illusion of reality; to take one simple example, characters in 'realistic' plays generally speak one at a time, and in grammatical sentences. The important point is that the illusion of reality is maintained; realism avoids gross violations of the laws of nature (People don't fly) or the introduction of purely symbolic characters or events (the Little Formless Fears of O'Neill's *The Emperor Jones*), while presenting characterizations and behaviour that are at least possible. And thus a realistic play asserts the claim that it speaks the truth, that what happens on stage is a reflection of the world its audience inhabits.

Historical plays were a mainstay of the nineteenth-century American drama, and continued to be written in the twentieth; most Broadway seasons in the 1920s saw at least one play about eighteenth-century France or nineteenth-century Mexico, and the 1930s had a thin but constant stream of plays about Washington, Lincoln or other American heroes. But the overwhelming majority of twentieth-century American plays are set in the present, again implying a close parallel to the real world. Moreover, a play set in the here and now is likely to reflect the external reality of the here and now, be it the Depression of the 1930s, the middle-class anxieties of the 1950s or the profound social changes of the 1960s and after; and one of the first important discoveries about domestic realism was that it could address large social and historical issues in theatrical terms.

Shakespeare wrote about kings, O'Neill wrote about Lazarus and Marco Polo, and Maxwell Anderson wrote about Elizabeth I and Mary of Scotland. But the overwhelming majority of modern American plays are about people from the same social and economic world as the playgoers – the urban middle class. That is not a narrow range, and can stretch from the barely-getting-by and underemployed to the comfortably well-off. But the extremes of the economic ladder, along with other fringes of society – blacks, rustics, etc. – are rarely represented before the 1960s, except as stereotypes, and infrequently thereafter except in plays specifically addressing minority subjects. Audiences for American plays are likely to see themselves or people like them, or people they might believably, with good or bad luck, be like.

Of all these adjectives being defined, 'domestic' may be the most significant. Not only are American plays about recognizable people in a recognizable world, but they are about the personal lives of these people. Whether a play is actually set in a living room, with a cast made up solely of family members, as an extraordinary number are, or whether the 'domestic' setting extends to an office and a circle of friends, the issues and events are presented in small and localized terms. Whatever the deeper meanings of an American play, on one solid level it is about love and marriage, or earning a living, or dealing with a family crisis.

Of course Americans did not invent domestic drama. Ibsen and Chekhov (to name just two) had written of realistic characters in domestic situations, and even hinted at larger social and moral issues through this mode. The gradual discovery of American dramatists, starting in the 1930s, was that domestic realism was their most effective vehicle for talking about larger issues – that the small events in the lives of small people could be presented so that they reflected the world outside the living room. Put another way, the insight becomes more than merely technical. Dramatists discovered that the real story of, say, the Depression was not in statistics and large social changes, but in the ways it affected a family in its living room. From there it was a small step to the discoveries of the 1940s and 1950s that purely personal experiences, even those without larger social implications, were valuable and dramatic in themselves. A national literature of plays set in living rooms is a deeply democratic national literature, one that assumes that the important subjects are those that manifest themselves in the daily lives of ordinary people.

And finally 'melodrama,' a word with unfortunate and undeserved negative connotations. Although 'melodramatic' is popularly used as a criticism of literature that invokes shallow or excessive emotional effects, the noun merely refers to a serious play with no pretension to tragedy. It is worth noting that few American dramatists aspire to high tragedy; indeed, the other characteristics already enumerated, particularly realism and the domestic setting, militate against any such ambition. Once again an essentially democratic impulse is at work, in the assumption that the important events of life, the things worth writing plays about, are the things that happen to the essentially ordinary, not the heroic.

As with the rebirth of serious drama at the beginning of the century, the discovery of domestic realism's power and potential was not the result of conscious artistic manifestos or collusion among writers. The form worked, and so individual writers were drawn to it; indeed, in a number of cases dramatists who began as non-realistic writers or who went through a period of experiment found themselves drawn back to domestic realism in later works: Eugene O'Neill most strikingly, but also Maxwell Anderson, Edward Albee, Sam Shepard and others. The remarkable accomplishment of domestic realism lies in its richness and adaptability, as writers as different in their styles, subjects and ambitions as Eugene O'Neill, Tennessee Williams, Arthur Miller and August Wilson have been able to say what they want to say in this mode.

Just as there have been few stylistic schools or movements in the American drama, so there have been few conscious agreements among writers about subject matter. Individual dramatists wrote about what interested or moved them, and cases of direct influence (e.g. of Tennessee Williams on William Inge) are far less common than situations in which playwrights responding to a similar reality independently found their way to similar subjects and approaches. Still, similar realities, or the discoveries of particularly strong or insightful writers, are likely to produce some similarities among contemporaries. It is both coincidental and (with hindsight) inevitable that the Great Depression of the 1930s would lead some dramatists to explore and even criticize the social and economic forces that were affecting the lives of millions; and the drama of social and political criticism remains a small but significant part of the American repertoire. (It may be the American theatre's commercial constraints, the need to sell tickets and make a profit, that kept political drama from becoming as dominant or as angry as it did in some European countries.)

Similarly, Tennessee Williams wasn't trying to move the American drama into new subject areas when his own personal empathy for the lost and weak led him to explore their experience in the 1940s and 1950s; it was the fact that these plays spoke to a perhaps unexpected hunger for counsel and reassurance in the hearts of his audiences that made them succeed, and allowed other writers with their own forms of counsel and reassurance to voice them in drama. Clear evidence that this was no conscious 'movement' is the fact that O'Neill's late and posthumous plays, some written as early as 1940 but not released until the 1950s, address some of the same emotional questions as the Williams plays that preceded them in the theatre but were written later; the two authors (and others) independently found their ways to subjects that only became a recognizable genre when audiences responded to them, making reassurance that the pains and insecurities of life can be endured a recurring theme in plays from the 1950s to the 1990s.

This, then, is a literary history with a plot. For various historical and artistic reasons the stylistic outline of the twentieth-century American drama has a clearly discernible arc. An art form that was essentially born afresh at the beginning of the century went through a period of exploration and experiment culminating in the discovery that one style was more amenable to American tastes and more adaptable to the demands that different writers made on it. That style – realistic contemporary middle-class domestic melodrama – was to become the dominant and artistically most fertile and flexible mode, the one in which the greatest American dramatists were able to create the greatest American plays, and in which writers with widely varying agendas could offer psychological insights, political criticism or spiritual counsel. So absolute was the superiority for American dramatic purposes of domestic realism that when, soon after the middle of the century, some significant changes in the theatrical structure led to another period of experiment and exploration, the centre held. Artistic discoveries were made – some of them greatly enriching the dramatic vocabulary – but domestic realism retained its place as the native and natural American dramatic style as new generations of dramatists continued to discover its flexibility and power.

Theatre history

By its very nature the drama – the literature of the theatre – is more closely bound to the marketplace than any other literary form. Poetry exists even if it is not published, and a novel may sell very few copies but remain on the library shelves to be discovered later. But a play simply is not a play until it is put on a stage, and it is likely not to exist for posterity unless it proves itself on a stage; it is the general rule that plays are not published unless they have had successful productions. So, since forces and events that are really more part of theatrical than literary history have a direct effect on the types of plays that are written and the types of plays that survive, a brief history of the twentieth-century American theatre is in order.

As mentioned earlier, there was professional theatre in colonial America and in the Spanish southwest before there was a United States. In the nineteenth century live theatre was a flourishing entertainment form, with more than two thousand resident professional companies across the continent, in almost every city of more than village size, each with a repertoire of classics and new plays. Though first Philadelphia, and later New York, was acknowledged as the cultural capital of the nation, audiences in Chicago, St Louis, Denver and dozens of other cities were just as likely to see new plays and leading performers in first-class productions.

This situation began to change in the last two decades of the nineteenth century, in part because New York-based producers began sending out touring companies of their biggest hits, drawing audiences away from the less glamorous local theatres. In the 1890s a cartel of producers bought up or sabotaged the competition in many cities, to give themselves a monopoly. In the new century most of the remaining resident companies succumbed to competition from vaudeville, movies and later radio. By 1920 at the latest, New York City had achieved an absolute dominance and virtually absolute monopoly of the American theatre, with the rest of the country reduced to local amateur fare and touring companies of last year's New York hits.

'Broadway' is the name of a street in New York City, and the label has come to be attached to the commercial theatres of that city (even though most Broadway theatres are in fact on the side streets intersecting Broadway), just as the term Wall Street has come to refer generically to the financial community. And for roughly the first half of the twentieth century 'Broadway' was for all intents and purposes the entire American theatre.

This was obviously an imperfect state of affairs. When virtually all the new plays, all the major playwrights, all the best actors, directors and designers were to be found in one square mile of one city, then the overwhelming majority of the population was being deprived of the opportunity to experience American theatre at its best. Meanwhile the intense competition for a limited audience meant that many talented artists were inevitably squeezed out or not given a chance. On the other hand, the concentration of the best and most ambitious in one place had some salutary effects. Writers and performers could be inspired and challenged by each other, and build on each other's accomplishments. It is surely not coincidental, for example, that the psychologically realistic performance style called 'Method Acting,' the emotionally evocative powers of such directors as Elia Kazan and Jose Quintero, and the major plays of Tennessee Williams and Arthur Miller all appeared at the same time, in the late 1940s and early 1950s. Each of these, along with the responsiveness of an experienced and sophisticated audience, fed on and nourished the others, helping to create an identity, a unified style and a very high standard of accomplishment that a more dispersed theatrical community would probably not have achieved.

Virtually all of the dramatists discussed in the second, third and fourth chapters of this book wrote for the Broadway theatre; and the overall arc of theatre history in the first half of the century – the period of experiment leading to the discovery of a dramatic mode that seemed most amenable to the task of addressing American concerns, and the subsequent development of this mode to its fullest potential – is itself clearly a product of a small community of artists observing and learning from each other's errors and accomplishments. It is quite likely that the shape of the American drama by mid-century would have been quite different, in unpredictable ways, if the performance arena of American theatre had not been so concentrated and localized.

This state of affairs began to alter around 1950, as a result of four distinct changes in the structure of the American theatre. The first was the appearance of an alternative theatrical environment within New York City, as young actors, directors and designers who could not find employment on Broadway formed their own shoestring 'Off- Broadway' companies. Motivated by the desire to be seen and to exercise their craft, they performed in lofts, unused theatres and various converted spaces, free to do uncommercial and experimental work simply because there was so little money or prestige to lose. And because there were so many good young actors, directors and designers in New York, some very good work was done Off-Broadway, leading critics and audiences to discover and appreciate the potential of this new venue.

With its lower budgets, Off-Broadway concentrated in its first few years on classics and revivals. By 1960, though, it found a new function, as a showcase for commercially risky new writers. Its generally younger and more adventurous audiences were open to new styles and subject matters, encouraging a period of experiment and stylistic diversity recalling that of forty years earlier. A generation of American dramatists had their first plays produced Off-Broadway, and the alternative theatre was soon recognized as a significant complement to Broadway, its new respectability inspiring a further expansion of the fringe and avant-garde to the ironically labelled 'Off Off-Broadway'.

About the same time that an alternative theatre was developing in New York, alternatives to New York theatre were being born elsewhere in America. In the late 1940s and early 1950s new professional resident theatre companies were founded in Dallas, Houston and Washington; and by 1960 there were a dozen such companies around the country. By 1966 there were thirty, with, for the first time in the century, more professional actors employed outside New York City than on Broadway. By 1980 there were more than seventy large, permanently established resident theatres in America and, since many of them awakened a hunger for theatre in their areas, at least ten times as many smaller professional companies functioning in local equivalents of Off-Broadway.

The opening of new theatres all over the country did not just mean that high-quality live drama was now available to more people than ever before in the century, though that was accomplishment enough. It meant more opportunities for new writers, particularly those from outside New York City and those doing challenging new work. Unlike the New York theatre, which was structured as a marketplace, with each production an independent commercial enterprise in competition with others, most regional theatres produced full seasons of from four to ten plays a year. This innovation, seemingly just a matter of logistics and organization, actually changed the way in which audiences, at least outside New York, experienced plays.

By buying tickets for the full repertoire in advance instead of waiting for each play to appear and be reviewed, audiences opened themselves to a variety of new theatrical experiences, taking the occasional Shakespeare or new American play along with more familiar fare. Many regional theatre audiences soon surpassed Broadway audiences in their willingness to experiment, and new American playwrights found their plays welcomed by producers in Minneapolis or San Francisco more readily than by their counterparts in New York. Many regional companies declared the encouragement of new writers, particularly from their region, as a central part of their mission; others, particularly the low-budget alternative theatres, were simply more able than commercial producers to take a chance on an unknown. By the 1980s there were routinely up to ten times as many new American plays produced Off- and Off Off-Broadway and outside New York as there were on Broadway.

The third development in theatre history that helped change the shape of American drama around mid-century was a revolutionary alteration in the financial basis of the American theatre. The marketplace structure of the Broadway theatre was in keeping with the American tradition of entrepreneurial capitalism; each production was a business intended to make a profit for its backers. From the start the alternative theatres offered a partial release from market pressures. Most of these companies, at least at the beginning, had such tiny budgets that profit or loss was not a significant consideration. Even more radically, virtually all regional theatres were established as non-profit corporations with (allowing for the constraints of their budgets) the quality of the work more important than its profit-making potential.

Until this point in its history, the United States was one of the very few countries in the world with no tradition of public support for the arts; those artistic institutions (museums, symphony orchestras, etc.) that could not support themselves had to rely on the generosity of individual philanthropists. But this also began to change in the 1950s. Earlier in the century the anomalies of American income tax laws had led the very richest families in America to establish charitable foundations; by giving away a lot of their money they actually got to keep more of it from the taxman. In the 1950s some of the largest charitable foundations in America – the Ford Foundation, the Rockefeller Foundation, etc. – turned their attention to the arts and began making large gifts to non-profit theatres, to fill the gap between ticket sales and production costs. In 1966 Congress created the National Endowment for the Arts to distribute government funds to arts institutions, and many individual states and cities followed with their own arts-funding arms. For the first time in American history the majority of theatres in the country were at least partially liberated from the pressures and constraints of having to make a profit or even to break even.

That, of course, opened new opportunities for new American dramatists. Even a masterpiece by an unknown writer is a risky proposition in the commercial theatre, but a theatre company that is not driven by the profit motive can take the risk. Moreover, such companies can afford to nurture and support promising writers through commissions, grants, or works-in-progress stagings. Even failure is not absolute in the non-commercial theatre; having a first play flop on Broadway can seriously damage a new writer's future; losing money in a theatre whose successes also lose money is no great shame.

This imbalance became even more marked as a result of the fourth development, the decline of Broadway in the last third of the century. Not only was its monopolistic hold on the American theatre audience broken, but many of the most talented theatrical artists and virtually all of the most talented new dramatists in the country were happily finding opportunities and building careers elsewhere. Broadway continued to be the home of musicals, light entertainments and star vehicles, but a loss of artistic energy and creativity was inevitable. Financial constraints also worked against Broadway: as production costs escalated by the natural forces of inflation, it became more and more risky for the commercial theatre to attempt anything but the safest, most conservative fare. By the 1980s Broadway had virtually abdicated the job of producing new American plays, instead becoming a central showplace for the best new plays produced elsewhere in the country.

Thus, while for the first half of the twentieth century the terms 'Broadway theatre' and 'American theatre' are virtually synonymous, the years after, say, 1960 can rightly be called the post-Broadway period. Important new plays and playwrights were as likely to appear in Chicago as San Francisco, Buffalo as Seattle, Louisville as Milwaukee; and many more new plays and playwrights were given the opportunity to appear. If some of the benefits of the Broadway monopoly – the intense concentration of talent, the artistic inbreeding – were lost, the openness of the now truly national American theatre to a variety of new voices and new styles made up for them. The fifth and sixth chapters of this volume show that Broadway's decline did not mean the decline of the American drama.

Outline of this volume

This is a study of American drama rather than of American theatre. Although, as the preceding discussion has implied, it is sometimes difficult to separate the two, the focus will always be on the literature rather than the production history. It is a study of the *staged* drama; and while some of the playwrights included in this book also wrote for television or film, only their theatrical plays will be discussed. It is also a study of the *spoken* drama; the Broadway musical is one of the American theatre's special accomplishments, but its serious consideration requires entirely different approaches that would confuse and diffuse the focus of this book. At the centre of any analysis in this volume is the text, the words of the play, written by an author with the intention of communicating something to an audience or affecting an audience in some way.

Literary study has, willy-nilly, certain prejudices, and one of these is a ranking of serious intention and accomplishment over popularity. It is not impossible for a great work to be popular or a popular work to be great, but in those cases where one must choose, the literary critic will opt for the artistic accomplishment over the commercial success. This volume unashamedly owns up to that prejudice, finding Eugene O'Neill's *Strange Interlude*, for example, more interesting than Anne Nichols's *Abie's Irish Rose*, even though the first ran for only 208 performances and the second for over 2300.

After this Introduction, the chapters of the book are divided chronologically, with all but Chapter 2 covering fifteen years each. (Chapter 2 covers the period from about 1890 to 1930, with its main emphasis, however, falling after 1915, so it fits the general pattern.) The dividing points are not arbitrary; each coincides closely with a historical or theatrical turning point that marks a discernible change in the nature of the drama: the beginning of the Depression in 1930; the start of Broadway's greatest period in 1945; the rise of Off-Broadway around 1960; and the establishment of a truly cross-country theatrical arena around 1975.

The division into periods means that some dramatists' work will be discussed in the course of two or more chapters, rather than all in one place. On the other hand, I have not hesitated to stretch the chapter boundaries when that simplifies things or helps the analysis: although Edward Albee's first plays were staged in 1959, he is clearly a dramatist of the 1960s and is discussed in that chapter; although Tennessee Williams's *The Night of the Iguana* was produced in 1961, it can most profitably be studied alongside his plays of the 1950s.

Finally, two technical points that may be violations of usual practice. Recognizing that plays exist first in the theatre, I date plays by their first production (not counting workshops or previews) rather than by their first publication. Generally the two are the same, but there may be a year's difference in some cases and a considerably longer gap in a very few. Unless otherwise indicated, the place of first production is New York City. And I do not identify quotations from the plays by edition or page number. Virtually none of these authors is published in a recognized Standard Edition, and students and casual readers are most likely to encounter the plays in paperback reprints, acting editions or anthologies of various sorts. Page references to unavailable editions would be useless at best and confusing at worst. In most cases the context of the quotation or the manner of its introduction should help locate it in the play.

Chapter 2
1890-1930: The Beginnings

Antecedents and roots

The nineteenth-century American theatre was an immensely popular and vital form of entertainment, with theatres and resident theatre companies in virtually every city of more than village size. But it was not the home of an artistically significant literature. Serious American drama at its most ambitious reached the level of blank verse pseudo-Shakespearean tragedy along the lines of George Henry Boker's *Francesca da Rimini* (1855) or large-scale costume melodrama filled with spectacle, like Steele MacKaye's *Paul Kauver* (1887); at its less ambitious it produced broader melodrama of the cheer-the-hero-hiss-the-villain kind, like George Aiken's enormously popular dramatization of *Uncle Tom's Cabin* (1852). Much of American comic drama was built on variants of the situation established in Royall Tyler's *The Contrast* (1787), the triumph of a supposedly uncivilized American (or Westerner, or Yankee farmer) over sophisticated Englishmen (or Easterners, or city slickers); among the many of this type were Samuel Woodworth's *The Forest Rose* (1825) and J.K. Paulding's *The Lion of the West* (1830).

The reasons for the general absence of literary depth or quality are many and not restricted to America; in Britain and on the European continent the eighteenth and nineteenth centuries were generally fallow periods for dramatic literature. It wasn't until late in the nineteenth century that serious literary writers, led by Ibsen in Norway, Chekhov in Russia and later Shaw in Britain, rediscovered the theatre as a vehicle amenable to ambitious literature, serious or comic. In America the inspiration was probably less the European dramatic models than the new realistic American novel as practised and championed by William Dean Howells and Henry James among others. Emphasizing depth of characterization and accurate depiction of social milieu, these novelists pushed fiction towards the close analysis of the ways in which individuals were affected by the world they inhabited, thus illuminating both psychology and social forces.

Howells and James both tried their hand at drama, generally unsuccessfully, and at dramatic criticism, with considerable influence on at least some of their contemporary playwrights. Their calls for a drama marked by believable and consistent characters, serious consideration of social and moral issues, and direct reflections of reality, along with the model of their fiction and the audience taste for realism which they developed, led to a variety of experiments in dramatic realism in the last decade of the nineteenth century. Much of this new realism was more a matter of surface than content. Some American theatrical producers, like some of their British colleagues, had already become fascinated with realistic stage settings as ends in themselves, putting real trees, real water, even real locomotive engines on stage. The realistic novel's emphasis on setting, with the story and characters clearly growing out of a specific time and place, was reflected in 'local colour' plays such as Augustus Thomas's *In Mizzoura* (1893) and *Arizona* (1899).

The most interesting and challenging sort of dramatic realism, however, was that which reflected the method of the realistic novel by placing characters in moral or practical dilemmas that required a direct acknowledgement and discussion of larger issues. 'Problem plays' such as Edward Sheldon's *Nigger* (1909), however melodramatic in structure or style, were clearly *about* something beyond their specifics; their action frequently stopped dead while characters debated the larger issues or at least made explicit connections between the specific and the general. It is noteworthy that in the history of the novel the Realism of Howells and James led to the Naturalism of Frank Norris and Theodore Dreiser, with its emphasis on the manner in which the environment – social, economic, biological – comes in conflict with human aspirations, because a similar development took place in the drama. A step beyond the play that raised issues was the one that saw that they were not easily resolved or escaped; and the plays of the twenty years surrounding 1900 that have the clear feel of a twentieth- rather than nineteenth-century sensibility about them are those that raise more questions than they answer and that leave their characters facing a future of continued dilemmas and partial solutions.

Generally cited as the first of these is James A. Herne's *Margaret Fleming* (1890). The title character, wife of a factory owner, discovers her husband's working-class mistress has died giving birth to his child. She adopts the baby and, while not completely breaking with her husband, leaves open the question of whether she will ever fully forgive him. Like other problem plays of the period, *Margaret Fleming* thus raises a number of moral and ethical issues; unlike the others, it quite consciously leaves many of them unresolved. Foremost among these is a significant questioning of received definitions of marital morality. In the early part of the play Fleming shows no particular remorse about his infidelity, only an annoyance at the inconvenience of his mistress's pregnancy, and later he apologizes to his wife with the evident expectation of being quickly forgiven. She shocks him by raising the spectre of the double standard:

> MARGARET Suppose – I – had been unfaithful to you?
> PHILIP (*With a cry of repugnance.*) Oh, Margaret!
> MARGARET (*Brokenly.*) There! You see! You are a man, and you have your ideals of – the – sanctity – of – the thing you love. Well, I am a woman – and perhaps – I, too, have the same ideals. I don't know. But, I, too, cry 'pollution.'

It is that – the realization that a wife might expect of her husband the kind of love and loyalty that he assumes from her – that moves Fleming to a deeper remorse and justifies Margaret's insistence that any reconciliation can only be partial and long in coming. *Margaret Fleming* also raises questions about sexual and economic exploitation, the relationship between class and moral obligation, and the nature of the maternal instinct; and it poses new answers, or at least partial dissatisfaction with old answers, to some of these while leaving others unanswered. It is easy to overstate the play's revolutionary nature; it is recognizably a product of its time, and has its share of coincidence and sensationalism – in the midst of all this conflict Margaret goes blind, and early versions of the text include a kidnapping. Still, as one of the first plays in the American repertoire to challenge received assumptions rather than affirming them, it recognizes and brings into the theatre a more complex image of reality than the American drama had previously sustained.

The double standard and socially defined sexual roles are the recurring subjects of the earlier plays of Rachel Crothers, which also reveal the limits that audience taste or authorial courage put on serious questioning of received values at that time. The heroine of the western melodrama *The Three of Us* (1906) is found in a compromising situation by the man she loves and is unable to defend herself because she is sworn to secrecy on another matter. In justifying her silence she makes a strong argument for a woman's right to a 'masculine' sense of honour, but then submits completely, apologizing for every sign of strength and independence she has shown, and admitting that all she really wants is the love and protection of a strong man. In *A Man's World* (1910) a bohemian feminist victimized by gossip discovers that the man she loves is unapologetically guilty of the sins he falsely accuses her of, and rejects him. But she cannot reject or answer his assertion that her values are irrelevant and impertinent in a world whose morality and assumptions are completely defined by men. *A Man's World* thus complains about an injustice while admitting that the injustice seems uncorrectable; in background action two other women are forced to forsake their attempts to define themselves outside masculine terms, one giving up a career for marriage and the other falling into the socially accepted role of spinster social worker.

He and She (1911) is Crothers' first really successful attempt to address these issues without sensationalism or surrender. A woman sculptor wins a valuable commission, but discovers that her teenage daughter has problems that will require her complete attention, and gives up the project. The difference from *The Three of Us* is that this is not a capitulation to social forces or traditional roles; the woman does not rush happily to find fulfilment as a traditional wife and mother, but regretfully interrupts her career because her personal values recognize a higher responsibility at this moment:

> I'll hate myself because I gave it up – and I'll almost – hate – her. You needn't tell me. . . . I've heard people say – 'A woman did that' and my heart has almost burst with pride – not so much that *I* had done it – but for *all* women. And then the door opened – and Millicent came in. There isn't any choice, Tom – she's part of my body – part of my soul.

In the subplot two other women pick futures – marriage in one case and a career in the other – that are right for them, and thus *He and She* does not settle for simple victory by either side, but rather presents life as requiring personal and difficult choices.

Admittedly these plays and the few others like them represent a small percentage of the drama of the late nineteenth and early twentieth centuries, and their hints of a new sensibility may be more apparent in hindsight than they were at the time. Contemporary audiences did not find the plays of Herne and Crothers significantly more advanced than, say, Eugene Walter's *The Easiest Way* (1905), in which a fallen woman is redeemed by love, only to fall again and be rejected by her lover, or Edward Sheldon's *The Boss* (1911), about an unscrupulous businessman's rise, fall and redemption. In both cases, and in dozens of others, the trappings of realism and contemporary relevance effectively disguised the conventional morality, contrived plots and cardboard characters. And even the most naturalistic plays of social commentary had a touch of the allegorical about them, with characters functioning more as representatives of The Modern Woman, The Downtrodden Poor and similar types than as individuals.

Meanwhile, some of the most popular playwrights of the period made little or no pretence of writing anything different from the kinds of plays that were the mainstays of late nineteenth-century drama. The light comedies and romantic melodramas of Clyde Fitch have only the surface details of contemporary life and manners to distinguish them from similar plays written twenty years earlier. Even less innovative were the melodramatic plots and striking stage tableaux around which actor-playwright William Gillette constructed starring roles for himself in such successful vehicles as the Civil War spy thriller *Secret Service* (1895) and *Sherlock Holmes* (1899). It was about this time that George M. Cohan began his career as playwright with such successes as *Little Johnny Jones* (1904) and *Forty-Five Minutes from Broadway* (1906), which are far more significant for their contribution to the development of the Broadway musical as an art form than for any special virtues in their romantic comedy plots and characterizations.

The more immediate roots of later twentieth-century American drama appeared in the second decade of the century, when the Little Theatre movement brought a new generation of writers to the theatre. Experimental theatre companies – what would later come to be called Off- and Off Off-Broadway in America and fringe theatre in Britain – had appeared in Europe towards the end of the nineteenth century and in Britain early in the twentieth, but it wasn't until after 1910 that the movement reached America as part of the broader social and intellectual ferment of the time. In a number of cities across the country young writers and political activists gathered with the self-conscious intention of creating new and experimental literature, and many of these young artists turned to the theatre. One such company, a loose collection of poets, artists and socialist activists based in New York City's Bohemian neighbourhood of Greenwich Village, declared themselves the Provincetown Players (after Provincetown, Massachusetts, where they spent summers and produced their first plays in 1915). A peripheral member of the group, Eugene O'Neill, had his first play produced by the Provincetown Players in 1916.

Eugene O'Neill: the early naturalistic plays

Before beginning a discussion of O'Neill's plays it is necessary to note that his career spanned more than fifty years, extending beyond his death with the release of several posthumous plays, and dividing into at least three distinct periods. So his work will be dealt with in segments spread over several chapters, with chronology occasionally violated for the purpose of discussing plays of similar style together. O'Neill's earliest plays, predominantly one act long or barely sketches, are all naturalistic and anecdotal, small or large episodes in the lives of unremarkable people. Many of the early plays deal with sailors on or recently off the sea; four of them – *Bound East for Cardiff* (1916), *In the Zone* (1917), *The Long Voyage Home* (1917), *Moon of the Caribees* (1918) – have essentially the same cast, the crew of the freighter *S.S. Glencairn*.

In the *Glencairn* plays, as in others of this period, O'Neill breaks with his immediate predecessors in the American drama by making no attempt to preach or to shape his characters or story to reflect some social issue or demonstrate some social thesis. The power of the plays lies in their unadornedly anecdotal quality: in *Bound East* an injured sailor dies and his mates are saddened; in *Voyage* a sailor looking forward to leaving the sea is robbed and shanghaied back onto a ship. By refusing to impose meanings on the anecdotes, O'Neill demands that attention be paid to these people's small adventures for their own sake, and thus announces a basic change, for America, in the definition of what drama does. These short plays are not totally free of thesis, but thesis is not what generates them. They are observational, putting on stage the dramatist's perceptions of reality rather than his moral or social opinions. His characters are conceived of as characters – that is, as presentations of human beings – rather than as issues or problems dressed in human clothing.

Paradoxically, this concentration on the characters as individuals and the plots as anecdotes gives the plays an unstated universality. By emphasizing the realistic and familiar in characterization, O'Neill makes his characters representative – not of social or moral types, but of humanity in general. The barely articulated camaraderie of the *Glencairn* crew, the boundaries of their imaginations and lives, the small but significant private secrets that separate them are all recognizably human, and become the true subjects of the plays. *Bound East* is about death, not as a moral issue but as an experience through which the dying man must find his way and to which the living must react. *Moon* makes us recognize that loneliness and the attempt to find pleasure in distractions are facts of life, things that people feel and do.

If the characters of O'Neill's early plays have anything in common it is obsessiveness, overt or hidden. A few of the plays – for instance, *Ile* (1917), about an Ahab-like whaling captain – are open studies in obsession, whose characters are destroyed by their inability to break free of some fixation. But almost all of O'Neill's characters are limited in their freedom by internal forces, if only the lack of the imagination necessary to make a change. It is not only because he had spent some time as a merchant sailor that he was drawn to what he would call in *Anna Christie* 'dat ole davil, sea' as a subject. Seamen share with farmers, another recurring group of characters in O'Neill's early plays, an emotional and spiritual attachment to their element that is beyond their conscious control. The hint of tragedy in *The Long Voyage Home* is not that the sailor Olson is forced back to the sea after trying to quit, but that he actually thought he could break away in the first place; the horror of *Where the Cross is Made* (1918) is not in the ghostly figures that appear in one of O'Neill's first experiments with non-realistic effects, but in the psychological study of a father and son both driven mad by a fantasy attached to the sea; the O'Henry-like twist in the plot of *The Rope* (1918) is built on the conflicting obsessions of land, family and revenge that drive a farmer and his prodigal son.

The undeniable strength of these early one-act plays made O'Neill's move to longer forms and to the mainstream Broadway theatre inevitable, and by the early 1920s he was firmly established as the most important new playwright in America. His first Broadway plays included a few more in the naturalistic mode of his beginnings, but signs of strain soon appeared. The slice-of-life mode of the shorter works did not lend itself easily to longer plays, and O'Neill began to reach beyond naturalism. As one of the most intellectual and consciously philosophical authors in American literature, he found himself moving more towards thesis drama, not with the political or social agendas of earlier playwrights, but with the desire to dramatize his philosophical and spiritual visions. As these ambitions became more clear to him, so did the apparent inability of realism to encompass them, and he soon rejected that mode for an extended period of stylistic and technical experimentation.

The full weight of philosophical content that O'Neill wanted the drama to bear will be analysed later in this chapter, in the section devoted to his non-realistic plays. The first signs of intentions more complex than the simple depiction of reality come in the extension of the early plays' sense of obsession-generated fate to a more fully developed tragic vision. In many ways his first Broadway production, *Beyond the Horizon* (1920), is a summation of his themes and concerns up to then, with his signature symbols of the sea and the farm, and his characters' uncontrollable ties to one or the other. What is added in *Beyond the Horizon*, and was soon to be seen as even more central to his vision, is an overriding sense of tragic doom that is not merely a function of obsession, as in the earlier plays, but an elemental fact of life.

Beyond the Horizon opens by introducing two brothers: the poetic dreamer Robert, drawn by the 'beauty of the far off and unknown, . . . beyond the horizon', plans to leave the family farm for the life of a sailor, while the more stolid, practical Andrew will stay on. But Ruth, the woman they both love, chooses Robert, and he decides impulsively to stay at home and marry her, while Andrew chivalrously goes to sea in his place. What follows is predictable because O'Neill makes it seem inevitable: totally ill-equipped to be a farmer, Robert fails as manager and husband, but fails himself more deeply by frustrating and repressing his *wanderlust*. Meanwhile Ruth finds that her somewhat whimsical choice of Robert over Andrew was a mistake that traps her in a stultifying, joyless life from which there is no release. And the wonders of the world beyond the horizon are wasted on the unimaginative Andrew, who has moderate success but none of the satisfaction he would have found within the bounds of the farm. The play ends with Robert, broken in body and spirit, dying with a vision of the afterlife that will be the only escape granted any of them: 'I can hear the old voices calling me to come – And this time I'm going! It isn't the end. It's a free beginning – the start of my voyage! I've won my right to my trip – the right of release – beyond the horizon!'

Beyond the Horizon is a play of unrelenting doom, in part the result of the characters' failure to be true to their own natures, but to a greater extent the workings of an ironic and cruel providence. It would be simplistic to say merely that they get what they deserve for not being true to themselves; each makes his or her fatal error for what seem at the time reasonable and honourable motives, and even at his most self-critical Robert insists that he and Ruth 'can both justly lay some of the blame for our stumbling on God'. The sunset that opens the play is as ironic as the sunrise that closes it; the first calls the characters' attention to the horizon at the moment that it is being closed, while the second offers an elusive hope for freedom in death that merely emphasizes the hopelessness of the characters who remain alive.

Anna Christie (1921) is a little lighter in tone, if only because its characters are allowed briefly to experience the fulfilment denied to those in *Beyond the Horizon*, before the gloom closes in on them again like the sea fog that dominates the final act of the play. The title character is a prostitute who is reunited in the first act with her sea-going and sea-hating father, who reluctantly takes her on board his coal barge where, to his horror, she comes under the spell of what he continually curses as 'dat ole davil, sea'. When the barge picks up the shipwrecked Mat Burke, who quickly falls in love with Anna, the play seems to be moving towards the hope of a new life. But Mat's proposal of marriage forces a confession of her past from Anna, leading to his curses and rejection. In the last act they are reconciled, slowly and painfully pushing past each practical and emotional obstacle to their getting together, but their earnest determination only has the effect of emphasizing the size of those obstacles in the audience's mind. The final images of the play are of futility and foreboding, despite the characters' attempts to believe in a happy future: 'Fog, fog, fog, all bloody time. You can't see vhere you vas going, no. Only dat ole davil, sea – she knows!'

A number of non-naturalistic plays (to be discussed later) came next, but the first period in O'Neill's career concluded with *Desire Under the Elms* (1924), a full-length tragedy of obsession and inevitability. Set in a barren New England farm in 1850, the play shows a conflict between father and son complicated when the patriarch marries a younger woman whose first line – 'It's purty – purty! I can't b'lieve it's r'ally mine' – establishes her as a dangerous rival to both. To ensure her dominance Abbie makes Cabot promise to leave the farm to her child if she bears him one, and then seduces Eben to produce the child. But in the process Abbie falls in love with Eben, killing the baby to prove her loyalty. Eben insists on being arrested with her, leaving the barren Cabot on his barren farm. *Desire Under the Elms* is, then, a drama of obsessions in conflict: Cabot's determination to live fully and to conquer his land; Eben's hatred of his father and need to lay to rest his mother's spirit, which he senses haunting the farm; and Abbie's compulsion to possess and control something after her years of poverty, compounded with the obsessiveness of her love for Eben.

But *Desire*, unlike the earlier realistic plays, reaches consciously for tragedy, invoking the myths of Oedipus (Eben's hatred of his father, and his confusion of his love for Abbie and for his mother), Phaedra (the stepmother's passion for the stepson) and Medea (the mother's sacrifice of the child); and successfully borrowing tragic stature through the resonances. It also uses Freudian psychology effectively in defining and explaining all three characters, most notably Eben, who early in the play attempts to declare his manhood by sleeping with his father's whore, and whose seduction by Abbie takes place in the room haunted by his mother's spirit:

> ABBIE Don't cry, Eben! I'll take yer Maw's place! I'll be everythin' she was t'ye! . . . Don't be afeered! I'll kiss ye pure, Eben – same's if I was a Maw t'ye – an' ye kin kiss me back's if yew was my son . . . .
>
> EBEN Maw! Maw! What d'ye want? What air ye tellin' me?
>
> ABBIE She's tellin' ye t'love me. She knows I love ye an' I'll be good t'ye. Can't ye feel it? Don't ye know? She's tellin' ye t'love me, Eben!

O'Neill's interest in dramatic imagery, apparent in the symbolic use of the horizon in *Beyond the Horizon* and 'dat ole davil, sea' in *Anna Christie*, continues in *Desire Under the Elms* in a much more sustained and developed way. Foremost among the play's symbols are the rocks that cover the barren farm, and the stone walls and fences built from them. Repeatedly invoked and cursed by each of the characters, they take on layers of meanings, suggesting first the difficult and Sisyphean life of the farmers, then the cold hardness of Cabot's religion and Eben's hatred (equating them), then the unending harshness of the characters' emotional lives and the fated boundaries within which they must live their tragedy.

It must be admitted that, for all their power, these early plays have flaws which themselves bear O'Neill's signature. Throughout his career he was to be convinced that the best way to show a character's concerns was to have him repeat them again and again, often in the same words. There is a sometimes numbing verbosity to *Anna Christie*, particularly in the scenes in which the essentially non-verbal characters struggle to express themselves; and each play's catch-phrase declares its significance by constant repetition: 'dat ole davil, sea', and *Desire*'s 'dead spit 'n' image', used variously to equate Eben and Cabot, Eben and his mother, the baby and each of them, even Eben and Abbie, thus stressing the tragic bonds that hold them all together. The dialogue in these plays also exposes O'Neill's tin ear for dialect and natural speech, with Anna and her father sounding at times like burlesque comic Swedes and Mat like a stage Irishman, while the Cabots waver between Yankee and hillbilly accents. This was a limitation O'Neill would never overcome; though at his best he could create a kind of stage poetry out of the eloquence or stammering of his characters, to the end of his career he would burden some with broad dialects or speech patterns never spoken by real human beings.

More significant than these technical flaws, and evidently more frustrating for O'Neill, is an obvious and seemingly unresolvable tension between the tragic ambitions of these plays and their realistic trappings. However effective the symbolism or classical allusions, there is something bathetic about the insistently realistic farmers of *Beyond the Horizon* and *Desire Under the Elms* speaking of fate and predetermination; and the gloomy fatalism of *Anna Christie* seems more an authorial vision imposed on the characters than a discovery growing out of the reality of the play. The plays are far from failures, but they were clearly too far from O'Neill's vision of what they should be to satisfy him. O'Neill would wrestle throughout his career with the double challenge of making his philosophical observations and tragic vision seem to grow naturally out of the events and characters of his plays rather than being imposed on them, and of making their darkness dramatically palatable to an audience without falsifying his vision. At this point he developed serious doubts about his ability to do this within the bounds of realism, or about realism's ability to carry the thematic burdens he wanted to put on it; and he began more than a decade of experiments in alternative forms before he finally returned to domestic realism in his last and greatest plays.

Other developments in realism 1915–1930

The mix of new American plays in the early 1920s included romantic comedies, costume dramas, mystery thrillers, local colour pieces and domestic melodramas with only superficial differences from similar plays written twenty or thirty years earlier. Nevertheless, among the young dramatists of the new century there were several others who, like O'Neill, were exploring the possibilities of realism and naturalism in the theatre. As will be seen later in this chapter, they represented only one strand of the various courses the American drama was taking, but it was the one that followed most directly the lead the previous generation had provided.

Susan Glaspell, one of the founders of the Provincetown Players, wrote a number of short satiric comedies and a few realistic dramas of some merit. *Trifles* (1916), her most successful work, is an efficient and effective one-act essay on sexual differences. A farmer has been murdered, possibly by his wife, and the men who come to the farmhouse to investigate leave their wives in the kitchen, on the unspoken assumption that there will be nothing of importance in that woman's room. As the wives idly tidy up, they discover and immediately understand the significance of things the men would never recognize as clues even if they found them: a piece of sloppy sewing, a broken birdcage, a dead songbird. Barely having to verbalize their insights, they reconstruct the story of a brutal husband and a wife driven to the end of her tether; barely having to verbalize their decision, they easily conceal these 'trifles' from the men.

The subject of *Trifles* is thus not just the differing perceptions and preconceptions of the men and women about what is significant, but also the women's growing sense of unity and sisterhood. Starting with sympathy for the murderess, they move through guilt at not having visited and helped her, to a sense of solidarity with her and each other, made concrete in the hiding of the evidence. One of the play's strengths is the elliptical nature of its language, the women acknowledging and expressing most of their discoveries and decisions through a silence and indirection that dramatize their empathy more than open discussion could. The special experience of women is the subject of Glaspell's other serious plays: *The Outside* (1917), about two alienated women forced to recognize their almost forgotten ties to the human world; *Bernice* (1919), about the effect of a woman's apparent suicide on the survivors; *Inheritors* (1921), which echoes *Antigone* in the conflict between an idealistic college student and a temporizing administrator; *The Verge* (1921), in which a woman striving for total, uncompromising liberation can find it only in insanity; and *Alison's House* (1930), in which the heirs of an Emily Dickinson-like poet struggle with their obligation to publish her embarrassingly intimate works.

The drama of unexpected moral complexities, in the tradition of Herne and Crothers, continued in early plays by Maxwell Anderson and Sidney Howard, though both moved to other styles later in their careers. Anderson's *What Price Glory?* (1924), written with Laurence Stallings, is a war story that resists all the received clichés and theatrical conventions about war: its soldiers are neither passionately dedicated patriots nor sentimentalized common men transfigured by the experience of battle, but ordinary Joes somewhat brutalized by their experience and merely trying to do their job and survive. Howard's *They Knew What They Wanted* (1924) gives unprecedented depth to what could have been a conventional romance, as the parties in a sexual triangle talk their way through the crisis to a compromise that each can live with. The conclusion is sentimental, but it is the sentiment inspired by practical little people making the best they can out of life, rather than triumphing over it in a fantasy happy ending.

To contemporary audiences these two plays, along with O'Neill's *Desire Under the Elms* the same year, were notable for another reason, their unusual frankness of language and characterization. As shocking as the coarse soldiers' language of *What Price Glory?* was, its unromanticized psychology was even more revolutionary; and the other two plays' depictions of sexual passions and unashamed violations of conventional morality were both hailed and deplored. This was the decade in which Freudian psychology became a subject of interest in America; and if an earlier generation's contribution to dramatic realism lay mainly in finding new complexities to moral and social issues, the 1920s saw an increasing interest in dramatizing the darker complexities of human psychology.

In some cases, of course, this new psychologizing was trivial, merely adding a thin Freudian gloss to a conventional murder mystery or drama of marital problems. But at its best the combination of Freudian psychology and dramatic realism could produce chillingly accurate depictions of characters and situations hitherto ignored or only glanced at by American drama. In George Kelly's *Craig's Wife* (1925) a cold and neurotic woman openly admits that she married for economic and social power rather than love. She defines her success by her elegant house, which she compulsively keeps in pristine order, screening out anything that could taint its perfection. When her husband finally realizes the extent of her obsession and leaves her, the pathetic woman is still tidying the picture-perfect living room that is all she has in life. Sidney Howard's *The Silver Cord* (1926) presents an even more disturbed woman, a mother so neurotically attached to her grown sons that she treats everything from their careers to their women as romantic rivals. She almost succeeds in destroying one son's marriage, defeated only by a wife more determined than she, and does end her other son's engagement, reducing him to infantile dependence on her.

The new psychological depth was not just a matter of presenting neurotic monsters on stage. Even a comparatively minor play such as *Dancing Mothers* (1924) by Edgar Selwyn and Edmund Goulding, about a wife who tries to reform her straying husband by acting wild herself, could gain resonances by showing that the self-indulgent and self-centred perceive themselves as normal, and cannot be changed overnight by simple stage conventions. And even comedy could occasionally support some psychological realism; the domineering mother of *Mrs Partridge Presents* (1925), by Mary Kennedy and Ruth Hawthorne, is not significantly different from the monster in *The Silver Cord*, and only her children's success in rebelling makes the play comic.

Yet another form of realism is represented by Elmer Rice's *Street Scene* (1929), a nearly plotless slice-of-life drama set on the sidewalk and front steps of a working-class New York City apartment building. In the course of two days we watch residents and neighbours go in and out, look out their windows, pause to chat or gossip on the steps and generally just carry out the very ordinary non-events of life:

> MRS FIORENTINO Good evening, Mrs Jones.
>
> MRS JONES (*stopping beneath Mrs Fiorentino's window*) Good evenin', Mrs F. Well, I hope it's hot enough for you.
>
> MRS FIORENTINO Ain't it joost awful? When I was through with the dishes, you could take my clothes and joost wring them out.
>
> MRS JONES Me too. I ain't got a dry stitch on me . . . . (*As Olga Olsen comes up the cellar stairs*) Good evenin', Mrs Olsen. Awful hot, ain't it?
>
> MRS OLSEN Yust awful. Mrs Forentiner, my hoosban' say vill you put de garbage on de doomvaider?
>
> MRS FIORENTINO Oh, sure, sure! I didn't hear him vistle. Don't go 'vay, Mrs Jones . . . . (*Willie Maurrant, a disorderly boy of twelve, appears at the left on roller skates.*)
>
> WILLIE (*raising his head and bawling*) Hey, Ma!
>
> MRS JONES If you want your mother, why don't you go upstairs, instead o' yellin' like that?
>
> WILLIE (*without paying the slightest attention to her, bawls louder*) Hey, Ma!

The bulk of the play is made up of scenes like this, pointless except that their depiction of uneventful reality is the whole point. In addition to the six families in this building and their visitors, the play calls for more than fifty passers-by, each briefly individualized by costume or manner, and for the continuous background sounds of the city – a carefully choreographed and orchestrated evocation of reality in whose context a skeletal plot of adultery and murder seems out of place. One is more likely to remember and believe in the smaller events and subsidiary characters: the young couple who are drunk and loving at midnight but bitter and abusive at dawn, the man who buys ice cream cones for everyone, the carefree children who do not understand the shame of being evicted. *Street Scene* builds on the innovation of Eugene O'Neill's *Glencairn* plays in presenting the facts of life as dramatic and interesting in themselves, and anticipates the Depression-era plays of Clifford Odets in using the domestic setting to suggest the larger world outside.

Comedy

Much nineteenth-century American comedy played on the idea of the artless American triumphing over the sophisticated Englishman or Bostonian, and variants on that convention survived into the new century in such plays as Booth Tarkington's *Clarence* (1919) and Don Marquis's *The Old Soak* (1922). The twentieth-century variant pitted old-fashioned values against new-fangled manners and morality, usually in a romantic setting, as in Langdon Mitchell's *The New York Idea* (1906), in which a divorced couple remain friendly and eventually remarry, shocking everyone but their equally sophisticated friends. Salisbury Field's *Wedding Bells* (1919) and Philip Barry's *Philadelphia Story* (1939) are among its direct descendants. A formula popular in the 1920s pitted young lovers against their more conservative parents; Anne Nichols's *Abie's Irish Rose* (1922), whose title says all that is necessary about its mix of romantic and ethnic humour, exploited the possibilities of the formula successfully enough to run for almost a decade. There was comparatively little pure slapstick or physical humour, that being the speciality of film comedians; and strikingly absent from the American tradition, with rare exceptions, was the quick-under-the-bed-it's-my-husband bedroom farce of the Feydeau school. As a result, there was little exploration of comedy's anarchic or critical potential in the American drama of this period – or, indeed, of any period except briefly in the 1960s. While audiences in the commercial Broadway theatre were surprisingly supportive of challenging serious drama, they came to comedies for relaxation and escape, and could be confident of leaving with their values and assumptions unshaken.

The two best-regarded comic playwrights of the period were George S. Kaufman and Philip Barry. Kaufman, usually writing with a collaborator, specialized in the satire of generally easy targets. *Dulcy* (1921, written with Marc Connelly) pokes gentle fun at a well-meaning but scatterbrained society matron, while *To The Ladies!* (1922, with Connelly) satirizes big business by having a clever wife guide her dull husband to a promotion. *Merton of the Movies* (1922, with Connelly), *The Butter and Egg Man* (1925), and *June Moon* (1929, with Ring Lardner) mock, respectively, Hollywood, Broadway and pop music. Many of Kaufman's plays are adaptations of short stories or developments of his collaborators' ideas, with his contribution being the comic structure and dialogue, paring the action to the bone and providing the fast-paced gags, wisecracks and clever repartee that gave the plays a knowing modern air. His only rivals in this particular genre were Ben Hecht and Charles MacArthur, who collaborated on two classic farces, *The Front Page* (1928), about unscrupulous newspapermen, and *Twentieth Century* (1928), about a Hollywood producer trying to seduce and sign an actress in the course of a train journey.

The overwhelming majority of Philip Barry's twenty-one plays were not humorous, but the overwhelming majority of those were failures, so it is on a handful of successful comedies that his reputation rests. And on that basis – on two plays in particular – Barry is recognized as the

American master of sophisticated drawing-room comedy. While Kaufman specialized in frenetic action and snappy repartee, Barry's was more a comedy of character and idea, with a philosophical and moral core beneath the light veneer. All his comedies are set in the world of the well-to-do; all see conventional morality and the business ethic as the enemy of the individual, especially the artistically inclined; all celebrate the instinct to violate convention, while keeping that rebellion well within the bounds of comedy. In *You and I* (1923) a sympathetic father, a frustrated artist himself, subsidizes his talented son's studies so that he can have both art and marriage; in *The Youngest* (1924) a young writer triumphs over his business-oriented family; in *Paris Bound* (1927) a self-consciously 'modern' couple find conventional morality difficult to escape. Typically with Barry, the humour in these plays lies not so much in single lines as in the twists of plot and the reactions of the conventional characters to the unconventional; and there is always an undisguised but controlled element of sentiment and seriousness, particularly in the recognition that happiness is a rare thing, to be grasped at in whatever form it takes.

It was with *Holiday* (1928) that Barry first found the perfect balance of these elements. The story itself is trivial and even contrived: a successful young businessman shocks his fiancée and her father by declaring his contempt for business values and his plan to make only enough money to allow himself to take a few years off to enjoy himself. These unconventional impulses are shared by the fiancée's ugly-duckling sister, and the play bides its time while the right couple discover each other. In the interim we get to watch the idle rich at play in the world of beautiful clothes, champagne suppers and sophisticated dialogue, while we enjoy the effect even mild rebellion has on the stuffy, cheer the growing independence of the sister, and finally delight in the plot twists that end one romance and start another. Determined to make his mark as a serious dramatist, Barry would not find the same comic success until *The Philadelphia Story* eleven years later.

Alternatives to realism

Although hindsight recognizes the beginnings of twentieth-century American drama in the most realistic plays, particularly those exploring social issues or psychological complexities, it would not have been quite so clear to contemporary audiences that this was the direction the drama was going to take. One of the most critically respected dramatists of the first two decades of the century, for example, was the now almost-forgotten Percy MacKaye, whose plays ranged in style from pseudo-Shakespearean verse tragedy to grand semi-operatic masques and pageants. MacKaye, whose father, Steele MacKaye, had been a master of nineteenth-century spectacle melodrama, inherited a sense of theatricality and ambitiousness, which in his case took the form of poetic dialogue, classical and mythical subjects, and spectacular staging. He experimented in almost every non-realistic mode – Shakespearean-style verse tragedy and comedy, Greek-style tragedy, pastoral American romance, *commedia dell'arte* – before turning his hand to pageants and community dramas, very large-scale outdoor performances on grand themes, often involving whole communities in their casts. A Massachusetts production of *The Canterbury Pilgrims* (1903) had a cast of 1500, and *St Louis, a Civic Masque* (1914) had 7500 principals, chorus members and extras on its vast playing area. Other titles – *Washington, The Man Who Made Us* (1916), *The Roll Call, a Masque of the Red Cross* (1918) – indicate both the grand subjects and the extra-dramatic mission of these pageants, dedicated as much to communal involvement and celebration as to literature.

MacKaye was not alone in any of these styles. Verse dramas, classical tragedies and pageant plays made up a small but respected and noteworthy part of the American dramatic repertoire in the early years of the century, and some contemporary critics thought they saw the future of American drama in these ambitious modes rather than in the quieter realistic plays. Meanwhile, other writers looked to other non-realistic forms. The social and political agendas of some little theatre groups led to clumsy attempts at allegorical and agitprop plays, while some more commercial writers continued to find vitality in historical and costume melodramas. Edward Knoblock's *Kismet* (1911), an Arabian Nights tale, was a big hit; and through the first thirty years of the century virtually every Broadway season had at least one exotic melodrama set in a Chinese opium den, one swashbuckler set in eighteenth-century France, one romantic adventure involving Mexican bandits, one biblical parable or epic, one pseudo-Shakespearean verse play, and so on.

European influences began to be felt in the American drama in the early 1920s, as the plays of Strindberg, Capek, Pirandello and the German expressionists and epic dramatists were read and produced in America. The Broadway season of 1922–23, for example, included productions of Capek's *R.U.R.* and *The Insect Comedy*, Pirandello's *Six Characters in Search of an Author*, Meinhard and Bernauer's 'fantastic melodrama' *Johannes Kreisler*, Claudel's *The Tidings Brought to Mary*, Andreyev's *Anathema*, and Ibsen's *Peer Gynt*, along with the Moscow Art Theatre's repertoire of Gorky and Chekhov, providing a particularly intense but by no means unique seminar in European drama. American experiments, particularly in expressionism and its offshoots – dream, memory, flashback and fantasy plays – were a staple of the 1920s, rarely accounting for more than a handful of plays a season but never absent.

As early as 1924 Kaufman and Connelly were able to use the form comically in *Beggar on Horseback*, in which a young musician has a nightmare vision of what a conventional job and marriage would be like for him; though the play's satire is primarily aimed at the business world, the audience could not see the joke if they were not familiar with expressionism and the parodic ways in which Kaufman and Connelly use it.

One of the most successful American expressionist plays is *The Adding Machine* (1923) by Elmer Rice, who had already broken new theatrical ground nine years earlier in *On Trial* (1914), the first appearance of the now-familiar device of using courtroom testimony to lead into flashbacks re-enacting the events leading up to the crime. Although Rice denied having read any European expressionist plays, he used many of the same techniques in *The Adding Machine* to portray visually and aurally, not objective reality, but reality as perceived and experienced by his central character, the identityless clerk Mr Zero. While early scenes only hint that something other than external reality is being presented, the play clearly moves into Zero's mind when his boss announces that he is to be replaced by an adding machine:

> *Soft music is heard – the sound of the mechanical player of a distant merry-go-round. The part of the floor upon which the desks and stools are standing begins to revolve.. . . . The platform is revolving rapidly now. Zero and the Boss face each other. They are entirely motionless save for the Boss's jaws, which open and close incessantly. But the words are inaudible. The music swells and swells. To it is added every offstage effect of the theater: the wind, the waves, the galloping horses, the locomotive whistle. . . . For an instant there is a flash of red and then everything is plunged into blackness.*

What has happened, we shall later learn, is that Zero has killed the boss, but we don't see that; in the expressionist manner we see and hear what goes on inside Zero rather than outside. Later, at Zero's trial, he pours out the emptiness of his life in an extended stream-of-consciousness monologue that is clearly more a depiction of his internal mental breakdown than of the external situation. The second half of the play, showing Zero's execution and afterlife, is completely stylized and unrealistic.

The Adding Machine's expressionism is not merely a technical experiment; Rice uses it as some of his contemporaries used realism, to dramatize his central character's psychology. Zero is so broken in mind and spirit that he cannot enjoy heaven, choosing to spend eternity serving a cosmic adding machine. The play includes some overt commentary on mankind's regression to 'a waste product, a slave to a contraption of steel and iron', but its theatricality is stronger than its politics; by taking us inside Zero, Rice gives us a direct experience of his loss of dignity and humanity.

Another alternative to realism during the 1920s was the folk play – a term describing the content or effect, and not suggesting a communal or non-literary creation. Though sometimes ostensibly realistic in the tradition of local colour drama, many of these plays used music, dance and other production elements to produce a dream-like or mythical effect. It was a function of the American social situation of the period that plays about southern blacks were likely to carry almost automatic associations of religious music, superstition and childlike fantasy, and two of the most successful folk plays were written by white authors about black characters.

Dorothy and DuBose Heyward's *Porgy* (1927), most familiar as the source of George Gershwin's opera *Porgy and Bess*, is a romantic fable of a crippled beggar's love for a whore. The play is almost as filled with music as the opera, either in the form of spirituals or of accompaniment to stage effects. It is clear, for example, that the end of Act One, a funeral service filled with singing, is meant to affect the audience on more than a merely realistic level:

> *The beautiful old spiritual beats triumphantly through the narrow room, steadily gaining in speed. Serena is the first to leap to her feet and begin to 'shout.' One by one, as the spirit moves them, the Negroes follow her example till they are all on their feet, swaying, shuffling, clapping their hands. . . . 'Allelujahs' and cries of 'yes, Lord,' are interjected into the singing. And the rhythm swells till the old walls seem to rock and surge with the sweep of it.*

Later a storm scene is orchestrated as a blend of wind and rain sounds outside and spiritual singing inside. To some extent scenes like these are merely local colour, attempts to capture the atmosphere of this unfamiliar and somewhat exotic culture. But *Porgy* is not, as some local-colour plays were, merely theatrical anthropology. For the play to work – for the love story to be more than pathetic or ludicrous, and for Porgy's search for Bess as the play ends to be felt as a noble romantic quest – the audience must relate to *Porgy* as to a romance or folk tale. The musical and staging effects, along with the melodramatic plot twists and even the racial stereotypes, are designed to induce the suspension of disbelief.

Marc Connelly's *The Green Pastures* (1930) is a *faux-naif* retelling of the bible from the supposed point of view of an uneducated southern black child translating the tales into the world of his experience. Thus all the characters are black, heaven is a continual fish fry occasionally interrupted for spiritual singing, and 'De Lawd' has the appearance and manner of an imposing country preacher. The world of the Old Testament is conflated into modern black America: the Flood is inspired by God's unpleasant encounter with some crap-shooters; Moses must deal with the Sublime Order of Princes of the House of Pharaoh, Home

Chapter; and Babylon is a flashy New Orleans night club. Only the discovery that some men have found through suffering the ability to love and forgive reunites De Lawd with his creation:

> GOD Through *sufferin'*, he said.
> GABRIEL Yes, Lawd.
> GOD I'm tryin to find it, too. It's awful impo'tant. It's awful impo'tant to all de people on my earth. Did he mean dat even God must suffer? . . . *(In the distance a voice cries.)*
> THE VOICE Oh, look at him! Oh, look, dey goin' to make him carry it up dat high hill! Dey goin' to nail him to it! . . . *(God rises and murmurs 'Yes!' as if in recognition . . . All of the angels burst into 'Hallelujah, King Jesus.' God continues to smile . . . .*

If one puts aside what is now likely to be seen as implicit racism in the play's patronizing picture of blacks, then the play's charm, humour and emotional power are apparent. *The Green Pastures* proves that the sophisticated tools of non-realistic drama can transport a play and an audience to a world of childlike faith and imagination, producing effects the naturalistic drama could not.

Eugene O'Neill: the experimental plays

The one dramatist who was most active and most successful in exploring alternatives to realism in the 1920s was Eugene O'Neill. Despite his beginnings in realistic drama, or perhaps because of them, O'Neill had already begun to turn away from realism by the time he wrote *Desire Under the Elms*. That play's conscious striving towards tragedy was one sign of O'Neill's dissatisfaction with the limits of realism; he spent the decade and beyond experimenting with almost every possible theatrical style and device: expressionism (*The Emperor Jones*, *The Hairy Ape*), fantasy and dream sequences (*The Emperor Jones*, *The Fountain*), Greek tragedy (*Mourning Becomes Electra)*, symbolic sets or lighting (*The Hairy Ape*, *Welded*, *All God's Chillun Got Wings)*, masks *(The Great God Brown*, *Lazarus Laughed*, *Days Without End)*, spoken thoughts *(Welded*, *Strange Interlude*, *Dynamo)*, pageant-like epic staging and spectacle *(Marco Millions*, *Lazarus Laughed)*. *All God's Chillun*, *The Fountain*, *The Great God Brown*, *Marco Millions* and *Strange Interlude* all cover fictional periods of twenty years or more, while *Mourning Becomes Electra* (a trilogy), *Marco Millions* (written as two plays, though cut for production) and *Strange Interlude*

(5½ hours) test the outer limits of playing time.

O'Neill was not experimenting for experiment's sake. He was driven during this period by a basic dissatisfaction with the limitations of conventional theatre. Every alternative he tried was essentially a stretching device, an attempt to make the drama as a literary form hold more, express more, cover more, go deeper, or show more of reality than it had been asked to show before, and more than O'Neill felt realism could capture. Before discussing the individual plays, then, it may be useful to summarize briefly the nature of the philosophical, metaphysical and psychological weight he was asking the drama to carry.

Any discussion of O'Neill's ideas must begin with the fatalism and sense of doom already apparent in his early plays. With rare exceptions, O'Neill's characters are trapped by some internal or external condition that limits their possibilities and frustrates their desires. In various plays this fate will take the form of biology (*The Emperor Jones*), economics (*The Hairy Ape*), social conditioning (*All God's Chillun*), individual limitations of mind (*Mourning Becomes Electra*) or spirit (*The Great God Brown*, *Marco Millions*), and metaphysics (*Strange Interlude*). In some cases the purpose of the play is to demonstrate the lack of freedom; in others, to explain its causes or to explore its effects. The assumption that life is defined by painful and inescapable limits is never far from the centre of O'Neill's work.

O'Neill once wrote half-seriously that the key to understanding all his works was to remember that he had been raised as an Irish Catholic. To him that meant an inheritance of guilt, repression and an instinctively pessimistic view of life – an inheritance he rejected as an adult but then spent his life fighting against while never completely escaping. His plays are filled with characters whose lives are desiccated and darkened by their religion, or who, like Anna in *Anna Christie* or the Tyrone family of *Long Day's Journey into Night*, are trapped in a conviction of guilt and unworthiness. These and other plays also illustrate O'Neill's rebellion against life-negating religion; *The Great God Brown*, *Strange Interlude*, *Lazarus Laughed*, *Dynamo* and *Days Without End* focus on characters who either succeed or fail in finding a life-affirming religious faith.

O'Neill was one of the most intellectual and well-read of all American dramatists, and the products of his thoughts and reading found their way into his plays. As a playwright he was directly influenced, at least briefly, by Strindberg and the German expressionists. As a thinker and analyst of reality he was strongly affected by his reading of Nietzsche, Jung and Freud, among others. From Nietzsche O'Neill took the view of life as a struggle between two contradictory forces: the instinctive, sensual, life-affirming Dionysian and the rational, inhibiting, life-denying Apollonian. Repeatedly in his plays, particularly in the 1920s, he dramatizes the conflict between these two forces, which exist both cosmically and internally, either as a struggle between characters (*Desire Under the Elms*), between individuals and social forces (*Mourning Becomes Electra*) or within the individual (*The Great God Brown*). O'Neill's sympathy, like Nietzsche's, was with the Dionysian impulse, and in *The Fountain* and *Lazarus Laughed* he attempts to celebrate the imagined triumph of life-affirmation. But far more often he recognizes the inevitable supremacy of the Apollonian as life-affirming characters are dragged under by the power of reason, convention and death.

In the work of the age's leading psychologists O'Neill found modern scientific explanations for the trapped, doomed quality he sensed in the human experience. Freud, after all, demonstrated that adults have virtually no free will, being almost inescapably the products of influences and traumas in their childhood. His explorations in the area of sex helped explain why that presumably life-affirming force so often turned against man. In several plays O'Neill gives his characters textbook Freudian neuroses, like the Oedipal complexes of Eben in *Desire Under the Elms* and Orin in *Mourning Becomes Electra*. In the latter play Freudian psychology is offered as a modern dramatic equivalent for the Fates and Furies of Greek tragedy, and elsewhere it is used to give verisimilitude to O'Neill's fatalism. O'Neill took from Jung the concept of racial memory – thoughts and instincts inherited biologically from ancient ancestors and still colouring and shaping our lives. The concept appears most overtly in *The Emperor Jones*, but it is also there in all the plays in which characters are driven by needs defined as inherently human rather than individual.

It should be obvious, even from this brief summary, that there are many parallels and connections among these various influences. The concept of inescapable fate runs through all of them; Catholicism and Apollonian life-denial are almost alternative terms for the same thing to O'Neill; and religious guilt, failed Dionysian impulses and Freudian sexual complexes offer non-contradictory explanations for the same observable behaviour. Philosophy, science and O'Neill's own instinctive sense of life reinforced and validated each other; and the burden he asked his plays to bear was not only to dramatize the truths he saw, but to provide one or more of these explanations. Indeed, O'Neill was inclined to mix his metaphors and offer more than one explanation at a time, sometimes to the detriment of a play's coherence. And, of course, like any dramatist, he wanted each time to tell a story that engaged an audience's emotions.

This, then, is some of the weight O'Neill was asking the drama as a literary form to carry. It is a greater burden than any previous American dramatist had put on the genre, and perhaps more than had been asked of it by anyone since the Renaissance. Even if all of O'Neill's plays had been failures, this demand alone would give him an important place in the history of American literature, for leading the drama, and other playwrights, towards a greater seriousness and artistic ambitiousness. And since he was asking such a young and relatively untried genre to carry such weight, it is not surprising that he felt the need to redefine its conventions and stretch its definitions to make it capable of the task.

O'Neill's non-realistic plays of the 1920s can be divided into two groups, not according to style, since virtually every play introduced new techniques and devices that were dropped after one experiment, but by the general thematic impulse that drove them. In keeping with O'Neill's need to explore the full depth of the fate-driven human experience, many of the plays look inward, employing one non-realistic technique or another to bring his characters' inner experience on stage. Meanwhile, the same desire to create works of stature that inspired his early forays into tragedy led him in other plays to reach outward to subjects whose importance could be measured by their epic size and scope. These two impulses were not mutually exclusive, and in fact the one play in which they come together most successfully, *Strange Interlude*, is probably O'Neill's most successful experiment of the decade.

Although there were hints of non-realistic experiment in the visionary sequence of *Where the Cross is Made* and the rudimentary symbolism of *Beyond the Horizon*, O'Neill's first big step into new forms came with *The Emperor Jones* (1920), the first of his inward-looking expressionist plays. *The Emperor Jones* shows the disintegration of an American black man who has used his sophistication to conquer the superstitious natives of a West Indian island. When a rebellion threatens, Jones escapes into the jungle, only to be met by a series of visions beginning with 'little formless fears' and going backward in his memory (a murder he committed, a chain gang) to other 'memories' that can only be racial (a slave auction, a slave ship), and finally to a nightmarish encounter with an African witch doctor and a crocodile god. Accompanying this backward movement are other symbols of Jones's internal collapse – the progressive stripping-away of his clothes until he is reduced to a loin cloth; the incessant beating of native drums, starting at normal heartbeat rate and gradually increasing in speed and volume; and the incremental reduction of the playing space as he moves to smaller and smaller clearings in the forest.

The Emperor Jones is about the essential human reality that lies behind any façade of civilization, and the impossibility of escaping it. In this case, the sophisticated black man, no matter how superior to the natives he considers himself, cannot escape his inheritance of primitive instinct and belief, which ultimately destroys him. The play is likely to offend modern sensibilities through its implicit racism, but certainly O'Neill chose a black man primarily because his story is so effectively dramatizable; it would be difficult to find as clear a set of symbols of historical regression for a white character. Jones's fall, like Icarus's, is a metaphor for the inability of any man to escape the limits imposed on him by the human condition. The expressionistic devices O'Neill employs make us experience this fall from within the character, and thus recognize a common truth, rather than allowing us the luxury of distance and dissociation.

In *The Hairy Ape* (1922), O'Neill's expressionism is more direct. Dropping the device of visions and dreams, he presents external reality in distortions that expose essence rather than surface. The subject of the play is at least two-fold: the social criticism of a capitalist world that not only lives off the labourer but dispossesses him, and the spiritual and psychological effects of losing one's sense of place and value. Yank, a stoker on an ocean liner, is first seen in the cramped, cage-like crew's quarters that reduce the men to stooping and moving about like apes, and in the stokehole in which they feed the monstrous furnaces. Despite the dehumanizing surroundings, Yank sees himself as triumphant man, source of the energies that move the world, until a chance encounter with a wraith-like rich girl from above decks makes him suddenly aware of a world of power that excludes him, and fills him with the raging determination to reassert his importance. Adventures in New York City – encountering the rich, who do not acknowledge his existence; being arrested; attempting to join the socialist IWW and being rejected as a spy – lead Yank to the realization that not only is he not at the centre of power, but he has no any real place in the human world. He finds more to empathize with in a zoo gorilla:

> So yuh're what she seen when she looked at me, de white-faced tart! I was you to her, get me? On'y outa de cage – broke out – free to moider her, see? Sure! Dat's what she tought. She wasn't wise dat I was in a cage, too – worser'n yours. . . .Yuh don't belong wit 'em and you know it. But me, I belong wit 'em – but I don't, see?

In his confused identification with the ape, Yank sets him free, only to have the animal's embrace kill him.

The play's method throughout is symbolic and expressionistic: the sets and stage directions of the scenes on board ship emphasize the workers' trapped, dehumanized condition, while contrasting their obvious vitality to the enervated rich above. The rich New Yorkers are puppet-like automatons, lifeless compared to Yank but invulnerable to his enraged attacks. Jewellery in shop windows is attached to over-sized price tags blinking their figures in electric lights, while prison is a row of cells receding into infinity. Both the imbalances of capitalism and Yank's disorientation at discovering that his energy is exactly the thing that disqualifies him from human society are thus presented as he experiences them, for us to experience with him.

Welded (1924), *All God's Chillun Got Wings* (1924) and *Desire Under the Elms* (1924) are essentially realistic plays in which O'Neill still strove to depict internal experience through non-realistic devices. *Welded* uses touches of expressionism and symbolism to explore a Strindbergan marriage: the couple are isolated from the rest of reality by spotlights, and join in a crucifixion pose at the end. *All God's Chillun* shows a doomed inter-racial marriage and dramatizes the inescapability of the husband's blackness by having a photograph and an African mask symbolically grow larger from scene to scene, while the room they are in seems to shrink. As shown earlier, *Desire Under the Elms* makes symbolic use of the stones and walls of the Cabot farm, while invoking both Freudian and Nietzschean explanations of the characters' passions.

For the most part, O'Neill's plays thus far use non-realistic devices intermittently or as supplements to conventional storytelling. In *The Great God Brown* (1926) he takes the leap of making the violations of convention central to the communication of his themes. The play is, characteristically, about several things at once. On one level it is an overt depiction and discussion of the Apollonian/Dionysian division of experience and its effects on man; on another it is a study of the gap, often ironic and sometimes tragic, between public face and private identity; on yet another it is an exploration of the female essence, the images by which women define themselves and are defined by men; and somewhere around the edges of all this it is an attack on the dehumanizing power of the business ethic. O'Neill's primary device for presenting these multiple themes is the use of masks. Each of the major characters in *The Great God Brown* wears a mask of his or her own face, representing the image seen by others, but removes it in private moments to let us see the real person beneath. As the play progresses, mask and face may grow further apart, so that we see truths about the character to which others are blind; at times the others, accustomed to a character's mask, cannot recognize the bare face. At a key moment in the play one character puts on another's mask and is accepted as the other person by everyone else.

At the centre of the play are two men, the successful but shallow William Brown and the tormented artist Dion Anthony, the one damning himself by settling for too little, the other destroying himself in his search for intense experience, both seeking comfort from the same two women. Dion represents the Dionysian impulse and Brown the Apollonian, and neither alone is sufficient, as Brown admits by taking Dion's mask when he dies and spending the second half of the play alternating between the two personalities. Compounding and to some degree confusing the central theatrical imagery of the play are other symbolic languages, some of them reaching more towards epic scope than psychological and spiritual insight. O'Neill calls for every set to resemble a courtroom, producing the almost subliminal impression of characters on trial for the manner in which they are leading their lives. One woman, identified symbolically with both the Virgin Mary and the redemptive female character in *Faust*, becomes a demonstration of Christianity's failure when she cannot save either of the men; and the other is presented as an instinctively pagan earth-mother who comforts each of the men as they are dying, with the Nietzschean assurance that the pain of life is the prelude to an eternal joy. And O'Neill expects his audience to be able to read subtle and significant character developments in every small alteration of a face or mask. Nevertheless, *The Great God Brown* was O'Neill's greatest commercial success up to that point; and despite the occasional murkiness of theme and symbol, audiences could see and appreciate that the play was addressing the tragic combination of torment and glory in the fully lived and fully self-conscious life, and that it was stretching the theatrical vocabulary to find new ways of expressing new thoughts.

O'Neill's epic impulse is best represented by *Lazarus Laughed* (1928), a play so huge in scope and so theatrically demanding that it has never received a full professional production. O'Neill takes up the story of the man Jesus raised from the dead where the bible leaves off, imagining Lazarus so transformed by his experience that he preaches a gospel based on the denial of death:

> Laugh! Laugh with me! Death is dead! Fear is no more!
> There is only life! There is only laughter!

With messianic and almost magical power, Lazarus inspires a cult of laughers-at-death, freed from fear through laughter and revelry. The play follows him through his progressive encounters with Jews, Greeks, Romans, finally the Emperor Tiberius. Each of the central scenes has almost the same outline, as initial scepticism or hostility is overcome by the irresistible power of Lazarus's preaching and laughter, and his converts join in orgiastic singing and dancing.

Lazarus Laughed is the culmination of O'Neill's fascination with Nietzsche, and is therefore significant despite its theatrical failure. This time O'Neill is dealing not so much with Nietzsche's Dionysian/Apollonian contrast as with the celebration of the god in man. Lazarus's gospel comes straight out of *Also Sprach Zarathustra*, insisting that if we recognize that man himself is God, then there can be nothing greater for him to fear. But that freedom carries new obligations with it, as Lazarus reminds the weak: 'That is your tragedy! You forget! You forget the God in you! You wish to forget! Remembrance would imply the high duty to live as a son of God – generously! – with love! – with pride! – with laughter! This is too glorious a victory for you, too terrible a loneliness!'

This, once again, is the sense of man's limitations that is inescapable for O'Neill: however he may wish to celebrate a Nietzschean godhead, his own instincts recognize that all but the sainted few are unable to rise above their human bonds. Lazarus's followers all revert to sobriety and fear of death when he leaves them; and Tiberius and Caligula, fearing

Lazarus more than they love him, have him put to death. Although Lazarus dies with laughter and the reassuring 'Fear not, Caligula! There is no death!', the last words are those of the despairing Caligula: 'All the same, I killed him and I proved there is death! (*Immediately overcome by remorse, groveling and beating himself*) Fool! Madman! Forgive me, Lazarus! Men forget!'

What makes *Lazarus Laughed* unstageable is O'Neill's determination to express this grandiose and ultimately tragic statement in grandiose terms. The script calls for a number of elaborate processions and spectacles and for a cast of 420, with as many as 100 characters on stage at one time. Moreover,

> *All of these people are masked in accordance with the following scheme: There are seven periods of life shown: Boyhood (or Girlhood), Youth, [etc.] . . .; and each of these periods is represented by seven different masks of general types of character as follows: The Simple, Ignorant; the Happy, Eager; the Self-Tortured, Introspective; [etc.]. . . . Thus in each crowd . . . there are forty-nine different combinations of period and type. Each type has a distinct predominant color for its costumes.*

Clearly this is in the realm of closet drama; even if O'Neill's vision were stageable, the information and impressions he is trying to communicate are beyond an audience's ability to absorb. Nevertheless, one can say in defence of *Lazarus Laughed* that O'Neill is trying to say something very large that almost justifies the package.

O'Neill had previously ventured into the realm of epic theatre with *The Fountain* (1925) and *Marco Millions* (1928). *The Fountain*, his story of Ponce de Leon and the search for the fountain of youth, spans twenty-two years and two continents, includes in its cast both Christopher Columbus and a symbolic beauty suggestive of Dante's Beatrice, and has as its theme nothing less than the essential unity of all religions and human experiences. As sometimes happens in O'Neill's more ambitious plays, the various strands and themes mingle to the detriment of the play's coherence and clarity. In the central scene his hero has a vision in which figures representing Buddhism, Islam, Christianity and the American Indian religion join together; age and youth are united; and Beatriz, the daughter of his lost love, is transfigured in an eternal flame. The symbolism is murky, and even his interpretation of it is more allusive than clarifying: 'Fountain of Eternity, Thou are the All in One, the One in all – the Eternal Becoming which is Beauty!' It is clear, though, that O'Neill is reaching for images of all-absorbing unity to communicate a somewhat oriental sense of an eternal flow that incorporates youth and age, life and death, man and nature. The final act of the play shows the effect of this vision on Juan: when the real Beatriz goes off with a young idealist, he sees his own youth repeated in an eternal cycle, and dies in peace.

Like *Lazarus Laughed*, *Marco Millions* is so grand in sweep as to be virtually unproducible as written: it was composed as two plays running for six hours, and was heavily cut by O'Neill at the demand of his producers. *Marco Millions* is built on a central satiric irony: the merchant and explorer Marco Polo travels from Venice through all the rich and exotic cultures of Asia; he lives for more than fifteen years in the court of the great Kaan of China, a philosopher-king who repeatedly attempts to engage him in religious and metaphysical discussion; and he is loved by the Kaan's daughter – and Marco, whose imagination cannot extend beyond commerce and the amassing of money, is totally unaffected by any of this. It takes little interpretation to see that the play is an attack on American materialism; O'Neill makes the connection explicit in a final stage direction:

> *The play is over. The lights come up brilliantly in the theatre. In an aisle seat in the first row a man rises, conceals a yawn in his palm. . . . In fact, it is none other than Marco Polo himself, looking a bit sleepy. . . . Arrived in the lobby his face begins to clear of all disturbing memories of what had happened on the stage . . . . His car, a luxurious limousine, draws up at the curb. . . and Marco Polo, with a satisfied sigh at the sheer comfort of it all, resumes his life.*

The satiric point of *Marco Millions* is a legitimate one, and the conception of Marco Polo as the quintessential American businessman is clever. What makes the play unmistakably part of O'Neill's experiments of the 1920s is his assumption that the irony requires an enormous gap between the reality Marco experiences and his limited perception of it. O'Neill repeatedly calls for overwhelming stage spectacles, such as the grand court of the Kaan, filled with warriors, musicians, courtiers, concubines and elaborate sets, just to make Marco's blindness all the more obvious: '*Marco stands, a sample case in each hand, bewildered and dazzled.*' In the end, though, *Marco Millions* is a mountain labouring to bring forth a mouse; unlike *Lazarus Laughed*, its ideas, however good, simply do not need or justify the size and weight of the production. If O'Neill was stretching the boundaries of the theatrical medium, he clearly reached one set of limits in these outward-looking plays: epic scope and sheer numbers were not the way to make the drama carry the burden he wanted.

The epic and psychological/spiritual impulses come together in *Strange Interlude* (1928), the most successful of O'Neill's plays of the 1920s, both artistically and commercially. The play covers twenty-five years and attempts a cosmic definition of humanity's relationship to God, but also uses all of O'Neill's Freudian knowledge and sense of fate to explore his characters' psychology and emotions. It is not a complete triumph; its extreme length and the extreme verbosity generated by its central technical device make it rather heavy going for most audiences. Still, in it he comes closest to the goal he had been striving for throughout the decade: stretching the medium so that it could carry greater dramatic and philosophical weight, while still producing an effective piece of theatre.

O'Neill avoids some of the diffuseness of the epic plays by limiting himself to a small group of characters in a domestic situation. Nina Leeds, mourning her war hero fiancé, marries his friend Sam Evans in the in hope of having a son to name after the dead Gordon. When she learns that hereditary insanity in Sam's family makes a child impossible, she turns to Dr Ned Darrell to father her son. As more than two decades pass, Sam dies, young Gordon grows up, and Ned finally breaks free, leaving Nina with the sexless and avuncular family friend Charlie Marsden. What gives this relatively simple story its depth, and the play its length, is a technical innovation designed to take us further into the characters than realism could. In *Strange Interlude* we hear each character's thoughts as well as his words, and a simple scene is thus textured by levels of passion, irony and insight that are not apparent on the surface. Here, for example, is the first meeting of Ned and Charlie (The ellipses are all O'Neill's, indicating pauses; some stage directions have been omitted).

> MARSDEN (*thinking sneeringly*) Amusing, these young doctors! . . . perspire with the effort to appear cool! . . . writing a prescription . . . cough medicine for the corpse, perhaps! . . . good-looking? . . . more or less . . . attractive to women, I dare say . . .
>
> DARRELL Here, Sam. Run along up the street and get this filled.
>
> EVANS Sure. Glad of the chance for a walk. (*He goes out, rear.*)
>
> DARRELL (*turning to Marsden*) It's for Nina. She's got to get some sleep tonight.
>
> This Marsden doesn't like me . . . that's evident . . . but he interests me . . . read his books . . . wanted to know his bearing on Nina's case . . . his novels just well-written surface . . . no depth, no digging underneath . . . why? . . . has the talent but doesn't dare . . . afraid he'll meet himself somewhere . . . one of those poor devils who spend their lives trying not to discover which sex they belong to! . . .
>
> MARSDEN Giving me the fishy, diagnosing eye they practice at medical school . . . like freshmen from Ioway cultivating broad A's at Harvard! . . . what is his specialty? . . . neurologist, I think . . . I hope not psychiatrist . . . a lot to account for, Herr Freud! . . . punishment to fit his crimes, be forced to listen eternally during breakfast while innumerable plain ones tell him dreams about snakes . . . pah, what an

> easy cure-all! . . . sex the philosopher's stone . . . 'O Oedipus, O my king! The world is adopting you!'

Here, in an essentially minor scene, the spoken thoughts tell us volumes about the characters, including insights that will reverberate ironically later. More significant than their insights into each other (though both are right: Ned's professionalism is an affectation and Charlie is sexless) is what each man exposes about himself: the mask of superiority Charlie uses to cover his inadequacy and fear of self-exposure, and the scientific detachment Ned will be unable to maintain under Nina's influence. The technique proves to be quite versatile, allowing not only this sort of insight and irony, but also moments of high drama, as in Nina's seduction of Ned, which begins with both maintaining a mask of scientific detachment, speaking in the third person, only to have the internalized passions overpower the restraints of speech:

> NINA Ned does not love her – but he used to like her and, I think, desire her. Does he now, Doctor?
>
> DARRELL Does he? . . . who is he? . . . he is Ned! . . . Ned is I! . . . I desire her! . . . I desire happiness! . . . But, Madame, I must confess the Ned you are speaking of is I, and I am Ned.
>
> NINA And I am Nina, who wants her baby. I should be so grateful, Ned. I should be so humbly grateful.
>
> DARRELL Yes – yes, Nina – yes – for your happiness – in that spirit! I shall be happy for a while! . . .

The centre of the play is Nina's relationship with the men in her life, and her attempt to control her own fate by controlling them. But the elaborate construct of husband, lover, son and friend she creates must constantly be maintained. Ned cannot be dismissed once he has served his biological function, but must be kept in thrall to her sexual power. Sam must be protected from the knowledge of his inadequacy, if only because his loyalty to the first Gordon has earned him the courtesy. Young Gordon must be shaped into the ideal form of his namesake, so that his mother can define herself through his glory. And even Charlie must be kept on a loose rein, as a tie to the simpler and safer past. But the child is never really hers, having a closer spiritual affinity to the man he thinks is his father; the husband is a burden; the lover constantly struggles to escape; and the passionless friend wins her in the end only because the struggle of life has worn her out, and his emptiness seems like a safe harbour.

So, once again, O'Neill is dramatizing the limits of man's – or, here, woman's – control over her fate. Nina is one of the strongest and most determined characters O'Neill had created up to this point; she is fully self-conscious and aware of her desire to shape the world to meet her needs; she spends twenty-five years in continuous and inventive stratagems to achieve her goals. And in the end she must not only acknowledge failure but give up the struggle in exhaustion. Her understanding of the struggle is, appropriately, expressed in sexual terms. Early in the play she tells Charlie, 'The mistake began when God was created in a male image . . . . We should have imagined life as created in the birth-pain of God the Mother.' Throughout the play she draws strength from her identification with this female God, but in the end she must acknowledge a force larger than hers that mocks her attempts to create and control her fate: 'Yes, our lives are merely strange dark interludes in the electrical display of God the Father!'

O'Neill would make only a few more attempts at new forms. His final play of the decade, *Dynamo* (1929), borrows an image from Henry Adams's philosophical autobiography *The Education of Henry Adams*, as O'Neill offers the electrical dynamo as the defining icon of the twentieth century, with his central character led to worship and ultimately sacrifice himself to the spirit of electricity. Thematically, *Dynamo* recalls elements of earlier plays: the conflation of Oedipal love, romantic love and religion in *Desire Under the Elms*, the search for a maternal saviour in *The Great God Brown*, the hunger for a new religion in *The Fountain* and *Lazarus Laughed*. But it is forced and clumsy – the scenes of Reuben Light worshipping his dynamo goddess are embarrassing when not comic, and O'Neill's perennial weakness for purple passions and purple prose (perhaps noticed, even if not pointed out, in some of the passages quoted earlier) is particularly apparent.

It is *Strange Interlude* that is the climax of O'Neill's stylistic experiments of the 1920s; and perhaps only with hindsight is it so strikingly apparent that, despite its epic scope and stylistic innovations, it is essentially a domestic melodrama. O'Neill was not giving up his artistic ambitions or the demands he was making on the drama as a literary form, but we can see in *Strange Interlude* the beginnings of a discovery that would generate his greatest plays: that he could say and do everything he wanted to through the mode of domestic realism.

In summary, then, the first thirty years of the century saw the American drama reborn as a serious literary form. Beginning with a tentative realism that resembled the naturalism of turn-of-the-century American novelists, dramatists discovered that their plays could acknowledge the existence of social and moral issues that did not have pat answers. Later generations of playwrights, finding audiences receptive to this new seriousness, shifted the focus from the theoretical to the personal, combining contemporary discoveries in psychology with the stage's ability to reflect reality, to present more fully rounded and fully explored characters than the American drama had previously supported. Indeed, in such works as Eugene O'Neill's early sea plays, Susan Glaspell's depictions of the female experience, and Elmer Rice's *Street Scene*, the convincing and insightful reflection of reality was a satisfying end in itself, without the need of thesis or social comment.

But realism was not the only mode being explored by artistically ambitious American dramatists, nor was it clearly the dominant mode. With its focus on the specific and its surface similarities to the shallower plays of the nineteenth century, it seemed to some writers and critics a limited and inflexible form. Some of the most ambitious, if not necessarily the most successful plays of the first thirty years of the century were experiments in forms that seemed inherently or potentially grander in scope than simple realism. Following the lead of European experimenters and staking out new stylistic and technical ground of their own, O'Neill and others redefined the boundaries of the medium, finding new artistic vocabularies with which to address new subjects. It was an exciting period of exploration during which even unsuccessful experiments were valuable in helping the theatrical community to discover what various styles of drama could and could not do.

By 1930, then, what had been merely an unpretentious medium of popular entertainment had become a vital art form that was still in the process of discovering its own potential. In any given year, the overwhelming majority of new plays were conventional potboilers of one sort or another, as was true in every art. But the possibility existed, as it really had not in the nineteenth century, for serious artistic work to be done in the dramatic mode. The next generation of American dramatists would explore that possibility even further, in the process discovering the American drama's native and natural voice.

Chapter 3
1930-1945: Reality and Realism

The 'plot' of Chapter 2 was the discovery by serious American writers that the drama was an amenable vehicle for their artistic ambitions, and the free-wheeling experiments in various styles and forms that ensued. The plot of this third chapter is two-fold: the irresistible intrusion of external reality into the world of literature, and the discovery that one of the many dramatic modes playwrights had experimented with in the first decades of the century was far more useful and natural to the American drama than any of the others.

The fifteen years covered in this chapter saw two extraordinary historical events – a decade-long worldwide economic depression, followed by a world war. It may be difficult for younger readers to understand how all-pervading and life-altering these events were. The Great Depression of the 1930s was not merely a matter of a stock market crash and an increase in unemployment. It was the cause of financial strain on almost every family in America, and directly affected and altered lives in ways that might not immediately be thought of as economic. Career choices were made or altered, dreams were abandoned, marriages were cancelled, crimes were committed, religious faith was lost or found – millions of people's lives were irreversibly altered by the economy.

There were also political and philosophical consequences. Whatever perspective hindsight has allowed since then, at the time the Depression did not seem like a natural, if extreme, part of a business cycle that would eventually correct itself. It had a cataclysmic, end-of-the-world quality that could be interpreted as the irreversible failure of capitalism and breakdown of the American economy. And that belief, or even suspicion, could lead to larger social and moral uncertainty. Was the American Dream – the belief that this was a land of opportunity in which hard work and talent would inevitably be rewarded – being proved invalid? If that national promise was being broken, what of the others – brotherhood, the Melting Pot, democracy? Not since the Civil War had the continued existence of America as a concept been so threatened. These were not the idle musings of philosophers; it was neither accident nor disloyalty that led a generation of young idealists to consider socialism or even Soviet-style communism as an alternative to an apparently dying system.

It is a simple measure of the importance and pervasiveness of this new reality that it became a recurring subject in the American drama, which had hitherto not been particularly quick to deal with historical events; even the First World War had appeared in only a handful of plays, and reflections of the social reality of the 1920s were generally limited to jokes about flappers and wild youth in escapist comedies. Alongside this discovery of new subject matter occurred a stylistic evolution. While some playwrights of the 1930s continued the experiments in dramatic form and method that had characterized the previous decade, a clearly discernible mainstream developed. By the end of the period covered in this chapter there could be no question that the dominant, natural voice of American drama was realism – realistic, contemporary, middle-class, domestic melodrama or comedy.

Each of those words was defined in the first chapter. At this point it is worth reiterating that the development was both technical and substantive. Domestic realism, which had shown itself capable of illuminating individual experience through psychological analysis, had seemed to Eugene O'Neill and some of his contemporaries ill-suited to the depiction of whole cultures or the exploration of metaphysical questions. The discovery by Clifford Odets and others that the great national traumas of the Depression and the war could be dramatized through their effects on a small group of people in a domestic setting meant that American dramatists did not have to wrestle with epic theatre or other non-realistic modes, but could present their observations in a theatrical form familiar to their audiences. Even more significant, however, was the cultural assumption that made the technical discovery possible: that the domestic experience, and not the financial statistics or mass movements, *was* the real story of the Depression or the war. A later generation of American dramatists, led by Tennessee Williams, would carry this assumption even further, recognizing that a group of people interacting in their personal, private ways could not only draw a picture of the world outside that was affecting those personal lives, but could have an inherent dramatic power and importance without need of a larger context. But for the dramatists of the 1930s the first step was a significant and exciting discovery in itself.

And it was indeed a discovery, made gradually and collectively, in the same erratic, unplanned manner as other developments in American dramatic history. The emergence of a dominant dramatic style was not the creation of a conscious 'school' or artistic movement. Such self-conscious direction or manipulation is actually quite rare in any of the arts in America; and, in fact, what little artistic doctrine existed in the 1930s – that influenced by the Communist Party – specifically rejected domestic drama, demanding literature about the masses. And, despite a growing consensus, some significant playwrights of the period, seeing that domestic realism had limitations as well as potentials, rejected the mode and continued to experiment in non-realistic styles. What happened in the 1930s was a legitimate evolution of the art form, as the experiments of the 1920s led writers, individually and collectively, to discoveries about what worked best in the theatre, and then to discoveries of what could be done within the dramatic mode most of them chose.

Drama of the Depression: Clifford Odets

One of the first playwrights to attempt to deal with the Depression in domestic terms was Elmer Rice, whose *We, the People* (1933) is an episodic panorama of domestic scenes reflecting the effects of the economy on more than forty characters. The central plot, about a family whose members must each give up a dream because of the Depression, is only the skeleton on which are hung scenes representative of the times: an immigrant who explains how empty the American Dream seems to the poor; a farm family broken by the economy; factory owners and bankers, oblivious to the suffering around them, buying fine art and politicians. The play ends with a public meeting at which characters from these various worlds give a series of speeches openly criticizing America for having lost its political, moral and economic way. *We, the People* is least successful on its domestic level, since Rice's attempt to include as many different examples of injustice as he can fit into the play reduces it to a series of sketches, too brief to allow any real empathy or identification by the audience, and almost of necessity drawn in the broadest black-and-white terms. By resorting to open speechifying in the final scene, and frequently before that, Rice implicitly admits a failure to dramatize his message. Still, the basic insight is an important one: the stories to be told about the Depression are set in people's kitchens and living rooms.

Clifford Odets borrowed the basic structure of *We, the People* two years later in *Waiting for Lefty* (1935), giving the episodes more of a focus by incorporating them in a frame plot. At a taxi-drivers' union meeting, with the theatre audience taking the role of the membership, various speakers address the question of a strike, their speeches fading into dramatizations of the events that have brought them to this juncture: a doctor victimized by anti-Semitism is reduced to driving a cab, another driver realizes that he can't afford to marry his girl, and so on. Episodes back in the union hall include the unmasking of a company spy in the audience and the climactic news that the union leader Lefty has been murdered, leading to the unanimous vote to strike.

Much that could be said in criticism of *We, the People* is equally true of *Waiting for Lefty*. The episodic structure allows little more than broadly sketched characterizations and almost forces the characters into undramatically overt statements of Odets's messages. Still, the union hall frame excuses much of this by presenting each scene openly as a vignette rather than as part of a continuous dramatic action, while the political speeches of the frame scenes are justified by the dramatic premise of a strike meeting. In other words, *Waiting for Lefty* defines itself as a kind of play that will operate in brief sketches and broad emotional effects, and audiences can relate to it on that level; at particularly successful performances the audience would rise and join in the final strike call.

Moreover, while the villainous characters tend towards caricature, the sympathetic figures are given some reality by the simple device of making them politically naive. Thus, for example, a wife nagging at her husband to fight for more money and dignity does not consciously spout Party doctrine; her motives are domestic, and she finds her way to politically correct insights and language by instinct and accident:

> I know this – your boss is making suckers outa you boys every minute. Yes, and suckers out of all the wives and the poor innocent kids who'll grow up with crooked spines and sick bones. Sure, I see it in the papers, how good orange juice is for kids. But damnit our kids get colds one on top of the other. They look like little ghosts. Betty never saw a grapefruit. I took her to the store last week and she pointed to a stack of grapefruits. 'What's that!' she said. My God, Joe – the world is supposed to be for all of us.

Waiting for Lefty represents an important advance over *We, the People*, not just in dramatic structure, but in the discovery that large political issues could not only be presented in domestic settings, but they could be expressed through characters thinking and speaking in domestic terms. The episodic structure of *We, the People* and *Waiting for Lefty* was used in some subsequent plays, notably the *Living Newspaper* series, but the logical next step was to maintain this method in a full-length, dramatically integrated domestic play.

Odets achieved this in his next and best play, *Awake and Sing!* (1935), which is on its surface merely the unremarkable domestic life of a New York City Jewish family. Father Myron Berger is a nonentity, and the family is ruled by mother Bessie, who coolly marries her pregnant daughter Hennie off to an innocent neighbour and as cold-bloodedly breaks up her son Ralph's romance with a poor girl who might be a drain on the family's finances. Grandfather Jake, an ineffectual socialist, falls or jumps from the roof, leaving his insurance money and ideals to Ralph, and Hennie deserts husband and baby to run away with the war veteran and minor gangster Moe Axelrod.

As the Bergers' lives seem frozen by the Depression, Odets lets us realize gradually that they are all honourably attempting to function in the ways they have been taught, but that they are living in a world in which the rules have changed. Myron believes in the American Dream, as expressed in quotations from his childhood hero Teddy Roosevelt, but he can do no more to achieve it than fantasize impotently about winning contests. Jake dreams of a better world, but cannot get beyond such platitudes as 'Life shouldn't be printed on dollar bills', and the books he bequeaths to Ralph have not been read. Ralph and Hennie want nothing more than what their culture has taught them young people should want – love, pleasure, the right to dream of a future – and are lost because they cannot have it.

Moe and Bessie are particularly victimized because their attempts to function do actual harm to themselves and others. Moe fought for his country and lost a leg in the process; he deals with the pain by adopting a persona of cynical realism and a determination to beat the system as the ultimate entrepreneur, a criminal. But his racketeering is stalled at the level of cheating Myron out of a couple of dollars on a bet, and his wise-guy mask leaves him unable to express his sincere feelings for Hennie in any other terms but 'I got a yen for her and I don't mean a Chinee coin'. In an important speech late in the play, Odets lets Bessie make it clear that she never acts out of malice or selfishness, but out of duty as she sees it:

> Ralphie, I worked too hard all my years to be treated like dirt . . . . Did I ever play a game of cards like Mrs Marcus? Or was Bessie Berger's children always the cleanest on the block?! Here I'm not only the mother, but also the father. . . . If I didn't worry about the family who would?

But this commitment to what she understands to be her moral obligation turns Bessie into the villain of the play, forcing Hennie into marriage, breaking up Ralph's romance, and later trying to steal Jake's insurance money.

Everyone is doing what they're supposed to do, and it isn't working. The economic collapse has changed the world so that all the old values and models for behaviour – responsibility, ambition, patriotism, the American Dream – are useless or actively dangerous. At the end of the play Ralph is inspired to the socialist cause by Jake's death, but his call to arms is clearly not Odets's solution. Ralph's socialism is as vague and utopian as his grandfather's ('Maybe we'll fix it so life won't be printed on dollar bills.'), with not much more promise of changing the world, and the play equates it with Hennie's decision to run off with Moe. The only thing – and thus the important thing – that her hedonism and Ralph's idealism have in common is a break with the values and patterns of their past. *Awake and Sing!* is not a call towards any specific path, but a call to any path that leads away from what the play has depicted as the deadly models of a dying culture.

After a brief detour into simple political drama with *Till the Day I Die* (1935), about the struggles of German communists against the Nazis, Odets spent the rest of the decade exploring the double discoveries that the most effective medium for discussing the Depression and the real stories to be discussed lay in domestic realism. *Paradise Lost* (1935) presents another family in its living room. Leo Gordon is cheated by his business partner; one son is tempted into a robbery during which he is killed; another succumbs to a debilitating illness; and a daughter sinks into near-madness when she and her fiancé cannot afford to marry. And yet these somewhat sensational events are not really the focus of the play. Long before the Gordons lose their money they have lost their moral and spiritual direction, their faith in themselves and the world around them. Watching his world disintegrate, Leo cries to his wife, 'What is happening here? Once we were all together and life was good.' When the financial crash finally reaches the Gordons it is almost anticlimactic; as the socialist Mr Pike points out, with all the attention paid to the economic Depression, 'No one talks about the depression of the modern man's spirit, of his inability to lead a full and human life.'

And here *Paradise Lost* parts company with *Awake and Sing!* because Odets does not put the blame for this malaise on the dead American culture, but on a temporary dislocation from what was best in the American culture. At the end of the play Leo Gordon rediscovers his belief in the American Dream and is moved to a climactic assertion of his faith:

> No! There is more to life than this! . . . Oh, if you could only see with me the greatness of men. . . . Yes, I want to see that new world. I want to kiss all those future men and women. . . . Oh, yes, I tell you the whole world is for men to possess. Heartbreak and terror are not the heritage of mankind! The world is beautiful.

Leo's vision may be as nebulous and Utopian as Ralph Berger's, but that is not the point. Just as the important thing for Ralph was to break with the past and set out on a new path – *any* new path – Leo's salvation comes from his regaining the ability to dream *any* dream. On one level *Paradise Lost* contradicts *Awake and Sing!* by embracing the values of the past rather than rejecting them; on another level it simply goes far beyond the earlier play, by seeing that the real Depression is spiritual rather than economic.

Golden Boy (1937), Odets's next and commercially most successful play, shifts the focus back towards the economic, while still speaking in personal terms. The central metaphor of a musician who becomes a prizefighter to earn a living clearly points an accusing finger at a world that forces a man to give up the most important part of himself in order to survive; the key to Joe's progress as a fighter is the development of his killer instinct, and the boy who could say 'There's no war in music' ends the play by killing an opponent in the ring. Underlining the importance of this personal cost, some of the background action is taken up with the buying and selling of shares in Joe's contract until the question can be unironically asked, 'How much does Joe own of himself?'

Odets reinforces his political point by carefully keeping any of his characters from functioning as simple villains. Joe's manager, the gangster who buys into his contract, and the girl who serves as bait to lure him away from his family are all shown to be victims of the Depression, so that the worst that can be said of them is that they are doing what they must in order to survive. The girl falls in love with Joe and, like Ralph Berger and Leo Gordon before her, is allowed a Utopian vision: 'Somewhere there must be happy boys and girls who can teach us the way of life! We'll find some city where poverty's no shame – where music is no crime – where there's no war in the streets – where a man is glad to be himself, to live and make his woman herself!' But this time Odets denies the validity of that hope. When Joe and Lorna run off together they are killed in an automobile accident. It is possible to interpret their deaths as suicide, indicating their realization that Lorna's fantasy is unattainable in life, or as a simple accident, suggesting a cruelly ironic God. Most likely, however, since Joe's fast car has been carefully identified with his growing commitment to boxing, is the conclusion that death is the inevitable extension of the self-destructive process Joe was forced to choose in his search for economic security.

Golden Boy is something of a throwback for Odets, his last play to focus on the economic effects of the Depression. His next play follows more directly the lead of *Paradise Lost* in finding deeper domestic and psychological losses that have little to do with the national economy. Ben Stark, the central character of *Rocket to the Moon* (1938), is a fairly successful dentist, and the important secondary figure of his father-in-law is actually rich; reminders that the country is still in the Depression are limited to background characters and passing comments. The centre of the play is both men's relationship to Ben's receptionist Cleo, an insecure, uneducated but intensely vital girl whose unformed quality is a large part of her attraction. Unhappy in his marriage, Ben has an affair with Cleo. Lonely in his riches, Mr Prince offers her a marriage of mutual exploitation, his money and worldly experience in exchange for her youth and passion.

The effect that Cleo has on both men is to make them aware of the spiritual emptiness of their lives. This feeling of inadequacy or incompleteness, resembling the 'depression of the spirit' that was the focus of *Paradise Lost*, is the true subject of the play, and is exposed in all the characters, even Ben's shrewish wife. More significantly, it is also there in Cleo, and is the force that drives her determination to live life more fully than the others. At the play's climax, when the men press her to choose between them, she astonishes both by choosing herself instead. When Mr Prince asks what she has to lose by choosing, she replies, 'Everything that's me'.

> I want a love that uses me, that needs me. Don't you think there's a world of joyful men and women? Must all men live afraid to laugh and sing? Can't we sing at work and love our work? It's getting late to play at life; I want to *live* it. Something has to feel real for me, more than both of you.

So once again Odets ends a play with a Utopian vision, but once again it is clouded by doubt. Cleo may just have the energy and courage to wrest her self-fulfilment from life, but Ben's sad little hope that he, too, has made a life-altering discovery is much less convincing: 'For years I sat here, taking things for granted, my wife, everything. Then just for an hour my life was in a spotlight. . . . I saw myself clearly, realized who and what I was. Isn't that a beginning? Isn't it?' The play's title refers specifically to the ability to break out of a spiritual inertia (Mr Prince to Ben: 'Why don't you suddenly ride away, an airplane, a boat! Take a rocket to the moon! Explode!'), but in Ben's case, at least, such an awakening is as unattainable a fantasy as space travel.

Other realists

Realistic plays about ordinary people coping with the Depression were a staple of the 1930s, with as many as half the serious new American plays in a Broadway season meeting this description. Few of these were noteworthy on their own merits; even Odets could not make a domestic story dramatically effective when its real subject was not larger than the specific anecdote. *Night Music* (1940), the romance of a boy and girl adrift in New York, does not succeed in evoking larger meanings, even though repeated references are made to the economic environment that constrains the lovers; and *Clash By Night* (1941) amounts to little more than its plot of a straying wife whose husband murders her lover. Domestic realism in itself was not the point; the important discovery of the 1930s was domestic realism as a means of depicting a larger reality.

Sidney Kingsley's *Dead End* (1935) uses the unlikely but striking setting of a New York City street on which a luxury apartment house stands next to a slum tenement to dramatize the seemingly unresolvable injustices of the Depression. Its central characters include a crippled and unemployed architect, a neighbourhood girl who has moved from tenement to apartment by becoming a rich man's mistress, a gangster who risks arrest for a sentimental visit to his mother and his old girlfriend, and the teenage boys of the slums. The girl loves the architect but leaves because she cannot accept a life of poverty; the gangster, after discovering that his old sweetheart is a whore and his mother curses him, is killed in a shoot-out with the police; and one of the boys is arrested and sent to a reformatory. Much of *Dead End*'s power comes from its *Street Scene*-like naturalistic depiction of the ordinary events of the day, particularly the activities of the boys, who wander aimlessly from innocent games to robbery and violence. But Kingsley shows how every facet of life on this street is shaped by the Depression, from the gangster's choice of crime – 'What da hell did yuh tink I wuz gonna do? Hang around dis dump waitin' fer Santa Claus tuh take care a me, fer Chris' sake' – to the girl's choice of lovers – 'I've known what it means to scrimp and worry and never be sure from one minute to the next. I've had enough of that.' – to the emptiness of the boys' days and hopelessness of their futures.

The Depression was not the only subject for realistic drama in the 1930s. Some writers used realism as an end in itself, in equivalents of the local colour plays of thirty years earlier: Louis Weitzenkorn's *Five Star Final* (1930) and Sidney Kingsley's *Men in White* (1933) both set melodramatic plots against almost documentary backgrounds, respectively a newspaper office and a hospital. Such private-lives-of-the-great plays as Robert E. Sherwood's *Abe Lincoln in Illinois* (1938), E. P. Conkle's *Prologue to Glory* (1938) and Sidney Kingsley's *The Patriots* (1943) employ domestic realism as a history lesson, betraying the unspoken assumption that the significant truths about America's cultural heroes are the domestic ones.

For other writers domestic realism could be the vehicle for presenting a world view not in itself particularly realistic. Robert E. Sherwood's *The Petrified Forest* (1935) and William Saroyan's *The Time of Your Life* (1939) are idealistic romances of different sorts that gain credibility by being presented realistically. Sherwood puts a wandering philosopher, a romantic young girl and a gangster into the same room and lets their unlikely conjunction dramatize the value of the fulfilled life and the capacity for the chivalric gesture, as the writer and gangster, doomed and spiritually lost, conspire to give the girl the future they can't have. Saroyan offers an openly sentimental and romantic picture of life as it might be if most men were inherently good and life rewarded them accordingly, celebrating the small adventures of the regulars in a friendly San Francisco bar:

> *The atmosphere is now one of warm, natural, American ease; every man innocent and good; each doing what he believes he should do, or what he must do. There is a deep American naivete and faith in the behavior of each person . . . . Each man is following his destiny as he feels it should be followed; or is abandoning it as he feels it must, by now, be abandoned; or is forgetting it for the moment as he feels he should forget it.*

The plot of *The Time of Your Life* is less significant than the milieu; like Sherwood, Saroyan uses dramatic realism to present his characters and their world as real, solid and ordinary, so that his rosy-coloured view of life becomes at least temporarily convincing.

Lillian Hellman's second and best play, *The Little Foxes* (1939), uses domestic melodrama as historical allegory. (Hellman's 1934 play *The Children's Hour*, about the tragedies caused by a malicious schoolgirl's lies, has a simple moral purpose, though some indecision about whether to focus on the sensational events or their psychological effects weakens it.) Set in the American South in 1900, *The Little Foxes* is the story of the Hubbard family, two brothers and a sister who have risen through skilful and ruthless business dealings to displace the old southern aristocracy in their town. Given an opportunity to become truly rich, they stop at nothing – theft, blackmail, double-crossing, murder – to succeed. The play works theatrically on its surface level of intrigue and betrayal, but it gains deeper resonances as Hellman repeatedly identifies the Hubbards as the spirit of the New South of the twentieth century, fated by historical inevitability to displace the Old South while still retaining a regional identity that distinguishes it from the commercial and industrial North. They explain this to their northern partner-to-be:

> MARSHALL It's very remarkable how you Southern aristocrats have kept together. Kept together and kept what belonged to you.
>
> BEN You misunderstood, sir. Southern aristocrats have *not* kept together and have *not* kept what belonged to them. . . . Because the Southern aristocrat can adapt himself to nothing . . . . *Our* grandfather and *our* father learned the new ways and learned how to make them pay.

And the play dramatizes this, characterizing the New South represented by the Hubbards as a mix of cold-blooded avarice and unhypocritical commitment to some higher values; at the end, when Regina cheats her brothers out of the lion's share of the new deal, Ben has the grace to accept with some philosophical resignation:

> I'm not discouraged. The century's turning, the world is open. Open for people like you and me. Ready for us, waiting for us.

> After all this is just the beginning. There are hundreds of Hubbards sitting in rooms like this throughout the country. All their names aren't Hubbard, but they are all Hubbards and they will own this country some day. We'll get along.

The play does not celebrate the Hubbards, nor does it leave them unambiguously triumphant. Regina's husband and daughter suggest a third, more moral stage in the development of the southern culture that is not yet strong enough to displace the Hubbards but is gathering its forces. As an historical allegory, *The Little Foxes* offers an explanation of how and why the South changed with the new century, and also a balanced evaluation of the costs and benefits of that change, all within the boundaries of a theatrically effective realistic domestic drama.

Plays of the Second World War

The Second World War affected every element of American life just as fully as the Depression had, so it may seem odd that it was not reflected as extensively in the drama. Of many possible explanations the simplest is probably that the war was more naturally the province of films, which could depict scenes of battle more realistically and effectively. While each Broadway season during the war saw three or four battlefield plays, virtually all were commercial and artistic failures. As with the subjects that had concerned the dramatists and audiences of the 1930s, the war story would be told on stage, when it was told at all, through small personal dramas.

A significant subgenre of pre-war and wartime plays uses domestic, personal stories to symbolize a national awakening or to invoke a call to moral arms; Lillian Hellman's *Watch on the Rhine* (1941) is one of the most effective of this group. On its surface it is an adventure story in a domestic setting: an American woman long resident in Europe returns to her mother's home with her German husband, an anti-Nazi underground fighter whose life is endangered by the spying of a Romanian guest in the house; the husband is eventually forced to kill the spy in order to be able to return to Europe to continue his work. But, while the surface drama is about the freedom-fighter, the play's subtext is centred on the Americans, the wife's mother and brother. Their story is one of awakening to the horrors and high stakes of the drama the Europeans are living, and of discovering in themselves the moral strength to share that commitment. Hellman never patronizes these characters; from the start they are shown to be open, gracious and honourable. But they are innocents just by virtue of being Americans, and the European characters, good and evil, are inclined to dismiss them as irrelevant. It comes as something of a surprise to everyone that they handle each new revelation as well as they do; and by the end, while they are still not as dangerously involved as the freedom-fighter, they have risen to the task of supporting him as much as they can, and to the knowledge that more will be asked of them in the future. On this level, then, *Watch on the Rhine* is an allegory of America's late discovery of the fascist menace, and a statement of faith in the nation's ability to react heroically once its eyes have been opened.

This is the pattern for the most successful wartime drama, very localized and personal stories that take on larger resonances through representativeness or implicit allegory. Similar in structure to *Watch on the Rhine* are Robert E. Sherwood's *Idiot's Delight* (1936) in which an American entertainer in Europe is politicized; Maxwell Anderson's *Key Largo* (1939), in which a fight against American gangsters is presented as a microcosm of the European struggle; Anderson's *Candle in the Wind* (1941), whose American heroine works to free her French lover from the Nazis, suggesting an America drawn into the European conflict only by its emotions but discovering an unexpected capacity for political and moral commitment; and Lillian Hellman's *The Searching Wind* (1944), in which a shallow American couple who spent the twenty years before the war in Europe must now face the judgement of their soldier son that they helped cause the war they did not help prevent. In *Tomorrow the World* (1943), by James Gow and Arnaud d'Usseau, an American family takes in a German orphan only to discover that the boy is a fervent Nazi, and the domestic adventure dramatizes both the American realization of the high stakes of the war and the faith that the German people can somehow be separated from their fascist ideology.

A cluster of plays of the 'plucky little Finland' type presents European civilians somehow fitting heroism into their daily lives: Robert E. Sherwood's *There Shall Be No Night* (1940), John Steinbeck's *The Moon is Down* (1942), Dan James's *Winter Soldiers* (1942). When actual soldiers do appear, they are likely to be off the battlefield: Paul Osborn's dramatization of John Hersey's novel *A Bell for Adano* (1944) shows an American major trying to establish democracy in a liberated Italian town, and being liberated in spirit by the townspeople; John Patrick's *The Hasty Heart* (1945) is set in a military hospital, as an international mix of soldiers work to befriend and soften an unlikable Scot who is dying; in Arthur Laurents's *Home of the Brave* (1945) a psychosomatically paralysed soldier is helped to realize that there is no guilt in being glad to have survived a battle in which his friend was killed. Each of these plays implicitly declares through its domestic focus that the real story lies in the personal experiences occasioned by the war, rather than in the war itself.

Alternatives to realism: Maxwell Anderson

Not every play of the 1930s was set in someone's living room. But most of the alternative modes of the 1920s had fallen by the wayside. Gone were the exotic melodramas – the Chinese opium dens, Mexican bandits and French swashbucklers. Expressionist plays, dream plays, open allegories and biblical parables virtually disappeared from the stage, and the very few that were produced were all failures. None of Eugene O'Neill's experiments in form in the 1920s bore fruit beyond the O'Neill plays themselves, as even he abandoned each alternative style after trying it. The few significant dramatic experiments of the 1930s were new responses to the limitations of domestic realism.

The dominance of domestic realism as a dramatic mode proved how flexible and effective in conveying an author's psychological, social and moral visions it was. Despite this striking effectiveness and the ease with which it seemed to adapt to various uses, domestic realism had its limitations. One problem of the genre which was noticed by contemporaries was that, when used as a reflection of larger issues, it implicitly went beyond realism to naturalism. That is, any playwrights who dealt with economic, social or moral forces by dramatizing their effect on individuals were accepting the same assumptions that had driven such nineteenth- and early twentieth-century novelists as Norris and Dreiser: they implicitly defined life as a struggle between the individual and the larger forces that made up his environment, and they implicitly conceded that the environment was stronger. Seen from this perspective, many plays of the 1930s seem to admit defeat for humanity before they begin: the characters of Lillian Hellman's *The Children's Hour* or Sidney Kingsley's *Dead End*, for example, have no chance against the social forces they fight, while the few of Clifford Odets's characters who have happy endings do so only through Utopian visions that the plays do not show them actually achieving.

One dramatist who reacted against this limitation was Maxwell Anderson. He began his career as a realist, with *What Price Glory*, and continued to write occasional realistic plays into the 1950s, but in the 1930s he became the modern American drama's foremost (almost by default) writer of verse tragedies in the Elizabethan mode. Anderson's break with convention, then, is not merely a matter of his writing in an unobtrusively irregular blank verse. Like O'Neill a decade earlier he was reaching towards tragedy, which is more than a literary mode, but a vision of the human condition incompatible with the implicit naturalism of his contemporaries.

However one chooses to define tragedy, a few characteristics are inescapable. Tragedy presumes a more exalted, consciously heroic picture of humanity than does melodrama; a character to whom a tragedy can occur is virtually by definition extraordinary. To the Greeks the tragic hero fell at least in part because his stature threatened the gods; to Shakespeare it was because the depth of his character and passions, and the extent of his virtues and vices, were too great for the earth to hold. Thus tragedy is anti-naturalistic; in the battle between man and the environment, tragedy is at least willing to raise the possibility that man could win. The tragic hero's defeat is ultimately at his own hands, as some inner flaw, division, conflict or excess brings him down; and the magnitude of that internal struggle is itself part of the tragic writer's exalted picture of humanity.

Thus, while another writer of the period might have presented the life of Elizabeth I of England as a domestic drama, emphasizing the ordinariness and universality of the private story, Anderson in *Elizabeth the Queen* (1930) presents a tragedy. He focuses on events late in Elizabeth's life, involving her love for the Earl of Essex. (Like any historical dramatist, Anderson is not too tightly bound by facts; the Elizabeth–Essex romance is more in the province of myth than history, and the key final meeting of the two after Essex's arrest never happened.) Falsely accused of treason, Essex recognizes in himself the capacity for treason, and accepts his death rather than pleading for mercy. Thus it is he, rather than Elizabeth, who is tempted by personal virtues and weaknesses to over-reach himself and thus bring on his inevitable destruction; and it is he who learns from the experience, achieves some self-recognition, and is changed for the better before his death. But Anderson is not a follower of the 'tragic flaw' school of tragedy. Essex is not a good man brought down by one small but fatal weakness such as vanity or ambition. Rather, like Shakespeare, Anderson sees that in certain contexts a man's virtues can become tragically self-destructive, that only the great are tempted to go too far. The qualities that doom Essex are those that make him great, as Elizabeth recognizes:

> You are young and strangely winning and strangely sweet.
> My heart goes out to you wherever you are.
> And something in me has drawn you. But this same thing
> That draws us together hurts and blinds us until
> We strike at one another. This has gone on
> A long while. It grows worse with the years. It will end badly.

If Essex is the traditional tragic hero in the play, brought down by the incompatibility of his greatness with the world around him, but granted the self-awareness to die a better man than he had lived, then Elizabeth is tragic in a more modern sense. Hers is the experience of making hard choices and then having to live with them, knowing that she has brought unending pain on herself by her decision. Like some of Shakespeare's heroes, she is forced to choose between her passions as an individual and her obligations as a ruler, voluntarily closing off some integral part of herself: 'he who would rule must be / Quite friendless, without mercy, without love.' But it is not until the final moments of the play that she fully realizes the price she is paying:

> Oh, then I'm old, I'm old!
> I could be young with you, but now I'm old.
> I know now how it will be without you. The sun
> Will be empty and circle round an empty earth . . .
> And I will be queen of emptiness and death.
> [ellipses Anderson's]

It is very much a modern sensibility that suggests that the one who lives may suffer more than the one who dies, and that a tragic heroism lies in choosing a path of pain.

Anderson's use of dramatic verse is skilful and impressive. The Queen's speech just quoted, for example, moves seamlessly from the uncensored and very dramatic outburst of the first line, through the simple statements of the next two lines, to the evocative imagery of the last two. Anderson never achieves, and rarely attempts, the linguistic complexity and density of Shakespeare, but he is able to use the occasional well-placed image or poetic device to intensify a moment or to underline a thought. At other times the verse is barely noticeable, merely a more regularly rhythmic prose, keeping the play from sounding too exotic or self-conscious.

Like most writers drawn to the story of *Mary of Scotland* (1933), Anderson sees the centre of her myth in the contrast and conflict with Elizabeth of England; and like most writers he does not resist the impulse to violate history by building his play around a face-to-face encounter between the two women who actually never met. In Anderson's version of the contrast Elizabeth is cold, intellectual and intensely political, and Mary sincere, naive and personally charming; the conflict is not so much between two queens as between a queen and a woman. In their climactic meeting Mary makes the mistake of greeting Elizabeth as a sister, 'a woman like myself, fearing as I do, / With the little dark fears of a woman,' only to be assured by Elizabeth that her power comes from having dehumanized herself: 'It's thus if you would rule; / Give up good faith, the word that goes with the heart, / The heart that clings where it loves.' And here Mary reaches the recognition of her failure – 'Oh, I'm to blame in this, too! / I should have seen your hand!' – and of her victory:

> MARY Still, STILL I win! I have been
> A woman, and I have loved as a woman loves,

Lost as a woman loses. I have borne a son,
And he will rule Scotland – and England. You have no heir!
A devil has no children.

ELIZABETH By God, you shall suffer
For this, but slowly.

MARY And that I can do. A woman
Can do that.

Mary will die as a woman, true to herself and tragically defeated only because the world of Elizabeth cannot support such a person. And Elizabeth, as in the earlier play, will triumph in the world's terms but live on with the recognition of her private defeat.

Winterset (1935) faces the challenge of applying this poetic style and tragic vision to a contemporary story, with mixed results. It seems on its surface to have much in common with the realistic melodramas of the Depression: its setting is a riverside slum alley much like the set of Kingsley's *Dead End*, and its characters are the barely surviving poor. But Anderson's interest in *Winterset*, as in his other poetic plays, is in his characters' inner torments, not their conflict with the world outside. Thirteen years before the opening of the play a man was falsely convicted of a murder, and three men now converge on a witness to the killing: the convicted man's son Mio, trying to clear his father's name; the actual murderer, determined to prevent a reopening of the case; and the judge, driven mad by the fear that he might have collaborated in an injustice. The play is built around obsessions: the murderer's damns him; the judge's drives him mad; and Mio's, which at first seems to be an ennobling quest, is revealed as a spirit-destroying prison resistant to the healing power of a woman's love:

When I first saw you,
not a half-hour ago, I heard myself saying,
this is the face that launches ships for me –
and if I owned a dream – yes, half a dream –
we'd share it. But I have no dream.

The climax comes, then, not with Mio's death, but a few moments earlier, as the woman's simple faith leads him to the light:

MIO Miriamne, if you love me
teach me a treason to what I am, and have been,
till I learn to live like a man! I think I'm waking
from a long trauma of hate and fear and death
that's hemmed me from my birth – and glimpse a life
to be lived in hope. . . .

MIRIAMNE He would have forgiven.

MIO He?

MIRIAMNE Your father.

MIO Yes. . . .
Then there's no more to say – I've groped long enough
through this everglades of old revenges – here
the road ends. – Miriamne, Miriamne,
the iron I wore so long – it's eaten through
and fallen from me.

Anderson tries to give tragic stature to this small story of redemptive but doomed love by surrounding it with the discussion of philosophical and metaphysical questions, but the attempt is misguided and the philosophy and metaphysics the weakest parts of the play. *Winterset* is most successful on the small, personal scale that Anderson's instinct towards tragedy seems to have mistrusted.

The remainder of Anderson's verse plays are less interesting. *Valley Forge* (1934) is notable for depicting a George Washington almost unprecedented in American drama, not the cardboard saint of tradition but a three-dimensional figure capable of rage and despair as well as idealism. The play makes no attempt at tragedy, being satisfied to use the power of verse to celebrate the common soldiers who were the true heroes of the American Revolution. *The Wingless Victory* (1936), a retelling of the Medea myth in American terms, wavers uncertainly between being an attack on intolerance and a tragedy based on its effects. *The Masque of Kings* (1937), Anderson's version of the Meyerling story, is too much a replay of *Mary of Scotland* to be satisfying, with Crown Prince Rudolph a Mary-like idealist and his father the Elizabeth figure. In *High Tor* (1937), the ghosts of seventeenth-century Dutch sailors interact with a modern American idealist and some low-comic criminals; Anderson tries to capture the tone of *A Midsummer Night's Dream*, in which high romance and low comedy can co-exist, but the play is an uneasy blend of romance, satire, fantasy, thesis drama and slapstick. *Key Largo* (1939), Anderson's drama of the battlefield deserter who finds the chance to redeem himself by standing up to a gangster, suffers from the same clashes of style and tone that weakened *Winterset*, as the author, evidently not trusting the inherent power of his domestic story, weighs it down with excess poetry and philosophy. His one post-war verse play, *Anne of the Thousand Days* (1948), is actually two-thirds prose, and again seems a return to old thoughts and patterns, with Anne Boleyn presented as an Essex-like tragic hero, instinctively passionate and ambitious, but less skilled at the cold-blooded games of power than her enemies.

In the end, the results of Anderson's experiment proved ambiguous. Without question, poetic drama was a viable medium, and at least a few of his verse plays were artistic successes. But Anderson's own later plays discovered nothing further in the alternative medium, and even receded from the new artistic ground staked out by his first explorations, proving most successful as domestic stories that might just as well have been written in prose. *Elizabeth the Queen*, *Mary of Scotland* and *Winterset* closed the same doors they opened by showing the limits as well as the possibilities of verse tragedy, and there seemed no place to go once the point had been made. Anderson himself returned to prose; and the few verse plays written in America in the years to come were more likely to be by poets trying their hands at drama (e.g. Archibald MacLeish and Robert Lowell) than by dramatists drawn to poetry.

On the other hand, Anderson's dissatisfaction with the prosaic language and naturalistic world view of much of American drama of the 1930s was a legitimate criticism of a real limitation of the realistic genre, and it is noteworthy that the next generation of essentially realistic writers would expand the definitions of realism beyond that limitation. The discovery that the evocative language and tools of poetry could be incorporated into a richer, heightened version of realistic prose would be one of the major contributions of Tennessee Williams; and while some dramatists, notably Arthur Miller,. would continue to look at domestic situations as the exemplifications of larger social forces, Williams and others in the 1940s and 1950s would, following Anderson's lead, find the value of such stories in the human interactions and adventures themselves.

Alternatives to realism: Thornton Wilder

The decade's other important stylistic non-conformist was, like Anderson, a conscious rebel against the limitations of domestic realism. To Thornton Wilder a play that was set too firmly in a specific time and place forsook the drama's ability to be universal, and a play that emphasized too assertively the realness of its setting lost the drama's ability to evoke a sense of magic and mystery. Moreover, like Anderson, Wilder had a greater interest and faith in the value of the individual adventure for its own sake than in its utility as a vehicle for political and social comment. Where Anderson's method for escaping the limitations of realism was to step into the clearly artificial through verse, Wilder's was to engage the audience's attention and imagination through the openly theatrical; where Anderson celebrated the value of the individual by raising him to tragic stature, Wilder sought out and illuminated the wonders of the most ordinary lives.

Until the mid-1930s Wilder was known primarily as a novelist, although he had published two volumes of short plays, *The Angel that Troubled the Waters* (1928) and *The Long Christmas Dinner* (1931). The second collection includes three plays that define the shape and purpose of Wilder's break with realism. In 'The Long Christmas Dinner' time is accelerated so that a single family dinner encapsulates the events of ninety years. Characters age and develop; some excuse themselves from the table to die, their seats filled by younger replacements; and the table talk describes the changes in the family's fortunes and surroundings as the years pass. 'The Happy Journey to Trenton and Camden' is assertively about nothing more than that: a family of four drives across New Jersey for a visit. In addition to condensing the several-hour trip into less than half an hour of playing time, Wilder focuses our attention by stripping away almost all distracting externals. The car consists of four plain chairs on a bare stage, the passing scenery is reflected only in the family's comments on it, and all secondary roles are played by a Stage Manager who also moves the few props. The small pleasures and adventures of the trip thus become remarkable because the play's unusual technique isolates them for appreciation. The third of Wilder's short experiments is the most ambitious. Like 'The Happy Journey', 'Pullman Car Hiawatha' has a Stage Manager, who marks the outline of a railroad car in chalk on the bare stage and cues the actors to carry on the chairs that will represent their berths and compartments. This Stage Manager is more obtrusive than the other, addressing the audience, controlling the action, instructing the passengers to let us hear first their small talk and then their private thoughts, and interrupting to call on a parade of allegorical figures who provide us with 'its position geographically, meteorologically, astronomically, theologically considered'.

The double time scale of 'Dinner' and the parade of allegorical figures in 'Hiawatha' would reappear in *The Skin of Our Teeth*, both of them in more sophisticated forms. The bare stage and minimal props, and the resulting focus on the universality of the human adventure, anticipate *Our Town*, as do the narrative and directorial figure of the Stage Manager, the impulse to bracket the dramatized event in its cosmic context, and even some specific language, including a reference to a town called Grover's Corners and a dead woman's farewell to the world.

Our Town (1938) is one of the best-known and most beloved of American plays. Its quiet story of the value of ordinary lives, told in a highly sophisticated style that has the deceptive appearance of simplicity, and its unashamed appeal to open and honest sentimentality make it perhaps the most accessible of great plays, and a staple of the amateur and student theatre repertoires. On its surface it is the story of thirteen years in the lives of a handful of small-town people; just below its surface it is a celebration of the holiness of the ordinary; and skilfully hidden by Wilder's art it is a carefully constructed exercise in non-realistic theatrical method.

Our Town opens in 1901 in the fictional small town of Grover's Corners, New Hampshire, where teenagers George Gibbs and Emily Webb are neighbours and school friends. Three years later we attend their wedding, and nine years later Emily's funeral. The play's central message is that we require no tragic heroes or allegorical interpretations to make ordinary life significant, because the simple and mundane facts are themselves almost too precious and extraordinary for us to absorb and appreciate; and the specialness of the ordinary is affirmed repeatedly throughout the play. Act One ends with news from George's younger sister Rebecca:

> I never told you about that letter Jane Crofut got from her minister when she was sick. He wrote Jane a letter and on the envelope the address was like this: It said: Jane Crofut; The Crofut Farm; Grover's Corners; Sutton County; New Hampshire; United States of America. . . . But listen, it's not finished: the United States of America; Continent of North America; Western Hemisphere; the Earth; the Solar System; the Universe; the Mind of God. . . . And the postman brought it just the same.

However tiny, however insignificant, however lost in the hugeness of the universe, Jane Crofut can still be found, and is still worth finding. Later, Wilder interrupts the preparations for the wedding to go back to the moment when George and Emily fell in love, only to discover that lives are not changed by grand dramatic events, but by moments so tiny and fleeting that they cannot be pinned down. All we can see is the moment the couple realize that the other moment has just passed:

> GEORGE Emily, if I *do* improve and make a big change . . . would you be. . . I mean: *could* you be . . .
> EMILY I. . . I am now; I always have been.
> GEORGE So I guess this is an important talk we've been having. [Ellipses Wilder's]

Two technical devices overtly announce the play's departure from the conventions of dramatic realism, and less obviously help the play to achieve the ends Wilder felt realism could not. The first is the almost complete absence of scenery or stage props. As in 'The Happy Journey' and 'Pullman Car Hiawatha', plain tables and chairs are used to imply all the other furnishings in the rooms around them, while most hand props are mimed by the actors. At the end of Act One, in the play's most famous visual effect, George and Emily sit on top of two ladders on a bare stage, creating the impression of their two houses and the upstairs windows through which they call to each other. Thus, as in the earlier plays, Wilder makes the ordinary significant simply by focusing on it in an unexpected way: when all we are allowed to see is a mother preparing breakfast and sending her children off to school, or a teenage boy puzzling over his algebra homework, or a girl wondering if she is pretty – each of these without even sets or scenery to distract us – then we can look at each of these afresh, and find the beauty in them.

The second break with realism is the use of a Stage Manager even more intrusive than the one in 'Pullman Car Hiawatha'. He provides continuity, background information and overt commentary, as in his Act Two salute to wives and mothers:

> I don't have to point out to the women in my audience that those ladies they see before them, both those ladies cooked three meals a day – one of 'em for twenty years, the other for forty – and no summer vacation. They brought up two children apiece; washed; cleaned the house, – and *never a nervous breakdown*.

He plays a variety of minor roles, and frequently interrupts and directs the action to control the flow and time scheme of the play. It is he who allows the dead Emily to revisit a day from her childhood before leaving the earth, and who counsels her on her return; and it is he who reminds us at all times that what we are watching *is* a play, an invented reality distinct from the one we inhabit, and that he can move back and forth between the two at will. He can also move backwards and forwards in time, and Wilder uses this magic to emphasize the play's pastoral and nostalgic atmosphere: 'Along here's a row of stores. Hitching posts and horse blocks in front of them. First automobile's going to come along in about five years.'

The play's emotional climax comes when the dead Emily insists on her right to relive a day from her past, a morning scene very similar to the one that opened the first act. She discovers that it is too painful to experience again:

> I can't. I can't go on. It goes so fast. We don't have time to look at one another. I didn't realize. So all that was going on and we never noticed. Take me back – up the hill – to my grave. But first: Wait! One more look.
> Good-by, Good-by, world. Good-by, Grover's Corners . . . Mama and Papa. Good-by to clocks ticking . . . and Mama's sunflowers. And food and coffee. And new-ironed dresses and hot baths . . . and sleeping and waking up. Oh, earth, you're too wonderful for anybody to realize you.
> [Ellipses Wilder's]

She asks the Stage Manager if any humans ever realize the preciousness of life. He answers, 'No. The saints and poets, maybe – they do some.' The purpose of the play, of course, is to number the audience among the exceptions.

Wilder's next play, *The Merchant of Yonkers* (1938), is a light farce loosely based on a nineteenth-century Austrian play about the clerks of an intimidating provincial shopkeeper, who steal a day off to go to the big city. Wilder adds the character of the local marriage broker, who encourages the boys in their liberation and catches the shopkeeper for herself. (A 1954 revision, *The Matchmaker*, was the basis for the very successful 1964 musical *Hello, Dolly!*) The play's enthusiastic employment of the artificialities of farce reflects Wilder's delight in open theatricality; and Wilder reaffirms his celebration of the simple pleasures of life, as the cold merchant is gradually humanized and the boys learn to appreciate both the value of adventure and the comfort of the safe and familiar.

The Skin of Our Teeth (1942) is a complex allegory saved from pretentiousness by an ironic attitude towards its own ideas and methods; almost everything Wilder says and does in the play is immediately made the subject of a joke, demystifying both philosophy and dramaturgy – this at the apparently acceptable cost of an ironic distance that prevents the degree of identification and emotional involvement that makes *Our Town* so powerful. It purports to be about a typical American couple who live in New Jersey with their children and maid. But Mr and Mrs Antrobus are also Adam and Eve after the fall, and also Noah and his wife, and also embodiments of the spirit of man and womankind from the cave days to the present, and also the actor and actress playing these roles; and Wilder breezily mixes these metaphors with no apparent concern for consistency. In Act One, Mr Antrobus returns home from a busy day at the office, where he has invented the wheel, the alphabet and the multiplication table, to discover that his family is threatened by approaching glaciers. In Act Two he addresses a convention of the Ancient and Honorable Order of Mammals before escaping from a sudden flood with some of the other delegates and their wives. In Act Three he returns from a world war against an enemy led by his son Henry (né Cain) and prepares to start up the work of civilization again.

The reassurance implied by the title, that humanity can and will survive anything, if only barely, must have been particularly welcome to its wartime audiences: despite moments of weakness, lapses of faith, and everything that an unpredictable cosmos can throw at them, Mr Antrobus's inventiveness and Mrs Antrobus's domesticity carry them through. Indeed, we are shown that survival and progress are a matter of instinct as much as conscious determination; even the wavering and sceptical maid Sabina is ultimately incapable of giving up: 'All right, I'll go on just out of *habit*, but I won't believe in it.' In each act of the play Mr Antrobus has a moment of despair or back-sliding from which he is able to rouse himself just in time: the inventor of the wheel is broken by the news that Henry-Cain has been throwing stones again until Mrs Antrobus's faith revives his spirits; the conventioneer is seduced by bathing beauty Sabina and ready to leave his wife, but her commitment to the family wins him back just in time for them to board the Ark; the war veteran's encounter with the unrepentant Henry again depletes his reserves of hope until he reminds himself that 'I have never forgotten for long at a time that living is a struggle. I know that every good and excellent thing in the world stands moment by moment on the razor-edge of danger and must be fought for. . . . All I ask is the chance to build new worlds and God has always given us that.'

Added to the play's anachronisms and mixed metaphors is its constant assertion of itself as a play. The actress playing Sabina repeatedly revolts against her part – 'I hate this play and every word in it' – and the lines she is forced to speak – 'I don't understand a word of this play.' These interruptions are startling and funny, disarm any audience suspicion that the play is over their heads, and in fact encourage the audience, who are rarely as confused as the actress claims to be. They are also used seriously; in Act Two the Sabina actress refuses to play the seduction scene 'Because there are some lines in that scene that would hurt some people's feelings and I don't think the theatre is a place where people's feelings ought to be hurt.' There's a joke there, about the popular belief that the theatre should be nothing but light entertainment, but also the opportunity to talk directly about the scene's implications. Sabina also interrupts the climactic scene between Mr Antrobus and Henry because last night the actors got carried away and actually fought; in the brief psychodrama that follows, the two actors find flaws in their offstage personalities that parallel and underline their characters' moral failings. And the play ends with a theatrical trick that is both joke and summing up: the action restarts with Sabina's opening lines, which she interrupts to address the audience: 'This is where you came in. We have to go on for ages and ages yet. You go home. The end of this play isn't written yet.'

Thornton Wilder's place in American drama, like Maxwell Anderson's, is an ambiguous one. There can be no question that *Our Town* is a play of major stature, one of a small handful of American dramatic masterpieces. But, as breaks with the mainstream of realistic drama, Wilder's formal and theatrical experiments proved to be as much dead ends as were Anderson's verse tragedies: he proved without question that they could be done, and done effectively; and, that having been proved, there oddly seemed little point in his or anyone else's doing them again.

And yet, like Anderson, Wilder did expose some of the failings of domestic realism as practised in the 1930s. In particular, he spoke of a hunger for more exploitation of the theatre's potential for imaginative magic, and also for consideration of domestic lives for purposes other than social or moral comment. *The Skin of Our Teeth* says that domestic melodrama, freed from the constraints of place- and time-specific social comment, can speak of the essential human condition, while *Our Town* insists that the domestic world is of interest and importance in itself. Two highly significant changes would be seen in the mainstream domestic drama of the 1940s and 1950s – a movement from political to psychological and spiritual subjects, and a loosening of the bounds of realism to allow the occasional dream sequence or expressionist effect. While neither of these can be traced directly to Wilder's example – just as the increasingly evocative and poetic prose cannot be said to come directly from Anderson, – in each case the later developments were manifestations of the same impulse to stretch the dramatic form that inspired Anderson and Wilder.

Eugene O'Neill

As he had in the previous decade, Eugene O'Neill spent the 1930s following his own drummer, paying little heed to what was going on in the mainstream or alternative drama around him. Only three new O'Neill plays were staged in the 1930s, each, characteristically, completely different in style from the others and from the string of stylistic experiments that had preceded them.

Mourning Becomes Electra (1931) is a trilogy based on the *Oresteia* of Aeschylus. On its surface, it is a domestic melodrama set just after the American Civil War, with none of the overt violations of realism that characterized O'Neill's plays of the 1920s. But it is significantly different from the domestic plays of O'Neill's contemporaries in more than its length (a total of thirteen acts, meant to be performed in a single night). Finding modern equivalents of the ancient Greek sense of Fate in a combination of mythical allusions, American Puritanism and Freudian psychology, O'Neill reaches in *Mourning Becomes Electra* for the stature and power of classical tragedy.

Aeschylus's trilogy traces the effects on the royal house of Atreus of a curse invoked by Atreus's brother Thyestes. Aeschylus joins the story at the end of the Trojan War, as Agamemnon, son of Atreus, is murdered by his wife Clytemnestra and her lover Aegisthus, son of Thyestes. In the second play Agamemnon's son Orestes, spurred by his sister Electra, avenges his father by killing the murderers; and in the third Orestes is haunted by the avenging Furies until the Olympian deities Apollo and Athena rescue him. O'Neill translates the story and characters bodily into an American setting. Following Sophocles and other writers, he focuses on Electra more than Orestes, but otherwise he makes few significant changes in the basic situation: General Ezra Mannon returns from the Civil War to be murdered by his wife Christine and her lover Adam Brant, Ezra's cousin. Lavinia Mannon (Electra) moves her brother Orin (Orestes) to kill Brant, after which Christine commits suicide. Driven mad by guilt, Orin kills himself; and Lavinia accepts the inescapability of the Mannon curse, withdrawing from the world to live out her days in the family mansion. The parallels to Aeschylus are open, with the title, the names (Ezra Mannon/Agamemnon), the set (the Mannon mansion is described as resembling the *skene* of Greek tragedy), and the use of a chorus of townspeople calling attention to them.

The power of Aeschylus's tragedy comes from three central sources: the concept of an inevitable Fate in the form of the curse, which separates the descendants of Atreus from the rest of humanity and marks them for pain and doom; the violations of profound taboos in the string of crimes and counter-crimes, particularly Orestes' matricide; and the allegory of a redeeming cultural evolution in the Olympian gods' defeat of the Furies. O'Neill's trilogy does not include the third of those elements, since he offers no escape or redemption for the Mannons, but he does reach for the dramatic power of the first two; and his biggest challenge in *Mourning Becomes Electra* was to find substitutes for Aeschylus's Fate and cultural taboos that would have an equivalent effect on twentieth-century American audiences.

In place of the curse O'Neill creates a Mannon temperament, a concentration of pride, intense passion barely held in check by Puritan repression, and susceptibility to feelings of guilt. The chorus's reaction to Christine's first appearance is that 'There's somethin' queer lookin' about her face. . . . Secret lookin' – 's if it was a mask she'd put on. That's the Mannon look. They all has it.' Each member of the family is described at some point as wearing a version of Christine's cold, emotionless mask: Ezra Mannon's voice '*has a hollow repressed quality, as if he were continually withholding emotion from it*,' and he admits in a weak moment that 'Something queer in me keeps me mum about the things I'd like most to say – keeps me hiding the things I'd like to show. Something keeps me sitting numb in my own heart – like a statue of a dead man in a town square.' The Mannons are not emotionless, but the only emotions that can break through their masks are the strongest, and those are likely to be warped by the strain. Disagreements of any sort instantly evoke intense, murderous hatred; uncertainties turn into panic or madness; regret becomes unbearable guilt. Their shared pain from this constant swing between emotional deadness and frighteningly uncontrollable passion is betrayed by their shared fantasy of escape to a South Sea island of natural, unperverted emotion. Orin and Lavinia do go to that island between the second and third plays, and Lavinia does find a shameless romantic and sexual abandon there, only to discover on her return that her experience and change are seen as depraved and sinful. Love, like other emotions, is only permitted to the Mannons in a pathologically intense and corrupted form.

As part of his modern image of Fate, and as an equivalent of the deeply frightening archetypal crimes of the *Oresteia*, O'Neill invokes Freudian sexual psychology, making the Mannon family a tangle of warped and uncontrolled sexual conflicts. The most centrally and explicitly presented are the Oedipal and Electra complexes of Orin and

Lavinia. Lavinia idolizes her father, rejects any suggestion of similarity to her mother (although O'Neill stresses that it is there), and is Christine's rival for each of the men: 'You've tried to become the wife of your father and the mother of Orin!' Orin not only has an unresolved Oedipal conflict – referring to Brant, he says, 'If I had been he I would have done what he did! I would have loved her as he loved her – and killed Father too – for her sake!' – but he is encouraged in it by Christine's excessive need for him: 'I feel you are really – my flesh and blood! She isn't! She is your father's! You're a part of me!' Though speeches such as these suffer from being a little too explicitly spelled out, the Freudian complexes do have the effect of exposing culture-threatening forces that can affect an audience with the emotional power of Aeschylus's mythic taboos. And, when seen as part of the more intricate and sophisticated Mannon temperament O'Neill has created, they do contribute to a sense of fatal forces beyond the control of the doomed characters.

Mourning Becomes Electra is as successful as one could imagine a modern version of a Greek tragedy to be. But, like the neo-Elizabethan tragedies of Maxwell Anderson or, indeed, most of O'Neill's other experiments, it is also a dead end. In proving that it can be done O'Neill also seems to have proved that this is as successfully as it can be done, leaving no room for any further exploration of the subgenre. It is worth noting that, its length and classical echoes aside, *Mourning Becomes Electra* is on its surface a realistic domestic melodrama. Perhaps unconsciously, O'Neill was discovering in his own terms what his contemporaries were discovering in theirs, that domestic realism was more adaptable to his ambitious visions than he had previously realized. O'Neill would certainly retain his tragic sense and his inclination to embody it in drama, and when he returned to that impulse near the end of his career, he would find shape for it entirely within the bounds of domestic realism.

O'Neill's next play, *Ah, Wilderness!* (1933), stands out even in the context of a career in which no two plays seemed to have been written in the same style. It is O'Neill's only comedy, with a tone and content that specifically deny the dark vision of human guilt and limitation that is O'Neill's hallmark – an elegiac, nostalgic piece set in an innocent small-town America of 1906 that probably never existed, in which an adolescent boy's rebellious fling amounts to little more than getting painfully drunk, and all evils are merely easily-rectified misunderstandings. It is almost as if O'Neill chose, as an act of will, to pretend he believed in everything that he disbelieved about life, and to write a play in that spirit. *Ah, Wilderness!* is pleasantly entertaining, in part because it is so bland and conventional; were it not by O'Neill there would be little to say about it, and what there is to say makes it of interest as a footnote to other plays rather than for its own merits. In particular, this is one of the first O'Neill's plays since some of the early one-acts to be drawn directly from his own life, though in a highly fictionalized and romanticized way; and it invites comparison with the later *Long Day's Journey into Night*, both for its similarities – the self-portraits have a similar ironic distance – and differences – the later play presents a far darker vision of the family and the world it inhabits.

Days Without End (1933) was intended as the second play of a trilogy on the theme of man's search for faith that began with *Dynamo*, but its failure led O'Neill to leave the third play unwritten. The hero of *Days Without End*, John Loving, is a man whose loss of faith has led to the disintegration of his personality, a condition O'Neill dramatizes by dividing the character between two actors, the good but weak John and the bitter and evil Loving. Only John and the audience see Loving, who wears a mask of John's face in a cynical sneer, but the other characters hear both figures as if they were one man speaking. *Days Without End* seems at times to be made up of leftovers from earlier O'Neill plays: the mask and divided personality from *The Great God Brown*, the desperate search for faith from *Dynamo*, the internal debates from *Strange Interlude*, with other echoes of *Lazarus Laughed* and *Welded*, and anticipations of *Long Day's Journey*. The play is clumsy in writing and construction, and O'Neill may have sensed that his experiments with expressionism and other non-realistic techniques had run their course; it would be thirteen years before his next play, which would mark a radical change in dramatic style, was produced.

Comedy

If one of the functions of drama is to reflect and illuminate the world of reality to its inhabitants, certainly another is to distract them from it; and it is not surprising that the majority of comedies produced during the Depression and war years either ignore the harsher facts of life, actively deny them, or deny their importance. Several distinct subgenres of comedy can be discerned in this period, and virtually all of them pretend that there is nothing more serious in life than whether or not a particular pair of attractive young lovers find each other.

Many comedies of the 1930s are virtually indistinguishable from similar plays of the preceding decade. Working with a new collaborator, Moss Hart, George S. Kaufman continued his practice of satirizing easy targets with fast-paced wit and humourous dialogue. *Once in a Lifetime* (1930) skewers Hollywood with many of the same jokes as the Kaufman–Connelly *Merton of the Movies* of eight years earlier, as does Samuel and Bella Spewack's Hollywood satire *Boy Meets Girl* (1935). Other successful comedies that might well have been written in any other decade include A. E. Thomas's *No More Ladies* (1934), about the wife who wins back her straying husband by pretending a dalliance of her own, and Kaufman and Hart's *George Washington Slept Here* (1940), about the predictable tribulations of a city man who buys a run-down country house.

Also similar to earlier plays are the 1930s versions of 'beautiful people' drawing-room comedies in the Noel Coward manner, whose only dramatic question is which rich and witty man will end up with which rich and witty woman. Essentially the same basic plot generates such plays as S.N. Behrman's *Brief Moment* (1931), in which a weak millionaire marries a poor girl who quickly makes a man of him; Philip Barry's *The Animal Kingdom* (1932), in which the man marries a conventional girl who emasculates him, but returns to his Bohemian mistress and happiness; and Samson Raphaelson's *Accent on Youth* (1934), in which a devoted secretary rejuvenates a middle-aged playwright. The best of this group alters the formula with a switch of genders: in Philip Barry's classic *The Philadelphia Story* (1939), icy heiress Tracy Lord is unforgiving of any imperfections in others, be they her ex-husband Dex's irresponsibility or her father's extra-marital straying, until an innocent slip of her own humanizes her and she finds happiness in remarriage to Dex. As with the other plays of this subgenre, the pleasure for the audience is provided, in almost equal parts, by witty banter, a slight touch of non-conformity, the inevitable pairing of the right couple, and a parade of expensive-looking costumes on elegant sets; characteristically, Barry adds just enough sentiment and emotional depth to give his variant a warmth many of the others lack.

Different in style, but appealing to the same how-the-other-half-lives escapism, is Kaufman and Hart's *The Man Who Came to Dinner* (1939) about curmudgeonly media celebrity Sheridan Whiteside, who breaks his leg on a speaking tour and becomes the involuntary houseguest of an Ohio family. He completely disrupts his hosts' lives with a steady stream of imperious demands and shameless insults ('My great-aunt Jennifer ate a whole box of candy every day of her life. She lived to be a hundred and two, and when she had been dead three days she looked better than you do now.'). Probably most theatre-goers were aware that the main character and his show business friends were based on real-life celebrities of the day; they certainly were titillated by a peep into a world in which names like Cary Grant, H. G. Wells, Lord and Lady Astor, and more than fifty others were dropped in casual conversation.

If the drawing-room comedies allowed the audience to imagine a world untouched by the Depression or the rumblings of war, another group encouraged the fantasy that life could somehow go its benign way in spite of outside events. Kaufman and Hart's *You Can't Take It With You* (1936) presents the eccentric Sycamore family: Grandpa quit work thirty-five years ago because he was not having any fun; his daughter writes unfinished plays and paints unfinished portraits; her husband manufactures fireworks in the cellar; and various friends and acquaintances add to the disorganized happiness. The one relatively ordinary member of the family loves an ordinary boy, and the plot turns on a disastrous meeting of the Sycamores and the boy's staid parents which threatens the romance until Grandpa's wisdom can begin to loosen up the in-laws-to-be. The faith that Grandpa's dinner table prayer expresses –

> Well, Sir, we've been getting along pretty good for quite a while now, and we're certainly much obliged. Remember, all we ask is just to go along and be happy in our own sort of way. Of course we want to keep our health, but as far as anything else is concerned, we'll leave it to You.

– is the fantasy the play allows the audience to enjoy for two hours: that love, tolerance and innocent eccentricity make the cold fact of the Depression irrelevant, and that the world is ultimately a good place in which only good things happen to good people.

The same fantasy of a loving providence watching over the innocent colours such comedies as James Hagan's *One Sunday Afternoon* (1933) and Maxwell Anderson's *The Star-Wagon* (1937), both of which allow their heroes to relive their lives and discover how satisfying they are. It is there in Paul Osborn's *On Borrowed Time* (1938), in which Death is magically trapped in a tree until the old man he has come for decides he is ready to go; and in Osborn's *Morning's at Seven* (1939), about four elderly sisters and their families, whose lives include problems no greater than wondering if a forty-year-old son will finally marry. It lies beneath *My Sister Eileen* (1940), the dramatization by Joseph Fields and Jerome Chodorov of Ruth McKenney's short stories about two naive Ohio girls who come to New York and find nothing harsher than romance, jobs and friendly neighbours. It infuses Joseph Kesselring's *Arsenic and Old Lace* (1941), in which a pair of little old ladies blithely murder lonely old men and let their mad brother, who thinks he's Teddy Roosevelt, bury them in the Panama Canal he's digging in the cellar. And it is certainly there in Mary Chase's *Harvey* (1944), about a gently potty man whose best friend is a six foot tall invisible rabbit. The same wilful blindness to reality may explain the popularity for several years of plays about the innocent adventures and small calamities in the lives of teenagers. Clifford Goldsmith's *What a Life* (1938), Chodorov and Fields's *Junior Miss* (1941), F. Hugh Herbert's *Kiss and Tell* (1943), and Norman Krasna's *Dear Ruth* (1944) all implicitly pretend that life holds no greater dangers than the school bully and no deeper emotions than puppy love. The subtext of all these comedies is the pretense that the more unpleasant facts of the outside world are not really there.

One effective escape from the unpleasantness of the present is a retreat into the past, and it is probably not coincidental that the most successful comedy of the 1940s is set in the 1880s. *Life With Father* (1939), drawn from the stories of Clarence Day by Howard Lindsay and Russel Crouse, celebrates an innocent past in which gruff and domineering father could easily be manipulated by his wife and children and the most dramatic event in their lives was his somewhat-delayed baptism. Much the same spirit permeates John Van Druten's *I Remember Mama* (1944), from Kathryn Forbes' stories of growing up in a Norwegian family in San Francisco, and the more acerbic *The Late George Apley* (1944) by John P. Marquand and George S. Kaufman from Marquand's novel: though the Boston Brahmin father in that play does turn his son into a stuffy duplicate of himself, the distance of history blunts any satire or suggestion that all is not for the best in this best of all possible worlds.

Still, a few comic writers of this period managed to acknowledge the world outside or the possibility of emotions deeper than love at first sight. S.N. Behrman gave the drawing room subgenre some additional weight by adding a bittersweet, elegiac tone in *Biography* (1932) and *End of Summer* (1936). In the first, a Bohemian artist writing her kiss-and-tell autobiography realizes it could harm an old lover; forced to recognize, perhaps for the first time, that her actions affect others, she destroys the manuscript and ends the play a deeper character than she began. *End of Summer* has a similar spirit, as an aging social butterfly comes to realize, and accept with some grace, the fact of her own inconsequentiality.

Other Behrman plays make a point of searching for comedy in the face of reality rather than through the denial of it. *Rain From Heaven* (1934) would be a typical drawing room comedy, even to the extent of being set in an English country house, were it not that the two men the heroine must choose between are an apolitical adventurer and a Jewish anti-Nazi fighter, so that scenes of typically brittle wit alternate with serious political and moral debates. In *No Time for Comedy* (1939), a Noel Coward-like playwright cannot bring himself to write drawing-room comedies in the face of the horrors of war, and the play reaches a comic resolution without pretending that the world outside its boundaries is also harmless and comic. Behrman is one of the very few wartime comic writers to acknowledge a reality harsher than the fact that Washington DC had a civilian housing shortage (the comic basis of Joseph Fields's 1942 comedy *The Doughgirls*) or that soldiers are attracted to pretty girls (John Van Druten's 1943 *The Voice of the Turtle* among others). In *Jacobowsky and the Colonel* (1944), adapted from a play by Franz Werfel, Behrman follows a desperate but resourceful Polish Jew as he stays one step ahead of the advancing Nazis in France.

It is striking how very few other comic plays in this fifteen-year period attempt a little extra depth of emotion, characterization or reality. Kaufman and Hart's *Merrily We Roll Along* (1934) traces the moral and artistic decline of a writer who sells out for success, but it is more notable for its narrative trick of telling the story in reverse chronology, starting with the writer's corruption and working back to ironic scenes of youthful idealism. Robert E. Sherwood's *Idiot's Delight* (1936), mentioned earlier in this chapter, would be a conventional light romance between sophisticated Americans in Europe were it not set on the eve of war, with the protagonists learning the importance and dangers of political commitment from the Europeans around them. Clare Boothe's *The Women* (1936) looks at first like a typical beautiful-people comedy, with the gimmick of an all-female cast only serving to permit lots of elegant gowns and catty repartee, but it confounds the conventions of the genre by presenting what should be a figure of fun, a deceived wife, as deeply unhappy. *Stage Door* (1936), by Edna Ferber and George S. Kaufman, is a sentimental celebration of the dreams and dedication of would-be actresses in a theatrical boarding house, but it offers a passing glimpse of reality when one of the girls kills herself, showing the toll that poverty and constant rejection can take.

By 1930, then, it was clear that drama in America had become a significant art form, capable of supporting ambitious literary intentions, both serious and comic, while still meeting its social function as a source of entertainment. Ironically, the theatrical style in which the twentieth-century drama had been reborn – realism – had begun to seem limited and limiting to some of the most ambitious playwrights, and Eugene O'Neill and others had rejected it in their search for other modes through which they might be able to dramatize subjects and artistic visions that earlier generations of playwrights would never have attempted. In this context, the fifteen years covered in this chapter represent an important turning point, the return to dominance of domestic realism with the realization by playwrights that it was capable of much more than had been asked of it previously.

The harsh and inescapable external reality of the 1930s demanded dramatic consideration in a way that the more benign 1920s had not, and the major discovery by Clifford Odets and other socially aware playwrights of the period was that the largest political and social issues could best be dramatized through their reflection in a domestic situation. This led in turn to the further perception that almost any large subject – social, philosophical, moral – could be addressed successfully in the same manner, and not only because realism was a more flexible and adaptable mode than it had previously seemed. Something in the instinctively democratic American perspective recognized that the important thing to be said about any large subject was the way it affected the lives of ordinary people, and thus the American drama found its native and natural voice in domestic realism.

Domestic realism was not infinitely adaptable, of course, nor had all its potential yet been realized; and the alternative styles of Maxwell Anderson and Thornton Wilder, and continued strivings of Eugene O'Neill, were generated by a legitimate reaction against some of the dominant style's apparent limitations. But the important discovery had been made: experiments in alternative styles such as expressionism were simply unnecessary. The occasional variation notwithstanding, the drama and dramatists of the Depression and war years made it clear that the typical American play was going to be set in someone's living room. The next step would be further exploration of just how rich a dramatic setting that could be.

Chapter 4
1945-1960: The Zenith of the Broadway Theatre

The era of Williams and Miller

The period covered in the preceding chapter saw the American drama finding its natural voice in domestic realism, and the exploration of some of the potential scope and power of that mode, particularly its ability to deal with large social, political and moral issues. In the fifteen years following the Second World War the next generation of American dramatists built on these discoveries, and on the psychological dramas of earlier writers, to carry the Broadway-based American theatre to its highest accomplishments. This period produced the greatest plays of America's greatest playwrights: the early works of Arthur Miller and Tennessee Williams, and the late and posthumous works of Eugene O'Neill. It was a golden age for Broadway in other ways as well, featuring the highest achievements of the musical theatre, most notably in the collaborations of Richard Rodgers and Oscar Hammerstein; the introduction and development of 'Method' acting; and the creative and illuminating accomplishments of such directors as Elia Kazan and Jose Quintero.

In one important way the drama of the 1950s turned away from the direction set in the 1930s. Although some playwrights, notably Arthur Miller, continued to address social and political issues by dramatizing their effect on the everyday domestic lives of ordinary people, others, led by Tennessee Williams (and anticipated by O'Neill), returned to the focus on psychological exploration that had been the concern of the realistic dramatists of the 1920s. Now, however, the concentration was not so much on explaining abnormal behaviour through Freudian analysis, as George Kelly, for example, had done in *Craig's Wife*, but on expressing and illuminating the emotions that lay beneath both normal and abnormal behaviour. Not the least of Method acting's contributions to the theatre was its making playwrights and audiences as well as actors aware of 'subtext', the emotional drama going on beneath the most mundane of conversations. For some of the best American dramatists of the 1950s and thereafter, the subject to be dramatized was their characters' spiritual and psychic adventure, what it felt like to be alive and trying to cope in the middle of the twentieth century.

More often than not, what they discovered was that it felt rather frightening, that even their most successfully functioning characters were actually being driven by feelings of inadequacy or inability to cope with the ordinary pains and pressures of life. Plays that dramatized this insight could give the audience a sort of comfort, in the demonstration that such insecurities were not individual failings but part of the human condition. Plays that showed realistically uncertain characters coping with or triumphing over their fears provided encouragement and inspiration. So the function and capacity of domestic realism was expanded once again, to include a spiritual content and purpose, in plays that not only explained or argued, but actually offered counsel, consolation and reassurance.

Just as the American drama of the 1920s was dominated by the figure of Eugene O'Neill and the drama of the 1930s given shape and focus in the plays of Clifford Odets, the mid-century drama can be defined by the work of, in this case, two dominant figures. (As always, Eugene O'Neill, though towering over the others, stood apart from them, with little effect on his contemporaries.) There is a difference, however, and it is not just a matter of number. O'Neill was without question the best and most ambitious American dramatist of the 1920s, but he was not a part of whatever mainstream there was. Odets helped to discover and shape the mainstream dramatic mode of his decade and after, but that is clearer in hindsight than it was at the time, when he seemed to be just one talented writer among many. The 1945–60 period was clearly dominated by its two leading dramatists; virtually every serious American play of the time functioned within artistic boundaries defined by the plays of Tennessee Williams and Arthur Miller.

It may be useful, therefore, to start with an overview and some generalizations, before looking closely at each writer. Williams and Miller had strikingly parallel careers. In each case some early writing led to an inauspicious New York debut: Miller's first Broadway play lasted four performances while Williams's closed out of town. Each followed with a second play that declared him to be a writer of importance, and with a third that would prove to be the finest play of his career. They went on to write some of the best American plays of the 1950s, and both ended their periods of greatest accomplishment in the early 1960s.

Beyond the parallels and the excellence of the individual plays, Williams and Miller had clearly developed dramatic styles and concerns that between them delimited both the manner and the matter of mid-century American drama. Following the lead of the previous decades, both wrote predominantly in the mode of contemporary, realistic domestic melodrama. But both helped to stretch that style a little through the introduction of symbolic or expressionistic elements that did not break or weaken the overall sense of realism. Miller followed in the direction begun by Herne and Crothers at the start of the century, and refined by Odets and the other realists of the 1930s, by using domestic stories to reflect larger political or moral issues. Williams, perhaps moved by an impulse parallel to those that led Maxwell Anderson and Thornton Wilder to other styles, helped to move domestic realism beyond this political focus towards its new function of illuminating psychological and emotional forces within his characters. Miller repeatedly took as his subject the ordinary person placed under extraordinary pressure by his society and either destroyed by it or triumphant over it; Williams wrote of extraordinary people, frequently the freaks and misfits of humanity, trying to survive the ordinary pressures of life emotionally and spiritually. Thus Miller's plays were, in effect, vertical in their thrust, celebrating the potential for stature and greatness in the Common Man (Miller's phrase), while Williams's were horizontal, speaking of the common humanity shared by the normal and abnormal. And while Miller, in keeping with his social agenda, strove for a style of clarity and direct statement, Williams brought the instincts of a poet to the drama, speaking and allowing his characters to speak in the language of symbols, emotionalism and rhetorical flourishes.

Miller and Williams were thus not merely the outstanding American dramatists of the period. Almost all their contemporaries also wrote predominantly in the mode of domestic realism, and chose as their subjects either the social/political or the psychological/spiritual. It would be tempting, therefore, to label each of them as being of 'the school of Miller' or 'the school of Williams' were the concept of such 'schools' simply not native to the American drama. While there are isolated cases of direct influence (Odets on Miller, Williams on Albee), what happened in the 1950s was as unplanned as what had happened in previous decades: individual writers followed their own muses, and their cumulative experience, along with audience response, gave a shape to the drama of the period. In this case, the styles and concerns of Arthur Miller and Tennessee Williams mark the ends of a very short continuum on which virtually every other serious American play of the 1950s can be placed, and thus define their age in a way that O'Neill and Odets did not define theirs.

Arthur Miller

Arthur Miller's first success, *All My Sons* (1947), introduces themes that would run through his work: the inescapable drive for economic success and public approval, the conflict between competing moralities, the temptation to do the wrong thing for the right reasons, and the identification of the political and economic system as the culpable cause of these torments and dilemmas. Its central character, Joe Keller, manufactured defective aircraft engines that killed some pilots during the war, but can come no closer to a confession than the assertion that he did what was necessary to support and protect his family, until the discovery that his son killed himself in shame leads him to recognize the higher moral obligation implied by the play's title. The play is told entirely in domestic terms: its fictional centre is the question of whether Joe's surviving son Chris should marry his dead brother's fiancée. The revelations about the past, and the direct discussion of the moral questions, come out in the course of resolving the family conflict; and Miller is scrupulously fair in presenting both the characterizations and the terms of the debate. Though Joe Keller is in the wrong, he is not the heartless capitalist of leftist propaganda, but a little man trying to do what seems to him to be the right thing:

> KELLER I could live on a quarter a day myself, but I got a family, so I . . .
> MOTHER Joe, Joe . . . it don't excuse it that you did it for the family.
> KELLER It's got to excuse it!
> MOTHER There's something bigger than the family to him.
> KELLER Nothin' is bigger!
> [Ellipses Miller's]

And while Chris is in the right, he has an unattractive element of self-righteousness in common with some of Miller's other 'correct' characters, and he is shown to be motivated more by his own psychological need for a world that makes moral sense than by any special purity of soul:

> Everything was being destroyed [in the war], see, but it seemed to me that one new thing was being made. A kind of responsibility. Man for man. . . . And then I came home and it was incredible. . . . Because nobody was changed at all. It seemed to make suckers out of a lot of guys. I felt wrong to be alive.

Chris *is* right, for whatever reason, and the discovery that the pilots Joe killed were his responsibility as much as his own sons leads to Joe's suicide. Miller has no doubt that Joe Keller is in the wrong for defining his moral accountability so narrowly and locally, and for losing sight of his connection to the larger human family. But the final moral judgement of the play is not Chris's 'There's a universe outside and you're responsible to it.' The play pities Joe Keller more than it condemns him.

It recognizes the sincerity of his values, inadequate as they were, and directs the blame at a world that would turn a man's love for his family into a crime and force him to choose between conflicting moral imperatives.

The same condemnation of an economic and social system that precipitates individual tragedy is the generating force behind *Death of a Salesman* (1949), Miller's best play and one of the finest and most important plays by any American dramatist. *Death of a Salesman* is the story of a small man driven to madness and despair by his inability to succeed economically in a system that seems to guarantee success; it is an exposé of the dark and dangerous side of the American Dream. To dramatize his hero's internal experience and focus attention on his spiritual and mental collapse, Miller inserts an expressionistic element in his essentially realistic play, as Willy Loman repeatedly withdraws into memories of the past and into imagined conversations with his brother Ben, his symbol of success. Thus, much of the play is internal; in key scenes it is not reality that we see, but reality as experienced, remembered or imagined by a man for whom reality is too frightening to face directly. Willy is a hard-working travelling salesman who has always tried to believe that he was more successful than he actually was. No longer able to keep up the pretence and yet unable to give up his dream, he puts all his hopes in one last fantasy of success for his son Biff, who tries to make his father recognize and accept their parallel failures. Determined to affirm his faith that Biff could succeed if he only had the starting capital Willy's insurance would provide, Willy kills himself.

Willy's life is defined by his absolute faith that success, measured in economic terms, is available almost without effort to those who deserve it, 'Because the man who makes an appearance in the business world, the man who creates personal interest, is the man who gets ahead. Be liked and you will never want.' The problem with that definition of how the world operates is that reality does not support it; Willy and Biff are well-liked and unsuccessful while their neighbours Charley and Bernard are unimpressive drudges who succeed. So Willy must work harder and harder to sustain his conviction that economic triumph is available to him and his sons, and to deny a reality that threatens that faith: thus the increasing withdrawal into the past, and thus a continuing conflict with Biff, who is slowly escaping the dream. Ultimately the preservation of the fantasy is more important to Willy than his own life, and he dies to affirm it.

The source of Willy's fantasy is so obvious that Miller does not even have to identify it. It is the American Dream – the cultural doctrine that America is by definition the land of opportunity, the place where any boy can grow up to be President, the place where success is a birthright waiting to be claimed, 'and that's the wonder, the wonder of this country, that a man can end with diamonds here on the basis of being liked'. Willy Loman's own experience proves that the dream is false: success is in fact not guaranteed to the well-liked, or even to the hard-working. But that is not the American Dream's greatest lie. The inevitable corollary to the belief that success is available to everyone is the conclusion that failure can only be the fault of the individual. Despite his growing realization that he does not want what the world calls success, Biff can only call himself a worthless failure for not achieving it. The possibility that Willy was simply not fated to be rich, or that he took the wrong route to it, is beyond Willy's comprehension. If they have not grasped the bounty that the American Dream insists is just lying there waiting for them, they must not be good enough, and that conclusion is so frightening to Willy that he will die rather than consider it. Even worse, the American Dream is exclusive; it will not allow its believers to acknowledge or even imagine alternative definitions of value or accomplishment. Willy has a loving wife, a supportive friend, and sons who try, within their limits, to love him. He enjoys looking at nature and working with his hands. But Willy cannot see the value in these things; the only definition of success available to him is measured in dollars, and in numbers that are always beyond his reach.

There are many resonances to Miller's choice of a salesman as the archetypal American, ranging from a reflection of the materialism that engulfed America in the post-war economic boom to the irony that Willy himself has been 'sold' the fantasies that trap him. One effect that the play underlines is the denial of any blame or criticism of Willy's blind faith in the American Dream; the fault is in no way his for believing what his culture gives him no option but to believe. A salesman's life is by its very nature constructed entirely of faith and fantasy: he trades the promise of delivery for the promise of payment. As Charley says at Willy's grave, 'Willy was a salesman. And for a salesman, there is no rock bottom to the life. . . . A salesman is got to dream, boy. It comes with the territory.' Even at the end of the play Biff (and only Biff; his brother reaffirms his commitment to Willy's doctrine) will only say of Willy, 'He had the wrong dreams,' acknowledging the possibility of alternative definitions of success and personal value, but not condemning his father for needing a dream.

The tragedy of Willy's commitment to his dream is highlighted by the contrast to his son, who painfully achieves a partial release never available to his father. The memory scenes show Biff being indoctrinated into Willy's creed, and as an adult he remains trapped by it. Early in the play Biff tells his brother about his recent experiences in the American West:

> This farm I work on, it's spring there now, see? And they've got about fifteen new colts. There's nothing more inspiring or – beautiful than the sight of a mare and a new colt. And it's cool there

> now, see? Texas is cool now, and it's spring. And whenever spring comes to where I am, I suddenly get the feeling, my God. . . .

It is obvious where that sentence wants to go; the logical continuation is something like ' . . . life is wonderful'. But that insight is not open to Biff:

> . . . I suddenly get the feeling, my God, I'm not gettin' anywhere! What the hell am I doing, playing around with horses, twenty-eight dollars a week! I'm thirty-four years old, I oughta be makin' my future.

A moment earlier Biff had confessed, as if it was a personal failing, 'I don't know – what I'm supposed to want.'

And yet Biff does escape his father's fate by choosing reality over the fantasy. A flashback shows the young Biff discovering his father with another woman and turning against him. Most critics agree that this is the weakest element in the play, offering an overly simplistic Freudian explanation for Biff's break from Willy. Fortunately Miller does not rely too heavily on this moment, instead identifying Biff's escape with a perceptual leap beyond the ability of anyone else in the family:

> I stopped in the middle of that building and I saw – the sky. I saw the things that I love in this world. The work and the food and time to sit and smoke . . . . Why am I trying to become what I don't want to be? What am I doing in an office, making a contemptuous, begging fool of myself, when all I want is out there, waiting for me the minute I say I know who I am!

The power of the dream is so strong that Biff cannot fully escape it; the only liberation Miller allows Biff is what feels to him like the painful acceptance of his own inadequacy. Biff never sees that Willy's doctrine is false, merely that it is not for him. He still judges himself a failure, but he accepts that failure, gives up the futile struggle, and looks for his fulfilment elsewhere, in the personal satisfaction of 'the things that I love in this world'.

Much ink has been spilled over the question of whether *Death of a Salesman* is a tragedy, a question raised by Miller himself in his extensive writings and interviews about the play. Certainly it has few of the characteristics of classical or Shakespearean tragedy. Miller's own justification for applying the label is based on the play's socio-political statement; he argues that, just as classical tragedy implicitly accused the gods of unfairly denying the tragic hero his full stature, *Death of a Salesman* points the accusing finger at the life- and dignity-destroying aspects of American culture. But few readers or theatre-goers are likely to share that specialized definition of tragedy. If the story of Willy Loman inspires pity and fear, it is in its personal and human level – in the way the fall of this unheroic little man reminds us of our own fragile mortality, and in Miller's success in convincing us that the pain and death of any human being is tragic. As Willy's wife says, in the play's most powerful lines,

> Willy Loman never made a lot of money. His name was never in the paper. He's not the finest character that ever lived. But he's a human being, and a terrible thing is happening to him. So attention must be paid. He's not to be allowed to fall into his grave like an old dog. Attention, attention must be finally paid to such a person.

Miller's next play, an adaptation of Henrik Ibsen's *An Enemy of the People* (1950), was not a commercial success, but it can be seen as an important prelude to the play that would follow, *The Crucible*. Ibsen's account of a doctor who insists on publicizing scientific findings that endanger his town's economy is complex and even ambivalent, but ultimately asserts the absoluteness of truth. Miller's adaptation changes the terms of the debate from a matter of true and false to a conflict of deeply held beliefs, and thus shifts the moral centre of the play to the individual's right and duty to follow his conscience even in the face of an overwhelmingly opposing social consensus.

Since that theme, to be further explored and affirmed in *The Crucible*, had very specific topical implications in 1950, a brief reminder of the historical context may be in order. Soon after the end of the Second World War, one part of America's politically conservative movement became deeply afraid of Communist Russia, and particularly of the prospect of Soviet espionage and subversion within the United States. This fear was not totally unfounded, of course, and the discovery of Soviet spies and agents, along with the exploitation of the issue by some politicians, fuelled the anxiety until it extended to almost every level of American society. The phenomenon, and the hysterical patriotic fervour it generated, are most strongly associated with the name of one of the first politicians to capitalize on the issue, Senator Joseph McCarthy, though in fact it was the Un-American Activities Committee of the House of Representatives, along with a number of self-appointed citizens' groups, that did most to spread the hysteria.

At its most extreme, this panic led to the suspicion of anyone whose patriotism was not absolutely unchallengeable – in particular, to anyone who had ever done anything that could be interpreted as un-American. Unfortunately that definition covered a large percentage of Americans who had come of age during the Depression and had joined, supported or at least agreed with some innocent cause which was now retroactively tainted because the Communists had also supported it. Careers and lives were destroyed because of petitions signed or meetings attended twenty years earlier; and the only defence that Congress or the civilian patriotic organizations would accept was abject confession and cooperation in the investigation of others. Miller himself was called before the House Un-American Activities Committee later in the decade, as were many of his friends and theatrical associates; like some and unlike others, he refused to name anyone else for the Committee to investigate, and was convicted of contempt of Congress.

In this light, Miller's version of *An Enemy of the People* is not merely an abstract speculation on the nature and power of truth, but a directly political assertion of the democratic right to freedom of belief. The issue is not whether Stockmann is correct, as it is in Ibsen, but whether the mere fact that his opinion is unpopular – indeed, is perceived as a threat to society – is sufficient cause to suppress it and to require him to repudiate it. Where Ibsen's Stockmann stands firm in his conviction that he is right, Miller's stands for his right to his convictions. Stockmann's final speech, which most translators give as some variant of 'The strongest man in the world is the man who stands alone', becomes in Miller's adaptation less a theoretical aphorism and more an acknowledgement of the price to be paid for freedom of conscience: 'We're the strongest people in the world and the strong must learn to be lonely!'

In *The Crucible* (1953) Miller addresses the subject as directly as the times would allow, through an easily decipherable analogy to the witch-hunt hysteria of colonial America. The play touches on every element of McCarthyism: how a legitimate cultural fear is exploited by those to whom such exploitation gives power; how innocent people can be suspected on minimal, manufactured or non-existent evidence; how growing hysteria can make suspicion tantamount to proof; and how well-meaning citizens and even public officials are unable to resist the self-perpetuating process or the deviousness of those manipulating it. The whole procedure, from tainted accusers to show trials and hypocritical offers of mercy in return for new names, is dramatized in a way that makes both its obscenity and its modern parallels clear. This encounter between the judges and a townsman charging an accuser with personal motives could easily have come from a congressional hearing of the 1950s:

> GILES If Jacobs hangs for a witch he forfeit up his property – that's law! And there is none but Putnam with the coin to buy so great a piece. This man is killing his neighbors for their land!
>
> DANFORTH But proof, sir, proof.
>
> GILES, *pointing at his deposition* The proof is there! I have it from an honest man who heard Putnam say it! The day his daughter cried out on Jacobs, he said she'd given him a fair gift of land.
>
> HATHORNE And the name of this man? . . . The man that gave you this information.

GILES, *hesitates, then* Why, I – I cannot give you his name.
HATHORNE And why not?
GILES, *hesitates, then bursts out* You know well why not! He'll lay in jail if I give his name! . . .
DANFORTH In that case, I have no choice but to arrest you for contempt of this court.

Remarkably, given the very specific contemporary phenomenon that generated *The Crucible*, the passage of time has not seriously dated it. Some of Miller's righteous indignation does seem strident and overstated when the opposition is no longer a real and present danger, but that loss is balanced by an increased awareness of his main character's spiritual struggle, which is not tied to the topical allusions. When John Proctor's wife is named by a girl who lusts after John, his attempt to expose the accuser results in his own arrest. Tempted to save himself by confessing, John decides that honour requires his death. On this level his story has Kafkaesque overtones: just as some of the weaker victims of the witch hunt come to believe in their own guilt because it is asserted so repeatedly, John is led to examine his life in the light of the accusations and to realize that, if he is innocent of witchcraft, he is not without other guilt. Confession of his adultery with the accuser begins his spiritual self-evaluation, which leads to the belief that even his execution for witchcraft would be unearned, since he would be dying in the company of the truly innocent.

So Miller shows that the witch hunters were right in an ironic way they could not have guessed: where there is suspicion there *is* guilt, but only because the true accuser is within. The human spirit must eventually be honest with itself, and what it finds when it is ready for self-examination is a profound sense of unworthiness. Yet that impulse to be unsparing is not just self-condemning but the means to salvation because, once acknowledged, it can be relied on as a guide to truth. John signs his confession because he believes himself too guilty to die with honour, but he recants when he realizes that his life would be more of a theft from the innocent than his death: 'Beguile me not! I blacken all of them when this is nailed to the church the very day they hang for silence!'

Again and again for Miller the key and the metaphor for self-discovery and self-acceptance is the individual's ability to name himself. Joe Keller insisted in *All My Sons* that 'I'm his father and he's my son, and if there's something bigger than that I'll put a bullet in my head' only to discover that those labels were insufficient. Biff Loman cried 'What am I doing . . . when all I want is out there, waiting for me the minute I say I know who I am!' and Eddie Carbone in *A View From the Bridge* will define the loss of his honour and self-respect in terms of the loss of his name. For John Proctor the final realization is triggered by the demand that his signed confession be displayed in public; he refuses

'Because it is my name! Because I cannot have another in my life! . . . How may I live without my name? I have given you my soul; leave me my name!' This discovery of his core of identity, his basic self that cannot be denied or found unworthy, enables Proctor to reject the lie of a confession and affirm both his readiness and his right to die.

Miller's next Broadway production, in 1955, was a double bill. *A Memory of Two Mondays* is a simple and realistic slice-of-life in an automobile parts warehouse, as observed by a college student temporarily working there. Its quiet social commentary lies in the sad irony that the boy sees the unpleasant working conditions and limited lives of his co-workers without realizing that such deprivations are an inevitable part of their social and economic caste – a caste he does not quite perceive because he is in the process of escaping from it. The companion piece, *A View From the Bridge*, came to have an independent life when Miller expanded it into two acts for the first London production in 1956; it is that longer version, whose major differences are the abandonment of awkward blank verse in favour of essentially the same dialogue printed as prose, and the addition of background action and detail, that is usually published and revived.

Much more than *Death of a Salesman, A View From the Bridge* represents a conscious attempt at classical tragedy. Like Eugene O'Neill in *Mourning Becomes Electra*, Miller uses Freudian psychology as a modern equivalent of inescapable fate, to give a tragic stature and intensity to the story of Eddie Carbone, whose love for his niece approaches the incestuous. When the girl falls in love with a cousin smuggled into the country, Eddie is driven to violate one of his culture's greatest taboos by reporting the alien to the immigration authorities; unable to admit his guilt, he dies defending himself against the charge. The dramatic centre of the play thus lies in Eddie's inner torment, and if the subject matter suggests a comparison to Shakespeare's *Othello*, a closer resemblance is to *Macbeth*, with the weak man driven by his internal demons ever deeper into his own damnation, while experiencing the additional agony of helplessly watching it happen to him.

But Miller's commitment to realism almost cripples the play, since it is essential to his conception of Eddie's character that he be unable to acknowledge, much less verbalize, what is happening to him. The nearest Eddie can come to expressing his feelings is to deny them – 'That's what you think of me – that I would have such a thought?' – and the treachery of informing is so beyond his capacity for self-awareness that he must fight to the death to deny having done it. So the burden of exposition and insight into Eddie's disintegration falls on other characters, particularly his wife Beatrice and the lawyer Alfieri. It is Beatrice who must, in two particularly clumsy scenes, confront Eddie with the complaint that he has stopped sleeping with her, and instruct Catherine to stop walking around in her slip, just to bring the sexual tensions in the household to the audience's attention. And it is Alfieri, in his role as chorus-narrator, who must repeatedly intrude on the realistic action with such self-consciously portentous statements as 'Eddie Carbone had never expected to have a destiny', and 'His eyes were like tunnels; my first thought was that he had committed a crime . . . but soon I saw it was only a passion that had moved into his body, like a stranger'. As this external commentary begins to become oppressive, the voice of an author insisting on the significance of his material rather than dramatizing it, it is ultimately self-defeating and unconvincing. The fact that Alfieri keeps telling us how meaningful Eddie Carbone's tragedy is almost forces the audience to resist the quiet pathos and horror inherent in the story.

A View From the Bridge was Miller's last play for eight years, and *After the Fall* in 1964 would represent a shift in his focus and style. His contribution in the 1940s and 1950s, aside from the merits of the plays themselves, was the reaffirmation and extension of the discoveries made by Clifford Odets and Lillian Hellman in the 1930s: that domestic realism could address such diverse and ambitious subjects as private morality, public policy, tragic destiny and the American Dream. Repeatedly he takes his protagonists and even tragic heroes from the ranks of the ordinary, affirming the very American concept that such people are worthy of attention. Never far from the focus of his study is the individual's relationship to the larger social and political community, and the evils that result from violation of the implicit social contract, either by the individual (*All My Sons*, *A View From the Bridge*) or by the society (*Death of a Salesman*, *The Crucible*). If there is a prosaic and sometimes preachy quality to his writing, that is almost inevitable, given the degree to which the plays are thesis-driven.

It is not accidental that Miller's critical and popular reputation is even greater outside the United States than within. Because they address social, moral and political issues in terms of their effect on ordinary Americans, his plays seem particularly and specifically American to foreigners, and are welcomed as accurate and evocative depictions of the American experience. But even American audiences can respond to that quality in Miller's plays; an important part of the continuing power of *Death of a Salesman*, *The Crucible* and the others is that they help to explain America, as a culture, to itself.

Tennessee Williams

In contrast to Arthur Miller, Tennessee Williams is an almost completely non-political writer. In fact, more than any other American dramatist, he helped move domestic realism beyond its accomplishment of reflecting large political and social issues through their effect on the domestic setting, and into the final stage of its evolution, the exploration of the emotional burdens of ordinary life. Though Williams might not have seen it this way, he and many of his contemporaries were returning to the modern American drama's roots, in the plays of Herne and Crothers that were the first to acknowledge that life was coloured by problems with no simple solutions. Playwrights of the 1920s had sought to explain the behaviour in such a world through Freudian analysis; the implicit question being answered in such plays as Kelly's *Craig's Wife* or O'Neill's *Desire Under the Elms* was 'Why did this happen?' For Williams, and increasingly for other American dramatists, the real question was 'What did it feel like to have this happen?' American plays began again to look inward rather than outward, as characters in the privacy of their domestic setting exposed their very private pains and joys. If Miller and others in the tradition of Odets were attempting to explain a culture to itself, Williams was exploring the workings of the human spirit and psyche.

And, in almost every case, the fragile and deeply wounded spirit and psyche. For much of his career Williams's popular reputation was as a sensationalist, with characters in various plays guilty of (or victims of) murder, rape, castration, cannibalism, alcoholism, promiscuity, homosexuality, and other shocking violations of moral and social norms. But these elements are not gratuitous in his plays. Like many other southern writers, Williams was drawn to extreme and even gothically bizarre stories and characters because he found in the fringes of acceptable human behaviour clearly dramatizable examples of universal experience. In Williams's eyes, everyday life is for most people an almost Sisyphean labour, the burden of a cold and threatening world borne by a lonely and self-doubting individual. His recurring themes are loneliness and insecurity, the fear of not being up to the task of living, and the grasping at any sustenance for the labour or distraction from the pain. Dramatizing this vision through explicitly melodramatic stories serves a double purpose: it underlines and clarifies the insights through intensity; and it also leads the audience to recognize the emotional experience they share with the characters, and thus to feel more sympathy and charity for those driven to aberrant behaviour.

One can find the seeds of many of Williams's later themes and characters in the volume of early one-act plays published in 1945 as *27 Wagons Full of Cotton*. Two patterns recur: the spiritual or emotional crippling of characters unable to cope with a cold and frightening world – a young man paralysed by the process of moving away from home in 'The Long Goodbye,' a weak-witted and weak-willed woman helplessly facing a summer of daily rapes in '27 Wagons Full of Cotton' – and the impulse of others to escape into a mythologized past – a whore fighting to retain the fantasy of an aristocratic heritage in 'The Lady of Larkspur Lotion', a travelling salesman regretting the good old days in 'The Last of My Solid Gold Watches', a mad spinster turning a brief encounter of twenty years ago into a dream lover in 'Portrait of a Madonna'. One sees unmistakably in these plays the sensibility that would create Amanda Wingfield, Blanche DuBois, Brick Pollitt, Chance Wayne and Williams's other victims of life.

Williams's first Broadway-bound play, *Battle of Angels* (1940) closed during its out-of-town previews; it would reappear in revised form as *Orpheus Descending* in 1957. His second attempt, *The Glass Menagerie* (1945) was much more successful, and announced Williams as an important new voice, not least because it stretched the boundaries of domestic realism in promising ways. Williams's method in *The Glass Menagerie* is as different from the realism of Odets and Miller as his purpose is. The central action of the play is, for the most part, a conventional domestic story, but it is filtered through the sensibility of a poet. Tom Wingfield, the narrator, defines the play's mode in his opening speech: 'The play is memory. Being a memory play, it is dimly lighted, it is sentimental, it is not realistic.' The non-realistic elements in *The Glass Menagerie* lie in the richly poetic language, the undisguised verbal and physical symbols, and above all the operative mode of suggestion – of allowing glimpses into thought processes and emotions that are too evanescent to be analysed or explicated logically. Williams does not violate realism as thoroughly or continuously as, say, O'Neill did in his expressionistic plays. But by presenting his domestic story as emotionally coloured memory he gives it the quality of a reality beyond mere verisimilitude. As Tom says in the play's opening lines,

> Yes, I have tricks in my pocket, I have things up my sleeve. But I am the very opposite of a stage magician. He gives you illusion that has the appearance of truth. I give you truth in the pleasant disguise of illusion.

The Glass Menagerie is about the Wingfield family, living in genteel semi-poverty in Depression era St Louis. Mother Amanda is a former southern belle inclined to live in the memories and values of her past; daughter Laura, hampered by a limp and an overpowering shyness, retreats from the world into her collection of glass animals; son Tom is trapped in a stifling job and aches to leave home, but Amanda insists that he first supply a man to care for Laura in his absence. The acquaintance Tom brings to dinner actually begins to bring Laura out of her shell, but the news that he is already engaged dashes any fantasies of romance. The play is narrated by Tom in the present, perhaps ten years after the dramatized action, and his relation to the story he is telling is a significant part of the play's subject.

In content as well as style, *The Glass Menagerie* is essentially different from the plays of Arthur Miller. It has virtually no larger political, social or moral agenda: though Tom as narrator repeatedly refers to the Depression and the warfare in Europe, it is always as things outside the play and separated from it, and Tom calls Jim 'an emissary from a world of reality that we were somehow set apart from'. The focus is on these three characters, on the internal qualities that created their shared adventure and on its effects on them. Williams presents all three members of the Wingfield family as unable to function in the world of reality, and presents that inability as a virtue more than a handicap. The characterizations and the play's symbolic language repeatedly identify Amanda's faded gentility and Laura's fragility as special qualities to be treasured and protected. Amanda's memories, even at their most foolish, have an evocative beauty; see, for example, her aria on jonquils in Scene Six. Laura is repeatedly labelled as special and precious rather than inadequate, through such symbols as the botanical hothouse in which she spends her days; the glass animals, particularly the unicorn; and Jim's nickname for her, 'Blue Roses'. Williams makes it clear that the cost of successfully functioning in the real world, if that were at all possible, would be the loss of everything that is special about these characters: the only glass animal Jim can take back to reality with him is the broken unicorn that is 'just like all the other horses'. The result is that, rather than rejecting Amanda and Laura as misfits, we come to cherish them as beautiful alternatives to the ordinary and to be grateful that they have found an environment in which, like the hothouse flowers, they can be protected and nourished. Some people are simply not equipped to deal with reality, says the play, and that may well be an indictment of reality.

So seductive and evocative is the alternative reality of Amanda and Laura that it is easy to forget that the central action of *The Glass Menagerie* is not the story being related but the relation itself; the reality of the play is Tom the sailor a decade later, reliving his past. His story is one of failing to recognize that his mother and sister were spiritually better off where he left them, and that he himself was more at home with them than he could be in the world of wars and depressions he keeps trying to convince himself is the real one.

> I traveled around a great deal. The cities swept about me like dead leaves. . . . I would have stopped, but I was pursued by something. . . .
>
> Perhaps I am walking along a street at night, in some strange city, before I have found companions. I pass the lighted window of a shop where perfume is sold. The window is filled with pieces of colored glass, tiny transparent bottles in delicate colors, like bits of a shattered rainbow.
>
> Then all at once my sister touches my shoulder. . . . Oh, Laura, Laura, I tried to leave you behind me, but I am more faithful than I intended to be!

Like that other tale-telling seaman, Coleridge's Ancient Mariner, Tom is driven by a sense of guilt, more for being untrue to himself than for deserting his mother and sister. And in the play's final moment he remains untrue to himself, trying to convince himself, in the face of all the evidence of the play, that the only reality is the one outside, and thus dooming himself to further torment: 'I reach for a cigarette, I cross the street, I run to the movies or a bar. I buy a drink, I speak to the nearest stranger – anything that can blow your candles out! – for nowadays the world is lit by lightning!'

A Streetcar Named Desire (1947) is without question Williams's best play, and shares with Miller's *Death of a Salesman* and O'Neill's *Long Day's Journey into Night* the very top rank of American drama. On its surface it is a simple, fairly sensational melodrama: frayed southern aristocrat Blanche DuBois visits her sister Stella in New Orleans and finds her married to the crude, intensely physical Stanley Kowalski. Shocked by Stella's new life, Blanche tries to take her away from it, but we gradually learn that Blanche is not as pure as she pretends and that she is struggling desperately to retain her emotional and mental equilibrium. The failure of her romance with Stanley's friend Mitch, and a violent and sexual confrontation with Stanley, finally break her; and the play ends with Blanche being led off to an asylum while Stanley and Stella remain together.

The emotional spine of the play is the decline and fall of Blanche DuBois, who enters the play in a state of physical and emotional exhaustion, and fights a tragically losing battle against insanity. She is a rich and complex character (one of the greatest acting roles in American theatre; Stanley Kowalski is another), made up of contradictory elements that she cannot reconcile. The last representative of an elegant southern aristocracy, she is committed to the values and moral imperatives of her class, values which the play gradually exposes as corrupted and barren. By nature unable to deal with reality, she escapes into a world of ideals; recognizing her own fragility, she desperately tries to make intimate contact with those who seem abler, confessing, in her final line of the play, 'I have always depended on the kindness of strangers.' When she is finally driven by her disastrous encounter with Stanley into the protective fantasies of psychosis, the play mourns her loss as well as the loss of the beauty and idealism she tried so hard to represent.

If the play's emotional centre is in Blanche's tragic experience, it is philosophically centred on Stella. Structurally, *A Streetcar Named Desire* resembles a medieval morality play, with Stella – and thus the audience – being pulled in two directions and having to choose between her tempters. Blanche stands for idealism, culture, purity, and the love of beauty, but also for falsehood, fantasy, weakness and the rejection of unpleasant reality. She is associated through verbal and physical symbols with whiteness, soft colours, a lost and romanticized past, cleanliness, art and the zodiacal virgin; and also with death, sterility, and debased or perverted sexuality. Stanley is essentially physical and sexual: *'Animal joy in his being is implicit in all his movements and attitudes. Since earliest manhood the center of his life has been pleasure with women. . . . Branching out from this complete and satisfying center are all the auxiliary channels of his life.'* He enters the play like a caveman, tossing Stella a package from the butcher with the cry of 'Meat!' His tastes are elemental and crude, and his manner brusque and violent, especially when he feels threatened. He is symbolically associated with Capricorn, gaudy colours and raw sexuality.

On this semi-allegorical level the play seems to be offering a choice between the higher and lower aspects of the human potential, and Stella – and thus the play – chooses Stanley over Blanche. It is important to see that Stella does *choose* Stanley. Blanche's report of being raped by Stanley shocks her, and she elects to believe that it is the fantasy of a broken mind. Blanche's mind *is* broken, but Stella would send her to the asylum sane rather than face the consequences of the truth; as she explains to her neighbour, 'I couldn't believe her story and go on living with Stanley.' And her friend approves of her decision: 'Don't ever believe it. Life has to go on.'

What justifies Stella's shocking choice is Williams's perception of the stakes. When Stella confesses to Blanche that she cannot bear life without Stanley – 'I can hardly stand it when he is away for a night. . . . When he's away for a week I nearly go wild! . . . And when he comes back I cry on his lap like a baby.' – she exposes an insecurity of her own that cannot be blamed on the values that handicap Blanche. When Stella temporarily flees to the neighbours after Stanley has hit her, Stanley cries desperately for her to return, in a scene of almost mythic power, and she does, recognizing the depth of his need for her. And when Mitch, stumbling towards romance, expresses himself in less than cavalier terms – 'You need somebody. And I need somebody, too. Could it be – you and me, Blanche?' – Williams demonstrates that Blanche is not the only one who finds that the task of living and coping demands the very utmost of her resources. In a world in which such insecurity is the norm, psychic and spiritual survival cannot be taken for granted. And however attractive may be the values of purity, culture and idealism for which Blanche stands, her own downfall demonstrates that they actively interfere with the ability to cope and function. Even before that is evident, Williams shows that Blanche has nothing to offer Stella; she can only make demands on her – demands of guilt, of moral obligation, and of desperation:

> Maybe we are a long way from being made in God's image, but Stella – my sister – there has been *some* progress since then! Such things as art – as poetry and music – such kinds of new light have come into the world since then! In some kinds of people some

> tenderer feelings have had some little beginning! That we have got to make *grow*! And *cling* to, and hold as our flag! In this dark march towards whatever it is we're approaching. . . . *Don't – don't hang back with the brutes!*
>
> [Ellipses Williams's]

What Blanche actually represents in the play's allegory are the human values and attributes that are life-draining rather than life-supporting, and they demonstrably do not work. Stanley, and the aspects of humanity that he represents, may be primitive and ugly, but they are eminently equipped to deal with reality. As he would frequently through his career, Williams uses sex as a metaphor for both vitality and spiritual sustenance: Stella explains that 'there are things that happen between a man and a woman in the dark – that sort of make everything else seem – unimportant'. In that context, Stanley's boast about having taken Stella from the house with white columns to the land of coloured lights becomes a valid claim to have saved her life. Williams can regret that the world is such that the Stanleys survive and the Blanches do not, but the most valuable counsel he can offer is to accept that fact and to choose life over death.

Williams's other plays of this period vary considerably in quality and commercial success. *Cat on a Hot Tin Roof* (1955) and *The Night of the Iguana* (1961), while not quite as overpowering as *Streetcar*, show him at the top of his talent. *Summer and Smoke* (1948), *Suddenly Last Summer* (1958) and *Sweet Bird of Youth* (1959) are flawed but effective. *The Rose Tattoo* (1951) and *Period of Adjustment* (1960) strive for less than the others and are less noteworthy; and *Camino Real* (1953) and *Orpheus Descending* (1957) are too basically flawed to succeed. The determining factor in each case is, to a great extent, Williams's success in balancing his impulse towards the poetic and symbolic with his need to ground the plays in a solid and humanly believable reality. Strikingly, the more successful his plays are as domestic realism, the more successful they are as poetry, allegory or universal statement.

Cat on a Hot Tin Roof (1955) is Williams at his near-best, with a strong, slightly sensationalistic domestic story dramatically effective for its own sake while also offering psychological insights and moral judgements. Like *Streetcar*, *Cat* has a double focus: Big Daddy Pollitt's need to decide which son to leave his plantation to, the successful but unloved Gooper or the loved but alcoholic Brick; and Maggie's fight to save her marriage to Brick. The who-will-inherit plot resembles the Blanche–Stanley conflict in forcing a choice on the audience – in this case between practical considerations, which clearly favour Gooper, and matters of empathy and morality, which favour Brick. Big Daddy is a Falstaffian figure, hugely and vulgarly alive. His one great hatred is of mendacity; he defines himself by his ability to deal with the truth, and has contempt for those who cannot. In this, Brick is his natural heir; he and Maggie share that hatred and that commitment to life, and part of the play's hope for Brick lies in the faith that those qualities are so integral to his nature that they cannot have been completely destroyed by his spiritual and psychological decline. So when Big Daddy decides in his favour, Williams is arguing that spiritual affinity is more important than practical concerns, while reaffirming his conviction that an enthusiastic immersion in life is to be celebrated and rewarded.

But, just as the *Streetcar* audience is more involved with Blanche's experience than with the play's metaphysics, any *Cat* audience will find the play in the emotional journeys of Maggie and Brick rather than in the question of inheritance. Brick, like Blanche, is trapped in a morality of purity that he cannot live up to, in his case an extremely limited definition of masculinity that forced him to reject his best friend. It is almost irrelevant whether Skipper or Brick actually had homosexual inclinations; for both, anything less than being a 'real man' was suspect, and neither could live with the suspicion. Skipper killed himself and Brick withdrew from the sexual arena; the trauma he is forced to go through during the play is not so much the acknowledgement of his role in Skipper's death as the acknowledgement of his own fears. He does not withdraw again from the truths he faces – indeed, he affirms his opposition to mendacity and his respect for his father by telling Big Daddy the truth about his health – and the play offers some hope of his recovery in this commitment to reality.

Maggie the Cat's story is also one of self-discovery and self-acceptance. She begins the play in almost manic desperation – much of Act One is an almost uninterrupted monologue, and the play's title is a symbol of nervous energy. She announces her determination to fight for the estate and for Brick's love, but also confesses that this struggle threatens her sense of herself. She thinks of herself as weak and feminine, and is frightened by what having to fight is doing to her: 'I've gone through this – *hideous!* – *transformation*, become – *hard! Frantic!*' But the truth is that Maggie is not inherently weak – she is, after all, Big Daddy's heir – and her spiritual adventure in the play is the process of recognizing and accepting this redefinition. At first she resists, instinctively separating herself from the role she is playing:

> *When she opens her eyes again, what she sees is the long oval mirror and she rushes straight to it, stares into it with a grimace and says: 'Who are you' – Then she crouches a little and answers herself in a different voice which is high, thin, mocking: 'I am Maggie the Cat!'*

But by the end of Act One she has integrated the new Maggie into her sense of self: 'I don't know why people have to pretend to be good, nobody's good . . . but I'm honest! . . . *I'm alive!* Maggie the cat is – *alive! I am alive, alive! I am alive!*' At the end of the play she has not only accepted the new definition of herself but also its implications for her marriage: 'Brick, I used to think that you were stronger than me and I didn't want to be overpowered by you. But now, since you've taken to liquor – you know what? – I guess it's bad, but now I'm stronger than you and I can love you more truly!'

That last passage comes from the 'original' Act Three of the play, as opposed to the 'Broadway Version', which requires some comment. Most editions of the play include two third acts, along with Williams's explanation that the first was the script he delivered to his director at the start of production, and the second represents changes he was advised to make – briefly, to soften Maggie, to bring Big Daddy back for another scene, and to offer more signs of Brick's spiritual recovery after the trauma of Act Two. All three were based on theatrical rather than literary considerations: the absence of the high-energy Big Daddy would make the last act anticlimactic, for example. Williams acquiesced in these changes, but chose to print his original text along with the revision when the play was published. It is worth noting that no new play has ever opened with exactly the same script the author first delivered to the producers; large and small changes are made before and during rehearsals as the needs of performance become clear. But, with rare exceptions, the published text is always the opening night text, and earlier versions are discarded with the notes and rough drafts. So the only unusual thing about the *Cat* text is not that there were revisions but that Williams saved his original version and published it. (Williams actually made a third version for a 1974 revival and published it in 1975; essentially it attaches the original ending to the Broadway act.)

We shall stretch the chronological boundaries of this chapter to include *The Night of the Iguana* (1961), since it deals with some of the same themes as *Streetcar* and *Cat*, and with some of the same power. *Iguana* does have a notable structural difference, in building towards an open, almost Shavian discussion of its issues; the minimal plot of the play clearly exists just to get two carefully delineated and contrasting characters to the same place so that they can meet and debate. The Reverend T. Lawrence Shannon, former minister reduced to leading third-rate guided tours of Mexico and in the process of one of his regularly scheduled nervous breakdowns, resembles Brick Pollitt of *Cat* and Chance Wayne of *Sweet Bird of Youth* in his self-destructive determination to deal with the world on his own ideal terms, without capitulation to its impurity. Hannah Jelkes is Amanda, Blanche and Alma Winemiller of *Summer and Smoke* reborn, but with one new and essential distinction: she has the strength and means to function in reality without being broken by it. The subject of their discussion is the question that drives virtually all of Williams's plays: how can the sensitive person survive the pain of existence?

For Shannon, the mere act of living is almost overpoweringly difficult

– 'It's horrible how you got to bluff and keep bluffing even when hollering "Help!" is all you're up to.' The earthy Maxine Faulk offers a sexual and financial partnership that she sees as an acceptable compromise with reality, but Shannon rejects that option as a capitulation to the world's corruption. Hannah dismisses his pride as pointless vanity, and insists that it is possible to engage the challenge of living. She rejects a moral or emotional fastidiousness that would lead one to spurn any source of assistance: 'Endurance is something that spooks and blue devils respect. And they respect all the tricks that panicky people use to outlast and outwit their panic. . . . Anything, everything, that we take to give them the slip, and so to keep on going. . . . '

> SHANNON You mean that I'm stuck here for good? Winding up with the . . . inconsolable widow?
>
> HANNAH We all wind up with something or with someone, and if it's someone instead of just something, we're lucky, perhaps . . . unusually lucky. [Ellipses Williams's]

The second-level plays of this period all reaffirm Williams's sympathy for those who find the task of living almost beyond their strength. He feels a particular affinity for the most sensitive, the Blanches and Bricks who commit themselves, however self-destructively, to their ideals, and yet he is forced to admire those with the courage to immerse themselves in life despite the pain; and he has the intellectual courage to recognize that neither way is easy, and that compromise and very modest victories may be all that are available. *Summer and Smoke* (1948) offers a milder version of the Blanche–Stanley choice and concludes, as *Streetcar* did, that life, in whatever form it takes, is preferable to the avoidance of life. Operating in broad strokes, as in *Streetcar*, he presents Alma Winemiller and John Buchanan as extreme opposites, she the sexually repressed daughter of a minister, and he a self-destructive Byronic hero. Much of the first half of the play is devoted to demonstrating their excesses: his whoring, drinking and gambling; her poetry circle and nervous attacks. Each of them is identified with a physical symbol, Alma with the statue of an angel and John with a medical chart of the human body. But the play is more concerned with their human story than with their allegorical significance; Williams is not presenting the clash of opposing forces as in *Streetcar*, but is dramatizing the dangers of the partial and unintegrated personality. In a key scene John forces Alma to examine the anatomical chart, demanding that she acknowledge that the sexual organs and their needs are part of being human, while she insists that life includes a spiritual dimension not to be found in a map of the body.

Each of them has been living a half-life, Alma driven to neurosis by the repression of her physical side, John reduced to self-disgust by his rejection of the spiritual. Thus, when the play brings them together again after a separation, only to discover that they have exchanged positions, she ready to accept her sexuality and he hungry for something deeper, Williams can regret the mistiming and still accept the alternative each finds. John chooses marriage to an ordinary girl whose main attraction is that she will not offer excesses of either body or soul, while Alma becomes a fallen woman, picking up strangers in the park. And for each the ending, while not the happiest imaginable, is offered by Williams as the happiest available. Once again, by starting from the assumption that life is a painful and lonely experience, Williams reaches the conclusion that any form of comfort is better than none. The weakest aspect of *Summer and Smoke* is its symbolism, which is awkwardly imposed on the human story. Perhaps sensing this, Williams wrote an alternative version of the play, *The Eccentricities of a Nightingale*, which was not staged until 1964; it cuts much of the symbolism and reduces the extremes of characterization, and the resulting quietly domestic play of mismatching and mistiming is in some ways more successful than the original, although it loses the larger resonances about the necessity of integrating soul and body.

Suddenly Last Summer (1958) is one of Williams's most gothic plays, and suffers somewhat from its excessive dependence on sensationalism, but at its core it is another exploration of one of Williams's recurring themes: the self-destructive danger of too great a commitment to purity. Its centre is the life and death of Sebastian Venable, whose extreme asceticism is a cover for homosexuality and whose death is the result of a cannibalistic attack by starving beggars. The dominant images of the play are of violent nature – a jungle-like garden, the attacks of birds on sea-turtles, a rape, the cannibalism – and of an imperfect purity built on the denial of nature – Sebastian's white suit and empty notebook, his pretence of asexuality, his mother's refusal to admit the truth of his life and death. Williams paints Sebastian somewhat more darkly than he did Blanche and Brick, recognizing that the search for purity, especially for the artist, can be selfish and cruel; and that uncharacteristic lack of sympathy gives an unpleasant tone to this reminder of the fate of those who would deny their connection to an impure and unforgiving world.

Williams's two images of the purist, the cruelly selfish artist and the helpless victim of life, reappear in *Sweet Bird of Youth* (1959), in which an aging gigolo tries vainly to hang on to symbols of his youthful purity. The sensationalistic plot turns on violence, political corruption, alcohol and drug abuse, sex for hire, racism, venereal disease, castration and hints of incest; and, as in *Suddenly Last Summer*, the gothic elements threaten to overwhelm the human story. But as in *Suddenly Last Summer* and, in a much more controlled way, *Streetcar*, Williams is using these symbols of violence and depravity to define the stakes of the game, to give a sense of the frightening and actively hostile world in which his characters must live out their adventures. Chance Wayne and his current patron Alexandra Del Lago are both fighting against the inevitable reality of time and their own moral and physical corruption. But she survives with the instinctive selfishness and strength of the egoist, while Chance, who never had any identity except his youth and sexuality, must go the way of Blanche DuBois, accepting fate with the passivity of the born victim: 'Whatever happens to me's already happened.' It is clear that, while Williams can admire the monstrous energy of Alexandra Del Lago as he did Stanley Kowalski, his sympathies are with the doomed Chance Wayne. The final line of the play is spoken by Chance directly to the audience: 'I don't ask for your pity, but just for your understanding – not even that – no. Just for your recognition of me in you, and the enemy, time, in us all.' In its bathos that may be the worst curtain line ever written by a great playwright, but its intent is clear: it is Chance's experience as a victim of life that Williams sees as universal, and that he wants to prepare the audience to face.

This discussion of Williams's second-level plays may have already given a hint as to the limits of his talent. It is certainly no criticism of any writer that he stakes out a particular aspect of the human experience and explores and re-explores it from slightly varying perspectives in his works. Williams's weaknesses were elsewhere – ironically in the area of his greatest strength, his poetic instincts and powers. At his best, Williams brought the sensibility and tools of a poet to the theatre more effectively than any other American dramatist, but he had to keep them under tight control or he would stray into distracting and obscuring excess. During his peak years he was usually able to anchor each play in a domestic realism solid enough that the imagery and poetic devices could resonate outwards without danger. But if he did not keep a very tight rein on these elements, they drew attention to themselves and away from the small and specific human story at the core of each of his plays. To the extent that *Suddenly Last Summer* is a play about cannibalism and not about the pain of living in a cruel reality, it fails; to the extent that the symbolism in *Summer and Smoke* seems imposed on a story that functions satisfactorily without it, it weakens the play. A playwright without a poet's instincts could not have written *The Glass Menagerie* or *A Streetcar Named Desire*, but throughout his career Williams would wrestle with the impulse to go too far.

Orpheus Descending (1957) demonstrates what happened when Williams was not in absolute control: an over-ambitiousness and over-complexity of symbolic and thematic intent that swamp the story and cripple its power to affect us on the human level. As the title makes clear, Williams intends mythic resonances for the play about an itinerant musician, Val Xavier, who comes to a southern town and stirs up the loneliness of the women and the enmity of their men. But the intended parallel to Orpheus, murky to begin with, is compounded with Christian imagery identifying Val with Christ, mystic themes invoked by the presence of an Indian Conjure Man, Williams's familiar symbolic identification of sex and life, and a variety of stray and disconnected symbols. All these levels of meaning, along with an excess of self-consciously poetic language, overwhelm a play whose potential strength would have been where Williams's strength always is, in the simply human story of lonely and frightened people reaching for anyone or anything that could ease their pain. Val is allowed some of the purest expressions of this experience in all of Williams: 'We're all of us sentenced to solitary confinement inside our own skins, for life!' and

> What does anyone wait for? For something to happen, for anything to happen, to make things make more sense. . . . Does everything stop because you don't get the answer? No, it goes right on as if the answer was given, day comes after day and night comes after night, and you're still waiting. . . .

But *Orpheus Descending* is just too symbolically and metaphysically cluttered to allow this human story to emerge.

Even less successful is *Camino Real* (1953) an openly symbolic and allegorical play without any anchor in domestic realism: a group of romantic characters drawn from literature (Don Quixote, Marguerite Gautier), history (Byron, Casanova) and myth (Kilroy) inhabit a decadent and inhospitable world that sneers at their values. The play's themes – a plea for compassion for the romantics wounded by an uncaring world, and a call for such victims to be true to their ideals and not surrender to despair – are all but lost when there is no recognizable human reality under the confused dramaturgy and symbolic excess. *The Rose Tattoo* (1951) and *Period of Adjustment* (1960) are minor comedies, both recognizing the spirit-crippling power of loneliness and withdrawal from life, and both offering immersion in sexuality as a life-affirming solution. *The Rose Tattoo* has an infectiously joyous air, not least in what seems almost like a self-parodying excess of symbols in the ubiquitous roses: rose tattoos, rose oil, a rose silk shirt, the names Rosario and Rosa delle Rose, and the flowers themselves. *Period of Adjustment* is quieter in tone and somewhat less successful, neither brainlessly funny enough to work on the level of empty romantic comedy nor developed enough to treat successfully the familiar Williams themes it hints at.

Williams continued to write new plays after 1961, but *The Night of the Iguana* was his last artistic or commercial success. Soon afterwards he went into a personal and artistic decline, triggered by the death of his longtime companion and exacerbated by alcohol, drugs and nervous breakdowns, from which he never recovered; the second half of his career would be marked by exactly the loss of control over his talents that had marred some of the early plays, producing a series of unfocused and overwritten failures. So the years and plays between *The Glass Menagerie* and *The Night of the Iguana* represent Williams's major contribution. They include some of the best plays in the American repertoire, and some of the most challenging roles for actors and actresses. Williams stretched the boundaries of domestic realism by approaching it with the instincts of a poet, making it support verbal and physical symbolism and a heightened, image-filled poetic prose. At the same time he reaffirmed its power as the native American dramatic form, by demonstrating that the dramatization of domestic events could illuminate the spiritual and emotional struggles of characters with an authenticity that made even the most bizarre convincingly human. His recurring observation – that ordinary, unheroic life was overwhelmingly challenging and painful for many people – proved to be such a powerful insight into the mid-century American experience that it was taken up by other writers, and came to be the dominant subject of the serious drama. Even more than Arthur Miller, whose work was really the culmination of earlier movements, Tennessee Williams redefined and reshaped the American drama in the 1950s.

William Inge

The one new dramatist of the 1950s who comes closest to being in the same league as Miller and Williams is William Inge, a protegé of Tennessee Williams's whose plays suggest a kind of domesticated Williams: sensitive and sympathetic studies of lonely and frightened people, but without the poetry or sensationalism of Williams. In each of his major plays Inge writes of small, ordinary people leading small, ordinary lives and frustrated by their inability to fulfil their romantic dreams. While he occasionally allows his characters a hope of happiness, he is more likely to offer them nothing more than the peace that comes from the resigned acceptance of their small, ordinary fates.

Come Back, Little Sheba (1950) is the story of a married couple, Doc and Lola, who have long since lost the romance and dreams of their youth, and are held together by little more than habit and practical need. Much of the first half of the play is given over to the exposition and demonstration of the many disappointments that make up their lives; even their dog Little Sheba ran away, and Lola's mourning turns Sheba into a symbol of all the other lost possibilities and hopes. Doc catalogues their disappointments and concludes, 'But we don't have any of those things. So what! We gotta keep on living, don't we? I can't stop just 'cause I made a few mistakes. I gotta keep goin' . . . somehow.' [Ellipses Inge's] The rest of the play is really about that 'somehow', as Lola and Doc try to keep going. Their successes, and even their efforts, are not consistent, and Inge does nothing to imply that Doc and Lola are extraordinary in any way; they are not even particularly attractive. Yet their unremarkable pains are real pains, and the task of coping with life, even on this level, is a heroic one. In the final scene Lola tells Doc of a dream in which she saw Little Sheba dead, and the play ends with their response:

> LOLA I don't think Little Sheba's ever coming back, Doc. I'm not going to call her any more.
> DOC Not much point in it Baby. I guess she's gone for good.
> LOLA I'll fix your eggs.

Lola and Doc do not change radically for the better as a result of their losses; nor do they kill themselves in despair; nor do they find new fantasies to build their hopes on. They go on, somehow. And Inge makes that seem remarkable.

Picnic (1953) shows another group of small-town midwestern characters forced to face the impossibility of their dreams and to make their peace with reality. In a plot situation that resembles Williams's *Orpheus Descending* but with none of that play's pretensions to symbolism, a handsome young drifter shakes up a neighbourhood of women, reminding them of half-forgotten fantasies and frustrations, and then leaves with the girl who has fallen in love with him. The end of the play, with Madge following Hal into an unknown future, has all the trappings of fairy-tale romance, the sort of happy ending that requires that the audience not ask what will happen next. But Inge raises the question, and makes it clear that Madge faces it, too:

> FLO He's no good. He'll never be able to support you. When he does have a job, he'll spend all his money on booze. After a while, there'll be other women.
> MADGE I've thought of all those things.

Madge and Hal are prepared to settle for considerably less than perfection because it is the best option available; they are not running towards a guaranteed happy ending but away from a guaranteed unhappiness. In the subplot a spinster schoolteacher willingly demeans herself by begging and finally trapping a man into marrying her; she does not love him and realizes he does not love her, but the alternative is unbearable. Each character in the play comes to recognize that the slim possibility of happiness is better than nothing – indeed, that something different, even if equally unhappy, is preferable to the unhappiness you know.

Inge shares with Tennessee Williams this conviction that no source of comfort, however fragile or ephemeral, should be disdained in a frightening and inhospitable world. It reappears in his next play, *Bus Stop* (1955), whose veneer of light romantic comedy only partially hides the recognition that life is lonely and any human contact is better than none. Bo, a simple cowboy, has abducted the saloon singer Cherie in the naive faith that if he loves her she must love him. The basic romantic convention on which that plot is built makes its happy ending inevitable, but Inge colours and surrounds the central action with reminders that the stakes are less happily-ever-after than anything-ever-after. Both Bo and Cherie are allowed quiet moments of reflection in which they realize that the lives they are leading are lonely and unpromising, and both readjust their expectations for the future accordingly. The small stories that surround the central action all serve to emphasize that Bo and Cherie's romance, even if imperfect, is better than most people get; the proprietor of the bus stop, for example, settles for brief sexual liaisons with the drivers. And the play ends, not with Bo and Cherie's triumphal ride into the sunset, but with Bo's friend, who has been left behind, discovering that the restaurant is closing and he has no place to go:

> GRACE Then I'm sorry, mister, but you're just left out in the cold.
> VIRGIL Well . . . that's what happens to some people.
> [Ellipses Inge's]

The Dark at the Top of the Stairs (1957), set in 1920s Oklahoma, is the story of Rubin and Cora Flood, whose basic differences almost destroy their marriage until both realize how much they need the other's support to face the pains and fears of life. The significance of the title is explained by ten-year-old Sonny, who is afraid to go upstairs to bed alone: 'You can't see what's in front of you. And it might be something awful.' At that point in the play his mother offers to go upstairs with him; in the final scene Rubin goes up first to light the way for Cora. The play thus continues Inge's exploration of the insecurities and fears of inadequacy that he, like Williams, sees as central to the twentieth-century American experience, and his analysis of the limited options open to ordinary people and the need to find – and settle for – imperfect solutions to inescapable problems. As in *Picnic* and *Bus Stop*, Inge surrounds the main characters with thematically supportive subplots, but somewhat less effectively, as the complexity of the secondary material threatens to upset the play's balance: the unhappy schoolboy and sexually frustrated sister whose lives teach Cora that her unhappiness is not unique actually take up more of the play's attention than Rubin does, and the complicated Oedipal relationship between Cora and Sonny seems grafted onto the play rather than integral to it. Apparently Inge was losing control of his craft; although he continued to write plays through the 1960s, this was his last commercial and artistic success.

Other new dramatists of the 1940s and 1950s

Miller and Williams (and, in a different way, Eugene O'Neill) so dominated the serious American drama mid-century that only a few other writers stand out, either for a body of notable work or for individual plays of special merit or importance. A survey of the Broadway seasons of 1945–1960 finds, even among the hits, a large number of one-play playwrights or genre plays that succeeded because they were that year's version of some type of play that Broadway always requires. That some were better than others is almost incidental to their effectiveness; each was at least skilfully assembled and better than any unsuccessful rivals, and met the audience hunger for that particular type of melodrama.

It is not surprising, for example, that the Second World War was followed by a string of plays set in the war or in the post-war experiences of returning soldiers; Miller's *All My Sons* fits partially in the latter category. Among the more interesting on their own merits are Arthur Laurents' *Home of the Brave* (1945), about a soldier dealing with the guilt of survival; Thomas Heggen and Joshua Logan's *Mr Roberts* (1948), a sentimental comedy about the Navy; William Wister Haines's *Command Decision* (1947), about Air Force officers making difficult choices in battle; *Stalag 17* (1951), by Donald Bevin and Edmund Trzcinski, a comedy in the unlikely setting of a prison camp; and Alfred Hayes's *The Girl on the Via Flaminia* (1954), in which an American soldier in post-war Rome learns that occupying forces can unwittingly injure and humiliate their hosts.

Another recurring subgenre is made up of dramatizations of fiction, history, biography and other non-dramatic sources: Ruth and Augustus Goetz's *The Heiress* (1947), from Henry James's *Washington Square*; William Archibald's *The Innocents* (1950), from James's *The Turn of the Screw*; *The Diary of Anne Frank* (1955), dramatized by Frances Goodrich and Albert Hackett; Ketti Frings's adaptation of Thomas Wolfe's *Look Homeward, Angel* (1957); Dore Schary's biography of Franklin Roosevelt, *Sunrise of Campobello* (1958); and William Gibson's drama of the youth of Helen Keller, *The Miracle Worker* (1959). Each of these plays is constructed with considerable theatrical skill, though their very nature keeps them subservient to their source material.

Among the other new dramatists of this period, Robert Anderson resembles Inge – and thus, by extension, Williams – in addressing the private unhappiness of life's victims and offering the counsel that no source of comfort or consolation should be disdained because it is imperfect. In his first Broadway success, *Tea and Sympathy* (1953), a prep schoolboy is suspected of being homosexual just because he is unusually sensitive, and the only person to offer him any sympathy is the unhappy wife of one of the schoolmasters. At the final curtain, with the oft-quoted (and parodied) line, 'Years from now – when you talk about this – and you will! – be kind', she offers herself to him sexually, both out of her own loneliness and as a way of reassuring the boy he is normal. Some critics have attempted to read *Tea and Sympathy* politically, as an attack on the mob hysteria and fear of the different that constituted McCarthyism. But Anderson's story is the personal one: of Tom, confused by the attacks and, in his innocence, almost convinced they are valid; of Bill, whose marital problems have the same roots as his compulsion to persecute the boy; and of Laura, whose sympathy for Tom's pain springs from her own unhappiness. The play does cheat a little to achieve its ends, making the villains all evil and the imminent sexual act unequivocally pure; while Anderson does not pretend that Laura's giving herself to Tom will be more than a very temporary interruption to her unhappiness, he does imply that the event will resolve all the boy's problems.

A similar quality might be said to mar *Silent Night, Lonely Night* (1959), about two unhappy people – her marriage is collapsing, his wife is insane – who allow themselves the comfort of a one-night affair. Anderson's point is that even an immoral act can be justified by its power to provide solace, but he does unfairly weigh the argument in his favour by setting the play on Christmas Eve, thus implicitly blessing the sin, and by rewarding both characters the next morning, her with a reconciliation with her husband, him with news of his wife's partial improvement. Like Inge, Anderson is sometimes charged with philosophical cowardice for pretending that complex problems have simple solutions, when he is really saying that even partial and imperfect solutions must be embraced gratefully when the alternative is unassuaged pain. And, like Inge, he is sometimes accused of introducing salacious material just for titillation, without facing its moral or psychological implications. But the moral simplifications that defang his shocking subjects are the very things that make the plays accessible to a mainstream audience, allowing it to see past the shock to the author's empathy for his main characters.

Lorraine Hansberry's *A Raisin in the Sun* (1959) is noteworthy both for its quality and for being the first play by a black woman to be staged on Broadway. The play's central action, about a black inner-city family who resist the attempts by racists to keep them from moving into a white neighbourhood, is actually not its moral centre. (It is noteworthy that the basic plot does not really depend on race; the Youngers might as easily be Italians, Irish, Jews or any group unwelcome in a restricted neighbourhood.) Its real insight is into a unique tragedy of black family life in America: that generations of racism have damaged the spirit of the men, forcing the women to assume responsibilities that only further the men's emasculation. The mother of this family has the power, and is always aware of her need to wield it without damaging the dignity of her adult son. She makes a point of giving him control of the family's money, even at the risk of his losing it:

> There ain't nothing worth holding on to, money, dreams, nothing else – if it means – if it means it's going to destroy my boy . . . . It ain't much, but it's all I got in the world and I'm putting it in your hands. I'm telling you to be the head of the family from now on like you supposed to be.

When he does lose it, the danger to his manhood is greater than the financial setback, and his mother calls on the other women in the family to help save his soul: 'What you tell him a minute ago? That he wasn't a man? Yes? You give him up for me? You done wrote his epitaph too – like the rest of the world? Well, who gave you the privilege?' They help him to find in himself the strength to reject the racists; and when the white man turns to the mother for help he gets only a joyful abdication of power: 'My son said we was going to move and there ain't nothing left for me to say.' The enduring power of *A Raisin in the Sun* lies more in its picture of the Youngers' domestic relations than in their encounter with racism.

Paddy Chayefsky began his career writing for television and ended it writing for film, and had his greatest successes (*Marty*, *Network*) in those media. But his stage plays of the 1950s offer an urban, Jewish complement to those of Inge and Anderson in telling small stories of small people with affection and respect. In *Middle of the Night* (1956) a middle-aged widower and a younger married woman fall in love, shocking family and friends on both sides; and the couple themselves begin to doubt their right to defy conventional mores. As Inge or Anderson probably would, Chayefsky allows his lovers to choose an imperfect happiness over a guaranteed unhappiness. Whether these ordinary people will have an ordinary chance at a future is treated with an unpatronizing respect for the reality of their emotions. In *The Tenth Man* (1959) the elderly members of a small Jewish congregation attempt an exorcism on a schizophrenic girl, and actually affect the dead-spirited young man they recruited to help in the ritual; with renewed faith and capacity for love, he proposes to her. Unfortunately a string of easy and patronizing jokes at the expense of the bored and confused old men almost overpowers the human story of the damaged young people reaching towards each other. In his later plays Chayefsky turned to a more intellectual, issue-oriented mode that proved more successful on film than on stage.

Between writing films and musical comedies Arthur Laurents offered several object lessons in the value of accepting imperfection in an imperfect world. *Home of the Brave*, in which a soldier forgives himself for what he took to be cowardice, has already been mentioned. In *The Time of the Cuckoo* (1952) an American woman in Venice has a brief romance with an Italian shopkeeper but is shocked by his moral casualness; she realizes too late that the price of her high moral standards is the loss of what would have been at worst a pleasant adventure. *A Clearing in the Woods* (1957), one of the very few non-realistic plays of the decade, confronts a woman with several girls and women who she realizes are herself at earlier ages. They help her see that her unhappiness has been caused by a perfectionism that led her to label each of them a failure; and by embracing them she learns to accept herself.

In a small but steady stream of plays by writers from other genres who were trying their hands at drama, Carson McCullers's *The Member of the Wedding* (1950) stands out as a sensitive portrait of a girl at the cusp of adolescence. Twelve-year-old Frankie vacillates between cuddling in the lap of her black maid and fantasizing about joining her brother and his bride on their honeymoon, between playing with her six-year-old cousin and discovering boys her own age. Her pre-teen angst, philosophizing and self-dramatizations are all treated with a respect that is not blind to their silliness. Archibald MacLeish's *J.B.* (1958) is a retelling of the Job story in modern terms, with J.B. a smug businessman who not only retains his simple faith in God through his ordeals, but develops humility and the capacity for love. MacLeish almost buries this modernization in additional and distracting symbolism, but the novelty of what was virtually the only verse play of the decade attracted and held audiences.

Eugene O'Neill

After *Days Without End* in 1933, Eugene O'Neill did not have another new play produced until *The Iceman Cometh* in 1946, and only one more (*A Moon for the Misbegotten*, 1947) before his death in 1953. But, although the last twenty years of his life were marked by extended periods in which physical and mental ailments kept him from writing, he was not as idle as the production record would seem to indicate. During this period he worked on a large number of uncompleted plays, most notably an eleven play cycle with the umbrella title 'A Tale of Possessors, Self-Dispossessed', in which he planned to follow a family through the two centuries of American history. He also planned a second cycle of eight one-act plays with the group title 'By Way of Obit', completed the somewhat autobiographical *Long Day's Journey into Night*, and had notes or outlines for other plays.

But O'Neill did not finish many of these projects, destroyed most of his notes and manuscripts, and ordered that *Long Day's Journey* be withheld from publication and production until twenty-five years after his death, to avoid embarrassment to his family. His widow overrode that request, and allowed the play to be published in 1956, and to be produced in Stockholm (in gratitude for O'Neill's having been awarded the Nobel Prize in 1936) and New York that same year. Meanwhile, a successful Off-Broadway revival of *The Iceman Cometh* in 1956 had reawakened interest in O'Neill, and gradually three more posthumous plays appeared: *A Touch of the Poet*, the only completed play of the 'Possessors' cycle (Stockholm 1957, New York 1958); *Hughie*, the only play from the one-act cycle (Stockholm 1958, New York 1964); and *More Stately Mansions*, a second play in the 'Possessors' cycle, surviving only in an overlong draft which had to be heavily cut for production (Stockholm 1962, New York 1967).

The last six plays have two extraordinary things in common. First, while they are not all of equal quality, as a group they are as superior to O'Neill's plays of the 1920s as those plays were to their contemporaries – which is to say that the most successful of them – certainly *Iceman*, *Moon* and *Long Day's Journey* – are among the very best plays ever written in America. Second, they contain none of O'Neill's characteristic experiments in technique, but are simple narratives of families or friends in the setting of their homes. After a lifetime of trying and discarding every dramatic and theatrical style he could adapt or invent in his attempt to stretch the genre and make it able to carry the weight and breadth of his concerns, O'Neill returned in his final masterpieces to the most conventional of forms. Perhaps the greatest artistic discovery of his life was that he could say what he wanted to say through domestic realism.

The Iceman Cometh is set in a cheap New York saloon populated by failures whose only activities are drinking until they pass out and comforting themselves with fantasies ('pipe dreams') of some day regaining their proper place in society. A friend talks them into actually looking for the job or contacting a friend as they had always promised, and shows no surprise when they all return in defeat, assuring them that admitting their worthlessness will free them to enjoy failure without guilt. In explaining how the process worked for him, he discovers an unbearable truth and retreats into the belief that he has gone mad. The others pretend to have been humouring him, and revert happily to their pipe dreams.

The play is about the need for fantasy – in particular, the ego-comforting fantasy that we are essentially better than we may appear to be at the moment – without which humanity cannot bear the pain of reality. The general tone of the first act, before Hickey's arrival, is of a contented, functioning community; as Larry Slade (one of the drunks, whose protective fantasy is that he does not care about life, but is just observing it with amusement) explains, 'To hell with the truth! . . . The lie of a pipe dream is what gives life to the whole misbegotten mad lot of us.' Larry calls it the Tomorrow Movement, since the pretence that each of them will achieve something tomorrow makes it easier to live with the failure of today. It is easy to miss the vitality and camer-aderie in the first act because the sordid and debased quality of the drunks' lives dominates our impressions. We notice it most when Hickey's campaign puts a damper on the party atmosphere, turning the friends against each other, driving them away from company to hide in solitude, and even depriving them of the solace of drunkenness. Larry identifies the change in the air: 'I'm damned sure he's brought death here with him. I feel the cold touch of it on him.'

Pipe dreams mean life and truth means death – one character kills himself after losing his fantasy – because the human animal is simply not equipped to deal directly with reality. The purest evidence of this is Hickey's own case. Hickey is honestly confused that the others do not find peace, and is compelled to tell his story, to show them how they should be reacting. What he wants to explain is that he gave up the fantasy that he would reform, and then killed his wife to save her from the pain of watching his surrender. But O'Neill makes brilliant use of a psychological insight he would also exploit in *Long Day's Journey*: that if a character talks long enough he will eventually say something he did not intend to say:

> Anyway, she forgave me. The same way she forgave me every time I'd turn up after a periodical drunk. You all know what I'd be like at the end of one. You've seen me. Like something lying in the gutter that no alley cat would lower itself to drag in – something they threw out of the D.T. ward in Bellevue along with the garbage, something that ought to be dead and isn't! . . . There's a limit to the guilt you can feel and the forgiveness and the pity you can take! You have to begin blaming someone else, too.

The momentum of his self-exposure brings him to the night of the killing: 'I remember I heard myself speaking to her, as if it was some-thing I'd always wanted to say: "Well, you know what you can do with your pipe dream now, you damned bitch!" '

The failure of the others to find peace in surrender to reality as Hic-key did is explained: that is not what happened to him. He did not kill the pipe dream; he killed a nagging wife, and then *created* the pipe dream that he had done it out of love. The truth about himself – that he was capable of such hate – was unbearable, and he built the whole fictional construct about giving up pipe dreams and finding peace, to protect himself from that truth. And O'Neill lets us watch him do it again: 'No!

That's a lie! I never said – ! Good God, I couldn't have said that! If I did, I'd gone insane!' Harry Hope senses that if Hickey was crazy, the others can then pretend that they were just humouring him, and thus did not really fail. Hickey realizes this: 'Now, Governor! Up to your old tricks, eh? I see what you're driving at, but I can't let you get away with – Yes, Harry, of course, I've been out of my mind ever since! All the time I've been here! You saw I was insane, didn't you?' Before our eyes, Hickey leaps into a new pipe dream, knowing full well that it is false, because it is easier to handle that contradiction than to face the truth. And the others (except for Larry, who lives with the truth and awaits the release of death) embrace Harry's fiction, rediscovering the community, the happiness and even the kick in the booze.

The return to realism in *The Iceman Cometh* marks a startling change from O'Neill's earlier styles, but there is a continuity of themes. The inherent limitation in the human capacity for dealing with reality that O'Neill discovers in this play is related to similar limits – of fate, genetics or psychology – that he had dramatized in such plays as *The Emperor Jones* and *Mourning Becomes Electra*. The important difference is in the tone of the play. There is an implied forgiveness offered to the audience: no one should feel guilty about relying on psychological and spiritual crutches if life is impossible without them. *The Iceman Cometh* does not bewail the fact that humanity needs pipe dreams to survive; it celebrates the good fortune that, needing such dreams, man also has the capacity to find and maintain them.

A Moon for the Misbegotten was actually written after *Long Day's Journey into Night*, though it was produced first, and is a kind of sequel to it, picking up one of the characters a decade later. The self-hating alcoholic James Tyrone, Jr enjoys the company of sluttish Josie Hogan, but Josie, who is actually a virgin and who loves him, realizes that he needs comforting more than sex, and gives up an opportunity to take him to bed. On its surface, then, the play is a simple and sentimental love story; its power and its stature come from the depth of its psychological insights and from the wealth of compassion those insights inspire. In this play O'Neill eschews all the metaphysical and psychological theses – tragic Fate, the Dionysian–Apollonian conflict, Freudian determinism, etc. – that generated his early plays, and dramatizes directly the painful process of living. Like most cynics, Jim Tyrone is an idealist and romantic at heart, and cannot forgive himself for his failure to live up to his own standards. He is defined by an emptiness of spirit, 'like a dead man walking slow behind his own coffin', interrupted only by flashes of self-loathing – 'You better watch your step. It might work – and then think of how disgusted you'd feel, with me lying beside you, probably snoring, as you watched the dawn come.'

O'Neill has not changed so much since *The Emperor Jones* and *Mourning Becomes Electra* that he thinks people can triumph over their limitations.

But he now feels that they can gain some peace by forgiving themselves for those limitations, and he shows this happening to Jim Tyrone. As Jim and Josie sit and drink together in the moonlight, he can speak openly about his crimes, notably his drunkenness and whoring during his mother's last illness and after her death. That the behaviour he describes is clearly a matter of weakness or even normal response to grief, not depravity, is no comfort; he can only condemn himself for it, and condemn himself further for exposing Josie to it. Josie's unselfish love enables her to give Jim what he cannot give himself: 'I'm proud you came to me as the one in the world you know loves you enough to understand and forgive – and I do forgive! . . . As *she* forgives, do you hear me! As *she* loves and understands and forgives!' He awakes the next morning, after sleeping in her maternal arms, 'Sort of at peace with myself and this lousy life – as if all my sins had been forgiven.' She, and the play, realize that this forgiveness is not a panacea. He remains dead in spirit, but at least he is freed from self-torment, and she prays only that he will be spared another relapse: 'May you have your wish and die in your sleep soon, Jim, darling. May you rest forever in forgiveness and peace.'

Josie also comes to a self-acceptance. She begins the play trapped in the role of bawdy slut, which is really a mask to cover her feelings of unworthiness – 'I'm an ugly overgrown lump of a woman, and the men that want me are no better than stupid bulls.' – but ends the play in peace, having sacrificed her one chance for romance, but appreciating the value of her capacity for giving. *A Moon for the Misbegotten* shares with *The Iceman Cometh* and, indeed, all of O'Neill's final plays this theme and tone of forgiveness. In *Iceman* O'Neill offered the audience a release from guilt by showing that some dependencies are part of the human condition, and not the result of individual inadequacy. In *Moon* he dramatizes how desperately important the capacity for self-forgiveness is, and offers the reassurance that such mercy, and the peace it provides, are available.

If there is such a thing as the single greatest American play, it is almost certainly *Long Day's Journey into Night*. In this unflinchingly probing exploration of what it is like to be human, O'Neill exposes some of the most frightening truths about what we do to ourselves and each other, and then probes even deeper to find, in the horrors themselves, the means of surviving the horrors. That he does so in a work that was a personal torment for him to write must increase the play's stature and our admiration for its accomplishment. *Long Day's Journey into Night* is a fictional portrayal of O'Neill's own family: miserly father, morphine-addicted mother, alcoholic and self-pitying sons. The Long Day of the play sees two crises: Mary, who has been off her drug for some time, starts taking it again; and the family learns that Edmund has consumption and must be hospitalized. Old grievances come to the surface with the new pains, and recriminations and accusations of blame are traded among the four members of the family. At the end of the play, with Mary lost in her drugged fantasies and the men paralysed by drunkenness, it seems impossible that the sun will ever rise again.

The first aspect of the play to note and then move beyond is the autobiographical element. The Tyrones are unquestionably based on the O'Neills, but one of the accomplishments of the play is O'Neill's transformation of the raw material of his life into something more. On the simplest level, he makes important changes in the characters, so they may better serve his symbolic and thematic purposes: Edmund is far more of an unformed adolescent than Eugene O'Neill was at that point in his life, for example; and the author conceals the fact that Mary O'Neill eventually overcame her addiction. More significantly, there is much more to the play than the telling of its story. The real importance of the autobiographical element is the fact that O'Neill had the courage and ability to take material that must have been painful for him to work with, to find a basis for hope and consolation in its apparently unbroken darkness, and to offer those gifts to the world through his art.

On the surface, there seems to be little basis for hope and consolation in the experience of the Tyrones. As a sunny day is darkened first by fog and then by night, we discover a network of sins committed by the Tyrones against each other. The father's miserliness and unsettled actor's life led to Mary's addiction and kept her from recovery; his constant criticism of Jamie alienated and embittered him; and his penny-pinching now threatens Edmund's life. Mary's addiction tortures all the men, as her repeated reminders of the sacrifices she made to marry Tyrone condemn him, and her obvious preference for Edmund contributes to Jamie's bitterness and Edmund's extended adolescence. Jamie actually killed a third son, by infecting him with measles, and has deliberately corrupted Edmund out of jealousy; and his dissipation and open antagonism are a continual source of pain to his parents. Edmund's birth led to Mary's addiction; his illness contributes to her relapse; his innocence is an affront to Jamie; and his petulance and self-pity impose guilt on all of them.

O'Neill offers no hope that these outrages could have been avoided, or that mutual atrocities are not inherent in man's fate. In fact, *Long Day's Journey* asserts more emphatically than any of his plays since *Mourning Becomes Electra* his lifelong conviction that human life is limited and controlled by forces outside our control. But, as he did in *The Iceman Cometh*, O'Neill forces fatalism to yield up a grain of reassurance. For one thing, the very complexity of the network of crimes relieves any one person from unique guilt. Exactly who, for example, is responsible for Mary's addiction? At various times the play blames the quack doctor, for first giving her morphine; Tyrone, for being too cheap to get a good doctor; Edmund, for being born; Eugene, for dying; Jamie, for killing Eugene; the parents, for preferring Eugene and inspiring Jamie's jealousy; Mary, for having an addictive personality; and so on, up to and including the economic forces of fifty years ago that led Tyrone's father to abandon his family. A similar list of causes could be assembled for every other offence in the play – Jamie's bitterness, Mary's relapse, etc. – leading inevitably to the sense that when responsibility is spread as widely as this no great burden of guilt can rest on any one person.

Even further, O'Neill recognizes that many of the crimes his characters have committed against each other were intended to be acts of love. Tyrone did not take Mary on the road with him to degrade her, but because he loved her too much to leave her behind. The end product was her degradation, but O'Neill makes us distinguish between intent and result. Mary does not baby Edmund to encourage his weaknesses, but because he is her baby. In the last act Jamie tries to warn Edmund against himself, but what slips out is a vicious expression of his resentment that he immediately regrets. It is a tragic fact of human interaction that sometimes an act or word of love gets inexplicably transformed in transit, and reaches its recipient as an attack. O'Neill cannot explain why that happens, and certainly cannot offer any means of preventing it. But he can help his audience to recognize that it does happen, and to learn to look past the received hate and honour the intended love; as Mary begs Tyrone, 'James! We've loved each other! We always will! Let's remember only that, and not try to understand what we cannot understand, or help things that cannot be helped – the things life has done to us we cannot excuse or explain.'

The grammar of that last phrase is significant, and is repeated elsewhere in the play (Tyrone: 'life had me where it wanted me'). In *Long Day's Journey*, O'Neill offers a psychological model that replaces free will with the limits imposed on a person by the events of his past. The Tyrones do not choose to behave as they do; they cannot help it, because their past experience has programmed them to behave this way. Tyrone is not miserly out of conscious malice; indeed, in his most self-aware moments he actually tries to be generous. But the extreme poverty of his youth warped him in a way he cannot reverse. In the last act he tells Edmund about his childhood, and then tries to fight his programming: 'If I took this state farm sanatorium for a good bargain, you'll have to forgive me . . . . You can choose any place you like! Never mind what it costs! Any place I can afford. Any place you like – within reason.'

The sad comedy of that reversal is repeated by each of the characters throughout the play, as they demonstrate O'Neill's thesis that even the occasional attempts at free will are doomed by this internal fate. O'Neill dramatizes this pattern most brilliantly in the last act, which contains some of his finest dramaturgy and dialogue. In a series of uncharacteristically eloquent and poetic speeches, the usually prosaic O'Neill allows each of the men to realize a truth about himself and then shows that each of them is unable to change in spite of the insight. Edmund admits that he doesn't have the poetic talent he wishes for, but then lapses into the bathetic imagery and posing he has just rejected. Jamie breaks through his wise-cracking persona to warn his brother against himself, but then returns to his viciously cynical mode. Tyrone talks about how his lust for money destroyed his talent as an actor, reaching the sorrowful conclusion that 'I'd be willing to have no home but the poorhouse in my old age if I could look back now on having been the fine artist I might have been.' But at the very moment that he is saying those words he is unscrewing light bulbs to save pennies.

How does a denial of free will offer any basis for hope? A recurring trope in the play, repeated at least ten times, usually by Mary, is some variant of 'It's not your fault':

> It's wrong to blame your brother. He can't help being what the past has made him. Any more than your father can. Or you. Or I.

If the psychological model O'Neill offers in this play is correct – if indeed we cannot control our actions – then, as O'Neill argued in *Iceman*, we cannot be held culpable for them. O'Neill is not saying that real crimes are not committed and real pain felt. He sees, however, that there is a second level of pain, built on guilt or recrimination, that piles additional torments onto the first level. He is too much of a fatalist to believe that the first level – the actual wounds we cause to each other, intentionally or not – can be avoided. But the second level can be. The same dark facts of reality that limit and pervert our attempts to be loving and moral are the reason we should forgive ourselves for our failures. Unlike the characters in *A Moon for the Misbegotten*, the Tyrones never achieve self-forgiveness, and that is why this play ends so very darkly. But the author has learned to view them with (in the words of his dedication of the play to his wife) 'pity and understanding and forgiveness;' and he offers the gift of that ability to the world.

A Touch of the Poet, the only surviving completed play from the proposed eleven-play cycle, is a slight digression in what was to be the saga of the Harford family, here kept mainly offstage; some of the play's secondary themes, notably the conflict between idealism and money-making, were evidently more important to the cycle as a whole than to this play. The focus here is on Con Melody, an army officer reduced to tavernkeeper, and on his daughter Sara's romance with Simon Harford; and the strongest themes are those that tie this play to *The Iceman Cometh* – the paradoxically life-giving nature of fantasy – and *A Moon for the Misbegotten* – the deep satisfaction women can find in unselfish love. Con Melody is a ridiculous figure, the village tavern owner whose cronies play up to his eccentricities just to cadge drinks. But O'Neill goes to great lengths to establish that there is also something imposing about this self-styled Byronic hero; like the pipe dreams of the bums in *The Iceman Cometh*, the persona of Major Melody is a self-delusion, but one that makes Con's disappointing life worth living, and gives him a vitality and dignity that reality cannot. Sara, who thought she had nothing but contempt for the fantasy, surprises herself with the depth of her disappointment when it is destroyed. Unlike the *Iceman* characters, Con will survive without his pipe dream, and on at least one level his life may be improved by its loss; he is finally able to look at his wife as his equal and tell her he loves her. But a glory has gone out of the man, and O'Neill leaves the audience wondering whether truth is really preferable to fantasy.

Although O'Neill's last two plays were not staged in America until after 1960, it seems appropriate to discuss them along with the other posthumous works. The one-act play *Hughie* is built on a rambling monologue by Erie Smith, small-time gambler, to the night clerk of his hotel, who responds only enough to indicate that he is really lost in his own thoughts. Erie reminisces about the previous night clerk, Hughie, who had lived vicariously through Erie's tales of the fast life, and whose appreciation had given Erie the confidence to keep going; since Hughie's death Erie has been on a losing streak. The clerk finally hears some of what is being said, and the play ends with the sense that he will replace Hughie as Erie's audience and ego-booster. Again there are echoes of *The Iceman Cometh* in the suggestion that fantasy can be more satisfying than reality (for the two clerks) and even a prop to support one through reality (for Erie). It is a slight piece, little more than a footnote to *Iceman* and *Touch of a Poet*, but it shows one direction O'Neill's exploration of realism was taking him, towards the exposure of inner experience through small talk.

It is difficult to analyse the final posthumous play, *More Stately Mansions*, very thoroughly, since it exists only in an unedited draft that is five times the length of an ordinary play. The two produced texts (Stockholm 1962 and New York 1967) and the published text represent different selections and ordering of scenes and speeches by different editors, none of them including more than half the manuscript. So one can only indicate some of the things O'Neill was considering putting into the play, without knowing for certain which would have survived his editing. The three texts have the same outline: Sara Melody has married Simon Harford, and competes with his mother Deborah for his love. The play is a series of power games, with the women occasionally uniting to manipulate Simon, and then splitting to fight over him. The struggle eventually takes its toll on Simon, who is in danger of losing his mind; and the women then outdo each other in sacrificing to save him, Sara giving up her ambitions while Deborah deliberately chooses insanity to free him from her spell.

The conflict between business and romanticism, with the dangers inherent in both, was evidently going to be the continuing theme of the cycle. But *More Stately Mansions* has other concerns as well, some of them, harkening back to much earlier periods in O'Neill's work. Simon Harford's disintegrating personality recalls *The Great God Brown*; the separation of woman into mother and whore, and then the confusion of the two, echoes both *Desire Under the Elms* and *Mourning Becomes Electra*; there is even a scene of spoken thoughts in the manner of *Strange Interlude*. In performance, the strongest aspect of the play is likely to be the domestic situation, with its horrifying picture of the power politics of marriage and family, filled with sudden reversals and shifting allegiances. Deborah Harford's conscious flirtation with insanity is also fascinating, and her deliberate entry into the summer house, which has been established as the place where she will go mad, is a powerful image of sacrifice. Still, the play cannot escape looking like the unformed and unwieldy draft that it is, and cannot be analysed or judged as a finished work.

Throughout his career O'Neill stood apart from the mainstream of American drama, not being influenced by what was being written around him and not having much direct influence on other dramatists, except by setting high standards. And yet the history of his career resembles that of the American drama as a whole, starting with small plays of psychological realism, moving through an extended period of stylistic experiment, and then returning to domestic melodrama, this time with the confidence that the conventional genre was far stronger and more flexible than it had seemed. He – and the American drama as a whole – also moved from overtly thesis-driven plays to those that at least give the impression of being generated by character, and from studies in such large subjects as Race and Fate to examinations of the psychological and spiritual experience of everyday life. And in his late masterpieces, all written before the major plays of Tennessee Williams and his contemporaries, he anticipated the discovery that the true calling of the mid-twentieth-century American dramatist lay in offering some hope and consolation to those for whom everyday life was a painful challenge.

Other older dramatists

Several playwrights whose careers began in the 1920s and 1930s remained active into the 1950s, though it is notable that few of them had more than one or two successes, which suggests that the public and the genre had moved beyond them. Maxwell Anderson, for example, attempted one more verse play, a couple of historical costume dramas, a musical and a domestic melodrama, and had his biggest hit with a popular adaptation of a suspense novel. *Anne of the Thousand Days* (1948) reaches for tragedy in presenting Anne Boleyn as an admirably intelligent and self-reliant figure tempted to abuse her power, but actually functions on a less heroic level, suggesting the ironic fate of an amateur politician defeated by those far more skilled at the game. The central question of *Joan of Lorraine* (1946) is whether Joan should have compromised her idealism to the inevitability of working with imperfect human associates. Anderson applies the question to the world war that had just ended, presenting Joan's story as a play being rehearsed by modern actors, who debate its interpretation. In *Barefoot in Athens* (1951), Anderson characterizes Socrates as a simple believer in democracy whose countrymen are unfortunately not as wise as he hopes; there are overtones of McCarthyism in his fate, and Anderson seems uncertain whether society's persecution of its inconvenient idealists is to be celebrated or deplored. *Bad Seed* (1954), an adaptation of a novel by William March about a murderous little girl who covers her complete lack of morality with a veneer of innocence, is the not negligible accomplishment of a proficient technician able to make a slight piece work in theatrical terms.

Lillian Hellman's post-war plays show her moving from the social commentary mode of the 1930s to psychological studies more typical of the 1950s – or, in other words, from the Arthur Miller end of the continuum towards the Tennessee Williams end. *Another Part of the Forest* (1946) returns to the characters of *The Little Foxes*, dramatizing events twenty years before the earlier play in which Ben blackmails his father into giving him control of the family business. It thus fills in some of the background to *The Little Foxes*, in particular the curiously friendly competition between Ben and Regina in the other play, both by showing one of their first clashes and by more explicitly identifying the qualities that they share and that they bemusedly recognize and appreciate in each other. But the play stands on its own as a fascinating study in familial politics and infighting, and as a chilling demonstration of the triumph of the ruthless and devious over the simple or the merely ruthless. Indeed, the biggest weakness of this play is that it does so very successfully what Hellman had already done so very successfully; except for the added insights into some of the characters, *Another Part of the Forest* is essentially a remake of *The Little Foxes* without the resonance of historical allegory.

The Autumn Garden (1951) is a much quieter play with overtones of Chekhov, as a group of old friends are led by the tactless comments of an outsider to recognize how completely they have wasted their lives. The discoveries serve no purpose, as they also understand that their inertia is too ingrained to alter; all that will happen is that they will now be more consciously aware of their own unhappiness, though an ironic resignation to the inevitable can temper and soften the regrets of age. As in Chekhov, very little actually happens, and much of that is offstage; life is not made of melodramatic climaxes, but of the gradual discovery that we are who we are, for better or worse. *Toys in the Attic* (1960) is a somewhat darker psychological study, as spinster sisters are so threatened by their brother's good fortune that they sabotage it to bring him back under their control. The plot involves incest, miscegenation, violence, neurosis and failure in a New Orleans setting, so a comparison to Tennessee Williams is inescapable; like Williams, Hellman is exploring the bargains and compromises people make with life in order to achieve something like happiness, and is interested in aberrant psychology primarily for what it can illuminate about normal life. But *Toys in the Attic* is unmistakably by the author of *The Little Foxes* as well, in its focus on the irresistible power of obsession and in an ironic distance that lets Hellman see that victories are sometimes Pyrrhic; an important secondary theme of the play is the danger of getting what you think you want.

Clifford Odets also turned from social criticism to psychological study after one last play of the earlier mode. Odets had spent much of the previous decade in Hollywood, and there are autobiographical overtones to his anger in *The Big Knife* (1949) at its corrupting seductiveness, but the Hollywood of *The Big Knife* is ultimately a metaphor for America and the drive for economic success, much like the boxing ring in *Golden Boy*. In both plays Odets shows that something must be sold to obtain a lot of money, and that is usually one's sense of self; this play's equivalent of Joe's violin is Charlie Castle's real name, which the studio made him change. *The Big Knife* is a weaker play than *Golden Boy*, in part because Odets's personal anger interferes with its focus, in part because it lacks the context of the Depression to establish universality. In the same Broadway season Arthur Miller offered Willy Loman as an example of the dangers of the American Dream, and Odets's Hollywood star proved harder to feel sorry for than the failed salesman.

Perhaps realizing that the time for plays like *The Big Knife* had passed, Odets turned to a more domestic analysis of relationships and psychological needs, combined with elements of the whodunit, in *The Country Girl* (1950). When a director casts a faded and alcoholic actor in a play, he is led to believe that the actor's dependent wife caused his downfall, and tries to keep her from interfering with the comeback. He ultimately learns that the opposite is true; the actor is self-destructive, and his wife's support is the only thing that has kept him going. At the centre of the play is an analysis of the politics of marriage, particularly the way in which the stronger party's willingness to serve the needs of the weaker actually puts the weaker in control. Another of Odets's observations is the human need for simple and schematic definitions, and the degree to which such definitions limit our perceptions. Because he wants a world made up of villains and victims, the director eagerly misinterprets all the wife's behaviour to make it fit his model; much of the play's theatrical power comes from the fact that we also misinterpret events and motives through much of its length. Odets's last play, *The Flowering Peach* (1954), is a retelling of the Noah story meant as an allegory of the modern conflict between faith and rationality, but defeated by stereotypical characterizations and weak jokes that are ineffective when not actually offensive. Despite the commercial success of *The Country Girl*, Odets remains a playwright – arguably *the* American playwright – of the 1930s.

Similarly Sidney Kingsley's one post-war success is not enough to move him out of his earlier place in history. *Detective Story* (1949) shares with Kingsley's earlier *Men in White* and *Dead End* an almost documentary background, this time of a police station, full of the routine comings and goings of officers and criminals. In the foreground is Detective McLeod, a man with an unequivocal sense of morality and a blind hatred of evildoers. When he discovers that his own wife has a tainted past, not even love can modify his absolute standards. He rejects her and, in torment, lets an escaping prisoner kill him. McLeod's drama implies a message about the need to balance morality with charity, but Kingsley does not press the generalization, focusing instead on the detective's psychological and spiritual pain.

Comedy

Although some of the important comic dramatists of previous decades – Kaufman, Behrman, Lindsay and Crouse – continued writing into the 1950s, none of them dominated the decade, nor did any newcomers. Broadway comedy at mid-century had evolved into a polished and somewhat homogenized commodity that could virtually be manufactured by formula, and an individual play's fortune was more likely to be a matter of how skilfully it resembled other successes than how original it was. A few subgenres, virtually blueprints for the construction of a comedy, account for most of the comedies, successful or not, of the fifteen-year period.

Not surprisingly, most comedies had an element of romance built into the plot, many relying on a comic staple that goes back at least as far as Shakespeare's *Much Ado About Nothing*, the couple who spend most of the play fighting until they finally realize what the audience has seen from the start, that they are in love. In Elmer Rice's *Dream Girl* (1945) the title character, escaping into various romantic fantasies, keeps fighting off the jocular cynic who eventually wins her. In Garson Kanin's *Born Yesterday* (1946) an uncouth businessman hires a reporter to give his dumb blonde girlfriend some class, and teacher and student spar through the play. Other romantic comedies let small plot complications loom large as temporary threats to a happy ending; the hero of Norman Krasna's *John Loves Mary* (1947) is a soldier who went through a marriage of form to help a friend's English girlfriend come to America and now must stall his eager fiancée until he can get a divorce. Still others built their suspense on whether the couple could avoid getting together: in George Axelrod's *The Seven Year Itch* (1952) and *Will Success Spoil Rock Hunter?* (1955) and Leslie Stevens's *The Marriage-Go-Round* (1958) the hero must resist the temptations of a desirable woman seemingly determined to seduce him.

Satire was fairly rare in Broadway comedy, and when it did appear it was usually directed at safe targets: government or military bureaucracy, stuffy businessmen, television, etc. Among the comedies that dared to suggest that the American military was not a thoroughly efficient and logically run operation, the strongest is the adaptation by Thomas Heggen and Joshua Logan of Heggen's book *Mr Roberts* (1948), about the crew of a Navy supply ship resisting the authority of their mad captain; it stands out also as one of the few comedies of the period to successfully mix low humour and sentimentality. In John Patrick's *The Teahouse of the August Moon* (1953, from the novel by Vern Sneider), soldiers trying to Americanize an Okinawan village find themselves going native instead; and in Ira Levin's *No Time for Sergeants* (1955, from the novel by Mac Hyman), a bumbling hillbilly is drafted into the Air Force, much to the regret of the Air Force.

The pattern of *No Time for Sergeants*, in which the innocent little person manages to wreak havoc on a mighty institution just by doing what comes naturally, worked just as well in other contexts. The dumb blonde in *Born Yesterday* ruins her sugar daddy's dishonest deals by applying the basic civics she learned from her tutor. *Life With Mother* (1948), Howard Lindsay and Russel Crouse's sequel to their 1930s hit, shows Mother once again manipulating Father simply by assuming she is in the right. In Samuel Spewack's *Two Blind Mice* (1949), two little old ladies run an obscure federal agency forgotten by the government and use the confusion and gullibility of the bureaucracy to keep from being shut down. In *The Solid Gold Cadillac* (1953), by Howard Teichmann and George S. Kaufman, another little old lady asks an innocent question at a stockholders' meeting and ends up running the company. Less innocently, the heroine of William Marchant's *The Desk Set* (1955) foils a corporate attempt to computerize her office by proving smarter than the computer.

Another recurring subgroup is made up of plays built around a magnetic and unorthodox character who deliberately and comically shakes up the conventional world. In Robert E. McEnroe's *The Silver Whistle* (1948) a vagabond philosopher challenges the demeaning routine of an old folks' home and rejuvenates its inhabitants. In S. N. Behrman's *Jane* (1952, from a Somerset Maugham short story), a plain-speaking country cousin plays the games of sophisticated society more skilfully than those born to it. In Gore Vidal's *Visit to a Small Planet* (1957) an uninhibited man from outer space causes trouble in Washington just for the fun of it. The hero of *The Pleasure of His Company* (1958), by Samuel Taylor and Cornelia Otis Skinner, is a charming socialite who returns from banishment on the eve of his daughter's wedding, determined to save her from a conventional life. The archetype of this pattern is *Auntie Mame* (1956), Jerome Lawrence and Robert E. Lee's adaptation of Patrick Dennis's novel about the life-loving and aggressively eccentric woman who drags her nephew through several lifetimes' worth of comic escapades with the rallying cry, 'Life is a banquet, and most poor sons-of-bitches are *starving* to death!'

With hindsight we can see that the Broadway theatre of the 1950s was the goal that the American drama had been moving towards for half a century. It was the full flowering of realistic contemporary domestic melodrama, a mode that proved itself to be the natural medium for the expression of the American experience in drama, by adapting to the very different styles and demands of Arthur Miller, Tennessee Williams and Eugene O'Neill. The important discovery of the previous generation, that realistic domestic stories could be used to reflect and comment on the world outside the living room, provided Miller with the dramatic tools for questioning the nature of the American Dream and the American system more acutely than any previous playwright; and such plays as *Death of a Salesman* and *The Crucible*, by showing how the injustices and failures of the system threatened the individual's sense of self, turned political and social theory into very human drama. Williams, starting from a sympathy for the outcasts and misfits of society, used domestic realism to present their experience with an intimacy that led audiences to recognize a shared humanity, and thus guided audiences to the recognition of their own feelings of inadequacy and failure, painful discoveries for which Williams then offered counsel and reassurance. O'Neill's career proved a microcosm of the history of the American drama as a whole, as he returned after years of experimentation to the domestic realism he had rejected in the 1920s and found that realistic plays with a domestic setting could carry the weight of his profound psychological and philosophical insights.

Even more striking than the divergence of these three leading playwrights, and the many second-level writers around them, is their convergence. Not only could the social critic, the poet and the philosopher all use variants of the same dramatic style, but they reached variants of the same dramatic statement. The Miller who argues in *Death of a Salesman* that failure in the world's terms is not necessarily a mark of personal inadequacy, the Williams who shows in *A Streetcar Named Desire* that everyone experiences life as painful and difficult, and the O'Neill who dramatizes in *Long Day's Journey into Night* his belief that our crimes against ourselves and others are not our fault, are ultimately offering the same gift of absolution and forgiveness.

The best plays of these three writers are the best American plays ever; and, along with their second best, and the best plays of secondary writers such as Inge and Hansberry, make up a body of work as yet unmatched in any other decade of the century. It might also be said that the American drama finally found its true subject in this period, in the realization that a significant portion of the audience found the task of day-to-day living an overpowering one, and was prepared to turn to the theatre for guidance and reassurance. The challenge of providing that almost religious service would continue to engage American dramatists for the rest of the century.

Chapter 5
1960-1975: The Post-Broadway Era

For most of the first half of the twentieth century the terms 'American theatre' and 'Broadway' were essentially synonymous, the commercial theatre centre of New York City being virtually the sole source of new American plays, and theatrical activity throughout the rest of the continent consisting of little more than revivals or touring productions of Broadway hits. As explained in the first chapter, this situation began to change around 1950, with the birth of a fringe or alternative theatre community within New York, and the establishment of the first resident professional theatres in other cities. Both Off-Broadway and the regional theatres began by staging classics and Broadway revivals, but both turned to new American plays in the early 1960s. Within a few years, as Off-Broadway and regional theatres became more established, the fringe moved outwards, creating an Off Off-Broadway of experimental companies in New York, and alternative theatres in cities around the country, all specializing in new work. Very quickly the lines dividing these theatrical worlds began to fade: by the 1970s a new young playwright might have his first play produced Off Off-Broadway, his second in Los Angeles or Chicago, and his third on Broadway; or he might make the same journey in reverse. Established Broadway dramatists were produced Off-Broadway or in regional theatres, while regional or Off Off-Broadway hits transferred to Broadway for commercial runs. The historical background of this chapter is the extraordinary expansion of the American theatre: as early as 1960 in New York, and by the late 1960s elsewhere in the country, new American plays were produced outside the realm of Broadway in unprecedented numbers, adding new dramatists and new points of view to the American dramatic repertory.

By its very nature, the expanded theatrical arena encouraged new experiments in form and content. In part because of the younger and more sophisticated audiences of the alternative theatres; in part a result of the discovery of challenging new European dramatists such as Samuel Beckett, Eugene Ionesco and Jean Genet, and a revival of interest in Bertolt Brecht; in part out of rebellion against the commercial theatre of

Arthur Miller and Tennessee Williams; and in part because the essentially non-commercial alternative theatres could more afford to take chances more than Broadway could – for these reasons and perhaps just because any art form that achieves its maturity invites challenge, some new dramatists of the 1960s rejected domestic realism and sought to expand the stylistic vocabulary of the American drama. The surprising result of this new experimentation would not become clear for another decade or so; at the time the 1960s resembled the 1920s as young playwrights, convinced that their new ideas or perspectives required new forms beyond the domestic realism of Miller and Williams, attempted to discover what else the drama could do.

This volume is not primarily a theatrical history, and generally has not discussed changes in acting or staging methods. Still, it is worth noting that a small but significant part of the alternative theatre movement, particularly in the late 1960s, was devoted to experiments in production and performance styles that had some effect on the literature of the theatre. Companies such as the Open Theatre and the Performance Group in New York, the Odyssey Theatre in Los Angeles, the Firehouse Theatre in Minneapolis, and the Living Theatre in self-imposed European exile shared an impulse to move away from author- and text-dominated drama to develop new forms. Along with other companies and directors, they treated a script as only the starting point for theatrical invention, relying extensively on improvisation and borrowing freely from such allied arts as mime, dance and religious ritual. Often their final product bore only a slight resemblance to the original text, while in some cases they dispensed with the playwright by composing the work communally, or reduced the author to the role of a scribe who recorded the final product of the group creation.

However anti-textual these approaches were, some playwrights found them amenable. Jean-Claude van Itallie and Megan Terry, for example, worked closely with the Open Theatre; and Sam Shepard and Lanford Wilson were strongly influenced by the company's style. In particular, all these writers found the company's rehearsal exercises, and the skills they developed in the performers, supportive of their own experiments in play construction and character development. In one such exercise, 'Transformation', actors playing a scene would be ordered suddenly to switch roles, jump to a different play, become animals, or otherwise make instant changes in the reality they were portraying. The purpose of the exercise was to stimulate the performers' imaginations and powers of concentration, as they created fully realized characterizations instantly. For the playwrights it provided a liberation from extended exposition and mood-setting, offered new and efficient means of characterization, and allowed for a cinematic structure of short scenes flowing or jumping into one another.

In *Interview*, one of the playlets of van Itallie's *America Hurrah* (1966), a number of disconnected scenes and events are tied together thematically through sudden juxtaposition, and the play is full of such stage directions as

> *The remaining actors return to the stage to play various people on Fourteenth Street: ladies shopping, a panhandler, a man in a sandwich board, . . . and so on. . . . Each time they approach the audience, they do so as a different character. The actor will need to find the essential vocal and physical mannerisms of each character, play them, and drop them immediately to assume another character.*

In a programme note van Itallie credits the original company with helping to shape his text through their rehearsal improvisations.

For *The Serpent* (1968), the Open Theatre explored improvisations on the theme of Creation for several months before van Itallie came in to observe and to combine the results of their work with material of his own in a final text, a series of scenes and interludes dramatizing or responding to the story of Adam and Eve. Enactments of the Temptation, the Fall and the murder of Abel alternate with symbolic stagings of the 'begat' genealogies of Genesis, the assassination of President John Kennedy, and the emptiness of twentieth-century life. The general style is non-realistic, incorporating mime, stylized and ritualistic movement, and distortions of time and space: the assassination is turned into a living newsreel film, which is re-run forward and backward, in slow motion and frame-by-frame, to capture the horrible fascination of the moment; and the genealogy is accompanied by the group miming of intercourse and birth. The governing idea of the play is that events precede the understanding of them, with humanity running a constant race to catch up to the implications of its own past.

Megan Terry's *Viet Rock* (1966) was also developed through improvisations which Terry moulded into a script. Earlier Terry had used Transformation as the structural basis for such plays as *Keep Tightly Closed in a Cool Dry Place* (1965), in which three prisoners act out their tensions by playing various roles, and *Calm Down Mother* (1965), in which three actresses portray the many aspects of woman. Lanford Wilson used Transformations in several early plays, to let a small cast fill the stage with characters, and to produce a cinematic flow of short scenes. Sam Shepard's characteristic juggling of reality levels and dramatic metaphors in such plays as *Action* and *The Tooth of Crime*, and his extensive use of magic, music and ritual also owe a great deal to the theatrical experiments he observed Off Off-Broadway. Though the fashion for improvisation, communal creation and de-emphasis of the text waned in the early 1970s, some of the techniques developed for acting and staging purposes have become staples of contemporary play construction.

Meanwhile, although the distinction between Broadway and the newer branches of the American theatre would begin to fade in the 1970s, it remained sharp in the 1960s; and it was still possible to identify individual playwrights as Broadway or non-Broadway, Edward Albee being one of the very few to move definitively from the alternative to the mainstream. On the other hand, no differences ever really developed among the kinds of plays written for the various alternative theatrical settings; and there was no clear distinction between, say, Chicago or Los Angeles playwrights and Off Off-Broadway playwrights, since in many cases they were the same people. In the absence of useful geographical divisions, it may be more profitable to classify the non-Broadway writers by genres and types of play; and here the distinction made in the previous chapter between the 'schools' of Tennessee Williams and Arthur Miller can still be of use. Many young writers of the 1960s followed the path of Williams and Inge, using the drama to explore and illuminate psychological states, particularly the burdens borne by those for whom day-to-day living was a difficult task. Others, in the very politicized national atmosphere of this decade, addressed social and political issues either directly or, in the tradition of Odets and Miller, by dramatizing their impact on the domestic lives of ordinary people; a significant offshoot of this group is made up of black playwrights dramatizing the black American experience to make moral and political statements.

Psychological and existential plays

One of the first original Off-Broadway hits was Jack Gelber's *The Connection* (1959), about heroin addicts idly waiting for their dealer. Gelber's fictional premise is a play-and-film-within-the-play in which actual addicts have been cast while a documentary film crew records their performance. As the 'actors' wander from or argue with the script, the supposed playwright, producer and film-makers are repeatedly drawn into the action, allowing Gelber to examine not only the nature of the heroin community but its connections to and impact on the 'straight' society. The addicts experience life as shapeless and disordered, something they must improvise their way through without a guide (as reflected by the presence on stage of jazz musicians who break into improvisations at seemingly random moments in the play), and from which heroin offers an escape. The straight characters sense the same frightening emptiness in existence, but frantically invent forms and labels to protect themselves from this realization. Thus the playwright and producer repeatedly insist that what we are watching is fiction, existing within a safely controllable frame, even though the addicts continually break through the frame; thus they constantly try to impose logic and meaning on the action, even when the addicts reject their definitions. Eventually the playwright confesses the inadequacy of his defences by joining the addicts and shooting up, allowing the play to wander shapelessly to its end.

The comic side of early Off-Broadway drama is best represented by Edward Albee's *The American Dream*, discussed later, and by Arthur Kopit's Dadaesque *Oh Dad, Poor Dad, Mamma's Hung You in the Closet and I'm Feelin' So Sad* (1962). Fabulously rich Madame Rosepettle travels the world, indulging herself in vengeance against everything she despises or resents, and dragging behind her – along with her husband's body, a talking fish and two giant venus fly-traps – her grown son Jonathan, whom she calls by whatever name comes first to mind, and whom she keeps in a state of arrested childhood to protect him from the world's corruptions. But Jonathan falls prey to the sexually voracious Rosalie, their bedroom encounter interrupted by the corpse of his father, which falls on them from its place in the closet. Jonathan panics and kills Rosalie, leading to the play's curtain line, when Madame Rosepettle returns:

> *Robinson!* I went to lie down and I stepped on your father! I lay down and I lay on some girl. Robinson, there is a woman on my bed and I do believe she's stopped breathing. What is more, you've buried her under your fabulous collection of stamps, coins, and books. I ask you, Robinson. As a mother to a son I ask you. *What is the meaning of this?*

From its title to its punchline, *Oh Dad* announces itself as a parody of absurd drama, or at least of the public's perception of plays that don't seem to follow traditional rules of dramaturgy. But behind the jokes is a view of life akin to that of Gelber and Tennessee Williams. Both Jonathan and his mother find the world frightening and hostile. Madame Rosepettle responds by attacking, making herself strong enough to shape her environment to her design, while Jonathan has no such resources and is left waving at a passing airplane in the futile attempt to make the universe acknowledge his existence.

A number of other Off-Broadway comedies touch on this basic existential insecurity, though generally without Kopit's bravura; in fact, their authors seem slightly disconcerted by their own insights, hurriedly retreating from them into easy and irrelevant jokes. In Bruce Jay Friedman's *Scuba Duba* (1967), a man whose wife has left him for a black lover finds himself adrift in a world of sexual freedom, psychiatric mumbo-jumbo and his own racist fantasies, and is unable to cope; in Friedman's *Steambath* (1970), death is a Turkish bath and God its Puerto Rican janitor, and the new arrival who tries to talk his way out comes to understand how empty and meaningless his life was. In Terrence McNally's *Next* (1969) a 48-year-old man called up by his draft board also tries to argue his way out and also gives up when he realizes how little he is fighting for. Elaine May's *Adaptation* (1969) presents life as a nightmarish TV game show that is impossible to win.

Paul Zindel's *The Effect of Gamma Rays on Man-in-the-Moon Marigolds* (Houston 1965) mixes black comedy with a domestic melodrama of survival in the face of despair. A neurotic and depressive mother and her disturbed and unhappy teenage daughters are barely able to cope with life's disappointments, but one girl's involvement in a school science project is her first tiny step towards self-acceptance and confidence. Zindel maintains a careful balance between light and dark: depending on whether one identifies more with daughter or mother, the ending is either optimistic or a futile fantasy of hope in the face of an overwhelmingly antagonistic universe.

The best of this genre is John Guare's *The House of Blue Leaves* (1971), mixing slapstick farce with pathos while demonstrating that coping with life is just beyond the ability of the ordinary person. A would-be songwriter saddled with an insane but sadly sweet wife plans an escape to Hollywood until a series of unlikely events somehow involving three nuns, a mad bomber, a deaf starlet and the Pope brings an end to his fantasies. In a mix of love, pity and desperation he kills his wife, and is left to face the world with no dreams at all. As with the other emptiness-of-life comedies, Guare seems vaguely embarrassed by his insight and his sympathy for his troubled characters, and hides his seriousness behind the mask of a *farceur*.

A similar mixture of tones characterizes Mart Crowley's *The Boys in the Band* (1968), one of the first commercially successful American plays to deal openly and sympathetically with homosexuality. A group of New York homosexuals, ranging from the very straight-acting Hank to the wildly effeminate Emory, gather at Michael's for a party, but the arrival of a (presumably) heterosexual friend of the host turns the revelry sour. Driven by his own insecurity, Michael is determined to prove that his friend is also gay, and forces everyone to play a self-exposing game that only ends in humiliating all of them, affirming the outsider's straightness, and revealing the depth of Michael's anguish. The image of homosexuals as self-hating neurotics may no longer be acceptable, but at the time Crowley's sympathetic presentation of them as feeling recognizable human pains, along with his salute to the courage with which they mask their insecurities with inventive humour, was a strong blow against current prejudices.

Edward Albee: the Off-Broadway plays

Of the many writers whose plays appeared in the first flourish of new Off-Broadway drama – from, say, 1958 to 1965 – some proved less interesting than novelty made them seem at first, some turned out to have only one play in them, some left the theatre for other callings. Among those who did produce one or more works of lasting interest, either for their inherent merits or as representatives of the time, only a very few were able or inclined to graduate from the fringe to the mainstream theatre. Most notable among these was Edward Albee, whose move to Broadway helped establish him as the dominant serious American dramatist of the 1960s.

Albee's first produced play, *The Zoo Story* (Berlin, 1959; New York, 1960), is essentially an extended monologue by a young New Yorker so alienated that any contact with another being, however one-sided or incomplete, is a relief:

> A person has to have some way of dealing with SOMETHING. If not with people . . . if not with people . . . SOMETHING. With a bed, with a cockroach, with a mirror . . . no, that's too hard, that's one of the last steps. With a cockroach, with a . . . with a . . . with a carpet, a roll of toilet paper . . . no, not that, either . . . that's a mirror, too; always check bleeding. You see how hard it is to find things? [. . .] with . . . some day, with people. [All but bracketed ellipses Albee's]

He chooses a conventional man he meets in the park to be his first attempt at human contact, but is unable to move his listener sufficiently by merely telling his story. Desperate for reassurance of his existence and his ability to affect another, he starts a knife fight and impales himself on his opponent's blade, dying with thanks for this ultimate proof: 'I came unto you and you have comforted me.'

The effect of this eloquent but tragically alienated character on the American theatre was of the same sort, if not the same magnitude, as the effect of John Osborne's Jimmy Porter on the British theatre. This was a new voice which had not been heard before in the drama but which was instantly recognizable as authentic, representing the experience of a hitherto ignored or unnoticed portion of the population. On the basis of a one-act play Albee was hailed as an important new writer and looked to for new insights into contemporary American life, an assignment he soon implicitly accepted.

After two brief and minor works, *The Sandbox* (1960), in which a dying woman meets a friendly Angel of Death, and *Fam and Yam* (1960), about the disconcerting effect of a young playwright (evidently Albee) on an older one (evidently Inge), Albee turned to an analysis of American racism in *The Death of Bessie Smith* (Berlin, 1960; New York, 1961), based on the traditional story that blues singer Bessie Smith died because she was refused admission to a white hospital after an automobile crash. Albee's central character is the admitting nurse at the hospital to which Bessie's driver brought her after being turned away elsewhere, and the bulk of the play precedes the arrival of the body. In the Nurse's bitter interactions with other characters, Albee shows how the emptiness of her life has produced a frustration and resentment for which racism is merely a convenient outlet. The climax of the play is not Bessie's death, but the Nurse's explosion a moment before it:

> I am *sick*. I am sick of everything in this hot, stupid, fly-ridden *world*. I am sick of the disparity between things as they are, and as they should be! [. . .] I am sick of going to bed and I am sick of waking up. . . . I am tired . . . I am tired of the truth . . . and I am tired of lying about the truth . . . I am tired of my skin . . . I WANT OUT!
>
> [All but bracketed ellipses Albee's]

This kind of pain must be avenged on somebody, and blacks are as convenient a scapegoat as any.

The American Dream (1961) is a darkly comic piece that combines echoes of Ionesco's *Bald Soprano* (in such toying with language as the extended discussion about whether a hat is wheat or beige coloured, and in the cartoonishly exaggerated characters) with a chilling condemnation of American values. On one level, then, the play is in the tradition of Arthur Miller's *Death of a Salesman*, using a domestic story, however unrealistically presented, to comment on the society outside the living room. But, where Miller condemned the American Dream for being impossible to achieve, Albee caustically depicts its dehumanizing success. Presenting what is implicitly a typical American family – selfish and catty Mommy, emasculated and almost invisible Daddy and wise-cracking Grandma – Albee satirizes and condemns a hollowness in American values that is not the result of inadequacy but of deliberate choice. Themselves almost totally inhuman, having withdrawn into a self-absorption from which they stir themselves only with great difficulty – it takes Daddy several pages to gather up the energy and determination to answer the doorbell – Mommy and Daddy *want* to be empty. Their comical abuse of language is not the result of linguistic incompetence, but a concerted effort to free themselves from the human obligations implied by communication. Grandma explains what happened to the child they once adopted in terms that explicitly identify misused words with acts of inhumanity:

GRANDMA One night, it cried its heart out. . . . Then it turned out it only had eyes for its Daddy.

MRS BARKER For its Daddy! Why, any self-respecting woman would have gouged those eyes right out of its head.

GRANDMA Well, she did. . . . But *then*, it began to develop an interest in its you-know-what.

MRS BARKER In its you-know-what! Well! I hope they cut its hands off at the wrists!

GRANDMA Well, yes, they did that eventually. But first, they cut off its you-know-what. . . . And then, as it got bigger, they found out all sorts of things about it, like: it didn't have a head on its shoulders, it had no guts, it was spineless, its feet were made of clay . . . .

The Young Man who appears at the end of the play is the twin of the unsatisfactory baby, who has been sympathetically affected by his brother's mutilation:

> Once . . . it was as if all at once my heart . . . became numb . . . almost as though I . . . almost as though . . . just like that . . . it had been wrenched from my body . . . and from that time I have been unable to love. Once . . . I was asleep at the time . . . I awoke, and my eyes were burning. And since that time I have been unable to see anything, *anything*, with pity, with affection . . . with anything but . . . cool disinterest [. . .] there are more losses, but it all comes down to this: I no longer have the capacity to feel anything.
> [All but bracketed ellipses Albee's]

Grandma recognizes immediately that this Young Man will satisfy Mommy and Daddy as his brother never did, because he is the American Dream. Mommy and Daddy did not mutilate their child to punish it, but to turn it into this; and the baby's only failure was in dying before it achieved the perfection of its brother. Mid-century America is characterized by an emptiness of spirit that requires emptiness around it, says Albee. A culture with waning values is not backsliding from what it really wants to be; it is actively striving for its complete loss of values, and will celebrate beautiful deadness as the achievement of all its desires. The artistic and commercial success of *The American Dream* prompted Albee's move to Broadway, where he became the predominant representative of the new generation of playwrights in the mainstream theatre; his Broadway plays will be discussed later in this chapter.

Sam Shepard: the early plays

Sam Shepard and Lanford Wilson, the two writers who would be recognized in the late 1970s as the leading American playwrights of their generation, both began Off Off-Broadway in the mid-1960s. Shepard wrote more than twenty-five plays between 1964 and 1975; not all of them were artistically successful, but the freedom provided by the noncommercial fringe theatre allowed him an extended apprenticeship during which he honed his craft and discovered his strengths. Shepard's earliest plays are almost spontaneous outpourings, rarely edited or rewritten after the first burst of inspiration. Almost inevitably, they are uneven and inconsistent, with radical shifts of character, tone and even dramatic mode – e.g., realism changing suddenly to expressionism or allegory and back again. (Shepard may have been influenced by the actors' exercise called Transformation, then popular in experimental theatre groups; see the discussion of experimental theatre earlier in this chapter.) Some are so personal that they almost require biographical footnotes: for example, *The Holy Ghostly* (1969), in which a son rejects and kills his father, is filled with extraneous details relevant only if one knows that the characters are based on Shepard and his father. Others make use of a private symbolism whose exact meanings are never explained: cowboys, rock musicians and supernatural monsters wander through his plays, often completely out of context, while in other plays characters suddenly become eloquent on such arcane subjects as urban planning (*Fourteen Hundred Thousand*, 1966) or dermatology (*La Turista*, 1967).

If, at their weakest, these early short plays are so personal in reference and symbolic vocabulary as to be almost unintelligible, at their best they have a seemingly improvised jazz-like rhythm that carries them theatrically, along with visual images of undeniable power, if unclear meaning, and self-contained speeches of instant characterization and convincing psychology. And if the plays are rarely linear and consistent enough to be said to make statements, some assumptions and observations about psychology, metaphysics and even morality do recur. One repeated theme is the need of the individual to create himself in a world that gives him no particular identity to start with. In such plays as *Cowboys* (1964), *Melodrama Play* (1967), *The Unseen Hand* (1969) and *Action* (London, 1974), characters repeatedly experiment with alternative identities and behaviours, trying to find something that will be 'real' to them. In *Action*, the purest but unfortunately least comprehensible presentation of this idea, four people at a Christmas dinner go through a series of seemingly unmotivated and unrelated behaviours, randomly dancing, telling stories, breaking furniture or busying themselves in incongruous domestic rituals. The play defies interpretation unless one takes special note of the fact that it opens with the line, 'I'm looking forward to my life', and closes with 'I had no idea what the world was. I had no idea how I got there or why or who did it. I had no references for this.' And midway through, one character says, '*quietly to himself*',

> Just because we're surrounded by four walls and a roof doesn't mean anything. It's still dangerous. The chances of something happening are just as great. Anything could happen. . . . You hunt for a way of being with everyone. A way of finding how to behave. You find out what's expected of you. You act yourself out.

Evidently what Shepard intends is a vision of humanity experimenting with identities and modes of conduct in an undirected and unguided attempt at self-definition, but in performance this theme is far less evident than it is on the page.

Slightly clearer are the plays in which invented identity is part of a process of escape, as characters for whom the world is too frightening or unpleasant try to create an alternative reality they can control. In *Chicago* (1965) Stu is unable to deal with the fact that his girl is leaving him, so he sits in a bathtub and pretends to be at the beach, describing his fantasy in elaborate detail to block out the reality. The characters in *Icarus's Mother* (1965) are oddly bothered by a passing airplane, and invent fictions about it, imagining its pilot and pretending to signal him. What they are doing, evidently, is mythologizing the source of their fear in order to domesticate and defang it. As happens in several Shepard plays, the fantasies actually displace reality; after they pretend the plane has crashed it really does, leaving them frightened by their power.

Something similar occurs in *Red Cross* (1966), in which three characters take turns exorcizing their fears through elaborately detailed fantasies until one of the stories seems to come true, causing the speaker's head to bleed – a visualization, perhaps, of the 1960s phrase, 'It blew his mind.' Indeed, one of the moral judgements Shepard implies in these early plays is that in some situations withdrawing from reality is dangerous or irresponsible. Among the very few characters towards whom he shows some anger or contempt are the father in *The Holy Ghostly* (1969), who lives entirely in a romanticized past; the revolutionary in *Shaved Splits* (1970), whose self-indulgent fantasies are equated with pornography; and the girl in *Cowboy Mouth* (1971), who uses her sexual power to try to force a man into her fantasy. On the other hand, Shepard makes no judgement of the condemned man in *Killer's Head* (1975), whose final thoughts in the electric chair are mercifully about the mundane pleasures of his ranch.

Another recurring element in the early plays is Shepard's conviction that the supernatural exists in the real world, in the form of magic, mysticism, religion or folklore; and that failure to acknowledge or respect its existence is perilous. The characters in *Icarus's Mother* and *Red Cross* are punished for toying with powers they hardly realized they had. In *Back Bog Beast Bait* (1971) a mythical two-headed swamp monster is proved to exist, and those who denied it are transformed into animals for their apostasy. The couple in *Cowboy Mouth* dream of a 'rock-and-roll savior,' and only interrupt their fantasy to torment a restaurant delivery man with the misfortune to be wearing a lobster costume. So they are unprepared when the lobsterman, in a symbolic evolutionary leap, cracks his shell and steps out as the rock superman; and, unappreciated, he is left to attempt suicide at the end. In *Mad Dog Blues* (1971) a rock singer and a drug-dealer somehow conjure up Marlene Dietrich, Mae West, Captain Kidd, Paul Bunyan and other characters from history and fable, who join them in a farcical search for a buried treasure that proves to be of less value than the comfort found in communion with these mythic figures.

In Shepard's eyes, alienation from the spiritual goes hand in hand with alienation from the past, and another of his implicit moral judgements, which would become very important in his later plays, is that a culture dissociated from its past is in serious danger. In *Fourteen Hundred Thousand* (Minneapolis, 1966) the emptiness of a couple building a bookcase for books they won't read is identified symbolically with the replacement of country villages by dehumanizing cities. In *La Tourista* (1967) an American in Mexico is repeatedly overpowered and displaced by figures from richer native traditions until he sees himself as Frankenstein's monster, an obscene parody of the human beings the others are. In *The Unseen Hand* (1969) a rebellious slave from a mechanized and regimented planet comes to earth for help in his revolution. He turns first to a trio of nineteenth-century Western gunslingers, whom he magically resurrects, but it is the unashamed patriotism of a high school cheerleader –

> I love Azusa! I love the foothills and the drive in movies and the bowling alleys and the football games and the drag races and the girls and the donut shop and the High School . . . and the YMCA and the Glee Club and the basketball games and the sock hop . . . and the county fair and peanut butter and jelly sandwiches and the High School band and going steady . . . and my Mom, I love my Mom most of all.

– that proves to be the key to his liberation by reminding him of his own cultural heritage.

These themes come together in *Operation Sidewinder* (1970), whose rather convoluted and melodramatic plot involves black revolutionaries, renegade Indians, a sightseeing housewife and a giant rattlesnake which is actually an experimental Air Force computer designed to contact UFOs. The Indians recognize the snake as the foretold sign of the apocalypse, and use it in a ritual that enables them to escape from earth just as the Third World War is about to begin. In this play Shepard finds almost every facet of American society – the military and the counterculture, black revolutionaries and the white middle class – inadequate because of materialism and alienation from spiritual values, while the Indians' religion has a stronger appreciation of truth than the white man's science.

Shepard remains challenging and difficult in his best play of this period, *The Tooth of Crime* (London, 1972), in which he freely mixes theatrical metaphors and even invents a new language for his characters. At the core of his plot is the cowboy movie staple of the aging gunslinger challenged by a younger sharp-shooter. But Shepard's Hoss and Crow are also an established rock star and the unknown musician threatening his popularity with a new style, a teen gang leader and a gypsy loner, a veteran automobile racer and a talented unknown, a Mafia godfather and a freelance hit-man, a champion boxer and his challenger, and pawns in an intergalactic game manipulated by unseen 'Keepers'. Shepard conjures up these images interchangeably and without explanation, in the faith that their accumulated evocative power will outweigh any confusion they cause.

In each of these definitions, Hoss is the aging establishment figure, strengthened by his knowledge of his art's history and traditions, but weakened by the cushioned inactivity of success. Crow is the self-sufficient loner, with no knowledge beyond his own limited experience and no sense of community with past or present, but with the totally amoral freedom to act that such alienation brings. At the play's centre is their inevitable duel, which takes the form of imposing alternative realities on each other. Crow attacks with images of failure designed to reduce Hoss to snivelling surrender:

> The dumb kid. The loser. The runt. The mutt. The shame kid . . . . Catch ya' with yer pants down. Whip ya' with a belt . . . . Hidin' dirty pictures. Hide 'em from his Ma. Hide 'em from his Pa . . . . Just gimme some head boy. Just get down on your knees. Gimme some blow boy . . . . Get down! Get down!

Hoss parries with images of Crow's lack of musical roots:

> Yeah, well I hear about all that kinda 'lectric machine gun music. . . . Fast fingers don't mean they hold magic. . . . You ain't come inside the South. You ain't even opened the door. The brass band contain yo' world a million times over. . . . You lost the barrelhouse, you lost the honkey-tonk.

The young man wins, his shallowness giving him the flexibility to dethrone the champ. He happily offers to teach Hoss how to be like him, wiping away all his personal history and cultural roots and programming him with a new, deliberately empty identity, but the new persona is too much of a leap for Hoss –

> Mean and tough and cool. Untouchable. A true killer. . . . No hesitation. Beyond pride or modesty. . . . Holds no grudge. No blame. No guilt. . . . Passed beyond tears. Beyond ache for the world. Pitiless. Indifferent and riding a state of grace. It ain't me! IT AIN'T ME! IT AIN'T ME! IT AIN'T ME!!

– and he both accepts defeat and reaffirms his own identity by killing himself with a stylishness Crow can only envy. There is a lot going on in *The Tooth of Crime*, but at its core is a view of the world similar to Tennessee Williams's in *A Streetcar Named Desire* or Edward Albee's in *Who's Afraid of Virginia Woolf?* Hoss, like Blanche DuBois, represents the world of the past, with its rich history and accomplishments, and with its weaknesses. Crow, like Stanley Kowalski, is the wave of the future, crude and shallow and strong. Like Williams, Shepard acknowledges sadly that Crow is destined to win out; unlike Williams, he has no grudging admiration for this raw power, sharing Hoss's view that dissociation from one's humanity is the equivalent of death.

Later in the 1970s Shepard would gradually move away from this style of mixed metaphors and sudden shifts in dramatic mode towards something more closely approaching domestic realism; and it is tempting to read *Geography of a Horse Dreamer* (London, 1974) biographically. Cody, who can foretell the outcome of horse races in his dreams, is held prisoner by gamblers. He is in a slump, no longer able to predict winners, until he shifts to greyhound racing, where his dreams once again prove profitable. A number of allusions identify Cody as a symbol of the artist exploited by commerce, so the play may be Shepard's acknowledgement of his own need to find a new artistic course. The biographical interpretation is not essential, however, since the play includes others of Shepard's familiar concerns. Several of his plays of the period (*Cowboy Mouth*, *The Tooth of Crime*, *Angel City*, *Suicide in* B^b) deal openly or symbolically with artists, labelling the artistic impulse as somehow magical. Cody is also identified with the old West and his captors with the city, and he is saved at the end by his cowboy brothers, so the play reiterates Shepard's idealization of the American past and his rejection of a present that does not respect it.

Lanford Wilson: the early plays

Like Sam Shepard, Lanford Wilson reached his full powers as a dramatist in the 1970s and 1980s; and like Shepard he used the freedom of the Off Off-Broadway theatre of the 1960s to explore themes and styles on his way to finding his true voice. Wilson's earliest plays are essentially character studies, using monologues and minimal interactions to provide insight into his generally unhappy characters. In *So Long at the Fair* (1963), an untalented would-be artist copes weakly with a seductive girl and the likelihood that he will settle into an ordinary life. In *The Bottle Harp* (1963), a boy from Nebraska visits his New York sister and decides to stay so they can comfort each other in their loneliness. *Home Free!* (1964) presents an incestuous brother and sister hiding from reality in a private world peopled by fantasy characters. In *Ludlow Fair* (1965) a woman talks at length about her melodramatic sexual past to her room-mate, who then speaks of her own empty and sexless life. A seemingly ordinary man in *Days Ahead* (1965) slowly betrays his madness in a monologue.

The best of this early group is *The Madness of Lady Bright* (1964). Leslie Bright, an aging homosexual, sits in his New York City room on a hot summer's night, trying to find some distraction from his isolation. In what is essentially an extended monologue, with an actor and actress playing minor choric roles, he ranges from despair to campy humour and back again, reminiscing about past lovers and fighting the awareness of his aging, loneliness and growing desperation ('Can't you see I'm going insane alone in my room, in my hot lonely room? Can't you see I'm losing my mind? I don't want to be the way I am.'). Finally he succumbs to all three in a wail of pain:

> I am young tonight. I will never be old. I have all my faculties tonight. . . . I am beautiful. I am happy! . . . I grow tired easily. *I grow brittle and I break. I'm losing my mind, you know. Everyone knows when they lose their mind. But I'm so lonely!* . . . Just take me home, please; take me home, please. Take me home now. Take me home. *Please take me home.* . . . TAKE ME HOME, SOME-ONE! TAKE ME HOME!

This play, like most of the others, is strikingly unjudgemental. With one or two notable exceptions, Wilson is generally charitable and sympathetic towards his unhappy characters.

While most of these plays are fairly conventional in form, except for their reliance on monologue, Wilson was also experimenting with non-linear and non-realistic modes. Like Sam Shepard, he was impressed by the then-popular rehearsal exercise in which actors practised jumping instantly from one scene and characterization into another; and in *Balm in Gilead* (1965), *This is the Rill Speaking* (1965), *The Rimers of Eldritch* (1966), and a few other short plays, he juxtaposes and overlaps very short scenes, sometimes out of chronological order, to create a musical or cinematic flow rather than a simple narrative. *Balm in Gilead* is set in an all-night New York coffee shop frequented by whores, drunks, addicts and other dregs of society. There is virtually no plot – one character is killed in a soured drug deal – but Wilson's purpose is to capture the milieu rather than to tell a story. His primary tool is overlapping dialogue, with several inconsequential things happening simultaneously:

> TIG You know in Egypt they had salves and things that could cure anything.
> DARLENE That's better than those other two creeps were acting. Did you see them?
> JOE They're just high. They're okay usually.
> TIG Cancer even! It says so.
> ERNESTO Show me where, you can't.
> TIG It does.
> DARLENE Why are you waiting for ten o'clock?
> TIG Hey, John; you ever read the Bible?
> JOE I'm meeting someone.
> JOHN What?
> TIG The Bible, stupid.
> JOE Like a business deal. A transaction.

If that's confusing to read, isolate three conversations, between Tig and Ernesto, Joe and Darlene, and Tig and John. This is actually a fairly simple passage, ignoring most of the more than thirty characters. Wilson also attempts stage equivalents of familiar cinematic effects: occasional freeze-frames and pin-spots focusing our attention on one character; the equivalent of background music and voice-overs by some choric figures; and a counterpart to the movies' use of slow motion in death scenes, as the moment of Joe's murder is repeated three times. *Balm in Gilead* is not really successful – there is too much going on at once, to too little apparent purpose, for the play to be clear in performance. Wilson would later learn how to moderate this chaos and use it to great effect in such plays as *The Hot l Baltimore* and *5th of July*.

In *The Rimers of Eldritch* Wilson employs a montage of short, disconnected scenes along with some of the cinematic effects of *Balm in Gilead* to capture the hypocrisies and ironies that hide behind the pastoral image of small-town America. The people of Eldritch guard their dark secrets with the same fervour with which they gossip about each others', coating everything with a frosty and hypocritical piety. At the centre of the plot is a murder trial; Nelly Windrod killed the town character, Skelly, when he seemed to be raping crippled teenager Eva Jackson. She is acquitted although we learn, and many in the town know, that Skelly was actually trying to save Eva from Robert Conklin, the real rapist. Wilson offers brief, almost subliminal excerpts of the trial early in the play, returning to it with longer excerpts, including the verdict, before finally showing the perjured testimony of Eva and Robert at the end. Meanwhile, in one subplot the happy news of an upcoming wedding is followed by revelations of the hidden violence and deceit that led to it, while another subplot shows a harmless romance destroyed by gossip; in each case the effect of witnessing events out of chronological order is to emphasize the corrosive power of the town's hypocrisy.

Almost uniquely in Wilson's canon, *The Rimers of Eldritch* is a play written in anger. Eldritch is so corrupted that it has no place for ordinary, healthy emotions and behaviour, or even for harmlessly illicit activity. What could have been an innocent romance between Eva and Robert can only find expression in rape. The fact that Cora sleeps with a younger man damns her as a whore and foils her attempts to expose the truth about Skelly's death. Skelly himself is a harmless old coot reduced to peeping in windows because the town has declared him untouchable. Nelly Windrod abuses her senile mother; Peck Johnson beats his daughter; Patsy Johnson forces an innocent boy to marry her. And through it all, the preacher and the townspeople mutter empty homilies. In his other plays Wilson generally paints an idealistic, pastoral picture of the inherent virtues of small-town America, but here he uncovers all its darkest secrets in an unrelenting assault.

Wilson also wrote more conventionally structured plays during this period. *Lemon Sky* (Buffalo 1970) is the young playwright's obligatory autobiographical play: a young man from the Midwest visits his estranged father and the father's new family in California. He embraces this household as the family he never had, but father and son slowly disappoint each other, and the son is left, some years later, still trying to pinpoint exactly what went wrong. Two essentially unoriginal plays are excellently crafted models of their genres. *The Great Nebula in Orion* (Manchester 1971) retells the old story of college room-mates meeting after some years and confessing, in the course of a drunken afternoon, how empty their lives are; Wilson enlivens the cliches and leavens the bathos by allowing each of the women a series of bitchy asides to the audience. And *Ikke, Ikke, Nye, Nye, Nye* (New Haven 1972) is a sex farce in which an obscene caller cannot handle actually being alone with a willing woman until he telephones her from another room, to the excitement and satisfaction of both.

Wilson's best play of this period is *The Hot l Baltimore* (1973). In a run-down residential hotel (symbolized by the title, a neon sign with one letter burned out), a group of society's discards and outcasts live in an unjudgemental and mutually supportive community. As happy and colourful prostitutes pass the time between clients, and residents tease and joke with each other, almost no actual plot takes place. The night clerk silently loves one of the whores; another whore leaves to move in with her pimp; a young man looks for record of his grandfather but gives up the search; and a couple of transients split up, one remaining to join the family. The play makes use of some of the techniques of *Balm in Gilead* – plotlessness, simultaneous and overlapping conversations, an imprecise and flexible time flow – in a much more controlled and successful way.

On one level *The Hot l Baltimore* is a very light-hearted version of Eugene O'Neill's *The Iceman Cometh*, leading us to re-evaluate what might at first seem an unattractive group of characters and to celebrate their fortune in having this nurturing and purifying environment. In this fantasy world the whores are delightfully bawdy, like April, or essentially innocent, like The Girl. Crotchety old Mr Morse is teased but protected; dotty old Millie is respected as a source of wisdom; Bill's chivalric adoration of The Girl is pure and beautiful; and ecstasy, in the form of a pizza eaten in the bathtub, is available with a phone call. As in *The Iceman Cometh*, this island of refuge from the judgements of the world is threatened, making our appreciation of it more intense. The scheduled demolition of the hotel is part of a larger pattern, a general rejection by American culture of the values and beauties of the past. The Girl mourns the decay of the hotel in the same spirit as her anger at the decline of the railroads, as a renunciation by a cold and practical present of a beautiful and gracious past. Wilson extends the symbolic meaning of the hotel to include all that is or was glorious about America, turning The Girl's attempt to guess Paul's home town into a kind of incantation ('Denver. Amarillo. Wichita. Oklahoma City. Salt Lake City. . . . ') and finally generalizing:

> GIRL Baltimore used to be one of the most beautiful cities in America.
>
> APRIL Every city in America used to be one of the most beautiful cities in America.

Everything that goes wrong in the play is caused by a short-sighted rejection of the past in the name of an empty future. Paul Granger's missing grandfather only went on the road because Paul's parents wouldn't let him live with them, and Paul foolishly gives up the search through hotel records because 'I got things to do'. Suzy leaves what she admits is her family for the false promises of a pimp. The transient Jackie justifies stealing from Mr Morse because she is young and he is old; and when her dream of a farm in Utah is shattered she abandons her brother as an unwanted burden and runs away. The others protect him from that realization and welcome him into their family, reminding us again at the end of the play how valuable this refuge from the uncaring modern world is. Earlier plays such as *Lady Bright* and *The Rill* had hinted at this reverence for the fading past, but *The Hot l Baltimore* marks the discovery of what would become a recurring theme of Wilson's mature work.

This sense of loss recurs in *The Mound Builders* (1975), set near an ancient Indian village in southern Illinois. Archaeologists trying to beat a dam project that will flood their dig come up against a local resident who anticipates the prosperity the dam will bring; when they use a conservation law against him, he wrecks the site, killing himself and one of the scientists, and rendering the research useless. Part of the play's point lies in the sad irony that these scientists searching for knowledge of the human past have so little awareness of the human character that they cannot see the effect their perfidy will have. But there is a broader and sadder message. Early in the play someone tries to explain why the Indians built mounds: 'Every society reaches the point where they build mounds. . . . For an accomplishment, honey, to bring me closer to Elysium; to leave something behind me for my grandchildren to marvel at. To say I'd built something!' That impulse to leave one's mark on the world, to defeat time and mortality, is universal. The Indians built mounds; Chad dreams of resort hotels; and the scientists fantasize about foundation grants and their pictures on the cover of *Newsweek*. And in every case the attempt at immortality is doomed. The mounds, the archaeological treasures, even the bodies of Chad and Dan are covered by the inexorably rising waters of the lake. Nothing is learned and nothing is saved. As the Aztec poet quoted by one character understood: 'This earth is only lent to us. / We shall have to leave our fine work. / We shall have to leave our beautiful flowers. / That is why I am sad as I sing for the sun.'

Political and social plays

The 1960s were a time of multiple political and social upheavals in America: the Civil Rights campaign (evolving later into the Black Pride and Black Power movements), the sexual revolution, the youth/drugs/rock-and-roll generation gap, the hippies, the anti-war movement, etc. Inevitably some of these subjects, and the social tensions they generated, found their way into the drama of the period. Since alternative theatre writers and audiences were, on the whole, young and liberal, their plays are generally critical of the status quo and sympathetic to the forces of change. Those dramatists who do take what might loosely be called the conservative position are often also critical of the present, calling for a return to the values or modes of the past. A very few are able to avoid taking sides, instead exploring with objectivity or sympathy the effects of the social and political changes on their characters.

To some extent, the distinction between these playwrights, who follow in the tradition of Odets and Miller by addressing social issues through domestic drama, and those already discussed, who focused on their characters' psychological and spiritual crises, is an imprecise one. Some of the plays discussed elsewhere in this chapter – for example, Albee's *The American Dream* and *Who's Afraid of Virginia Woolf?* and Wilson's *The Rimers of Eldritch* and *The Hot l Baltimore* – have a social/ political content, openly or implicitly indicting contemporary American culture as the source of the insecurity and pain that is their focus. Many plays of the period, however, are primarily about the larger issues, using the individual stories as representative or metaphor. In Israel Horovitz's *Line* (1967), for example, a group of people use intimidation, trickery and sex as they jockey for position in a waiting line, clearly a symbol of the struggle for success in a capitalist society that only rewards the ruthless. Horovitz's *The Indian Wants the Bronx* (1968) and *The Primary English Class* (1975) address the racism just below the surface of American culture: an East Indian lost in New York City is tormented and brutalized by a pair of young hooligans, for no other reason than that he is an outsider and thus fair game; and a teacher's inability to communicate with her students leads her to a violent outburst that betrays the xenophobic core beneath the very thin veneer of liberalism.

Terrence McNally's *Where Has Tommy Flowers Gone?* (New Haven 1971) is a character study of a child of the 1950s adrift in the 1960s, with no connection to anything beyond the icons of television and popular culture, and nothing to moderate his demand for instant gratification of every passing impulse. Like the characters in Albee's *American Dream* and Horovitz's *Line* he has not fallen away from the highest values of American culture but completely fulfilled them, and the play eventually recoils in horror at the world that created this amoral monster. In Jason Miller's *That Championship Season* (1972), the many mutual betrayals of a group of supposed friends are shown to be the direct result of the lesson they learned from their basketball coach twenty years earlier, to play dirty and win at any cost; athletic achievement reflects the American success ethic, and the men's moral and spiritual decay is an inevitable product of blind adherence to it.

Taking an approach closer to that of Albee's *Virginia Woolf*, Robert Patrick's *Kennedy's Children* (1973) portrays the emotionally wounded victims of a corrupted culture. In a series of parallel monologues, five characters who were shaped by the idealistic dreams of the early 1960s – a Vietnam veteran, a political activist, etc. – show how they have been maimed by the failure of the rest of the decade to live up to its promises. Each speaker's frame of reference is limited, even trivial, but each one's suffering is real; and by the play's end the sheer quantity of pain on stage is overwhelming. This is the true legacy of the 1960s, the play says: not the dreams or the hopes, but the disappointments, and a generation broken in mind and spirit. Jean-Claude van Itallie's *America Hurrah*, an evening of three one-act plays, also condemns the spiritual emptiness of the 1960s. In *Interview*, the demeaning rituals of a job interview lead to more nightmarish examples of American impersonality and disregard for the individual, while *TV* depicts the artificial emotions of television drama as more meaningful than a desiccated reality. In *Motel* van Itallie relates spiritual bankruptcy to aesthetic barbarianism, concluding that a culture without taste will create humans without humanity, and dramatizes that dehumanization by peopling the stage with actors encased in larger-than-life papier-mâché dolls; while a motel owner proudly describes the vulgar furnishings of a characterless room, her guests wantonly vandalize and destroy it.

In *Indians* (London 1968; Washington 1969), Arthur Kopit analyses the American psyche through the symbolic figure of 'Buffalo Bill' Cody, an actual frontier hero who later became a showman. Kopit shows Bill discovering to his horror that his Wild West Show actually contributes to the destruction of his Indian friends, by giving white America a distorted view of the cowboys-and-Indians world which then becomes the basis for national policy. Like some plays already mentioned, *Indians* is implicitly about Vietnam, showing how American myth-making, particularly the need to see ourselves as the 'good guys,' inevitably distorts our perception of other cultures; and offering in Bill Cody, the essentially good man who does the wrong thing for the right reasons and then does not have the moral courage or the ability to reverse himself, a not wholly unsympathetic portrait of President Lyndon Johnson.

The Vietnam War, on the battlefield and in its effects back home, was the single most significant social and political fact in American life in the late 1960s and early 1970s, and some dramatists did not disguise their criticisms of it in metaphor. Megan Terry's *Viet Rock* is a phantasmagorical picture of the war's atrocities, with scenes of basic training and battle combined with attacks on politicians who ignore all criticism in a blindly patriotic fervour. Barbara Garson's *MacBird* (1967) is a cheerfully libellous Shakespearean pastiche casting President Lyndon Johnson as Macbeth, who engineers the assassination of 'John Ken O'Dunc' and carries on the war to protect his power against the slain king's brother Robert. Kenneth H. Brown's *The Brig* (1963) is more broadly anti-military, overpowering the audience with a high-volume depiction of the incessant humiliation and brutality of military discipline.

David Rabe

The one American dramatist who most successfully captured the story of the Vietnam War is David Rabe; in three independent but complementary plays – *The Basic Training of Pavlo Hummel* (1971), *Sticks and Bones* (1971) and *Streamers* (New Haven 1976) – Rabe addresses horrors not normally thought of as part of a war story and thus illuminates the central fact of this particular war: that not all the casualties and atrocities were on the battlefield. Like Brecht's *Man is Man*, which it resembles in outline, *Pavlo Hummel* shows the transformation of a hapless young man into an ideal soldier. But there are two bitter differences: the intended end product is not a fighting machine, as in Brecht, but a dead body; and Pavlo does not passively allow himself to be changed, but races eagerly to his destruction.

In Rabe's eyes, the Army preys on weak, insecure adolescents and turns them into well-trained cannon fodder by offering the sense of identity, home and manhood the young men lack:

> NOW YOU ARE TRAINEES, ALL YOU PEOPLE, AND YOU LISTEN UP. I ASK YOU WHAT IS YOUR FIRST NAMES, YOU TELL ME 'TRAINEES'! . . . AND YOU LIVE IN THE ARMY OF THE UNITED STATES OF AMERICA. . . . WITH BALLS BETWEEN YOU LEGS! YOU HAVE BALLS! NO SLITS!

The Army thus promises everything an American boy is conditioned to want and to fear he does not have, and it lies. Pavlo is turned into a soldier but still cannot find friends or impress girls or avoid being wounded in battle. Vietnam is not a place of manhood, but a place where you haggle with whores over pennies and where the wounded beg to die. Pavlo is killed in a drunken fight, no closer to the secret of manhood than he was as a civilian, and with just the barest awareness that this irrelevant and pointless death, not heroic triumph, was what he was trained for. Something unhealthy in the American psyche confuses personal worth with victory in battle, Rabe charges, and something evil in the American system exploits that confusion to feed its cannon and fill its body bags.

Rabe brings the war home in *Sticks and Bones*, in the form of a blinded and guilt-ridden veteran who alternately mopes about, tries unsuccessfully to communicate the horrors he witnessed, and lashes out in rage at the offensive ignorance of those who were not there. But the play is not so much about him as about his family; and by naming his characters Ozzie, Harriet, David and Ricky, after the all-American family of a 1950s television series, Rabe makes them symbolic of the whole nation. David's family cannot or will not accept the horrible truths he brings home from the war – Harriet reacts to his pain by offering him fudge, and Rick works very hard at acting as if nothing is wrong – and the depth of their need to deny is the play's real subject. The presence in their home of a son who is imperfect, who has killed and still seems capable of killing, who has loved a woman of an alien race, calls into question every assumption by which they define their family and themselves. Rather than answer or even acknowledge those questions they reject the questioner; the play ends with the family killing David. His attempt to share his experience with those at home, his very presence as a reminder of the war, threatens an entire way of life, and to defend that way of life the family, and the nation, will do things that debase everything they are meant to represent.

Streamers, the strongest of the three plays, is written in sympathy rather than anger. Without pointing an accusing finger anywhere, Rabe depicts the horror of a war that seems to have become a self-perpetuating natural force. In an army camp in America several young soldiers await reassignment, possibly to Vietnam. Each has his own small neuroses. Roger, an educated black man, seems lost between his roots and his class. Ritchie a homosexual, keeps the others off balance by joking so much about his gayness that they do not know whether to believe him. Billy, who may be a repressed homosexual, is holding some emotional tension under tenuous check. Carlyle, an uneducated black, treats everything with the suspicion that street life has taught him. The play's title is macabre military slang for a snarled parachute hanging uselessly above a man plunging to his death, as described by an older soldier:

> This guy with his chute goin' straight up above him in a streamer . . . . He looks right at me. Then he looks up in the air at the chute, then down at the ground . . . . He started going like this. (*Cokes reaches desperately upward with both hands and begins to claw at the sky while his legs pump up and down.*) Like he was gonna climb right up the air.

That, of course, is the central image of the play. Each of the young men feels himself being pulled towards death by forces out of his control, and each is slowly going mad under the pressure. Roger has had one breakdown already, and spends the play trying to make himself invisible through perfect soldiering and practised ordinariness. Billy denies and represses, insisting that Ritchie cannot be gay, and compulsively washing and polishing everything in sight. Ritchie's fear and loneliness force him to be ever more overt in his appeals for sexual comfort. And Carlyle, exhausted by trying to apply street logic to an inherently illogical situation, is driven to angry rebellion. Finally there is violence – barely motivated and almost accidental – and death – random and irrelevant

– leaving the survivors with more confusion and tension to be unable to cope with. The play barely mentions Vietnam, but the war defines the lives of these characters, just as it did for every American male of draft age for more than a decade; the question was never *whether* it would affect their lives disastrously, but only *when*.

Black dramatists

The Civil Rights movement of the early 1960s, and the Black Pride and Black Power movements that evolved from it later in the decade, changed the lives of black Americans more than anything since the Civil War. The heightened self-awareness that came from these changes demanded artistic expression; and Off-Broadway and its equivalents around the country, with their educated and liberal audiences, proved more receptive to black playwrights than the commercial theatre had ever been. Some writers used the opportunity to vent their anger at racism, others to express hope for the future, still others to examine the nature of the black experience and its effect on individual black Americans.

In LeRoi Jones's *Dutchman* (1964) a middle-class black man is accosted by a seductive white woman who taunts him for denying his blackness until he reacts in rage, at which point she calmly murders him and prepares to begin again with another victim. Jones's despairing view of the possibilities of racial harmony is voiced in Clay's angry outburst, which stresses the fact that hatred of whites is *the* single defining element in black psychology: 'You don't know anything except what's there for you to see. An act. Lies. Device. Not the pure heart, the pumping black heart. You don't ever know that. And I sit here, in this buttoned-up suit, to keep myself from cutting all your throats. I mean wantonly.' The only thing that keeps blacks from wanton murder, he says, is their fear of that unlimited passion within them; they have not learned the white trick of hating calmly, and so the whole race consciously chooses to make itself neurotic by repressing and displacing its anger. This insight makes Lula's earlier taunting newly frightening; we see now that it was calculated to produce the anger that would justify her killing him. Murderous racial hatred is the defining emotion in whites as well as blacks, says Jones, the only difference being that whites have found ways of rationalizing and living with their genocidal impulses that blacks have not yet mastered. There can be no hope for racial harmony in these circumstances, only the prospect of war when blacks learn to kill as calmly as Lula.

Jones's *The Toilet* (1964) offers an alternative basis for the same conclusion, as a teenage gang leader takes part in an attack on an outsider whom he secretly loves. Race is barely mentioned in the play, though the victim is cast as white and the gang members as black, but this is clearly an allegory of race relations. There may be isolated individuals in any group who are capable of amity, Jones admits, but the social roles imposed on them will force them to act as enemies, whatever they feel. *The Slave* (1964) is set in the racial civil war of the near future, as the leader of the black forces explains to his white ex-wife that it hardly matters whether he is right or whether he particularly believes in his cause. The war, the destruction and the waste are foregone conclusions, and Jones can offer only resignation without even the comfort of outrage or anger.

The Slave was one of Jones's last plays for white audiences. Within a few years he left the theatre altogether, taking on the Muslim name of Amiri Baraka and devoting himself to social and political action. The few plays he wrote as Baraka were deliberately simplistic and one-sided propaganda works intended for theatrically unsophisticated black audiences. Like similar agitprop plays by other black writers of the period, they have little literary merit as measured by traditional standards – plots are mechanical and arbitrary, characters one-dimensional, and messages spelled out in overly simple and undramatic ways – but of course such judgements are unfair, since these playwrights were not trying to write traditionally literary drama. (Baraka himself follows the convention of retaining the name Jones on the earlier plays.)

Not all black dramatists followed the lead of Amiri Baraka. Douglas Turner Ward was one of the very few to find humour in the black experience; in his double bill of *Happy Ending* and *Day of Absence* (1965) he reminds both black and white audiences that the current condition of blacks is not without its power and rewards. In *Happy Ending* a young black man is offended by what he takes to be his family's subservient grief at their white employers' divorce, only to learn they are really mourning the end of their systematic embezzling. In *Day of Absence* black performers in whiteface burlesque the confusion and dismay of a town of white southerners on the day all the blacks magically disappear and there is no one to cook the food, care for the children and do all the other menial jobs without which society collapses. Charlie L. Russell's *Five on the Black Hand Side* (1969) offers another essentially comic view by translating the civil rights movement into domestic terms, a downtrodden wife using sit-ins and picket lines to win concessions from her husband and finally to liberate his spirit.

But the same observation, that the rapid social changes of the 1960s were forcing even the most conservative and unpolitical blacks to redefine themselves, had serious and even frightening implications to other black writers. Charles Gordone's *No Place to be Somebody* (1969), about small-time criminals trying to make it big; *Ceremonies in Dark Old Men*

(1969) by Lonnie Elder III, about a family caught up in crime and revolution; and Joseph A. Walker's *The River Niger* (1972), about former gang members facing adulthood, all present black characters who find neither their old ways, the model of younger revolutionaries nor the imitation of white behaviour fully satisfactory. In each play at least one character dies while others are left to wrestle with the challenge of finding new identities, and the only reassurance any of the plays can offer is one that faintly echoes Clifford Odets in *Awake and Sing!*: the previous condition of blacks in America was a living death, so that any change will be for the better.

The black playwright who proved most successful in exploring and dramatizing the effects of the 1960s and 1970s on black Americans was Ed Bullins. Although he wrote a few 'revolutionary' plays in the manner of Amiri Baraka, his more characteristic plays focus on domestic life within the black community rather than on black–white relationships. Bullins is fascinated and disturbed by the very thin line that separates ordinary life in black America from sudden violence, despair or madness. Structurally, his plays are misleadingly loose and seemingly aimless in construction, wandering plotlessly through images of life in the urban ghetto until an abrupt and unexpected moment of violence not only accelerates the pace, but also forces a reappraisal of the earlier non-action as carrying in it the seeds of the sudden change.

In *Clara's Ole Man* (San Francisco 1965), a young black man spends an afternoon idly drinking and chatting with a girl he has just met and her mannish friend. Only when he repeats an overheard reference to 'Clara's ole man' does he realize that the women are lesbians; he is punished for his ignorance by being beaten up. In *Goin' a Buffalo* (1968), small-time criminals are sidetracked from their plans to leave town and start fresh by an unexpected quarrel and murder, and an outsider takes advantage of the confusion to steal both their money and their women. *In the Wine Time* (1968) is almost actionless through most of its length, merely depicting a poor black family in its daily routine of lounging, drinking, arguing and making up. When the boy of the family is attacked by a gang and his uncle is forced to kill to save him, we realize that ordinary ghetto life has crossed the line into dangerous anarchy and that these characters' lives had been moving inexorably towards some catastrophe from the start. A sequel, *In New England Winter* (1971), presents the uncle after he has left prison and joined with others planning a robbery. The personalities prove too different for the project to work, one man's obsessive order and meticulousness clashing with another's impulsiveness, and the play ends with a sudden fight and murder.

A recurring observation in Bullins's plays is that black Americans are often trapped more by the limits of their personalities and imaginations than by the external forces of racism. A central metaphor in *Goin' a Buffalo* is chess, a game that the gang members play in their idle moments; ironically they are unable in real life to plan several moves ahead or adjust their strategy when obstacles appear in their way. In *In the Wine Time* Cliff Dawson floats aimlessly through what he assumes is a life without possibilities, and as a result his life is altered by a decision he must make at the spur of the moment. Yet, as Cliff's relatively calm acceptance of his wife's infidelity in the sequel shows, this same limited vision can ironically save Bullins's characters from the despair that would come if they could fully appreciate the extent of their failures or losses. *The Taking of Miss Janie* (1975) explores this paradox by following a group of black and white characters from the late 1950s through the early 1970s, showing how they both fail to live up to their youthful dreams and adjust to that failure through compromise and rationalization. The central irony of the play is that none of the young people can imagine that their lives and identities will ever change, while from the perspective of the future none can see anything unpredictable in what actually happened. The same limitation that keeps them from imagining alternative futures saves them from the pain of imagining alternative pasts. On its surface the play seems to lament the lost idealism of the sixties, but it actually accepts the inevitable compromises, recognizing that they are more apparent to the observer than to the participants.

Broadway

Virtually all of the pre-war playwrights still writing in the 1960s ended their Broadway careers with anticlimactic disappointments: Sidney Kingsley's *Night Life* (1962), Lillian Hellman's *My Mother, My Father and Me* (1963), S. N. Behrman's *But For Whom Charlie* (1964). Even the younger writers of the 1950s had trouble in the new decade. A string of critical and commercial failures in the 1960s contributed to William Inge's suicide in 1973. A similar succession of failures led Paddy Chayevsky to desert the theatre for a more successful career in film. Lorraine Hansberry's only other completed play, *The Sign in Sidney Brustein's Window* (1964), closed three days after her death in 1965; and Robert Anderson's only hit was the relatively trivial *You Know I Can't Hear You When the Water's Running* (1967), four comic sketches with light hints of pathos. Even Tennessee Williams had an uninterrupted string of flops after 1962. Meanwhile, with more and more of the creative energy of the American theatre located beyond Broadway, only a very few new writers found their artistic home there; and the two most important new Broadway playwrights of the 1960s could hardly have been less alike. Edward Albee, the first dramatist from the alternative theatre to move successfully to Broadway, wrote complacency-challenging dramas about the emptiness of American culture and the uncertainties of human existence. Neil Simon, a former television joke writer, proved a master craftsman of light, unthreatening and highly entertaining comedies.

Edward Albee: the Broadway plays

Who's Afraid of Virginia Woolf? (1962), Albee's first Broadway production, is without question his best play, ranking just behind the very best of O'Neill, Williams and Miller in the American repertoire. It is the picture of a seemingly hellish marriage whose positive and supportive aspects we only gradually appreciate, as we understand how frightened and alienated the characters are, in an America similar to that of *The Zoo Story* and *The American Dream*. George and Martha (the names suggest an archetypal American couple) subject each other and another couple to a long liquor-filled night of humiliation and exposure, culminating in the revelation that their adult son is a fantasy they have maintained for twenty-one years to comfort themselves in the face of life's disappointments. Chastened by this exposure to a desperation far beyond their shallow experience, the other couple leave George and Martha to face the future without one of their most depended-on emotional crutches. The play has nothing to do with Virginia Woolf; its title, a parody of a song from a Disney cartoon, 'Who's Afraid of the Big Bad Wolf', begins as a characterizing device, identifying George and Martha as intellectual and verbal games-players, and later comes to symbolize the unnamed fears and insecurities of modern life.

Albee has three purposes in the play: to show that George and Martha's experience of life, like Jerry's in *The Zoo Story*, is dominated by an insecurity and alienation that make the task of coping with day-to-day disappointments overwhelming; to argue that their commitment to survival, even if the only means is degrading or abhorrent behaviour, is heroic; and, slightly peripherally, to expose the cause of this painful experience as a dehumanizing force in the American culture, similar to that revealed in *The American Dream*, and identified in this play with the second couple, Nick and Honey. The emphasis on depicting and celebrating characters who find daily life overwhelming puts Albee squarely in the tradition of Tennessee Williams, a debt Albee acknowledges in the play with brief allusions to *A Streetcar Named Desire*.

Throughout the play George and Martha let slip veiled or open admissions of their sense of weakness and failure, as in Martha's third-act confession of her need for him:

> George who is good to me, and whom I revile; . . . who can hold me, at night, so that it's warm, and whom I will bite so there's blood; who keeps learning the games we play as quickly as I can change the rules; who can make me happy and I do not wish to be happy, and yes I do wish to be happy . . . . who has made the hideous, the hurting, the insulting mistake of loving me . . . who tolerates, which is intolerable; who is kind, which is cruel; who understands, which is beyond comprehension.

Torn between fear of their inadequacy and refusal to avoid the struggle for existence, they live lives defined by pain. George offers as a self-description the declension 'Good, better, best, bested', and Martha admits to the portrait of her father, 'I cry all the time too, Daddy. I cry alllll the time; but deep inside, so no one can see me.'

Like many of Williams's characters and like Jerry in *The Zoo Story*, George and Martha seem less bizarre when we interpret their behaviour as reactions to a world that offers them no support. Their games are attempts to cope with the pain, either as distraction – Martha calls them 'the refuge we take when the unreality of the world weighs too heavy on our tiny heads' – or as a way of keeping themselves in shape for the battle: George assures Nick, 'Martha and I are merely . . . exercising . . . that's all . . . we're merely walking what's left of our wits.' [Ellipses Albee's] A particularly skilful blow is likely to be met by congratulations or mutual celebration, and they refuse to let each other relax and coast: when George scores points off the completely outmatched Nick and Honey, Martha's reaction is contempt for his 'pigmy hunting;' and when George decides on the final battle over destroying the fantasy son, he insists that Martha be at full strength to fight him.

If George and Martha must use every tool at their disposal to help them fight the dehumanizing forces pitted against them, then the loss of any tool must seriously endanger their struggle. The imaginary son compensated for their sterility, helped to hold them together and gave them a constant in their lives, but he has become a weapon they use against each other, and therefore must be sacrificed. After the initial trauma Martha accepts this judgement, and the play ends with them facing an uncertain future with a bravery that consists of admitting their fear but not turning back.

Albee forces us to recognize this immersion in life as heroic, particularly in contrast to the bland, evasive and ultimately cowardly Nick and Honey, who consciously avoid the challenges of living. Nick keeps what emotions he has under control, tries not to get involved in unpleasantness, and admits there is no passion in his marriage or in his pursuit of Martha. Honey drinks until she passes out, decides not to remember unpleasant things, and takes secret precautions to avoid having children. They do not even have real identities: 'Nick' is the name used in the stage directions, but it is never spoken in the play; and 'Honey' is merely his perfunctory endearment for her. Such lifelessness is not merely evasion of responsibility; Albee sees it as a cause of the central couple's pain. The main reason for setting the play in a university is the ease with which the two men can be characterized by their specialities. In Act One George defines the contrast in an extended mockery of Nick's genetic research:

> It's very simple, Martha, this young man is working on a system whereby chromosomes can be altered.[. . .] We will have a race of men . . . test-tube-bred . . . incubator- born . . . superb and sublime.[. . .] *But!* Everyone will tend to be rather the same.[. . .] I suspect we will not have much music, much painting.[. . .] There will be a certain . . . loss of liberty. [. . .] Cultures and races will eventually vanish . . . the ants will take over the world. [Unbracketed ellipses Albee's]

George's academic discipline symbolizes his commitment to life: 'And I, naturally, am rather opposed to this. History, which is my field [. . .] will lose its glorious variety and unpredictability.[. . .] the surprise, the multiplexity, the sea-changing rhythm of . . . history, will be eliminated. There will be order and constancy . . . and I am unalterably opposed to it.' The force that causes life to be painful and difficult for those committed to living it is the irresponsibility of those who attempt to find security by rejecting their humanity.

One important quality of *Who's Afraid of Virginia Woolf?* to which this analysis has not done justice is its passion. George and Martha experience and express extreme and uncensored emotions, and the play celebrates those passions, good and bad, as part of the couple's commitment to the challenge of living. Oddly, that passionate quality, which had also driven *The Zoo Story*, begins to disappear in Albee's later plays, displaced by a dry intellectuality that dooms some to failure and deprives even the most successful of the theatrical power of the early plays. *Tiny Alice* (1964) is the story of a religious man faced with what he has wanted all his life, the opportunity to sacrifice himself to an impossibility that can only be accepted through faith. But Albee presents this adventure as a series of gratuitously mystifying puzzles and paradoxes, more an intellectual construct than a living drama. The mystery Julian is asked to worship is not God but Alice, an essence who may be nothing more than a mouse in a dollhouse, and who is represented by a woman named Alice who seduces him and then orders his death. The play ends with Julian's long dying monologue, in which he either rambles incoherently or accepts Alice as his God:

> How long wilt thou forget me, O Lord? Forever? How long wilt

> thou hide thy face from me? [. . .] Oh, Alice, why hast *thou* forsaken me? [. . .] Alice? . . . God? SOMEONE? Come to Julian as he . . . ebbs. [. . .] Is that the humor? THE ABSTRACT? . . . REAL? THE REST? . . . FALSE? It is what I have wanted, have insisted on. Have nagged . . . for. [. . .] COME, BRIDE! COME, GOD! [. . .] I accept thee, Alice, for thou art come to me. God, Alice . . . I accept thy will.
>
> [Unbracketed ellipses Albee's]

What makes *Tiny Alice* difficult is Albee's conviction that what cannot be understood through any means other than faith must be kept outside the audience's understanding; he simply fails to make the story clear much of the time. Meanwhile, the general air of mystification is maintained through secondary puzzles and red herrings, and through the general refusal of any of the secondary characters to act in recognizably human ways. Given its ambitions, it may be remarkable that *Tiny Alice* is as nearly comprehensible as it is, but it certainly tests the outer limits of a theatre audience's ability to grasp abstract ideas.

A Delicate Balance (1966) returns to some of the same concerns as *Who's Afraid of Virginia Woolf?*, though in a considerably subdued manner. When friends unexpectedly demand refuge with Tobias and Agnes, their presence disrupts the tenuous stability of the household, inspiring alcoholic sister Claire to ever more outrageous conduct, and driving daughter Julia to territorial hysteria. But it is their departure, when they decide they are imposing, that finally panics Tobias, who cannot bear the discovery that even the closest friendship has limits.

The title thus refers to the fragility of both social constructs and individual psychology. Agnes opens the play by entertaining herself with the prospect of going mad, but Julia's leap from childish petulance to the murderous defence of her turf, and Claire's from mildly anti-social to aggressively bizarre behaviour, show how thin the line of sanity is. Even more disturbingly, the carefully maintained balance of peace and civility in the household, the preservation of the illusion of normality, is shown to be very easily disrupted. The first half of *A Delicate Balance* has some parallels to the opening of Eugene O'Neill's *The Iceman Cometh*, in that an apparently diseased social milieu becomes somewhat more attractive in retrospect when a disrupting influence produces something much worse.

But more central than that, and tying this play to *Virginia Woolf*, is the recognition that social constructs, family rituals and even aberrant behaviour are defence mechanisms against an unnameable existential fear. Claire's drinking, Julia's regularly scheduled returns to her childhood, Agnes's reduction of every unpleasant thought to the level of abstract intellectual speculation, and Tobias's constant placating and smoothing-over of awkward moments are all ways of protecting themselves from what Agnes calls a plague of communicable terror. As Claire reminds Tobias, whole lives can be built on the fantasy of order: 'We have our friends and guests for patterns, don't we? – known quantities. The drunks stay drunk; the Catholics go to Mass, the bounders bound. We can't have changes – throws the balance off . . . . Just think, Tobias, what would happen if the patterns changed: you wouldn't know where you stood.' For everyone there is the unbearable reality that an entire life must be constructed to blot out. For Tobias it is the possibility that no commitment is absolute and there is no promise one human can make to another that has no bounds. So, at the end of the play, when Harry and Edna decide that they have indeed imposed, their attempt to spare their friend the terror they carry actually brings it on.

The abstractly intellectual quality of Albee's later plays is seen in its purest form in *Box* and *Quotations from Chairman Mao Tse-Tung* (1968), an interrelated double bill sometimes called *Box-Mao-Box* because the first play is repeated at the end of the second. In *Box* we see only a hollow cube, while the recorded voice of a woman muses on its philosophical and moral implications. In *Mao* Mao Tse-Tung lectures on the class struggle, a Long-Winded Lady describes her fall overboard from an ocean liner, and an Old Woman recites a Robert Service-like poem about being rejected by her children in her old age. About midway through, the voice from *Box* joins the rotation of monologues, speaking a line or two from that play; at the end, the live actors are silent while the voice finishes her reprise.

The theme of both plays is cultural decay; the voice moves from an admiration of the workmanship of the box to a recognition that all standards are declining, and that we have gone beyond noticing the loss. The three monologues of the inner play demonstrate this decline through Mao's dependence on empty rhetoric, the Lady's self-centredness and the Old Woman's bathetic verse. But *Box-Mao-Box* is as much an intellectual exercise as a dramatic statement, a Beckett-like experiment in how little can be put on a stage and still have a play, and an attempt to structure the interlocking monologues musically. With *Tiny Alice* before it and *All Over* after, it represents Albee's furthest swing of the pendulum away from the passionate psychological realism of his early plays.

All Over (1971) is set in a more recognizable reality than *Box-Mao-Box*, but it is no less desiccated and lifeless. Its dramatic situation is in fact a deathwatch: an important man is dying, and his Wife, Mistress, Son, Daughter and Best Friend (who are given no other names) wait out his final moments. They show very little emotion; the few brief lapses, generally by the Son and Daughter, are treated as distasteful self-indulgences. Only at the very end does the Wife allow herself an open expression of grief – not for her husband's death, but for her lifetime of unhappiness. Like Honey in *Virginia Woolf* and Agnes in *A Delicate Balance* these characters distance themselves from experience to save themselves from the pain of having it. The difference is that Albee seems to approve, and to offer such deliberately chosen lifelessness as an acceptable model. That this seems to contradict everything his earlier plays stood for is a minor difficulty – authors have the right to change their minds. But a play full of characters repressing all emotions and keeping all interactions at a safe distance runs the risk of being as anaesthetized as they are; and the biggest problem with *All Over* is that it is dull.

In contrast to the dry plays that preceded it, *Seascape* (1975) is a charming and happy fable, using an entertaining bit of fantasy to make a positive if not particularly original statement. Nancy and Charlie, a middle-aged couple relaxing on a beach, encounter Sarah and Leslie, two human-size lizard-like amphibians who have chosen this moment to evolve out of the sea. After some comic shock on both sides, the newcomers prove to be very much like the humans, with middle-class amphibian values and parallel tensions in their marriage. But the introduction to life on shore involves discovering some painful new sensations, such as human emotions, and the lizards waver in their decision to evolve. Nancy and Charlie beg them not to go back into the sea, and the play ends with the humans' promise to help their new friends through their great adventure.

Sarah and Leslie are delightful characters, and their little breakthroughs – learning to shake hands, marvelling at Nancy's breasts, experiencing sorrow for the first time – almost steal the show. But the fact that they do not appear until the end of Act One is a clue that the play is actually about Nancy and Charlie, the lizards' evolutionary leap a metaphor for the humans' uncertainty about what to do now that the main business of their lives seems completed. Charlie's commitment to helping the newcomers find their way is also his recognition that, for humans as well as lizards, the end of one stage in life is the cue to evolve into the next. *Seascape* is a small play, and one might even feel that Albee wasted a particularly clever invention on a fairly clichéd message. But it makes its small point effectively, and with a liveliness and lightness of touch that had been missing from Albee's plays for over a decade.

Neil Simon

Neil Simon is, in commercial terms, the most successful dramatist in the American theatre, and probably in the history of the world. Beginning with his first Broadway success, *Come Blow Your Horn* (1961), he has turned out comedies at the rate of almost one a year; since most of these were successes, there has been hardly a season in thirty years without at least one Simon play – and sometimes as many as four – being performed on Broadway. And yet Simon has been accorded the least critical and academic study of any major American playwright, for a reason that says more about literary criticism than the plays. Literary analysts have never been comfortable with skilled craftsmen who make few if any pretensions to high art. Writers of detective novels, for example, may do what they do very well, but what they do is not what most literary critics are trained (or inclined) to analyse and judge. Much the same is true of the authors of stage comedy; criticism that presumes some value to the illumination of existential truths or the exploration of the deeper nuances of the human condition finds itself with little to say about writers who merely want to make the audience laugh, even if those writers are remarkably skilled at doing that.

The legitimate observation, for example, that almost none of Simon's early plays has more than a skeletal and mechanical plot must be accompanied by the recognition that the story lines are merely frameworks on which to hang comic situations and dialogue. Simon's general pattern is to juxtapose fairly simple opposites and then reconcile them or just enjoy the conflict. *Come Blow Your Horn* is about a boy who leaves home to move in with his playboy older brother. In *Barefoot in the Park* (1963) newlyweds go through a period of adjustment while his conservative nature gives way to her free-spiritedness. In *The Odd Couple* (1965) two divorced men rooming together prove to be comically incompatible. *The Star-Spangled Girl* (1966) brings two would-be revolutionary writers in contact with an all-American girl, with romance inevitably triumphing over politics. *The Sunshine Boys* (1972) reunites two feuding ex-vaudevillians and lets them renew their fighting. Other Simon plots of this period are about victims of their own limitations. *Plaza Suite* (1968) is three one-act portraits of middle-aged New Yorkers facing or denying the passage of time; *The Last of the Red-Hot Lovers* (1969) shows an unadventurous middle-aged man trying very unsuccessfully to commit adultery; and *The Prisoner of Second Avenue* (1971) watches a New Yorker being driven crazy by the comic pressures of city life. *The Good Doctor* (1973) is a collection of sketches adapted from Chekhov, generally of the small defeats of small people; *God's Favorite* (1974) is a modern-dress comic retelling of *Job*; and *The Gingerbread Lady* (1970), Simon's one serious play of the period, shows an alcoholic trying to recover.

A more legitimate criticism is that virtually none of these plays uses the implicit opportunity to explore character. The couple in *Barefoot in the Park*, for example, function as Beautiful Young People obviously fated to live Happily Ever After; the personality difference is not even introduced until midway through the play, after some earlier comic material has been exhausted, and is resolved without any soul-searching by either side. Only two of the early plays offer real psychological insights. In *The Odd Couple* prissy, meticulous Felix and aggressively sloppy Oscar may be static and somewhat one-dimensional characters, but their adventure rings true: two divorced men living together will recreate their failed marriages, each playing exactly the same infuriating role that drove his wife crazy, and each being driven crazy just like the wife he replaces. *The Sunshine Boys* recognizes that a partnership of fifty years is very much like a marriage, most dangerously threatened by petty exasperations rather than major incompatibilities, and ultimately surviving through a habit that is indistinguishable from love.

One reason why Simon's characters show little depth or individuality is that they all sound alike. At this stage in his career Simon was essentially a joke-writer, and his characters are all compulsively clever and wise-cracking; *The Gingerbread Lady* failed partly because the characters sound like they're making jokes even when they're being serious, and audiences were disoriented. Still, joke writing is an art, and so is the construction of a scene or a whole play built on jokes. Simon does not just throw a string of jokes together, but assembles them so that they feed on each other, the whole totalling more than the sum of the parts. Consider this exchange from the first scene of *The Odd Couple*:

> OSCAR Felix and Francis! They broke up! The entire marriage is through . . . .
> SPEED After twelve years?
> VINNIE They were such a happy couple.
> MURRAY Twelve years doesn't mean you're a *happy* couple. It just means you're a *long* couple.

The basic joke is in Vinnie's line, with its incongruity. It is a small joke, but Murray's repetition underlines it, and then his reversal tops it. Simon's plays are filled with such progressions, several small jokes producing more laughter together than they would alone. A related skill is the manipulation of running gags; each Simon play works inventive changes on a basic pattern, such as Oscar's exasperation in *The Odd Couple.*

Nevertheless, however impressive Simon's craft is, it is limited. It is no criticism of a comic writer to point out that he ultimately affirms conventional values, but Simon rarely even questions them, and there is no edge of danger in his early plays. The playboy in *Come Blow Your Horn* marries and settles down, thereby assuring us that his brother, after sowing some harmlessly wild oats, will do the same. The strangely Bohemian neighbour in *Barefoot in the Park* proves a perfect gentleman; the would-be lecher in *Last of the Red-Hot Lovers* is just too nice a guy to stray; and so on. The occasional attempts at seriousness or sentimentality, as in *The Gingerbread Lady*, are wooden or bathetic. Simon's milieu is middle-aged, middle-class New York, and he builds much of his humour on the familiarity of that world to his audience, citing currently fashionable brand names, for example, but he is lost when he strays too far from that territory; *The Star-Spangled Girl* does not work because he really doesn't know what young California rebels act and sound like. Not until the 1980s would Simon prove able to transcend these limitations.

On the other hand, if there is any doubt about Simon's supremacy in the limited realm of Broadway comedy, one need only consider the other successful comedies of the period, most of which do essentially the same things his do, though rarely with as much skill. Following in the tradition of the 1950s, the majority are built on the romance of a couple who spend the play either discovering the attraction that was obvious to us from the start or overcoming obstacles that usually turn out to be little more than misunderstandings. Among those that rise above the formula, through particularly witty dialogue or unexpected touches of characterization, are Jean Kerr's *Mary Mary* (1961), about an almost-divorced couple who rediscover their love just in time; Bill Manhoff's *The Owl and the Pussycat* (1964), about a stuffy author and a prostitute who evolve from feuding neighbours to lovers; Abe Burrows's *Cactus Flower* (1965, adapted from a French play), about a philandering dentist who involves his plain nurse in his subterfuges, only to have her blossom into a beauty and catch him in the end; Bob Randall's *6 Rms Riv Vu* (1972), about apartment-hunters who have a brief romance before returning to their spouses; and Bernard Slade's *Same Time, Next Year* (1975), about a couple whose affair consists of twenty once-a-year meetings.

Arthur Miller

When Arthur Miller returned to the theatre in 1964, after an absence of eight years, it was with a new subject and purpose. Where his earlier plays had all used domestic stories to raise and explore larger social and political issues, he was now to focus on the individual experiences themselves. His particular subject was guilt – or, rather, guiltiness: the sense of having failed some abstract or specific moral obligation and thus of having proved oneself inadequate or unworthy. To be sure, that subject was present in such early plays as *All My Sons* and *Death of a Salesman*, but as a means of criticizing the social situation that produced such feelings. Now Miller turned his attention to dramatizing the pain of feeling inadequate and to offering counsel and consolation to help alleviate that pain. In effect, then, he left the tradition of Odets to join the camp of

Williams, Inge and the later O'Neill. To the extent that such plays as *After the Fall* and *The Price* offer an explanation of life that frees the individual from unnecessary self-condemnation, they resemble both *The Night of the Iguana* and *Long Day's Journey into Night.*

In *After the Fall* (1964), Quentin faces his third marriage with the fear that his history makes him unworthy of the opportunity for happiness. He finds only failures and betrayals in his life, not only in his first two marriages but in his relations with friends and family, and in their relations with each other, so that he is convinced he has no right to impose himself on another victim. He finally realizes that such defects are part of the human condition, not his personal failing, and that hope for the future can be built on the ability to accept and forgive the inevitable lapses, and to carry on in spite of them. To dramatize this wholly internal experience, Miller uses a method even more expressionistic than the fantasy and memory sequences in *Death of a Salesman*: the fictional premise of *After the Fall* is that Quentin is talking to a friend represented by the audience, and the events he remembers are acted out on the stage behind him, Quentin moving back to join them and then returning to the conversation. Thus the bulk of the stage becomes a representation of Quentin's mind, with memories played out in various spaces as Quentin recalls them, and he is not always in complete control: some characters appear unbidden or through a chain of association to interrupt his talk. Through this device Miller can provide insights into Quentin's thinking that even he is not aware of, and thus keep us a step ahead of him in his discoveries.

Quentin is haunted until the final minutes of the play by the conviction that he is morally responsible for all the failed relationships in his life, but Miller shows that the guilt frequently lies elsewhere. Quentin's first wife Louise is neurotically jealous and unbending in her self-righteousness. His second wife Maggie, in her more than neurotic insecurity, makes ever-increasing demands Quentin cannot possibly meet (summed up in her simple requirement, 'Love me, and do what I tell you. And stop arguing.'), and yet he feels guilty for not meeting them. Quentin is so prepared to accept blame that he finds ways of condemning his innocence – 'Is it altogether good to be not guilty for what another does?' When a woman he has helped thanks him, he feels unworthy; when a friend's wife tries to seduce him, he feels ashamed for not even realizing what was happening; when he visits a concentration camp and recognizes the humanity of the builders, he fears that this empathy is proof of his baseness: 'I can easily see the perfectly normal contractors and their cigars, the carpenters, plumbers, . . . good fathers, devoted sons, grateful that someone else will die, not they, and how can one understand that, if one is innocent? If somewhere in one's soul there is no accomplice?'

A potential danger of the play is that Quentin's virtual lust for guiltiness and atonement – twice in the play he actually strikes crucifixion poses – will alienate the audience, but it is necessary for us to see the excesses of his self-condemnation before he does, so that we can be prepared for his climactic discovery. For theatrical reasons, Miller makes the realization come upon Quentin in a rush, beginning with the memory of his final quarrel with Maggie. To save himself from her madness, Quentin attacks her, and is once more almost drowned in guilt: 'What love, what wave of pity will ever reach this knowledge – I know how to kill?' The various betrayers and victims of the play gather on stage, and the concentration camp site is lit.

> Or is it possible – that this is not bizarre . . . to anyone? And I am not alone, and no man lives who would not rather be the sole survivor of this place than all its finest victims? What is the cure? [. . .] No, not love; I loved them all, all! And gave them willing to failure and to death that I might live, as they gave me and gave each other.

Some of the play's early detractors condemned Miller for excusing himself (i.e. Quentin) by putting the blame on everyone else, particularly Maggie, a character clearly based on his late wife Marilyn Monroe. But the accomplishment of *After the Fall* is that Miller offers Quentin a basis for self-forgiveness that does *not* transfer the guilt elsewhere, but rather provides a greater charity towards others. If, after the Fall, we are all imperfect, then perhaps we can accept and forgive our imperfections and build our lives on that foundation.

> To know, and even happily, that we meet unblessed; not in some garden of wax fruit and painted trees, that lie of Eden, but after, after the Fall.[. . .] And the wish to kill is never killed, but with some gift of courage one may look into its face when it appears, and with a stroke of love – as to an idiot in the house – forgive it; again and again . . . forever?

And if that sounds more like something O'Neill or Williams or Inge would say, *After the Fall* is further vindication of those writers' discovery that the most-needed gift of mid-century American drama was consolation and reassurance.

Miller's next play, *Incident at Vichy* (1964), offers little in the way of consolation, but it is, like *After the Fall*, an analysis of guilt and responsibility. In a group of Jews and other racial undesirables rounded up by the Gestapo in 1942 is an Austrian count, Von Berg, obviously arrested by mistake. The others, led by the Jewish psychiatrist Leduc, force Von Berg to recognize his complicity, if only passive, in the Nazi atrocities; and when Von Berg is given a pass to leave, he hands it to Leduc, staying to die in his place. On one level the play is a celebration of that inexplicable act, an affirmation of man's capacity for good set against the almost overwhelming contrary evidence of the Nazis. Beyond that, it is an analysis of social responsibility that recalls Miller's *All My Sons*. Von Berg, whose hatred of the Nazis never really extended beyond a distaste for their vulgarity, and who could overlook a cousin's participation in atrocities because he was of his class, is forced to recognize his larger obligation as Leduc makes him see that silence, even disgusted silence, is an acquiescence.

Had Miller written this play ten or twenty years earlier, its meanings would have stopped here, with Von Berg's discovery of his moral failure and his acceptance of responsibility and punishment. But *Incident at Vichy* is not really about Von Berg; it is about Leduc, whose experience raises issues larger even than Nazism and identifies guilts that even the seemingly just and innocent can bear. In the middle of the play Leduc tells a German officer that his life is worth more than the German's 'Because I am incapable of doing what you are doing. I am better for the world than you.' But when he later tries to convince Von Berg of his culpability, we sense, perhaps a few seconds before he does, the applicability of his words to himself: 'Jew is only the name we give to that stranger, that agony we cannot feel, that death we look at like a cold abstraction . . . the man whose death leaves you relieved that you are not him, despite your decency.'

So at the end of the play, when Leduc takes the pass and escapes, it is not Von Berg's fate that is meant to haunt us. By his own logic Leduc has declared himself the moral equivalent of the Nazis just by defining someone else's life as of less value than his own; and that he is passively accepting Von Berg's sacrifice is no more an excuse than Von Berg's allowing his cousin to do the dirty work for him. On this point, then, *Incident at Vichy* contradicts *After the Fall*: those who survive at the expense of others, or even those who accept the deaths of others as less tragic than their own would be, must carry a burden of responsibility and guilt for the rest of their lives; and the fact that this capacity for selfishness is universal does not make it forgivable.

The Price (1968) is more in the spirit of *After the Fall*, going beyond even forgiveness to offer the capacity for pride and a sense of accomplishment. The story is of two brothers, Victor and Walter Franz: when their father was ruined in the Depression Victor stayed to support him, giving up his chance for an education, while Walter went off to college and a lucrative medical practice. When they meet to sell their father's furniture, Victor presses the charge that Walter deserted them and ruined his life. His need to make sense of the past fuels his attack even more than the old resentments: an affirmation of his image of himself as victim of his brother's selfishness will at least give a shape to his wasted life. But Walter refuses to accept the role of villain, slowly forcing Victor to the realization that his father had manipulated him; in a powerful theatrical moment Walter points to the furniture and reminds Victor – and the audience – that the whole premise for their being in this room is the sale of these valuable pieces that could have been sold back then.

This discovery, which threatens to make Victor's whole life meaningless, pushes him a step further into repressed memories. He knew his father was using him, but felt an obligation to remain loving and supportive: 'But you're brought up to believe in one another, you're filled full of that crap – you can't help trying to keep it going, that's all. I thought if I stuck with him, if he could see that somebody was still . . . I can't explain it; I wanted to . . . stop it from falling apart.' [ellipses Miller's] 'It won't work, Vic,' says Walter.

> Is it really that something fell apart? . . . What was here to fall apart? Was there ever any love here? When he needed her, she vomited. When you needed him, he laughed. What was unbearable is not that it all fell apart, it was that there was never anything here . . . . It's that there was no love in this house. There was no loyalty. There was nothing here but a straight financial arrangement. That's what was unbearable. And you proceeded to wipe out what you saw.

This final revelation, which Walter thinks is an accusation, is actually the vindication of Victor's life that the play has been striving for. Earlier in the play Walter said people create themselves to escape the horror they fear most. The idea of a world without love was unbearable to Victor, and he denied it by making himself into a loving person. Without support, without any precedent in his family, without even his conscious awareness, he created out of himself the capacity for love, sacrifice and service; and that is an explanation of himself he can live with.

Affirming the value of even modest human accomplishment is also the goal of Miller's next play, the somewhat less successful *The Creation of the World and Other Business* (1972), a partially comic retelling of the story of Adam and Eve in which God makes absolute and inexplicable moral demands while Lucifer promotes anarchy. In this vacuum Adam must find his moral values and his capacity for goodness within himself, and the play ends with his struggling to guide Cain towards repentance and Eve towards forgiveness. The play is crippled by Miller's obvious uneasiness with both allegory and comedy, but its generating idea is clearly of a piece with *The Price*: the reassurance that man has the capability to create in himself the virtues and life-justifying qualities that the world outside does not provide him.

Tennessee Williams

The first two decades of Tennessee Williams's career represent an unmatched contribution to the American drama, not only in the individual plays but in the irreversible shift in the intention and effect of domestic realism, from the representation of larger issues to the exploration of the psychological and emotional burdens of everyday life. The second two decades of his career offer the much less happy picture of an artist with the misfortune to outlive his talent. Beginning in 1962, a series of personal crises and professional setbacks, exacerbated by alcohol and drug abuse, produced an artistic decline from which Williams never recovered. Although he continued to write for most of the next twenty years – it seemed to be a symptom of his emotional state that he felt compelled to produce new work even when craft and inspiration failed him – virtually all of his plays after *The Night of the Iguana* were artistic and commercial failures. (The following analysis will stretch the chronological boundaries of this chapter to include Williams's plays after 1975.)

His plays of this period are not significantly different in kind from those of the 1950s, and almost every one has one character or scene or speech that recalls Williams at his best. The otherwise undistinguished *The Mutilated* (1966), for example, includes a Christmas carol that is the purest Williams:

> I think the strange, the crazed, the queer
> Will have their holiday this year
> And for a while, A little while,
> There will be pity for the wild. . . .
> I think the mutilated will
> Be touched by hands that nearly heal,
> At night the agonized will feel
> A comfort that is nearly real. . . .
> I think for some uncertain reason
> Mercy will be shown this season
> To the wayward and deformed,
> To the lonely and misfit.

The generally moribund memory play *Vieux Carré* (1977) is enlivened by one of Williams's greatest comic characters, the mad landlady Mrs Wire; and some of the same grotesque comic energy appears in *The Gnadiges Fraulein* (1966) and *Kingdom of Earth* (1968). But one scene or speech or character does not make a play, and in every case the hint of Williams's earlier power either exists in a vacuum, unintegrated with the rest of the play, or lapses into excess. This speech from *The Milk Train*

Doesn't Stop Here Anymore (1963), for example, would not be out of place in *Streetcar* or *Summer and Smoke*:

> Have you ever seen how two little animals sleep together, a pair of kittens or puppies? All day they seem so secure in the house of their master, but at night, when they sleep, they don't seem sure of their owner's true care for them. Then they draw close together, they curl up against each other. . . . We're all of us living in a house we're not used to. . . . We have to creep close to each other and give those gentle little nudges with our paws and our muzzles before we can slip into – sleep and – rest for the next day's – playtime.

But it is completely out of place in *Milk Train*. Self-contained arias in the earlier plays – Blanche's 'Don't hang back with the brutes' speech in *Streetcar*, for example – were similarly self-conscious and even strained in their imagery and ideas, but they put into words impressions that the rest of their plays were dramatizing. But *Milk Train* is about the need for assistance in dying, not living; and that theme is never addressed with the poetic intensity Williams gives to this really irrelevant thought.

Another symptom of Williams's decline is the compulsive reworking of flawed material. Previously Williams was able to adapt and revise successfully: *Menagerie*, *Cat*, *Summer and Smoke* and *Iguana* were all based on earlier short stories; and *Eccentricities of a Nightingale* is a valid alternative version of *Summer and Smoke*. But now the tinkering proved fruitless. *Vieux Carré* recycles earlier autobiographical material with no advance in technique or insight, except for the depiction of Mrs Wire. *Milk Train* was produced in a preliminary version a year before its Broadway opening, revised for Broadway, revised again for the published text, and revised once again for a second Broadway production in 1964, with no significant improvement. *The Red Devil Battery Sign* closed out of town in 1975, was rewritten for a 1977 London production and revised again for publication in 1979. It seems to have begun as a play about conspiracy, the first version being set on the day of President Kennedy's assassination, but by the final version that element has been cut so clumsily that the central action, about a frightened woman's frantic romance with a dying musician, makes little sense.

One of Williams's biggest personal disappointments was the play variously known as *Out Cry* and *The Two-Character Play*, which he worked and reworked through at least five versions between 1967 and 1975 without conquering its basic flaw, which recalls the earlier *Camino Real*: the play operates in almost purely metaphoric terms, without the anchor in domestic realism that characterized his best work, and the result is too lifeless and fragile to hold an audience. A brother and sister acting team rehearse a play about an agoraphobic brother and sister, the distinctions between performers and roles gradually disappearing. The inner play, essentially an extended metaphor of madness and the capitulation to madness, is sensitively balanced by the extended metaphor of the outside play, which is of resisting madness and trying desperately to function in reality even when the capacity for coping has been exhausted. *Out Cry* is thus one of the purest expressions in the Williams canon of some of his recurring themes: the pain of existence, the feelings of inadequacy, and the seductive power of madness, along with Williams's admiration and affection for those who fight, even (or especially) if they lose. But its purity is its downfall. The characters in *Out Cry* have no reality as human beings; they are merely the voices speaking Williams's words. Samuel Beckett at his best could create living drama out of virtually disembodied voices, but that is beyond Williams's scope.

This play also exposes another, particularly sad symptom of Williams's decline, his growing inability to separate himself from his plays. There was never any question in his prime that some of his characters were projections of aspects of his own personality; and indeed a large part of the achievement of a Blanche or Brick was that Williams was able to transmute his own emotional experiences into fully realized dramatic creations. But in *Out Cry* the expressions of stage fright, the temptation to withdraw from the world rather than face its ridicule, and the compulsion to carry on in spite of waning powers are all clearly the author speaking, without translation into art. The same embarrassing self-exposure cripples *In the Bar of a Tokyo Hotel* (1969), whose secondary character of a painter in the midst of a mental and artistic breakdown (expressed in a style of unfinished sentences everyone in that play affects) is so obviously a self-portrait that he breaks the bounds of the play:

> I've understood the *intimacy* that should, that has to exist between the, the – painter and the – I! It! Now it turned to me, or I turned to it, no division between us at all any more! The one-ness, the! . . . There was always a sense of division till! Gone! Now absolute one-ness with! . . . This work is hard to confine to. . . . Always before I felt controllable limits, I.

And Scott and Zelda Fitzgerald in *Clothes for a Summer Hotel* (1980) are used as almost mythic examples of burnt-out artists who outlived their limited talents; that they are hardly-disguised projections of the author's own self-pity is obvious from the unsubtle ways Williams betrays his identification, repeatedly and pointlessly questioning Scott's heterosexuality, and once again making them speak for him more than themselves: 'The losses accumulated in my heart, the disenchantments steadily increased. . . . Wouldn't accept it. Romantics won't, you know . . . . I went back to the world of vision which was my only true home.'

Another form of falling-off of powers is evident in two plays of the period which work as domestic melodramas though they have little new to say, *Small Craft Warnings* (1972) and *A Lovely Sunday for Creve Coeur* (1979). In the first, the denizens of a neighbourhood bar rearrange their pairings and their lives, settling for the best options available; in the second, a schoolteacher has her fantasies of romance and social advancement destroyed, and accepts a less glamorous fate. What we have here, in short, are two William Inge plays, about little people surviving little disappointments and settling for little satisfactions. Both might in fact have been more successful had Inge written them, without the awkward attempts at poetry in the first and at grotesque humour in the second.

Williams spent the last twenty years of his life trying to say the same sorts of things he had said in his best plays, using the same tools of poetic imagery, grotesque humour, intense sympathy for human suffering, and frequently sensationalistic plots. It was not that he could not do these things any more, but that he could not make them work. Even at his most successful, he had been a dramatist of broad strokes and extreme images, but he was able to control his rich poetic imagination and produce powerfully effective art. With the loss of that control, the plays of his final period either seem empty compared to the emotional resonances of his great plays or approach self-parody in their undisciplined excess. The poetry becomes disconnected and unmanageable, as in *Milk Train*; the style becomes self-conscious and precious, as in *In the Bar*; the characters and emotions become too personal, as in *Out Cry* and *Summer Hotel*; the plot structure becomes disjointed, as in *Red Devil*; the plays become either all symbol and no reality (*Out Cry*) or mere narrative without resonance (*Vieux Carré*). At his peak Williams was a playwright of unrivalled imagination and poetic power; we can see now that what really made those great plays work was his ability to discipline, control and shape the outpourings of his genius. The brilliance of Tennessee Williams's masterpieces was always obvious; the particular nature of his decline tells us to go back and appreciate their craftsmanship.

The period in the history of the American drama covered in this chapter is less clear in definition than some of the others. Unlike the 1930s or 1950s it is not dominated by one writer or style; it seems, like the 1920s, to be more a period of transition or exploration. In retrospect, one can see a clear and straight line from O'Neill's early experiments, through Odets, to the triumphs of Miller and Williams. But, having reached that zenith, the Broadway theatre seemed to be marking time; of the Broadway playwrights of the 1950s, only Arthur Miller continued to grow and produce effective new work in the 1960s. Meanwhile, the face of the American theatre changed in the 1960s, inevitably producing changes in the American drama. As new theatres welcomed new writers, the dominance and homogeneity of Broadway-based drama was broken. Inevitably, some of these new writers experimented with alternatives to domestic realism, as O'Neill and others had done in the 1920s, and perhaps with the same motivation, the belief that the new statements and points-of-view they wanted to bring to the drama required new forms.

Still, as new and diverse as the playwrights of the 1960s were, what they had to say proved not all that different from what previous generations of dramatists had offered and previous generations of audiences had responded to. The social criticism of writers such as LeRoi Jones, David Rabe, Lanford Wilson and the Edward Albee of *The American Dream* and *Who's Afraid of Virginia Woolf?* stemmed from the same impulse that had driven Clifford Odets and Arthur Miller: the need to warn Americans against the failings and dangers inherent in their own culture. The existential pain dramatized by Albee in *The Zoo Story* and *Virginia Woolf*, by Jack Gelber in *The Connection*, by Sam Shepard in *Action* and *The Tooth of Crime*, and by many other playwrights treated in this chapter is not radically different from the loneliness and insecurity that were the continuing subject of Tennessee Williams and William Inge.

Moreover, despite the wide-ranging experiments in alternative styles, some of which proved particularly useful to the more radical new writers and would eventually affect the mainstream drama, the fact remains that the majority of the plays described in this chapter are set in someone's living room. It was noted in the preceding chapter that the artistic voyage of American drama before 1950 – from realism, through experiment, back to domestic realism – was replicated in the individual case of Eugene O'Neill. While it may not yet be wholly apparent, the next chapter will show that the same is true of the careers of Albee, Wilson, Shepard and many of their contemporaries. Having proved its effectiveness and versatility in the 1950s, domestic realism was challenged by a new generation of writers in the 1960s. It would withstand that challenge successfully in the following decade, confirming its power as the natural voice of American drama. Meanwhile, for the first time in more than half a century the terms 'Broadway' and 'American theatre' were not synonymous, and the range of American drama became broader, richer and more varied.

Chapter 6
1975-1990: A National Theatre

New theatres, new plays

By the mid-1970s Broadway's domination of the American theatre was decisively over. Off- and Off Off-Broadway theatres routinely produced six or more times as many shows as Broadway did each season; and the hundreds of professional theatres around the country together staged many more productions than all the New York City theatres put together. This imbalance was particularly extraordinary in the case of new American plays, which were simply no longer the domain of Broadway. For various reasons, including financial constraints as well as the artistic flourishing of other venues, by 1980 the Broadway repertoire was made up almost entirely of plays first produced elsewhere.

In some cases, to be sure, this was merely a modification of Broadway's old out-of-town-trial system, with a commercial producer placing a new script in a regional theatre for an inexpensive test run before bringing it to Broadway. But in the majority of cases new plays were produced independently in regional or fringe theatres, only attracting the attention of Broadway producers when their success suggested commercial potential. The originating theatres could then participate in any commercial profits, and the plays would benefit from the publicity and 'seal of approval' Broadway still had the power to bestow. One way or another, the best new American plays of this period usually did get New York productions eventually, but many bypassed Broadway.

This change in the theatrical balance of power had direct and indirect effects on the plays and playwrights themselves. First, there were many more opportunities for new plays to be produced. Even in its most active years the Broadway theatre had rarely introduced more than fifty new American plays a season; by the early 1980s the combination of regional and New York fringe theatres could easily produce six or eight times as many. It also meant an opportunity for greater variety. Having at least partly escaped the inhibiting shadow of high budgets and the need to show profits, regional and fringe theatres could be far more open than the commercial theatre to playwrights representing minority groups, addressing minority interests, dealing with uncommercial subjects or employing challenging and unconventional styles. Plays by women; plays by black, Hispanic and Asian dramatists; plays by and about the concerns of feminists, homosexuals and Vietnam veterans; plays reflecting regional subjects and cultures; plays in styles ranging from the mime/dance/ritual of performance artists to the most conventional domestic realism – for the first time in the twentieth century the American theatre had the capacity to stage such a broad representation of the American experience. Inevitably, many of these new plays and playwrights were unsuccessful or unimpressive, but the sheer quantity meant a greater opportunity for plays of merit or potential to be seen.

Another by-product of the new theatrical set-up was the chance for a new play to be staged several times, in several versions, on its way to its final form. Most regional theatres, and many New York companies, had new play workshops and staged readings of works-in-progress built into their structure. By the 1980s it was the rule rather than the exception for a playwright to see a new work performed in a workshop or staged reading, then in one or more full productions, and perhaps then a commercial New York staging, with the opportunity to revise and polish the script based on what was learned each time. August Wilson's *Joe Turner's Come and Gone*, for example, had workshop stagings at New Dramatists in New York in 1983 and 1984, a staged reading at the O'Neill Theatre Center in Connecticut in 1984, a full production by the Yale Repertory Company in 1986, and further productions in Boston, Seattle and Washington on its way to a Broadway run in 1988. (One minor problem generated by this pattern is deciding which date to assign to such plays; in this chapter the opening date of the production that corresponds most closely to the published text will be used.)

Despite the increased variety of voices, styles and subject matter represented by the new American drama of the 1970s and 1980s, a clear mainstream remained; and the discovery by previous generations of dramatists that the natural American dramatic mode was domestic realism was continually affirmed. Whatever the subject, however dominant the social or political agenda, however particular or unconventional the life experience being explored, the overwhelming majority of American dramatists continued to find that their concerns could best be dramatized through the everyday, personal experiences of ordinary characters. There were, to be sure, some noticeable alterations in this dominant mode; and two stylistic innovations in particular became so widespread and so generally accepted that they practically became absorbed into the definition of realism.

One was a general replacement of traditional act and scene structure by something considerably more fluid. While many dramatists continued to write thirty- or forty-minute scenes separated by closed curtains and set changes, many others wrote in five- or ten-minute units, with brief scenes flowing into each other or leaping over time and space in a cinematic way. Indeed, it is likely that film and television were the inspirations for this structure, with a generation of dramatists and audiences raised on those more fluid media able to absorb information offered in short bursts and to make the imaginative leaps from one to the next. (A somewhat more cynical way of looking at it would be that a generation raised on American commercial television could think only in ten-minute units.) But it is worth pointing out that Shakespeare did much the same thing, and this fluid structure may have been made possible, as Shakespeare's was, by the nature of the theatres in which the plays were performed. For various historical reasons, the majority of regional theatres had arena or thrust stages rather than the conventional proscenium stage, while many fringe companies, in New York and elsewhere, performed in converted spaces that were essentially bare rooms. These settings, along with the relative poverty of most companies, encouraged the use of bare stages or minimal, impressionistic sets; and these theatrical factors combined to allow and even encourage a more cinematic flow of scenes.

A second violation of traditional play structure that came to be familiar enough that it was hardly noticed was the breaking of the fictional frame to allow direct audience address by one or more characters. Of course this, too, had been done before, by Shakespeare among others, but always in plays that were consciously declaring themselves as non-realistic through the device – Wilder's *The Skin of Our Teeth*, for example, or Williams's *The Glass Menagerie*. The difference now was that a character, frequently a narrator figure like Williams's Tom Wingfield, could step outside the play for a moment without seriously breaking the illusion; and the audience would not be disconcerted, but would experience the play as continuous and realistic. A play that combined these two innovations, such as Mark Medoff's *Children of a Lesser God* (Los Angeles 1979), would have been considered highly experimental thirty or forty years earlier, but fits comfortably within the expanded definition of realism.

So, with this expanded definition, it is possible to say that the dominant mode of American drama in the 1970s and 1980s remained the realistic, contemporary, middle-class domestic melodrama discovered by Odets and others in the 1930s and refined by Miller and Williams in the 1950s. Both Miller and Williams continued to write into this period, though Miller produced only a few new plays and Williams was in severe decline. (Williams's plays of the late 1970s were discussed in the previous chapter.) From the generation of the 1960s Edward Albee and Neil Simon continued to define the extreme boundaries of the commercial drama; and Sam Shepard and Lanford Wilson came into their full powers, fulfilling their early promise with mature works. The closer we come to the present, the more difficult it is to distinguish the major writers from the merely interesting or trendy, but David Mamet and August Wilson would seem to be unquestionably in the first category; and such others as A. R. Gurney, Beth Henley, David Henry Hwang, Terrence McNally, Marsha Norman and Wendy Wasserstein are certainly worthy of attention.

Arthur Miller

Miller's few plays of this period are an even mix of the predominantly social and thesis-driven drama of his early years and the psychological and spiritual explorations he had begun in the 1960s. *The Archbishop's Ceiling* (Washington 1977) is fully in the tradition of *The Crucible* in condemning an evil society for its crimes against individuals. Writers in an unnamed eastern European country must function with the suspicion, but not the certainty, that their rooms are bugged and that one or more of them may be government agents. So the most innocent conversations have multiple levels, as they perform for an audience that may or may not be there while simultaneously trying to be open and honest with each other. Deciphering the ambiguities of reality is not a theoretical puzzle for these characters; the society that controls their lives forbids them to live naturally. Every conscious moment must be spent in trying to interpret, decode and clarify not only each other's words and actions, but their own.

As in *The Crucible*, where false accusations led John Proctor to discover real guilt in himself, the constant monitoring of themselves and each other forces the characters of *The Archbishop's Ceiling* to face their private, unpolitical ambiguities and duplicities. The American visitor Adrian sincerely wants to help his friends, but he is writing a novel about them, and he and they can never be sure whether he is thinking of them as people or characters. Even if they are not spies, Marcus and Maya enjoy the small luxuries of officially approved writers, and cannot themselves be sure that the desire to retain those does not lead them to do the government's work unbidden. Here, even more than any acts of direct repression, are the social atrocities Miller condemns: first, the theft not only of privacy but of personality. It is not only in Adrian's novel that they will be fictionalized; when everyone must constantly play a role, no one can be himself. Even worse is the forced imposition of an awareness of one's own moral failures; being obliged every day to search out the truth about oneself, and thus every day to face oneself, is the most exquisite torture a diabolical state can impose.

The American Clock (1980) is an attempt at social epic, a history of the Depression of the 1930s in both national and personal terms, narrated through flashbacks by two survivors. Lee Baum's memories are all domestic, of his middle-class Jewish family's decline into genteel poverty, while Arthur Robertson, a financier who anticipated the crash and saved his riches, observes as society crumbles around him. Miller is clearly uncomfortable with the epic mode, and the play has some of the same flaws – a disjointed, episodic structure and scenes which are too obviously striving to be archetypal to have much reality – as *We, the People*, Elmer Rice's early attempt at reflecting an entire culture through domestic realism. But if the play's dramaturgy is clouded, its point of view is not, and Miller succeeds in stating a judgement of the Depression even if he fails to dramatize it. His premise is that, for all its destruction, the Depression ultimately affirmed and strengthened the American experiment. Baum and Robertson both repeatedly express their astonishment that the apparent breakdown of the system did not lead to revolution from the left or right, and find the explanation in a redefinition of the American Dream; seeing his mother as representative, Lee realizes that

> She was so like the country; money obsessed her but what she really longed for was some kind of height where she could stand and see out and around and breathe in the air of her own free life. With all her defeats she believed to the end that the world was meant to be better.

In an early version of the play Robertson reached a similar conclusion, which the London text gives to another character: 'Roosevelt gave them back their belief in the country. The government belonged to them again! . . . The return of that belief is what saved the United States, no more, no less!'

Miller's next plays are as different from these – and, indeed, most of his earlier work – as could be. The one-act *Elegy for a Lady* (New Haven, 1982) is a delicate study in intimate human relationships and a teasing exercise in ambiguity. An older man enters a women's shop to buy a gift for his somewhat younger lover, and enlists the help of the proprietress in making his selection. This involves some explanation of the relationship: although accepting him as her lover, the woman has maintained a certain independence and distance; now she is dying, and the man fears that too intimate a gift will violate her boundaries while one too impersonal will misrepresent his feelings. The proprietress, first guessing and then speaking with some confidence about the woman's feelings and thoughts, is able to reassure him that his lover understands his love and has been totally satisfied by the form their relationship took.

The power of the play comes from the mystical and dream-like tone that develops as the proprietress becomes progressively identified with the dying woman, speaking for her and allowing the man to speak to his lover through her. At first she merely sympathizes, but soon she is describing the woman's feelings with insight and authority, and then speaking for her:

> MAN But her independence means more to her than any relationship, I think.
> PROPRIETRESS How do you know? – You were the one who ordered her not to love you.

Eventually the two women blend into one, so it is not clear who is speaking or being addressed:

> PROPRIETRESS I never condemn anyone; you know that . . . .
> MAN That's your glory, but in some deepest part of you there has to be some touch of contempt.

And thus the proprietress's eventual assurance that he has not failed his lover and can honour her best by respecting her privacy provides the comfort that the man needs.

Miller invites speculation: has the proprietress had a similar experience and relationship in her life, that she has such insight into this one? Is she graciously offering herself as a surrogate, through whom the man can work out his feelings? Does she know the woman for whom she speaks so confidently, and does that mean she also knows this man who appears to be just a chance customer? Does he know her, so that he is slowly opening up to a friend, rather than to a stranger? Is this in fact the woman, with both parties delicately speaking in the third person to protect themselves from the painful reality of her dying? Or is this indeed a dream, a projection of the man's internal process of acceptance and resolution? This uncharacteristic excursion into ambiguity is completely successful, the various possibilities resonating against each other to deepen the emotional effect.

Its companion piece, *Some Kind of Love Story* (1982), has similar ambiguity, though without the same resonance and delicacy. A private detective has been questioning a prostitute about a murder for years; at each meeting she offers another tantalizing hint of what she knows, but if he pushes too hard she withdraws into schizophrenic fantasy. He has come to realize that she will never tell him everything, and even to suspect that she has no information of value and is just stringing him along to keep their relationship alive, but he also realizes that he has too much invested in the search, and in her, to give up. During the play she metes out one more clue, keeping him on the string, but bringing him no closer to the solution of the mystery or the resolution of the strange hold they have on each other. As the title makes clear, Miller is suggesting that this kind of mutual need and mutual exploitation has the effect, if not the familiar face, of love; and that both parties are getting more from the continued cat-and-mouse game than they could from the achievement of their ostensible goals. But these two characters are much less insightful and eloquent than the pair in *Elegy for a Lady*, and the play must operate in broader, cruder strokes that present the ambiguities in their feelings without evoking the feelings themselves.

The same imbalance exists in Miller's other double bill of the 1980s. *I Can't Remember Anything* (1987) is an affectionate but external picture of two elderly people, showing how the banes of old age – physical frailty, fading memory, alienation from friends and family – are blessedly accompanied by philosophical resolutions, quieted passions and a focus on the smallest and most immediate pleasures and pains: with every reduction of power comes a matching reduction of desire, so the loss is not felt and the quiet passage towards death is eased. It is a quiet, pleasant play, with the elegiac tone of *Elegy for a Lady*, if not its emotional depths. The second half of the bill, *Clara* (1987), is a much more ambitious play, an attempt to address moral and psychological issues akin to those of *The Crucible* and *After the Fall* in the context of this new intimate style Miller was exploring in the 1980s. A man whose daughter has been murdered is questioned by a sympathetic but determined detective who suspects that guilt for not protecting her has blocked the memory of the key suspect's name. But rather than releasing guilt, the questioning revives the man's faith in his daughter's essential goodness and his own. The play thus explores two familiar concerns of Miller's. The first is connection and responsibility, the need to recognize one's ties to and effects on others. As far back as *All My Sons* Miller saw dissociation as a moral lapse, and he introduces Albert Kroll as a man so cut off from humanity that he has difficulty carrying on a coherent conversation. Under Fine's relentless guidance, Kroll remembers the ways he helped to shape his daughter through his stories and behaviour, and accepts responsibility for the woman she became. The second and decisive step is the reaffirmation of one's own highest virtues. Like Victor in *The Price*, Kroll has lost sight of his own achievements, but as he remembers his former idealism, so different from his present acquiescence to minor corruption, he is gradually reintegrated with that purer self; and it is when he can embrace his younger self with joy that he can declare his love for his daughter and, incidentally, remember the name of the probable murderer.

These late plays provide a quietly satisfying if slightly uncharacteristic coda to Miller's long career. Certainly his most significant contributions to the American repertoire lie in his plays of the 1940s and 1950s, and in his reaffirmation that the American drama was able to address social and political issues without losing the power and immediacy of domestic realism. If in such later plays as *The Price* and these one-act sketches he acknowledges that small, personal epiphanies or accommodations are also of importance, and worthy of dramatic exploration, he does not deny his earlier work but supplements it. If the dramatists of the 1950s can be divided into the Miller and Williams camps, Miller alone lived long enough and evolved enough to write effective plays of both types.

Neil Simon

Having established himself as the pre-eminent craftsman of joke-filled urban comedy, Neil Simon continued to dominate that branch of the American drama through the 1980s. Such plays as *California Suite* (1976), *Chapter Two* (1977), *I Ought to Be in Pictures* (1980) and *Rumors* (1989) fit comfortably in the genre of *Barefoot in the Park*, confidently working minor variations on the formulas Simon mastered in the 1960s. Tentative steps beyond those limits, particularly in the direction of deeper characterization or more serious subjects, repeatedly produce the weakest moments in the plays. Well into the 1980s, then, it appeared that Simon was limited to excellence in a very narrow field – not just comedy, but the comedy of topical references, upper-middle-class New York City characters, and unpausing gags. But in 1983 Simon discovered a capacity for presenting real sentiment, believable characters and even social and historical insight, without wandering too far from the type of comedy that was his natural mode.

Brighton Beach Memoirs (1983) is a sentimental and semi-autobiographical comedy of life in Depression-era New York. Jack Jerome is overworked to the point of a heart attack trying to support his wife Kate and sons Stanley and Eugene, and Kate's widowed sister Blanche and her two daughters. In the course of a week there is a variety of large and small crises: Jack's heart attack, Blanche's failed first attempt at dating, fights between Kate and Blanche and between Blanche and her daughter Nora, Stanley's chagrin at losing his paycheck in a poker game. The story is narrated by Simon's fictional self, thirteen-year-old Eugene; and for virtually the first time in Simon's career the comedy is based on character rather than isolated jokes, the central comic figure being Eugene himself, caught between boyish enthusiasm for baseball and adolescent discovery of lust, and constantly feeling put upon for being the family errand boy and scapegoat. A typical joke, part of a furtive discussion of sex with his older brother, has the rhythm of a standard Simon two-liner:

STANLEY How horny can you get?
EUGENE I don't know. What's the highest score?

But that line is clearly Eugene's; he has his own voice, which is a function of his age, his precocious self-awareness, and the conflicting forces pulling at him. This character may be a familiar one, but Simon gives him a reality and individuality, and he provides a strong centre and point of view for the play.

But the real strengths of the play, and innovations for Simon, lie in the background characters, who are not only allowed to be real and individual, but to feel actual, unmawkish emotions and to express them dramatically. The central quarrel between Kate and Blanche is triggered by a specific event, Blanche's failed date, but expands movingly and believably as two lifetimes of repressed pain come to the surface:

JACK You want to get it out, Blanche, get it out! Tell her what it's like to live in a house that isn't yours. To have to depend on somebody else to put the food on your plate every night . . . . Tell her, Kate, what it is to be an older sister. To suddenly be the one who has to work and shoulder all the responsibilities and not be the one who gets the affection. . . .

BLANCHE I love you both very much. No matter what Kate says to me, I will never stop loving her. But I have to get out. If I don't do it now, I will lose whatever self-respect I have left. For people like us, sometimes the only thing we really own is our dignity.

'For people like us. . . . ' In this scene, in the earlier scene in which Jack must painfully guide Stanley towards compromising his ideals in order to keep his job, in Stanley's deep shame at realizing that the boyish indiscretion of gambling becomes adult irresponsibility when his contribution to the family income is involved, and in the later scene in which Blanche fights to make contact with her alienated daughter, the play rises above Simon comedy and indeed above comedy. *Brighton Beach Memoirs* actually has less in common with Simon's previous work than it has with the Depression-era plays of Clifford Odets. Like them it uses the native American dramatic form of domestic realism, here tempered with warm and unobtrusive humour, to depict a society in crisis and transition, through its reflection in the everyday lives of ordinary people.

A sequel, *Biloxi Blues* (1985), is not nearly as strong, but represents a consolidation of Simon's new-found powers. It amounts to a very conventional military comedy, with Eugene Jerome experiencing the adventures and misadventures of basic training, but if there is little new, there is also a clear escape from Simon's old vices. The characters have some individuality; the warm humour of Eugene's embarrassed visit to a whore is not spoiled by gratuitous jokes; and the climactic encounter between the rebellious soldier Epstein and the dedicated Sergeant Toomey resolves one of the play's few dramatic subplots satisfactorily. This may seem like faint praise, and indeed the play would hardly attract notice if written by someone else. But all these things had once been beyond Simon, and his casual competence here demonstrates that the superior *Brighton Beach Memoirs* was not a fluke.

(Biographical and psychological critics will note that the character of Arnold Epstein, a bitter New York City Jew whose experience of the army is of humiliation and anti-semitic prejudice, appears to represent a side of himself Simon was unwilling to put into Eugene. There is in fact a pattern of displacement running through all three Eugene Jerome plays. The wise and loving father who guides the Jerome family and supports his in-laws in *Brighton Beach Memoirs* was absent from the Simon family, who actually had to live with their relatives. And a key speech of anger at the father in *Broadway Bound*, a speech that clearly represents the author's feelings, is given to Stanley rather than Eugene.)

Broadway Bound (1986) picks up the story after the war. Stanley and Eugene, now aspiring comic writers, sell their first sketch to a radio programme, but discover that jokes drawn from their family life are painful to the loved ones being ridiculed. Their father leaves the family for his long-time mistress; their socialist grandfather rejects his daughter Blanche for marrying a rich man; and Eugene cajoles his mother into telling the story of the romantic highlight of her youth, the night she danced with movie star George Raft. Even more than *Brighton Beach Memoirs*, *Broadway Bound* is a melodrama with humour rather than a comedy with serious moments; Eugene's parents are not the comically squabbling couple of his radio sketch, but two mismatched and emotionally repressed people with the not-uncommon experience of having drifted apart. The play sympathizes more with Kate than with Jack, but it recognizes that her single-minded commitment to home and family has turned her, like Odets's Bessie Berger, into a repressive, life-draining figure from whom Jack can reasonably feel a need to escape.

It is that double perspective that gives Kate's scene of reminiscence such power, as she briefly becomes (and Eugene is allowed to see) the open, exciting girl she was before she chose to turn herself into a wife and mother. The same complexity is seen in the other characters: grandfather Ben rejects his rich daughter and lives apart from his wife in the name of his socialist principles, but it was he who taught his daughters how to be loving even in the face of rejection. Blanche wears her new prosperity with the same mix of pride and pain with which she wears her mink coat: 'It keeps out the cold, but it also stops my father from reaching out and holding me.' And even Simon's on-stage surrogate becomes more than merely an attractive narrator when his experiences shake him. Stanley's explosion of anger at Jack leads Eugene to realize his own emotional repression –

> I was the one who should have had the fight with him. Only I didn't know I was so angry. Like there's part of my head that makes me this nice, likeable, funny kid . . . and there's the other part, the part that writes, that's an angry, hostile real son of a bitch. [Ellipses Simon's]

– just as Kate's exposure of her secret, romantic self both thrills and embarrasses him – 'Dancing with my mother was very scary. . . . And holding her like that and seeing her smile was too intimate for me to enjoy. Intimacy is a complex thing. You had to be careful who you shared it with.'

Lost in Yonkers (1991) is a return to more familiar Simon territory, as two precociously witty boys endure the comic frustrations of living with their tyrannical grandmother. But a subplot shows his new mastery of seriousness and psychological depth as the boys' mentally retarded aunt stands up against her mother and claims her measure of dignity and independence. The three Eugene Jerome plays, and *Lost in Yonkers* represent a significant advance for Neil Simon and a valuable contribution to the American repertory. Without departing too far from his strengths as a joke-writer and easy entertainer, he moves into the realm of Clifford Odets by dramatizing the social effects of first the Depression and then the post-war prosperity through their impact on the domestic lives of ordinary people, and into the realm of William Inge by sympathetically and respectfully observing these ordinary people as they make the small compromises and adjustments necessary for survival.

Meanwhile, with Simon continuing to define and dominate the realm of Broadway comedy, few new writers made much impression in this field. Larry Shue's farces provide a satisfying number of laughs in a totally unthreatening environment; in *The Nerd* (Milwaukee 1981) a man is tormented by an unbearably dull and boorish visitor he cannot get rid of, while *The Foreigner* (Milwaukee 1983) presents a man who tries to assure himself a quiet vacation by pretending not to understand English, only to discover that his deception elicits the solicitude and confidences of others. Some of Christopher Durang's plays have a benignly satiric tone: *A History of the American Film* (1978) takes its characters through parodies and pastiches of every classic film from *Intolerance* to *Dr Strangelove*, celebrating in the process the way in which films have guided and reflected American culture; and *Beyond Therapy* (1981), in which two neurotic lovers turn for help to their psychiatrists only to find them even crazier, pokes easy fun at easy targets. Other Durang plays are marked, and possibly marred, by a much darker tone: *Sister Mary Ignatius Explains It All for You* (1979) angrily satirizes a nun whose religion and values are restricted to the most repressive, irrational and life-denying elements of Catholicism; and *The Marriage of Bette and Boo* (1985) is a nightmarish account of the doomed marriage of two sadly trivial people, narrated by their son, who seems to feel that their neuroses and weaknesses were all deliberately adopted with the express purpose of warping him.

Edward Albee and other dramatists of the 1960s

Edward Albee's rate of production, like Arthur Miller's, slowed in the 1970s and 1980s, in his case through an admitted frustration with a theatre and audience that seemed unable to appreciate his work; his last commercial success was *A Delicate Balance* in 1966. The general trend of cold intellectuality and lack of dramatic vitality in his plays that was noticed in the last chapter continued to the point almost of self-parody, as Albee seemed more and more like an author talking to himself in a private vocabulary. *Listening* (British radio, 1976; Hartford, 1977) presents three people in the decaying garden of a mental hospital, two staff members and a patient who make fragmented and unsuccessful attempts to penetrate each other's self-absorption. The Man begs The Woman to acknowledge that they were once lovers; The Woman goes through the rituals necessary to break through The Girl's catatonic withdrawal; and The Girl, in her occasional lucid moments, must constantly interrupt the others' conversations to be heard. In the end The Girl slits her wrists and, although The Man is frightened, The Woman seemingly approves and allows her to die.

As early as *The American Dream* and *The Death of Bessie Smith* Albee had made the Miller-like judgement that self-absorption was immoral and could be murderous. But just as *All Over* seemed to be celebrating the emotionless, life-avoiding characters he had condemned in his earlier plays, so *Listening* seems to accept behaviour he had previously rejected. The play offers no particular sympathy to The Man, who seems most interested in making human contact with the others, and it makes no real criticism of their self-absorption. When The Man finally explodes and cries to The Woman, '*Reveal* yourself!' she replies, 'There is no revelation *in* me'; and a moment later her only response to The Girl's suicide attempt is an aesthetic approval: 'Done beautifully.' In *The Zoo Story* Albee presented the inability to communicate with others as tragic, and in such plays as *Virginia Woolf* and *A Delicate Balance* he stressed the human need for integration and involvement with others. But in *Listening* he seems to be postulating, without emotion, a view of human existence as so empty and without external value that self-referential detachment is an appropriate behaviour.

Oddly, a companion piece, *Counting the Ways* (London, 1976; Hartford, 1977), seems to take the opposite position, celebrating romantic love with a benign and comic awareness of its foibles and imperfections. Beginning when She asks He, 'Do you love me?' it is a series of sketches and blackouts presenting situations in which love may be called into question: in the midst of unsuccessful sex, after an argument, at a time of grief, while consulting flower petals, etc. Nothing really unpleasant is discovered, so the last scene, in which He finally asks if She loves him and She replies, 'I don't *know*. I *think* I do.' is a reassuring joke, suggesting the occasion for another round of considerations, not a negative. Albee labels the play 'A Vaudeville', and it functions as an occasionally thoughtful but never threatening light entertainment.

The Lady from Dubuque (1980) is Albee's most ambitious and significant play of the period, ultimately a failure of execution rather than of conception. A stranger appears in the home of a dying woman and claims to be her mother, here to help her through the process of dying. (The title refers to a comment made in the 1920s about the sophisticated magazine *The New Yorker*, that it was not edited for the archetypal midwesterner; Albee seems to use it to identify Elizabeth as the ultimate Other.) Whether she is meant to be the angel of death or just a guide along the way, the play has echoes of both Thornton Wilder's *Our Town* and Tennessee Williams's *The Milk Train Doesn't Stop Here Anymore* in the idea that dying is a weaning away from life that requires some assistance. But one of the problems of the play is that this process is never really dramatized; Jo's departure seems quick and easy, and is seen entirely from the outside, while her husband's feeling of desertion and exclusion is most strongly dramatized, even though the play keeps declaring it to be irrelevant: 'What does Sam know? Sam only knows what *Sam* needs. . . . And what about what Jo needs? What does what Sam needs have to do with that?'

There are other structural flaws to the play. Since the supernatural visitor does not appear until the very end of the first act, there is no clear focus to the first half, and the audience must spend Act Two recalling and reinterpreting earlier events while keeping up with the new action. Meanwhile, Elizabeth is so pointlessly and cruelly coy about her identity and mission through much of the second act that the audience is likely to turn against her before her role as comforter and guide is finally made clear, and thus to sympathize with Sam's outrage more than the play desires. The potential power of *The Lady from Dubuque* lies in its insight that death is a process the dying must be guided through, even at the cost of excluding the living. The failure of *The Lady from Dubuque* lies in the basic error, far too representative of Albee's later work, of presenting this as an intellectual concept rather than as the play's dramatic and emotional core.

The Man Who Had Three Arms (Chicago, 1982) is a failure of another sort: like some of Tennessee Williams's late plays, it seems to reflect an author unable to separate himself from his own creation. The play takes the form of a lecture by the title character, whose physical abnormality made him a celebrity until the third arm withered and the public lost interest in him; he is reduced now to being the substitute speaker on a minor lecture circuit, telling his story and venting his anger at the fickleness and cruelty of the world. (An Albee play titled *The Substitute Speaker* was repeatedly announced as in progress during the 1960s.) There are some small indications that Albee intended the play as a study in the harmful effects of celebrity on the celebrated, so that Himself's petulance, self-pity and lashing-out would be seen as signs of his spiritual decay. The problem is that Himself's story, of instant elevation to stardom on the basis of his curiosity value and just as instant rejection when he no longer continued to amuse, is very close to Albee's own experience; and Albee was never reticent about expressing his resentment towards the critics and audiences that forsook him after his early triumphs. In the absence of any effective distancing of the play's voice from its central character's, it must inevitably be experienced as Albee's own uncensored demand that the world apologize. Albee's few short plays since 1982 have been kept away from the commercial theatre and neither published nor revived after their brief runs, so, like Williams's late self-pitying works, *The Man Who Had Three Arms* makes an embarrassing coda to a career that began with such accomplishments.

Few other dramatists of the 1960s and early 1970s continued to produce work that fulfilled their early promise. Arthur Kopit's only important play after *Indians* is *Wings* (radio, 1977; New Haven, 1978): through a collage of distorted sounds and fragments of language and non-language, he captures the experience of a woman having and then recovering from a stroke; the rather static play is better suited to its original medium than to the stage. David Rabe proved to be a one-subject playwright with nothing to say after his three powerful Vietnam plays. *In the Boom Boom Room* (1974), an attempt to turn the sordid and pitiful life of a go-go dancer into a metaphor for the American spiritual condition, does not make the evocative connection; and *Hurlyburly* (1984), about a group of men whose treatment of each other and particularly of their women is marked by an infantile need for immediate gratification and a Mamet-like ability to convince themselves of their depth and sincerity, remains on the rather unattractive specific level, with no resonances at all.

Terrence McNally had turned from social commentary to simple farce in the early 1970s, and *It's Only a Play* (1985) seemed to continue that process, skewering the pretensions, despairs and hypocrisies at a Broadway opening night party with delightfully bitchy humour. But McNally's next two plays have more substance. *Frankie and Johnny in the Clair de Lune* (1987) is one of the best of a subgenre of plays about unconventional but deeply sincere lovers slowly breaking through the defences of emotionally wounded women, in what is something of a *leitmotif* in American drama of the 1970s and 1980s; others include Edward J. Moore's *The Sea Horse* (1974), Lanford Wilson's *Talley's Folly* (1979) and *Burn This* (1987), Mark Medoff's *Children of a Lesser God* (1979), William Mastrosimone's *The Woolgatherer* (1980), Gardner McKay's *Sea Marks* (1981), Sam Shepard's *Fool for Love* (1983), and John Patrick Shanley's *Danny and the Deep Blue Sea* (1984). We meet McNally's couple in the middle of the sexual act on their first date, and then watch as the man tries to change the tone of the evening from lust to love. At first she is incredulous, thinking he's just making ritual sweet talk; then annoyed, thinking he is patronizing her; then annoyed at the discovery that he is serious; then threatened by his unrelenting attack on her scepticism; then frightened by the exposure of her own loneliness and need for love; and finally won over. The play is alternately amusing, touching and emotionally disturbing but, like others of its type, it ultimately affirms the romantic fantasy that even in the twentieth century sincere, honest emotion, however unpoetically expressed, will eventually lead to happily-ever-after.

In *The Lisbon Traviata* (1989), McNally seduces us with a lovingly amused look at a harmless eccentricity, only to shock us later with the exposure of its underside. Stephen and Mendy are homosexual opera buffs of the sort that combine flamboyant, self-dramatizing displays of emotion with real obsessiveness, collecting obscure bootleg recordings of Maria Callas and delighting in their erudition in this very narrow field. But a first act of almost pure comedy is followed by the discovery that Stephen's lover Mike is leaving him for another man; Stephen's obsessive hobby is also a way of distancing himself from real life and real emotions, and Mike is willing to settle for a lesser man who will be emotionally present for him. The shock and pain almost draw Stephen back into reality, but in the end he is left alone with his music. *The Lisbon Traviata* had a particularly complex production history, beginning as only the first act, and going through several significant revisions afterwards; and McNally never really succeeded in integrating the two very different acts into one play. And yet, even as a loosely interrelated double bill of one comedy and one drama, it presents a sympathetic reminder that even the most apparently happy in the gay world can be lonely and frightened.

Lanford Wilson

Lanford Wilson's plays after 1975 build on the strengths he had discovered through his earlier experiments, particularly the ability to depict the complex emotions and relationships of a group of characters through a domestic realism given a lyrical tone by a musical and poetic use of language. Wilson is something of an heir to Tennessee Williams in this regard, though both his prose poetry and his view of life are generally much more subdued and gentle. Themes introduced in such earlier plays as *The Hot l Baltimore* are developed further, particularly Wilson's respect for love and commitment in whatever form they take and his awareness of the folly and danger of discarding the past.

5th of July (1978), one of Wilson's best plays, mixes humour and sentiment in the Chekhovian mode of *The Hot l Baltimore* and *The Mound Builders*, filling the stage with characters each of whom is following his own agenda, and finding that the most dramatic effects come from small commitments and betrayals. Kenneth Talley, legless Vietnam veteran, has returned to his family estate in Missouri with the plan of selling it and moving on. Also in residence are his lover Jed, his sister June and her precocious daughter Shirley, and his eccentric Aunt Sally. They are visited by Ken's former student friends John and Gwen, who may buy the house to convert into a recording studio for Gwen's as-yet-imaginary career as a singer. Each of these characters is the centre of his or her own drama: June is sorting out her unresolved sexual tensions with John; John is cold-bloodedly exploiting his rich wife; Gwen is in the overdue process of discovering an identity and purpose for herself, somewhat in parallel with Shirley; and so on. Wilson keeps all these individual stories going and makes it clear that none of the characters would admit to being a background figure in someone else's drama, while still guiding our focus onto Ken. His desire to sell the house is a symptom of a central void in his character, the inability to accept responsibility for his own life and the fear of commitment to anything because it may not be perfect; June accuses him of being 'the only person I know who can say "I'm not involved" in forty-five languages'.

Ken's experience in the play is the slow acceptance of his need to stop drifting and define himself, and of his need for, and ability to rely on, Jed's love and support. Each of the characters and subplots is able to serve him in this process. From his anger at John's perfidy he discovers his ability to feel a commitment to his values; from Sally he learns the strength of character that is his inheritance; in Gwen and Shirley's comic attempts at self-definition he sees challenges being faced without the fear of being ridiculous. And when he falls down and Jed is there instantly to protect and help him, he accepts his dependence with the security that it will not be exploited.

Wilson confirms Ken's discoveries with several evocative symbols and images. The garden Jed is planting is a commitment that will take years to complete, and the lost species of rose he has rediscovered proves the Talley land a source of life and hope. Aunt Sally scatters her dead husband's ashes in the rose garden, offering the future nourishment from the past. Ken himself recognizes that the mildly scatological myth told by Gwen's musician, in which Eskimos die because their meat is spoiled, is wrong: 'Don't choke on it, don't turn up your nose, swallow it and live, baby.' And it is he who reads the end of a student's science fiction story:

> After they had explored all the suns in the universe, and all the planets of all the suns, they realized that there was no other life in the universe, and that they were alone. And they were very happy, because then they knew it was up to them to become all the things they had imagined they would find.

That the task of becoming yourself is a precious opportunity rather than a burden is a very positive statement for a play about cripples and misfits to make; that the strength to meet this challenge can come from the past and from an unjudgemental acceptance of support in any form is an addition particularly characteristic of Wilson.

The Aunt Sally of *5th of July* and the husband whose ashes she scattered are the subjects of *Talley's Folly* (1979): in a boathouse on the Talley estate in 1944, Matt Friedman courts Sally Talley, and the two emotionally damaged people discover that their common neuroses and fears paradoxically seem to guarantee their happiness. Matt opens the play by asking the audience for its support in his quest for Sally's heart – 'If everything goes well for me tonight, this should be a waltz, one-two-three, one-two-three; a no-holds-barred romantic story, and since I'm not a romantic type, I'm going to need the whole valentine here to help me' – and from then on this little play pretends to be no more than the small, sweet love story it is. It fits comfortably in the context of Wilson's other work in both its quiet poetry and its conviction that in a sometimes cruel world the less-than-perfect is not to be disdained as a refuge.

In *Angels Fall* (1982) Wilson uses an openly artificial device, an accident at a uranium mine, to bring a disparate group of characters into the same place so they can react to each other: a priest, an Indian doctor, a professor recovering from a nervous breakdown, and a rich woman and her lover-protégé. In the course of an afternoon they each slowly emerge from their self-absorption to recognize similarities to the others, and to rediscover their own strengths through the comparisons. Niles's breakdown came when he realized that he did not believe the absolute judgements on which his books and lectures were based, and Father

Doherty recognizes this as a loss of faith. The tennis player Zappy's hypochondria and pre-tournament nerves are a trivial and comic counterpart to Don's fear that Indian health problems are too great to be tackled and Niles's conviction that his life is falling apart. Niles's despair that the academic world is no longer an ivory tower protected from reality is reflected in Doherty's accusation that Don is drawn to laboratory work as an escape.

In trying to convince Don that he has the equivalent of a priestly calling to serve his people through medicine, Doherty evokes an image with resonance for all of them. Niles's commitment to teaching is a calling, and Doherty points out that he never actually quit his job and will almost certainly return to it. Even Zappy can respond to the word, in an eloquent account of how the discovery of his athletic talent changed and defined his life. The final realization is Doherty's as he admits that the calling he is trying to force on Don is his own, and that the doctor must be free to define himself. These rediscoveries of the meaning in their lives do not radically change any of the characters; nobody leaves the play cured or significantly stronger for the experience. All they can really hope for is the ability to carry on, and late in the play Niles and his wife discover in the middle of a witticism that they have it:

> NILES I seem to be coming apart at the seams on you, don't I? Everything is coming unglued all at once.
> VITA Not quite.
> NILES Not quite all at once, or not quite everything?
> VITA Not quite unglued.

'The only good thing that comes from these silly emergencies, these rehearsals for the end of the world,' concludes Doherty, 'is that it makes us get our act together.' Like many previous American plays, *Angels Fall* searches for reassurance that life can be endured, not triumphed over; and like some, it finds that capacity within the ability of the ordinary person.

Talley & Son (1985; an earlier version was staged in 1980 as *A Tale Told*) returns to Missouri on the same night as *Talley's Folly*, letting us see what is going on back at the house while Matt is courting Sally: essentially a lot of mutual backstabbing among the members of this power- and money-hungry family that resembles Lillian Hellman's Hubbards, though without their dramatic vitality or allegorical power. Wilson seems to have written this play for his own benefit, to give an off-stage reality to the other two Talley plays, without realizing that the story had little to say on its own. Most of the characters, particularly the women, are underwritten; and a comment that the soldiers who will soon be returning from the Second World War will not recognize America because it has changed so much does not really define the play, since the Talleys have not been identified with past or future. Sally's appearance at the beginning and end of the play helps make the rather weak point that she is wise to leave her family for Matt, but that confirms this play's status as little more than a footnote to *Talley's Folly*.

Burn This (Los Angeles, 1987) is an uncharacteristic departure for Wilson, a venture into the world of violent passions and bizarre behaviour more natural to Sam Shepard or David Mamet, and his success in this unfamiliar territory is only partial. The story of the uncontrollable sexual attraction between a somewhat repressed dancer and a near madman suffers from Wilson's inability to make them believably part of the same reality. Anna is a collection of attitudes and clichés (the sensitive but insecure artist, the brittle New Yorker, the sophisticate conquered by a man's tears, etc.) rather than a recognizable person. Pale is a more consistent and individual character, if only because he is defined by his violent self-contradictions: the passionate man driven mad by urban life. Even his most disconnected behaviour, such as interrupting a string of obscene curses at a half-dozen targets to take off his trousers because he is afraid of wrinkling them, is believable if you acknowledge the existence of such people; and certainly his uncontrollable careering about the stage has a theatrical excitement to it. But, for all their typical New York City neuroses, Anna and the secondary characters function on a moderate emotional plane, with no evidence of a capacity for intensity or obsession on Pale's level. Thus *Burn This* does not convince as the story of a mutual surrender to boundless passions, because no one in the play has boundless passions except Pale, and he is clearly an aberration. Instead, the audience is likely to experience it as the story of an emotionally vulnerable girl 'pillaged and raped' (her words) by the Byronic magnetism of an ultimately dangerous alien, and to believe her expressions of helpless foreboding at the end more than his.

It might be instructive to experience this play alongside Sam Shepard's *Fool for Love* which, despite flaws of its own, does succeed in depicting two people trapped by their obsessive emotions. Overpowering passion and aberrant behaviour are not really within Wilson's natural scope; his quietly poetic and Chekhovian style is better suited to capturing the chivalric yearning of Bill for The Girl in *The Hot l Baltimore*, the fairy tale romance of Matt and Sally in *Talley's Folly*, or the understated steadfastness of Jed's devotion to Ken in *5th of July*. Even more, the value of his most successful plays is in the way they gently remind us of truths and values we have lost sight of: the beauty of the past, the comfort of friendship, and the folly of rejecting any source of emotional support because it is imperfect.

Sam Shepard

During the 1970s Sam Shepard gradually abandoned the collage-like style of his early plays for a somewhat more consistent domestic realism, in which bizarre events might take place but the dramatic modes or reality levels would not change or overlap without warning. Among the remaining plays of the earlier style, *Angel City*, *Suicide in* B^b and *Seduced* are the most interesting. Each is at its core a criticism of some aspect of American culture, and each uses symbolism or violations of reality to dramatize its attack. In *Angel City* (San Francisco, 1976) Hollywood producers reaching for the ultimate special effect succeed only in turning themselves into monsters; it thus follows such earlier plays as *Red Cross*, *Icarus's Mother* and *Back Bog Beast Bait* in warning against the irresponsible use of the imagination. In *Suicide in* B^b (New Haven, 1976) a musician symbolically kills off the various false identities others have imposed on him, the play condemning the public's tendency to project its own needs, fears and fantasies on the artist. The central character of *Seduced* (Providence, 1978), based on the billionaire recluse Howard Hughes, is ancient, unkempt, half-mad, and living only in the fantasy-memories he pays others to supply for him. Like the other two plays, *Seduced* has no real plot or character development, but is an extended metaphor for the state of America ('I'm everywhere!. . . I'm all over the country . . . . A phantom they'll never get rid of . . . . Everything they ever aspired to. The nightmare of the nation!'), an image of the pioneering and entrepreneurial spirit turned into a lifeless but indestructible obscenity.

Curse of the Starving Class (London, 1977) is the first play in what would become Shepard's mature style; there are no magical phenomena, sudden character transformations or mixtures of reality and myth, just a family in their home, experiencing the events of a two-day period. Granted, family, home and events are all unusual, but they are not impossible; and when the play offers a metaphor for the human condition, it is through the representative power of realism, the same method used by Clifford Odets, Eugene O'Neill and the mainstream of American drama. Its subject, implied by the title, is a gnawing hunger, but not for food; Shepard repeatedly jokes with that image by bringing abundant quantities of food on stage. His characters are starved for some modest level of fulfilment, for just the chance at happiness, self-definition or self-esteem. Everyone in the family suffers from a sense of incompleteness, which Wesley blames on the faceless 'they' who run the world. Weston identifies them more explicitly with the institutions of American capitalism: 'I figured that's why everyone wants you to buy things. . . . They all want you to borrow anyhow. Banks, car lots, investors. . . .

So I figured if that's the case, why not take advantage of it? Why not go in debt . . . ?' So, on one level, *Curse of the Starving Class* is a variant on *Death of a Salesman*, an indictment of the American system for awakening a hunger it does not fill, and for then punishing the hungry.

The curse seems inescapable: Ella follows the empty promises of a trickster and is cheated; Weston dreams of a new start in Mexico, but the debt collectors are close behind; Emma's fantasized life of crime is just a further commitment to capitalism ('the perfect self-employment') and therefore doomed. Since this is a Sam Shepard play, it is not surprising that the one faint and ultimately chimerical hope lies in a reintegration with tradition and the land. Weston explains how a return to the farm gave him back his sense of self: 'Then it struck me that I actually was the owner. That somehow it was me and I was actually the one walking on my own piece of land. And that gave me a great feeling.' But Weston must run away again, to escape his creditors, and Wesley can only remain by putting on his father's clothes and identity; 'And every time I put one thing on it seemed like a part of him was growing on me. I could feel him taking over me.' Losing one's identity is not a solution to the problem of achieving one's identity; and Shepard can only end the play with the story of a fight between a cat and eagle, an unending and unwinnable struggle to the death.

Arguably Shepard's greatest play, *Buried Child* (San Francisco, 1978) offers a powerful dramatic metaphor for both the corruption of the American spirit and the hope for its salvation. In an assertively ordinary midwestern farmhouse, what would otherwise be an all-American family is marked by spiritual and physical decay. Grandfather Dodge is sick and virtually immobile; his wife Halie flirts drunkenly with the local priest; one son is crippled and the other brain-damaged. Grandson Vince is frightened by the family's bizarreness, but he is unable to deny his ties to them, and at Dodge's death Vince assumes his persona. Ominous hints are dropped through the play of a hidden scandal, eventually revealed as the incestuous child of Halie and her son Tilden, which the family killed and buried; in the final moments Tilden enters with the exhumed body as Halie describes a miraculous abundance of vegetables springing up from the hitherto dead farmland.

Except for those magical crops, the play is wholly realistic in style; these may be odd characters behaving oddly, but they could exist. But the play does not use domestic realism in the manner of Odets and Miller or of *Curse of the Starving Class*, addressing a larger subject through demonstrating its effects on a specific domestic situation. Rather, Shepard here adapts the method of *Operation Sidewinder*, *Seduced* and others of his non-realistic plays, making the specific stand directly for the general through the power of metaphor. This household is not America typified, but America personified: Tilden is identified as a former All-American athlete; the midwestern farm setting automatically suggests a universal;

and Shepard directly invokes the names of Norman Rockwell (the early twentieth-century illustrator who specialized in heartwarming Americana) and the 'Dick and Jane and Spot' of children's books.

If this family is what Norman Rockwell's America has become, it is as the direct result of a loss of connection to the past and a refusal to acknowledge its own past failings. The family exists on denial; Dodge will not admit he recognizes himself in old photos, Vince tries to run away from the family that does not live up to his fantasies, and Bradley insists, 'Nothing's wrong here! Nothin's ever been wrong! Everything's the way it's supposed to be! Nothing ever happened that's bad! Everything is all right here! We're all good people!' The central symbol of both dissociation and denial is the dead child: 'It wanted to grow up in this family. It wanted to be just like us. It wanted to be part of us. We couldn't allow that to grow up right in the middle of our lives. It made everything we'd accomplished look like it was nothin'. . . . I killed it.' One might hear echoes of David Rabe's *Sticks and Bones* in that speech, and Shepard does not have to press the metaphoric meaning any further. Dozens of examples will come to mind of the American cultural compulsion to bury our sins and to rewrite history so that we are always the good guys: slavery, the Indians, Vietnam. To preserve the family's sense of itself as pure, the family does the most un-family thing possible, killing a child, and thus destroys its capacity to be a family. By committing atrocities to deny its capacity to commit atrocities, America has decayed into an obscene parody of itself.

But the process is not irreversible. Vince runs away, but discovers that the past cannot be denied: 'I drove all night. . . . I could see myself in the windshield. . . . As though I was looking at another man. . . . And then his face changed. His face became his father's face. . . . And his father's face changed to his Grandfather's face. And it went on like that.' Through courage, honour or madness – and Shepard does not really seem to care which – Vince accepts his connection to the past and returns, declaring himself the new owner of the farm. At virtually the same moment Tilden enters with the exhumed child, and Halie's offstage voice ends the play:

> I've never seen such corn. . . . Tall as a man already. This early in the year. Carrots, too. Potatoes. Peas. It's like a paradise out there. . . . It's a miracle, Dodge. I've never seen a crop like this in my whole life. Maybe it's the sun. Maybe that's it. Maybe it's the sun.

That last word is a pun, of course, referring to both Vince, who is going to save the family by drawing on his spiritual inheritance (earlier Halie called him 'an angel. A guardian angel. He'd watch over us. He'd watch over all of us.'), and to the baby, who will now finally be acknowledged and dealt with. Neither offers an easy or costless salvation; Vince appears to sacrifice his youth and even his sanity, sinking into his grandfather's immobility, while the baby will bring the shame and scandal it was killed to avoid. But this difficult, painful and uncertain path is the path to spiritual rebirth, as the sudden flowering of the farm makes clear. And on the metaphoric level Shepard repeats the assertion that is a *leitmotif* of his career: that the future of America lies in a reintegration with the best, and an acknowledgement of the worst, in its past.

In *True West* (San Francisco, 1980) Shepard invokes mythic imagery and power without once violating the bounds of domestic realism. Aspiring screenwriter Austin and his criminal, semi-literate brother Lee are trying to write a script based on an idea of Lee's. They ultimately fight, and Austin is trapped in the position of victor, unable to kill his brother but afraid to let him go. Much of the theatrical power comes from the simple display of animation; as in some other Shepard plays, notably *Fool for Love*, the characters spend a good part of the play careering violently about the stage, unable to control the raw energy within them. But beneath this surface the play is an inquiry into the subject of its title: is the uniquely American myth of the West valid and, if not, what is the reality that displaces it? (Clearly this question is of more than passing significance to Shepard, whose reverence for the mythic past drives so many of his plays.) Lee's script idea is a pure chase story, one man following another across country in a drive for vengeance with no clear beginning or end:

> So they take off after each other straight into an endless black prairie. . . . What they don't know is that each of 'em is afraid, see. Each one separately thinks that he's the only one that's afraid. . . . And the one who's chasin' doesn't know where the other is taking him. And the one who's being chased doesn't know where he's going.

That account, which ends the first act, is eerie and powerful, but the question remains whether it is 'true'. The answer comes in the final fight between the brothers, when Austin is temporarily the victor and then tries to find a way out of the impasse: 'I'll make ya' a deal. You let me outa' here. Just let me get to my car. All right, Lee? Gimme a little headstart and I'll turn you loose. Just gimme a little headstart. All right? . . . ' At that moment, without violating any of the rules of dramatic realism, *True West* leaps from domestic melodrama into myth and metaphor. What will follow this tableau, it is perfectly clear, is the real-life enactment of Lee's script. The fiction has come alive, and the living people have become characters in the fiction. Myth and reality combine, the True West encompassing both of them, as once more Shepard asserts his conviction that the dream of the American West is a truth that it is dangerous folly to deny.

Fool for Love (San Francisco, 1983) is a theatrical study in inescapable and uncontrollable human passions, as Eddie and May, long-time lovers and, we eventually learn, half-brother and sister, are trapped in a can't-stay-together-can't-stay-apart obsession. Even more than *True West*, the play is driven theatrically by the high energy Shepard demands in performance; his opening stage direction stresses that '*This play is to be performed relentlessly without a break*', and Eddie and May do nothing at half-steam. At various times she '*erupts furiously*', '*stares at him wild-eyed*', gives an '*agonized scream*', '*slams wall with her elbow*', '*pins him with her eyes*', is carried '*screaming and kicking*', and '*starts lashing out*'. He '*bangs his head into wall*', '*moves violently toward her*', '*crashes into stage-right wall*', '*drags chair violently*', and '*crashes to the floor*'. Both of them slam all doors repeatedly, and they take turns moving around the edges of the room, hugging the walls as if testing the power of the physical boundaries to hold them in. Quiet, controlled emotions are not within their repertoire; a '*Long, tender kiss*' is followed immediately by a vicious knee to the groin. *Fool for Love*, like many of Shepard's other plays, functions primarily as metaphor; it is not the narrated story of Eddie and May that the play is about, but their dramatized passion – and, by extension, any passion. The image of the motel at the edge of the desert, the pitifully small and fragile enclosure in the vastness with people bouncing off its inner walls, stands for the inevitably doomed attempt to keep infinite human emotion within the finite human frame. To be human, says the play, is to be made of forces so large they cannot be controlled, so essential they cannot be escaped, so extreme and contradictory that tenderness cannot be separated from violence or love from hate, so inherent in the fact of being human that their story can have no real beginning or end.

A Lie of the Mind (1985) has an almost O'Neill-like theme, the superiority of fantasy over realism as a means of coping with life. Insanely jealous Jake has viciously beaten his wife Beth, leaving her brain-damaged. He retreats to his family and to madness, while Beth slowly recovers in the bosom of her own dysfunctional family. When Jake's brother Frankie visits, Beth fantasizes him into a kinder, repentant Jake. Jake believes that Beth and Frankie are lovers, but in his madness he finds an escape from his jealousy, and he gives them his blessing. Once again Shepard builds his play on a metaphor, this time Beth's brain damage, which stands for the dissociation from self and distortion of relationships that is everyone's condition: Jake and Mike both see Beth only as extensions of themselves (Beth to Mike: 'You have a feeling I'm you.'); Lorraine, Sally and Meg all impose fictional identities on others and then relate to them (Beth to Meg: 'You don' know. Only love.'); Baylor withdraws ('This is my father. He's given up love.'). Thus, just as Beth must fight her way back to the use of language and coherent thought, everyone else in the play must find some means of defining self and surroundings; and for virtually all of them the process paradoxically involves a fantasy or escape from reality. By imagining Frankie to be a kind and loving Jake, Beth can find peace; by withdrawing into madness, Jake escapes his jealousy. Mike takes comfort in the role of protective and avenging brother; Sally and Lorraine run away to an imagined Utopia; and Meg and Baylor find serenity in the rituals and habits of a long marriage. Only Frankie, the one character trying to deal with the problems rationally and realistically, is left confused and alone, unable to take Beth's counsel: 'Pretend. Because it fills me. Pretending fills. Not empty. Other. Ordinary. Is no good. Empty. Ordinary is empty.'

It is striking that Shepard capped a long period of stylistic experiment with a return to domestic realism in his later plays. Of course his brand of realism is not the same as anyone else's, with unusual and even bizarre characters acting in ways that are not fully explained. But, with the occasional small exceptions, such as the magic vegetables in *Buried Child*, his best plays since *The Tooth of Crime* do not reach for new theatrical vocabularies, mix metaphors, offer isolated images in place of continuity, violate the laws of reality or change inexplicably from the tone and dramatic genre in which they begin. If the best of them – *Buried Child*, *True West*, perhaps *Fool for Love* – achieve mythic resonances, it is through exploiting the power of realism, not escaping from it. Like Elmer Rice, Maxwell Anderson, Eugene O'Neill and others before him, and like Lanford Wilson alongside him, Shepard found his way back to the American drama's natural mode.

David Mamet

David Mamet brings to the American theatre a particular sensitivity to the ways in which language is used, apart from its communicative function, to give shape and meaning to its users' lives. His characters tend to come from closed worlds with their own cultures, myths, codes and jargon – old men, swinging singles, minor criminals, high-pressure salesmen, Hollywood hustlers. Their lives are marked by frustrations and feelings of impotence that are likely to explode in bitter displays of rage unless they are controlled or diverted through something that gives the characters a reassuring sense of dignity: affectations of wisdom, of honour, or of moral superiority. In most cases, all it takes to achieve this comforting sense is language that *sounds* wise, honourable or moral; for Mamet's characters form is more convincing than content. Early in his career Mamet was celebrated for the accuracy of his ear and the perfection with which he captured the speaking styles of each subculture. Later it became apparent that Mamet-language is as artificial and stylized in its own way as, say, Tennessee Williams's lush prose poetry: a heightened style that captures the essence of its speakers through intensification rather than stenographic transcription.

The importance of self-definition and the use of language to achieve it are both apparent in Mamet's first successful play, the brief *The Duck Variations* (Plainfield, Vermont, 1972). In a series of short scenes two old men sitting in a park exchange inanities and misinformation on a variety of subjects, including ducks. Though virtually everything they say is empty or wrong, the ability to express themselves in the rhythms of epigrams or the vocabulary of profound thought is sufficient to give them a sense of value and dignity:

> The battle between the two is as old as time. The ducks propagating, the Herons eating them. . . . Each keeping the other in check, down through history, until a bond of unspoken friendship and respect unites them, even in the embrace of death. . . . Survival of the fittest. The never-ending struggle between heredity and environment. . . . Who can say to what purpose? . . . We do not know. But this much we *do* know. As long as the duck exists, he will battle day and night, sick and well with the Heron, for so is it writ.

Of course that does not mean anything, but it sounds as if it did, and that is enough to satisfy both speaker and listener.

In *Sexual Perversity in Chicago* (Chicago, 1974) this kind of meaning-in-empty-language is shown to have a dark side. The perversity of the title is a normal, healthy sexual romance which is broken up when the couple's friends invoke the codes of their respective societies. Bernie is a self-proclaimed swinger whose reduction of women to sex objects barely disguises a fear and hatred of their sexual power ('Coming out here on the beach. Lying all over the beach, flaunting their bodies.'). Joan is a feminist who is afraid of sex ('Look at the incidence of homosexuality . . . the number of violent, sex-connected crimes . . . all the anti-social behaviour that chooses sex as its form of expression.'). Separately they use their mastery of their jargons to lure Dan and Deborah away from their rather charming romance and back into their old habits of thinking:

> BERNIE You getting serious? I mean she seemed like a hell of a girl, huh? The little I saw of her. . . . So what can you tell from seeing a broad one, two, ten times. You're seeing a lot of this broad. . . .
>
> JOAN . . . the fact is if you take a grown man whose actions and whose outlook are those of a child, who wants nothing

> more or better than to have someone who will lick his penis and grin at his bizarre idea of wit. . . .

Eventually what *sounds* right becomes more real to the lovers than what feels right, and the play ends with their separation, Deborah blaming herself for falling, while Dan joins Bernie in ogling and cursing the girls on the beach.

That language creates identity and also helps protect one from feelings of inadequacy and rage is central to *American Buffalo* (1977). Three minor criminals planning a robbery are so insecure in their relationships to each other, and so incompetent as criminals, that the jockeying for power and the sheer logistics of the job overwhelm them. One finally blows up in rage and frustration, but then finds his way back, with the others, to the level of calm and friendship at which they began. The title refers to a valuable coin that triggered the plan; it may suggest that these characters, and their view of themselves as outlaws, are pitiful anachronisms. The men speak a *patois* that resembles the dialect invented by Damon Runyon for his stories of the New York underworld in the 1920s: almost consciously artificial, combining extra-correct grammar and courtesy with vulgarity, obscenity and a certain amount of verbiage designed merely to fill the empty spaces. For those who really have nothing, it is essential to *sound* as if they have values ('You're right, and you do what you think is right, Don.'), dignity ('No, I am sorry, Don, I cannot brush this off. They treat me like an asshole, they are an asshole.'), and wisdom ('I want to tell you something. . . . Things are not always what they seem to be.'). Language gives them the confidence that they are capable of pulling off this job, even as their bumbling proves that they are not – that is to say, language gives them the confidence that they are worth something, even if the facts of their lives suggest otherwise. And, significantly, when things fall apart it is the language of forgiveness and reconciliation that leads them to the belief that they have achieved reconciliation, and thus to peace: 'I fucked up.' – 'No. You did real good. . . . ' – 'Thank you.' – 'That's all right.'

As with some of Sam Shepard's plays, it is the danger of barely repressed violence that drives *American Buffalo* and several other Mamet plays in the theatre, but what lingers in the audience's mind is likely to be the hint of Tennessee Williams and Eugene O'Neill in the picture of characters frightened by life and unsure of their ability to cope with its ordinary demands. Several of Mamet's lesser plays have this same underlying observation: *A Life in the Theater* (Chicago, 1976) traces the small rise of one actor and the small, sad decline of another through the subtle power shifts in their backstage conversations; *The Woods* (Chicago, 1977) shows a clearly incompatible couple driven by their fears to stay together; *Edmond* (Chicago, 1982) shows a man conspiring in his own degradation because the only shelter from the pain of existence is a complete loss of ego. Other minor plays draw their energy from surprising and enlightening juxtapositions of language. In the radio play *The Water Engine* (1977) the exalted and optimistic language of a World's Fair guide and a riches-promising chain letter clashes ironically with the sordid tale (and classic American legend) of big business suppressing a lone inventor's discovery. *Prairie du Chien* (radio, 1979) juxtaposes a ghost story of high passions and violence with the more mundane emotions of reality; *Lakeboat* (Milwaukee, 1980) lets the inarticulate talk of a shipload of sailors create a sense of the milieu; and *The Shawl* (Chicago, 1985) enjoys the gap between a charlatan fortune-teller's professional patter and his plain-spoken explanations of the tricks of the trade.

The subculture of *Glengarry Glen Ross* (London, 1983) is that of high-pressure land salesmen. Two familiar Mamet themes are reiterated: the degree to which their language and thought processes limit and control them, and the importance to such marginal men of the fantasy that they have a calling and a code that set them apart from others. The play opens with brief scenes that constitute a virtual textbook lesson in selling modes: begging his boss for better 'leads', Levene ducks and weaves, trying to anticipate and counter every objection – ' . . . all I'm saying, put a *closer* on the job. There's more than one man for the . . . Put a . . . wait a second, put a *proven man out* . . . and you watch, now *wait* a second' [ellipses Mamet's] – while Moss, luring a colleague into a criminal plot, uses the trick of making him think he has already agreed; and Roma disguises his pitch so well that we think he is just rambling aimlessly to a stranger until 'You get befuddled by a middle-class morality' leads somehow to 'I do those things which seem correct to me *today*' and thence somehow to 'This is a piece of land. Listen to what I'm going to tell you now.'

But these skills are all that these men have, and they are forced to make them a basis of pride and self-definition, for want of any other material. When through the office manager's interference Roma loses a sale, the identification of salesmanship and manhood bursts out:

> ROMA You stupid fucking cunt. . . . Whoever told you you could work with *men*? . . . What you're hired for is to *help* us . . . to help *men* who are going *out* there to try to earn a *living*. You *fairy*. . . .
>
> LEVENE You have to learn it in the streets. You can't *buy* that. You have to *live* it. . . . 'Cause your partner *depends* on it . . . . you're *scum*.

And yet, despite this pride and sense of comradeship, these men cannot help cheating, lying and manipulating; it is what they do. The lead Williamson makes Levene demean himself for is a fake; Moss pushes his robbery plan until he finds someone he can con; Roma and Levene instantly invent a complex lie to befuddle a wavering customer; and

Roma, after declaring undying admiration for Levene, tries to cheat him out of commissions. The skills and instincts that give these characters their sense of identity and worth also cripple them, making them incapable of honesty, honour or even a simple, unpremeditated conversation.

Speed-the-Plow (1988) tells much the same story, this time in the somewhat less life-or-death context of Hollywood, as a producer and studio executive con each other with unctuous flattery and insincere intimacy. When a secretary tries to advance herself by joining the game, the producer whose project she threatens uses his superior skill to destroy her. Many of Mamet's familiar themes are here: the comradeship of fellow professionals and contempt for the outsider; the use of language that has the sound and rhythm of meaningful statement as a substitute for meaningful statement; the inability of the born conman to stop conning, with the paradox that the most skilled conmen may be the most easily cheated; and the need of these shallow characters to believe they are deeper and more sensitive than they really are. What is new, or has not been so apparent since *The Duck Variations*, is the humour; it is not only psychologically insightful that the men can convince themselves that they are deep thinkers and sensitive human beings –

> GOULD Charl, I just hope.
> FOX What?
> GOULD The shoe was on the other foot, I'd act in such a . . .
> FOX . . . hey . . .
> GOULD Really, princely way towards *you*.
> FOX: I *know* you would, Bob, because lemme tell you: experiences like this, *films* like this . . . these are the films . . .
> GOULD . . . Yes . . .
> FOX *These* are the films, that whaddayacallit . . . [*long pause*] that make it all worthwhile. [Ellipses Mamet's]

– it is also devastating satire. And still there is a point being made: if even these characters must play psychological and linguistic tricks on themselves to give themselves a sense of worth, then that need and that method must indeed be universal. Mamet may have a tendency to tell the same story over and over, but it is a good story, and he tells it with style, insight and theatrical vitality.

August Wilson

The most important new American dramatist of the 1980s is August Wilson, a black writer who has taken upon himself the task of dramatizing the experience of black Americans in the twentieth century through a series of plays, one for each decade (though not offered in historical order). It is part of Wilson's conceit that each play is written from the point of view of that decade, without the comforts or ironies hindsight might provide. So successful has Wilson been in this project that each of the four plays produced in the 1980s is both a powerful drama and a startling illumination of the black experience.

From a later perspective, the 1920s might be seen as a significant turning point in black American life, as the last remnants of the slavery world slowly gave way to the promises of the new century. But Wilson's historical insight is that only one of those movements was apparent at the time: clearly something was ending for black Americans, but there was no sense of anything replacing it. *Ma Rainey's Black Bottom* (1984) is set in 1927, as the blues singer Ma Rainey (a historical figure, fictionalized in the play) records some songs, and her musicians rehearse and kill time. One of them, Toledo, expresses the era's sense of black history:

> Everybody come from different places in Africa, right? Come from different tribes and things. Soonawhile they began to make one big stew. . . . Now it's over. . . . The colored man is the leftovers . . . . Done went and filled the white man's belly and now he's full and tired and wants you to get out the way.

The music that fills the play (which is not the familiar my-baby-left-me slow blues but the earlier, very earthy and life-affirming style of the 1920s) is a powerful identity-defining force, and the source of Ma Rainey's power:

> I always got to have some music going on in my head somewhere. It keeps things balanced. . . . White folks don't understand about the blues . . . You don't sing to feel better. You sing 'cause that's a way of understanding life. . . . The blues help you get out of bed in the morning. You get up knowing you ain't alone.

But it is also a commodity being bought and abused by the white world. Ma is aware of this, and has the intelligence to make sure she is paid well for selling her birthright: 'As soon as they get my voice down on them recording machines, then it's just like if I'd be some whore and they roll over and put their pants on.' The younger musician Levee is too blinded by ambition to realize that his music will be taken from him; and when the white man steals his songs and denies him the chance to record, he has no philosophy or historical perspective to fall back on. He experiences in an overpowering instant what Ma and Toledo had seen and accepted: the apparent end of a culture and a race.

Lashing out in frustration, he kills Toledo, ironically and tragically destroying his last connection with his racial heritage. If history has shown that things actually (if slowly) improved for black Americans after the 1920s, Wilson makes us understand that an unforeseeable future offered no comfort to a people whose only experience was the loss of a rich and life-defining past.

In *Fences* (1987), set in 1957, Wilson again chooses a historical moment which hindsight recognizes as coming just before a period of positive change for black Americans, and again he makes us see that the future was not apparent at the time. Troy Maxton represents the first generation of black Americans to progress into the middle class through pride and determination, but his instinct is to consolidate and protect what he has rather than risk it with further ambition; he sabotages his son Cory's hopes of college because they seem dangerously over-reaching. Troy *has* accomplished a great deal, and one of the play's purposes is to remove our memory of the even greater advances to come in the next decade so we can appreciate the plateau black Americans reached in the 1950s. But those who could only see how far they had come would have difficulty imagining an even grander future; and the inevitable clash between father and son becomes especially painful when both have reason to believe themselves in the right. One of the sad ironies of the 1950s that Wilson dramatizes is the fact that those who should have been heroes of black history, the first generation to achieve security and dignity, became villains in the eyes of the next generation because their fear of losing what they had gained made them conservative. That is the purpose of the coda-like final scene, in which Cory is made to understand that his generation's victories will be achieved because of those who came before, not in spite of them.

The second irony in the play is that Troy himself is unable to appreciate and enjoy his own accomplishments. Because he can only see it as something to lose, Troy's success weighs him down with responsibility, and this responsible family man is driven to violate the family in search of some respite from the burden: 'She gives me . . . a different understanding about myself. . . . I ain't got to wonder how I'm gonna pay the bills or get the roof fixed. I can just be a part of myself that I ain't never been.' Since at least as far back as *Death of a Salesman* dramatists have found a particularly American tragedy in the morality or social condition that forces a man to deny himself in order to do right. Troy Maxton fails the test – he violates his own principles in the name of self-fulfilment – and he pays the price. But like Arthur Miller, though somewhat less angrily, August Wilson recognizes that the fault lies in the times rather than the man; and, like most other plays on this theme, *Fences* will not condemn its hero-victim.

With *Joe Turner's Come and Gone* (1988), Wilson returns to the theme of lost identity, this time in personal terms. In 1911 any black American under the age of 45 was born into freedom, and even those older had spent most of their lives free. Wilson makes us see that this extraordinary advance had its price; as he explains in an introductory note, 'Isolated, cut off from memory, having forgotten the names of the gods, . . . they carry as part and parcel of their baggage a long line of separation and dispersement . . . as they search for ways to reconnect.' While the younger characters in the play function in an empty and hedonistic present, those with experience and memory look for some continuation or replacement of their lost cultural identity. The metaphor that Bynum, the conjure-man, offers for this missing element is the private song each person must create for himself out of himself and sing as a declaration of himself. Loomis, recently freed from the private chain gang of Joe Turner, has lost his: 'What he wanted was your song. . . . Now he's got you bound up to where you can't sing your own song. Couldn't sing it them seven years 'cause you was afraid he would snatch it from under you. But you still got it. You just forgot how to sing it.' Loomis rediscovers his selfhood almost by accident, in a moment of frustration at his wife's attempts to win him to her religion:

> MARTHA Jesus bled for you . . . .
> LOOMIS I don't need nobody to bleed for me! I can bleed for myself . . . .
> (*Loomis slashes himself across the chest. He rubs the blood over his face and comes to a realization.*)
> I'm standing! I'm standing! My legs stood up!

It is a significant dramatic weakness of this scene that Wilson must explain the meaning of the moment in a stage direction ('*Having found his song, the song of self-sufficiency, . . . he is free to soar. . . .*') rather than making it fully clear on stage. That, along with the heavy reliance on mysticism, makes *Joe Turner's Come and Gone* the least effective of his plays so far. Nevertheless, the central image of the personal song is powerful; and the experience being described, of the first black generation faced with the task of self-definition, is a valuable chapter in the enriched history of America Wilson is writing.

The Piano Lesson (1990) also combines a resonant and evocative central imagery with an obscuring mysticism. In 1937 a black family owns an old piano filled with reverent associations: their slave ancestors were sold for it, family portraits are carved in it, and their father died stealing it from the white family that had owned it and them. But now that man's son has the opportunity to buy the farm his family had been slaves on, and, despite his sister's opposition, wants to sell the piano to raise the needed money. The audience must respond to the piano as an almost holy relic, and share Berniece's conviction that 'Money can't buy what that piano cost', but Boy Willie's dream is equally compelling; his quietly understated comment that 'This time I get to keep all the cotton' makes it clear that his triumph will be achieved in the name of those who preceded him.

The conflict between sister and brother presents one part of Wilson's definition of the black American experience of the 1930s: an opportunity, for the particularly fortunate, to have either a past or a future, but not both. To achieve his ambition, Boy Willie must, it seems, turn his back on the sacrifices of his ancestors ('You can always get you another piano'); in order to honour those ancestors, Berniece must not only block her brother but deny the validity of his ambition ('You right at the bottom with the rest of us'). Boy Willie hints at a way out of this impasse when he shows that he does recognize the piano's symbolic value, insisting only that the way to show respect to such an inheritance is to make use of it:

> The only thing my daddy had to give me was that piano. And he died over giving me that. I ain't gonna let it sit up there and rot without trying to do something with it. . . . Alright now, if you say to me, Boy Willie, I'm using that piano. I give out lessons on it and that help me make my rent or whatever. Then that be something else. I'd have to go on and say, well, Berniece using that piano. She building on it.

At the end, when Boy Willie must fight a vengeful ghost from the white family (a theatrically weak climax, since the mystical element seems imposed on the essentially realistic play, and the fight is off-stage), Berniece saves him by playing a song calling on the spirits of their ancestors to intervene. Now that Berniece has made use of the piano, Boy Willie acknowledges her right to keep it, and leaves to look elsewhere for his money; and Wilson ends this play with the most hopeful and reassuring note of his four plays so far: if the past is not just honoured but invoked, not just respected but used to serve the present, then it need not stand in opposition to the future.

Women dramatists: Wasserstein, Henley, Norman

Almost certainly more plays by women were produced in the 1970s and 1980s than during any other period, possibly more than in all other periods combined. Those of Wendy Wasserstein, Beth Henley and Marsha Norman stand out and complement each other. Wendy Wasserstein's plays focus on the stressful and confusing experience of the generation of American women who came of age in the 1970s, the first to experience the opportunities and frustrations of the feminist era. *Uncommon Women and Others* (1977) follows a group of over-achieving young women through their years in an exclusive college as they try to sort out the conflicting signals their culture, their families and their own emotions are sending them. The college repeatedly reminds them of their opportunities and of their obligation as an intellectual élite to be in the forefront of social revolution – 'Am I saying that anatomy is destiny? No . . . . Today all fields are open to women' – but at the same time offers courses in Gracious Living and suggests that worldly success 'can happen without loss of gaiety, charm or femininity'. The highest achiever in the graduating class is applauded for choosing law school, while the weakest is advised to take a typing course. Buffeted by these contradictions, given models ranging from wife-and-mother to bluestocking to professional to 'girl friday for an eastern senator', the young women flounder helplessly. One retreats into the brainless cheerleader mode of an earlier generation; one cuddles her dolls while dreaming of the handsome doctor who will save her from having to make choices; one retreats into near-catatonia; one moves forward, without enthusiasm or real direction, along the professional path: 'Just once it would be nice to wake up with nothing to prove. . . . But if I didn't fulfill obligations or weren't exemplary, then I really don't know what I'd do.' In a frame set six years later, some of the women meet to discover that nothing has become clearer; they still think of themselves as in transit, and the rallying cry of their college days – 'When we're thirty we're going to be pretty fucking amazing' – has merely been adjusted to' . . . forty-five . . . '.

Though it has an entirely new set of characters, *Isn't It Romantic?* (1983) essentially picks up the story where *Uncommon Women* left it. Struggling to begin her career, Janie Blumberg must juggle the distractions of equally struggling and frantic friends, the intrusive and smothering love of her parents, and the allure of what once might have been every woman's dream, a rich and loving doctor who wants nothing more than to marry her and absorb her into his life. Wasserstein's point is that choices have become harder, not easier; even in the post-feminist world the prospect of surrender to a protective man is very enticing, and all the things modern women have been taught to demand as their right are draining. Janie's friend Harriet asks her apparently successful mother whether the whole feminist agenda is just a fantasy: 'Mother, do you think it's possible to be married or live with a man, have a good relationship and children that you share equal responsibility for, build a career, and still read novels, play the piano, have women friends, and swim twice a week?' Her mother's answer is that it is impossible, but that a woman who chooses one or two items from that list and is willing to forgo the others without regret can find satisfaction. Janie and Harriet make their choices, career for one and marriage for the other, and begin the move forward with very unsteady optimism.

Wasserstein's *The Heidi Chronicles* (1988) continues the saga with a new group of characters. The life of evidently successful art historian Heidi Holland is told in flashbacks starting in the 1960s: self-consciously 'sensitive' adolescence, college idealism, seduction by an attractive sexist, feminism, watching activist friends burn out or sell out, and so on. Wasserstein shows that the problems that confused women in 1970 and 1980 are still unresolved; making what is supposed to be an inspirational speech at her school reunion, Heidi confesses, 'I'm afraid I haven't been happy for some time. . . . It's just that I feel stranded. And I thought the whole point was that we wouldn't feel stranded.' Particularly depressing is the discovery that her generation of women, those who began the struggle in the early 1970s, not only accomplished little for themselves, but merely made it easier for their younger sisters to conform, accepting the material benefits of their career opportunities with no sense of debt or of responsibility to continue the effort. And yet Heidi's story, like the others, ends with a quiet optimism. Heidi has adopted a baby girl, and she closes the play with the most traditional of comforts – that she has helped make the future a little better for her daughter: 'And [a man]'ll never tell her it's either/or, baby. And she'll never think she's worthless unless he lets her have it all. And maybe, just maybe, things will be a little better. And yes, that does make me happy.'

Beth Henley's plays are generally about ordinary southerners who turn out to have unexpected quirks or askew values, so that mundane events turn into bizarre adventures and bizarre adventures into unnoticed trivia. *Crimes of the Heart* (Louisville, 1979) is the story of the MaGrath sisters of Mississippi: Babe has just shot her husband, after he found her with a fifteen-year-old black boy; Lenny has decided on the life of an old maid; and Meg has come home from her failed attempt at a singing career for an affair with her old boyfriend. Filling any gaps in this already complicated situation are their antagonistic cousin, their dying grandfather, Babe's amorous lawyer, some compromising photos, a horse struck by lightning, and the memory of their mother, who committed suicide by hanging herself and the family cat. Clearly, Henley is drawing on the same tradition of gothic and grotesque that has inspired southern writers from Faulkner to Williams, though with a particularly benign and loving humour that assures all will be well.

The basic joke of *Crimes of the Heart* is a moral and emotional displacement that makes the sisters' response to events always somehow just off-centre:

MEG So, Babe shot Zackery Botrelle, the richest and most powerful man in all of Hazlehurst, slap in the gut. It's hard to believe.

LENNY It certainly is. Little Babe – shooting off a gun.

Repeatedly a fact or plot twist that should distress them is met with utter calm or hardly absorbed, while smaller things distract their attention or engage their emotions. Childhood grudges are more real than present dangers; and grief over their mother's death is overshadowed by curiosity about why she killed the cat. The waxing and waning tensions of the moment lead them to respond with hysterical laughter to the news that their grandfather is in a coma, while the thought of the lightning-struck horse moves them to tears. But beneath these jokes is a serious vision. Meg faces her failure as a singer and her inability to win her old swain back; Lenny overcomes the embarrassment that has stifled her romantic life; and Babe, driven to a failed suicide attempt, realizes why her mother killed the cat: because death might be lonely, and she needed the company. When Meg sums up the situation by saying, 'But, Babe, we've just got to learn how to get through these real bad days here, I mean, it's getting to be a thing in our family', Henley shows that the MaGrath sisters' curiously displaced reactions to reality actually protect them from the pains of reality. Unlike their mother, they will never find life too painful, and unlike her they will never be lonely. The play ends with the sisters laughing over the pleasure of eating birthday cake for breakfast, their inherent innocence guaranteeing them happiness.

Henley's next play has much the same quality, with slightly skewed characters having bizarre experiences but ultimately protected by their inability to distinguish between the trivial and the serious and by their inherent good natures. In *The Miss Firecracker Contest* (Buffalo, 1981) a poor girl looks for some vindication for her indignities in the local beauty contest while her rich cousins cope with their separately tangled love lives. She loses the contest, the love lives are not sorted out, and a few broken hearts seem possible in the future. But all is purified by the complete absence of malice, and the characters end the play innocently enjoying a fireworks display. *The Debutante Ball* (Costa Mesa, 1987) lays the gothic element on somewhat more thickly: a mother who has lived in scandal since killing her husband plans to use her daughter's formal presentation to society as the occasion of her own social rehabilitation. But baby is panicked by the approaching ball, and it is eventually revealed that she was the murderer. Secondary characters more than fill the play with additional bizarre touches and plot twists, but the saving innocence of the earlier plays is absent here, and the play is more grotesque than comic.

Indeed, it is true of both these plays that, at their weakest, they seem made up of leftover pieces of *Crimes of the Heart*, revisiting the same world and recycling the same jokes with decreasing effectiveness. Perhaps sensing this, Henley turned in *Abundance* (Costa Mesa, 1989) to a new setting and subject, tracing the lives of two mail-order brides in the nineteenth-century West. One meets nothing but hardship and disappointment, but attains a certain moral stature by resolutely honouring her commitment to her brutal and unproviding husband, while the other rises above her difficulties to achieve security and soçial success, but at a price to her soul. The play's title takes on multiple and sometimes ironic resonances as the promise implied in the myth of the American West is examined through the experience of women.

[The regional local-colour quality of Henley's earlier plays is present in several other successful dramas and comedies of the period. Both Alfred Uhry's *Driving Miss Daisy* (1987), about the friendship that develops between a southern gentlewoman and her black chauffeur over twenty-five years, and Robert Harling's *Steel Magnolias* (1987), about the mutual support provided by the women in a Louisiana beauty parlour, have some of the flavor of *Crimes of the Heart*. Earlier, Preston Jones's 'Texas Trilogy' of *The Last Meeting of the Knights of the White Magnolia* (Dallas, 1973), *The Oldest Living Graduate* (Dallas, 1974) and *Lu Ann Hampton Laverty Oberlander* (Dallas, 1974) offered a similar sentimentally comic picture of small-town Texas life.]

The third notable woman dramatist of the period is Marsha Norman, whose early plays show little people taking little steps in their search for happiness, steps that the plays cannot see as offering a great deal of promise or hope. *Getting Out* (Louisville, 1977) finds Arlene, released from prison after eight years, attempting to adjust to spiritual as well as physical freedom. The guarded, self-protective persona she had adopted in order to survive the brutality of prison is difficult to give up, especially since there are those, like her former pimp, ready to exploit any sign of softness in her. Overlapping and sometimes simultaneous scenes show her younger self, Arlie, entering prison as a dynamic and self-reliant, if undisciplined, girl and slowly learning to repress her personality. Norman thus shows Arlie turning to Arlene at the same time that she shows Arlene trying to find the remnants of Arlie in her; by the end Arlene has begun the reintegration of the two personalities, but there is clearly a long and uncertain way to go.

Third and Oak (Louisville, 1978) is made up of two slightly connected plays, 'The Laundromat' and 'The Pool Hall'. In the first an embittered middle-class widow and an unhappy working-class girl break through their reticence and the distance between them just enough to recognize some shared pain, but with no real hope that the discovery will change their lives. In 'The Pool Hall' a rebellious young black man and the older man who has been his responsible but intrusive father figure reach some respect for each other's values and boundaries, but again one senses a temporary truce rather than a resolution of their ongoing difficulties. *The Holdup* (San Francisco, 1983) uses an encounter between two young men and an older outlaw in 1914 to depict the passing of the Old West and the need to accept new realities; and *Traveler in the Dark* (Cambridge, 1984) shows a man who has tried to reduce life to a matter of fact and rationality tentatively rediscovering his capacity for faith.

Norman's most powerful play, *'night, Mother* (Cambridge, 1982) is about fortyish Jessie Cates, who announces to her mother Thelma that she intends to commit suicide later that evening, and then methodically goes through her chores, putting the house and her affairs in order while her mother moves from incredulity through attempts at reasoning, comforting, bargaining and begging. Jessie's reasons for ending her life are clearly elucidated, all the more forcefully for obviously being the result of cool and rational consideration. Life simply has nothing to offer her, and it makes more sense to end it than to endure: 'I'm just not having a very good time and I don't have any reason to think it'll get anything but worse.' It is not, as her mother guesses, that some specific disappointment or pattern of disappointments has depressed her; it is that there is no convincing argument in favour of life. What she is, what her life is, and what is likely to come are all just not enough: 'So, see, it doesn't much matter what else happens in the world or in this house, even. I'm what was worth waiting for and I didn't make it. Me . . . who might have made a difference to me . . . I'm not going to show up, so there's no reason to stay.' [Ellipses Norman's] She has no fantasy that things will be better after death; they will just be over: 'Dead is everybody and everything I ever knew, gone.'

'night, Mother is, however, less an argument for the legitimacy of suicide than it is an assertion of the individual's right to choose and control her own destiny. Even more satisfying to Jessie than the release death offers is the experience of taking possession of herself. When her mother cries that Jessie does not have to kill herself, she replies that having the choice is the whole point: 'No, I don't. That's what I like about it.' For the first time in an essentially passive life, Jessie is taking control – 'It's all I really have that belongs to me and I'm going to say what happens to it' – and that argument proves to be the unanswerable one. After the off-stage suicide Thelma can only acknowledge her daughter's right to make the decision: 'Forgive me. (*Pause*) I thought you were mine.'

Minority voices

The mainstream American theatre seems unable to accept more than one major black dramatist at a time so, with Ed Bullins dominating the 1970s and August Wilson the 1980s, only a few isolated plays by black authors reached the larger audience. Ntozake Shange's *for colored girls who have considered suicide/when the rainbow is enuf* (1975) is less a play than a staged reading of poems and set pieces, spoken by black women discovering their feminism and sisterhood through their relationships with men. Samm-Art Williams's *Home* (1979) tells of a black North Carolina farmer who leaves the land and loses his moral bearings until he returns home; an evocative mix of realism and folk styles expands the play into an allegory of the race's disconnection from its roots. Charles Fuller's *Zooman and the Sign* (1980) depicts the decay of middle-class black neighbourhoods as the residents lose the will to fight or save a lost generation of frighteningly amoral black youths. In Fuller's *A Soldier's Play* (1981), the murder of a black soldier in a southern army camp in 1944 leads to the discovery that racial pride can sometimes be an inadequate mask for racial self-hatred.

Of course many black writers, along with those of other ethnic or special interest groups, chose to write to and for their own limited audiences. Theatres such as the Negro Ensemble Company in New York and the East West Players in Los Angeles specialized in ethnic drama (and, not incidentally, gave employment to ethnic actors), only occasionally attracting the attention of the general public. Still some non-profit theatres, and the occasional commercial producer, attempted to bring these voices into the mainstream. Luis Valdez's *Zoot Suit* (Los Angeles, 1978) depicts a 1942 murder case tinged by racism and its effects on the Mexican–American community in a style that is a salute to the rhythms and vitality of that community. In David Henry Hwang's *FOB* (1980) the contempt of the Americanized second-generation Chinese for those Fresh Off the Boat is shown to cover an insecurity about the success of their own assimilation and an envy of the newcomers' living connection with Chinese tradition and culture. Hwang's *The Dance and the Railroad* (1981) corrects the popular image of the passive, exploited coolie workers in nineteenth-century America by showing how a continuing bond with their roots gave them the strength to survive and fight for better conditions.

That a minority writer can bring a special perspective to bear on the majority culture is demonstrated by Hwang's *M. Butterfly* (1988), based on the unlikely but true story of a French diplomat who carried on a twenty-year affair with a Chinese actress without knowing that she was both a spy and a man. Hwang's insight is that the story is one of cultural stereotyping more than of political or sexual confusion. In his version Rene Gallimard is so enraptured with the fantasy that he is acting out the archetypal story of *Madame Butterfly*, as the sexually irresistible Pinkerton worshipped by the submissive and self-effacing Butterfly, that he wilfully blinds himself to any evidence of the contrary in his affair with Song Liling. The westerner, insecure in his manhood and his culture, needs to believe in a woman from an exotic culture who is weaker than he and who will find her fulfilment in being dominated and exploited; and when he thinks he has found such an ideal, he will give her anything she asks just to keep up the illusion. Thus, Hwang sees, it is ultimately he who willingly submits to her domination; and the play ends with Gallimard's discovery that all along he was the Butterfly and Song the Pinkerton. In his courtroom testimony when the two are finally caught, Song makes the play's larger implications explicit:

> As soon as a Western man comes in contact with the East – he's already confused. . . . The West thinks of itself as masculine – big guns, big industry, big money – so the East is feminine – weak, delicate, poor. . . . Her mouth says no but her eyes say yes. The West believes the East, deep down, *wants* to be dominated.

The events of the play coincide with the period of the Vietnam War, and Hwang offers a metaphor for how completely America misjudged its role and power in Asia.

Minority drama is not merely a matter of race. As the subject of homosexuality came to be more openly discussed in the 1970s and 1980s, and as the homosexual community became more demanding of attention, several plays with homosexual characters or themes found their way into the mainstream theatre. The breadth of homosexual drama in the 1980s ranges from the casual inclusion of homosexuals among straight characters, in such plays as Lanford Wilson's *5th of July*; through plays in which the characters' sexuality is incidental to other issues, such as Terrence McNally's *Lisbon Traviata*; to plays specifically addressing gay issues like AIDS, such as William M. Hoffman's *As Is* (1985) and Larry Kramer's *The Normal Heart* (1985); to the flamboyantly campy drag extravaganzas and pastiches of Charles Ludlam (*Bluebeard*, *The Mystery of Irma Vep*) and Charles Busch (*Psycho Beach Party*, *Vampire Lesbians of Sodom*). Harvey Fierstein's *The Torch Song Trilogy* (1981), a long evening of three plays originally written and produced separately, operates fully within the mainstream tradition of domestic realism, using the life of the author's alter ego, Arnold Beckoff, to chart American homosexuals' evolution in the 1970s from impersonal promiscuity and self-depreciating humour to mature self-acceptance. The trilogy deals openly, if frequently comically, with the emotional difficulties of the homosexual life, showing how an unbearable loneliness drives Arnold to convince himself he enjoys the demeaning anonymous sexual encounters of the first play, while social prejudice makes him accept the patronization of heterosexuals who are shallower than he in the second; in the third play he mourns a lover killed by gay-bashing hoodlums and hears his mother deny the validity of his emotions. As he learns to honour himself and demand respect from others, he discovers that he is really a very ordinary and conventional person, whose dream is to have exactly what his mother had – a marriage, a family, and the right to mourn his lover openly. The trilogy reaches its resolution when that desire is honoured, as his mother accepts Arnold's widowhood as the equivalent of her own.

Mark Medoff's earlier plays had some theatrical vitality but little weight; *When You Comin' Back, Red Ryder?* (1973) is a thin reworking of *The Petrified Forest*, with a criminal holding several people hostage in a roadhouse, giving them the opportunity to display their true characters. But *Children of a Lesser God* (Los Angeles, 1979) is strong drama as well as good theatre. James Leeds, teacher of the deaf, falls in love with the proud and intelligent student Sarah Norman, and prods her to learn to speak rather than using sign language. But Sarah refuses, choosing to use a language in which she is eloquent rather than one in which she will always be inferior, and the conflict destroys their relationship. Some of the play's theatrical strength comes from the clear necessity to cast an actual deaf actress as Sarah, thus giving the events on stage a special reality. Medoff creates a multi-levelled stage vocabulary as James must sign all his dialogue to the deaf characters while speaking it aloud for the audience's benefit, and then speak the other characters' lines as they sign them. In a powerful emotional climax Sarah cries out in the barely intelligible sounds of the untrained deaf speaker, to demonstrate the violation of her dignity that James is demanding. Dignity is what the play is ultimately about; beneath the debate over speaking or signing is the hearing world's assumption that its ways are best. James's teaching is essentially paternalistic and patronizing, much the equivalent of racism; and Sarah's refusal to trade her proud and satisfying identity as a deaf person for the indignity of attempting to pass as hearing makes this play by a white male writer as powerful a repudiation of prejudice as any by minority dramatists.

An ethnic minority of a special sort is the subject of A.R. Gurney's plays. Gurney wrote some promising short plays in the 1960s, notably *The Golden Fleece* (Los Angeles, 1968), a retelling of the Medea myth in modern dress, with the off-stage tragedy reported and misinterpreted by Medea's very ordinary American neighbours. In the 1980s Gurney focused on the lives of American WASPS, well-to-do White Anglo-Saxon Protestants from old families, the Brahmin class that once ruled the country but now found itself reduced to a powerless and anachronistic minority. *The Dining Room* (1981) is a series of short sketches set in a variety of Brahmin dining rooms, each offering an insight into the culture: an adulterous couple are moved to discretion by the room's sense of history; a marriage breaks apart; an adult daughter tries to retreat into the past but is not welcomed; a father gives funeral instructions to his adult son. In *The Perfect Party* (1986) a man's attempt to give his life meaning by engineering a perfect social event is made, a little too explicitly, into an allegory of the fading American culture's attempt to retain its greatness through sheer will power. In *The Cocktail Hour* (San Diego, 1988) a playwright asks his Brahmin family's permission to produce a play about them, and the resulting debate exposes the culture's decay while still lovingly saluting its faded greatness. *Love Letters* (New Haven,

1988) is a reading for actor and actress of a lifetime's correspondence between a boy and girl from this community who are held together by a bond that can only be recognized afterwards as love. In one scene of *The Dining Room* a teenager treats his aunt's elegant dinner service and customs as quaint and primitive tribal artifacts, and that is essentially the point of view of Gurney's plays, more respectful than the boy but anthropological in nature, capturing and saluting a dying culture.

During the 1970s and 1980s Arthur Miller wrote a play indicting the totalitarian regimes of eastern Europe for invading the innermost core of their citizens' privacy; Sam Shepard offered a bizarre farm family as a microcosm of America's moral decay; Lanford Wilson showed a legless homosexual Vietnam veteran rediscovering the ability to love and hope; Marsha Norman presented suicide as the ultimate symbol of self-control; August Wilson showed a black family able to move forward with pride in their past; Neil Simon looked back on the Depression with warm and respectful humour – and every one of those plays was set in a living room.

The twentieth-century American drama was born in realism, as a few playwrights tentatively and almost accidentally brought a little extra substance to what had been a lightweight entertainment form by attempting to reflect the world as they saw it: with human behaviour that was sometimes unpredictable, and problems that could not all be resolved by the final curtain. That the drama's future would lie in realism was not at all clear from the start. In two particular periods – the years before 1930, when the American drama was young enough that no one could be sure what style fitted it best, and the 1960s, when cultural and artistic revolution was in the air – some of the century's most able and serious dramatists, from Percy MacKaye through Eugene O'Neill to Sam Shepard, tried to push the drama away from realism into other forms that would be more able to fulfil their artistic ambitions. Virtually all of them succeeded, at least in individual plays of merit and power, as did Maxwell Anderson, Thornton Wilder and other non-realistic playwrights.

And yet, in spite of all the experiments and all the individual successes, the backbone of American drama remains in realism. The continual discovery and rediscovery and re-rediscovery of this fact *is* the history of the twentieth-century American drama. It is in fact two parallel discoveries, of realism and of domestic drama. Realism – the use of art and artifice to create the effect of a direct reflection of the world outside – has proved remarkably adaptable. It is able to carry a thesis, as in the problem plays of Herne and Crothers; to depict life without comment, as in the earliest O'Neill plays; to introduce audiences to unfamiliar characters and perspectives on life, as Albee and various minority writers can; to let us see past the unfamiliar surface to a recognizable core, as Williams does; to capture, explain and even judge the largest social and political forces, as Odets and Miller do in their best plays. It is equally flexible, able to stretch its definition and rules to fit the styles and sensibilities of writers as individual and as different as O'Neill, Miller, Williams and Shepard. Above all, it works in the theatre: American audiences respond to it, it is understandable, and it meets an audience hunger to have their experience of life reflected and clarified through art.

Even more striking is the continuing power and adaptability of the domestic setting. Twentieth-century American playwrights did not invent domestic drama, but they – and their audiences – took to it from the start, recognizing not only that a variety of subjects could be addressed through the domestic focus, but that in fact it was the way these subjects affected the daily lives of ordinary Americans that was most worth addressing. Odets discovered in the 1930s that the true story of the Depression was not to be found in statistics and mass movements, but in how it permeated every aspect of everyday life. Miller in the 1940s and 1950s, Albee in the 1960s, and Shepard in the 1970s identified and condemned the failures of the American Dream by dramatizing the damaged lives of its victims. But American domestic drama is not limited to political and social comment. The special insight shared by Williams and O'Neill, in his last plays – that many Americans found the challenges of ordinary life almost beyond their capabilities – gave domestic realism a new focus that was to dominate the American drama in the second half of the century. O'Neill, Williams, Inge, Albee, Wilson, Wasserstein, Mamet, even Miller offer their audience spiritual comfort and counsel by depicting characters facing their own insecurities and self-condemnations. Realistic plays with a domestic setting can illuminate the inner world as well as the outer.

And so, while the twentieth-century American drama has had its talented mavericks and profitable diversions, the overall arc of its history is clear: a literature born in the impulse to describe the real world has always returned to that impulse – as, indeed, have its individual practitioners, as O'Neill, Rice, Anderson, Albee, Williams, Shepard and almost every other writer who challenged the hegemony of domestic realism eventually found it the most amenable vehicle for their individual styles and concerns. As the drama approaches the end of the century, in a theatrical context far more open and varied than it had at mid-century, it is inevitable that new writers, voicing new concerns in new styles, will appear; and there is every reason to believe that they will be able to say what they want to in plays with a realistic, contemporary, middle-class domestic setting.

Chronology

DATE	AMERICAN DRAMA	WORLD DRAMA	OTHER LITERATURE AND ART	HISTORICAL/CULTURAL EVENTS
1890	Herne *Margaret Fleming*	Ibsen *Hedda Gabler* Strindberg: *The Stronger*	Howells *A Hazard of New Fortunes* Wilde *The Picture of Dorian Gray*	Ellis Island opens DAR founded Frazer *The Golden Bough*
1891		Wedekind *Spring's Awakening*	Garland *Main-Travelled Roads* Hardy *Tess of the D'Urbervilles* Kipling *The Light that Failed* Shaw *The Quintessence of Ibsenism* Mahler: Symphony #1	First international copyright law Zipper invented
1892		Ibsen *The Master Builder* Shaw *Widowers' Houses*	Doyle *The Adventures of Sherlock Holmes* Kipling *Barrack-Room Ballads*	Grover Cleveland elected US President Lizzie Borden case

		Wilde *Lady Windermere's Fan*	Dvorak: 'New World' Symphony Tchaikovsky: 'The Nutcracker'	
1893	Thomas *In Mizzoura*	Pinero *The Second Mrs Tanqueray* Wilde *A Woman of No Importance*	Crane *Maggie*	World Exposition in Chicago Dreyfus trial in France Henry Ford builds first car
1894		Ibsen *Little Eyolf* Shaw *Arms and the Man*	Kipling *The Jungle Book* Twain *Pudd'nhead Wilson* Debussy: 'Afternoon of a Faun'	Sears-Roebuck begins mail-order business
1895	Gillette *Secret Service*	Wilde *The Importance of Being Earnest* Wilde *An Ideal Husband*	Conrad *Almayer's Folly* Crane *The Red Badge of Courage* Wells *The Time Machine* Yeats *Poems* Tchaikovsky: 'Swan Lake'	Roentgen discovers X-rays Gillette invents safety razor Freud *Studies in Hysteria*

DATE	AMERICAN DRAMA	WORLD DRAMA	OTHER LITERATURE AND ART	HISTORICAL/CULTURAL EVENTS
1896		Chekhov *The Seagull* Ibsen *John Gabriel Borkman* Jarry *King Ubu* Wilde *Salome*	Hardy *Jude the Obscure* Puccini: 'La Bohème'	McKinley elected president Nobel Prizes established First modern olympics, Athens
1897		Rostand *Cyrano de Bergerac* Schnitzler *La Ronde* Shaw *The Devil's Disciple*	Conrad *The Nigger of the Narcissus* James *The Spoils of Poynton* Kipling *Captains Courageous* Sousa: 'The Stars and Stripes Forever'	Queen Victoria's Diamond Jubilee Gold Rush in Alaska
1898		Pinero *Trelawney of the 'Wells'* Strindberg *The Road to Damascus*	Crane *The Open Boat* James *The Turn of the Screw*	Curies discover radium Spanish–American War Hawaii annexed

			Wells *The War of the Worlds*	
1899	Gillette *Sherlock Holmes* Thomas *Arizona*	Chekhov *Uncle Vanya* Ibsen *When We Dead Awaken* Shaw *You Never Can Tell*	Norris *McTeague*	Boer War begins Veblen *Theory of the Leisure Class*
1900	Fitch *The Climbers*	Wedekind *The Marquis of Keith*	Baum *The Wizard of Oz* Conrad *Lord Jim* Dreiser *Sister Carrie* Puccini: 'Tosca'	Planck formulates quantum theory Freud *The Interpretation of Dreams*
1901		Barrie *Quality Street* Chekhov *The Three Sisters* Strindberg *The Dance of Death*	Kipling *Kim* Norris *The Octopus* Washington *Up from Slavery*	Queen Victoria dies; Edward VII king McKinley assassinated; Theodore Roosevelt elected president

DATE	AMERICAN DRAMA	WORLD DRAMA	OTHER LITERATURE AND ART	HISTORICAL/CULTURAL EVENTS
			Rachmaninoff: Piano Concerto #2	
1902	Fitch *The Girl With Green Eyes*	Barrie *The Admirable Crichton* Gorki *The Lower Depths* Shaw *Mrs Warren's Profession* Strindberg *The Dream Play*	Doyle *The Hound of the Baskervilles* James *The Wings of the Dove* Kipling *Just So Stories* Potter *Peter Rabbit* Wister *The Virginian*	Boer War ends W. James *The Varieties of Religious Experience*
1903		Gorki *Summer Folk* Synge *In the Shadow of the Glen*	James *The Ambassadors* London *The Call of the Wild* Norris *The Pit* film: 'The Great Train Robbery'	Wright brothers fly Ford Motor Company founded First baseball World Series

1904	Cohan *Little Johnny Jones*	Barrie *Peter Pan* Chekhov *The Cherry Orchard* Molnar *Liliom* Shaw *John Bull's Other Island* Synge *Riders to the Sea*	Conrad *Nostromo* James *The Golden Bowl* London *The Sea Wolf* Puccini: 'Madame Butterfly'	Panama Canal begun Abbey Theatre founded, Dublin
1905	Belasco *The Girl of the Golden West* Fitch *The Woman in the Case* Walter *The Easiest Way*	Shaw *Man and Superman* Shaw *Major Barbara*	Wharton *The House of Mirth* Lehar: 'The Merry Widow'	Einstein formulates Special Theory of Relativity
1906	Cohan *Forty-five Minutes from Broadway* Crothers *The Three of Us*	Shaw *Caesar and Cleopatra* Shaw *The Doctor's Dilemma*	Galsworthy *The Man of Property* Sinclair *The Jungle*	San Francisco earthquake

DATE	AMERICAN DRAMA	WORLD DRAMA	OTHER LITERATURE AND ART	HISTORICAL/CULTURAL EVENTS
	MacKaye *Jeanne d'Arc* Mitchell *The New York Idea*			
1907	MacKaye *Sappho and Phaon*	Strindberg *The Ghost Sonata* Synge *The Playboy of the Western World*	Conrad *The Secret Agent* Picasso: 'Demoiselles d'Avignon'	Boy Scouts founded First Ziegfeld Follies Adams *The Education of Henry Adams* W. James *Pragmatism*
1908		Barrie *What Every Woman Knows*	Bennett *The Old Wives' Tale* Grahame *The Wind in the Willows*	Taft elected President First Model T Ford General Motors Corporation founded
1909	MacKaye *The Canterbury Pilgrims* MacKaye *The Scarecrow* Sheldon *Nigger*		Stein *Three Lives*	Peary reaches North Pole

1910	Crothers *A Man's World*	Molnar *The Guardsman* Shaw *Misalliance* Synge *Deirdre of the Sorrows*	Forster *Howard's End* Stravinsky: 'The Firebird'	Edward VII dies; George V king
1911	Crothers *He and She* Knoblock *Kismet* Sheldon *The Boss*		Wharton *Ethan Frome*	Amundsen reaches South Pole Chinese revolution ends Manchu dynasty
1912			Grey *Riders of the Purple Sage*	Wilson elected President Titanic sinks F. W. Woolworth Company founded
1913	MacKaye *A Thousand Years Ago*	Shaw *Androcles and the Lion* Shaw *Pygmalion*	Cather *O, Pioneers!* Lawrence *Sons and Lovers* Mann *Death in Venice*	US institutes income tax

DATE	AMERICAN DRAMA	WORLD DRAMA	OTHER LITERATURE AND ART	HISTORICAL/CULTURAL EVENTS
			Proust *Swann's Way* Stravinsky: 'The Rite of Spring'	
1914	Mackaye *St. Louis* Rice *On Trial*	Molnar *The Swan*	Burroughs *Tarzan of the Apes* Joyce *Dubliners*	First World War begins Panama Canal opens First Charlie Chaplin films
1915			Conrad *Victory* Lawrence *The Rainbow* Masters *Spoon River Anthology* Maugham *Of Human Bondage* film: 'Birth of a Nation'	Einstein formulates General Theory of Relativity

1916	Glaspell *Trifles* MacKaye *Washington* O'Neill *Before Breakfast* O'Neill *Bound East for Cardiff*	Gorki *Enemies* Pirandello *Liola*	Joyce *Portrait of the Artist as a Young Man* Sandburg *Chicago Poems* Tarkington *Seventeen* film: 'Intolerance'	
1917	Glaspell *The Outside* O'Neill *Ile* O'Neill *In the Zone* O'Neill *The Long Voyage Home*	Barrie *Dear Brutus* Pirandello *Right You Are!* Yeats *At the Hawk's Well*	Eliot *Prufrock and Other Observations*	US enters First World War Russian revolution Jung *Psychology of the Unconscious*
1918	MacKaye *The Roll Call* O'Neill *Moon of the Caribees*	Joyce *Exiles*	Cather *My Antonia* Hopkins *Poems*	First World War ends Wordwide influenza epidemic kills 22 million Theatre Guild founded

DATE	AMERICAN DRAMA	WORLD DRAMA	OTHER LITERATURE AND ART	HISTORICAL/CULTURAL EVENTS
	O'Neill *The Rope* O'Neill *Where the Cross is Made*		Tarkington *The Magnificent Ambersons*	
1919	Field *Wedding Bells* Glaspell *Bernice* Tarkington *Clarence*		Anderson *Winesburg, Ohio* Maugham *The Moon and Sixpence*	League of Nations founded 'Black Sox' scandal in baseball
1820	O'Neill *Beyond the Horizon* O'Neill *The Emperor Jones*	Capek *R.U.R.* Shaw *Heartbreak House*	Eliot *Poems* Fitzgerald *This Side of Paradise* Lawrence *Women in Love* Lewis *Main Street*	Harding elected President Prohibition begins Women get the vote Joan of Arc canonized

			Wharton *The Age of Innocence*	
			Holst: 'The Planets'	
			film: 'The Cabinet of Dr Caligari'	
1921	Glaspell *Inheritors*	Capek *The Insect Play*	Dos Passos *Three Soldiers*	First scheduled radio broadcasts, KDKA, Pittsburgh BBC founded Sacco and Vanzetti case
	Glaspell *The Verge*	Maugham *The Circle*	Moore *Poems*	
	Kaufman & Connelly *Dulcy*	Pirandello *Six Characters in Search of an Author*	Robinson *Collected Poems*	
	O'Neill *Anna Christie*		Picasso: 'Three Musicians'	
		Shaw *Back to Methuselah*		
1922	Kaufman & Connelly *Merton of the Movies*	Brecht *Baal*	Cather *One of Ours*	Irish Free State proclaimed Mussolini forms Fascist goverment in Italy Emily Post *Etiquette*
	Kaufman & Connelly *To the Ladies!*	Brecht *Drums in the Night*	Eliot *The Waste Land*	
	Marquis *The Old Soak*	Pirandello *Henry IV*	Joyce *Ulysses*	
	Nichols *Abie's Irish Rose*	Witkiewicz *The Water Hen*	Lewis *Babbitt*	

DATE	AMERICAN DRAMA	WORLD DRAMA	OTHER LITERATURE AND ART	HISTORICAL/CULTURAL EVENTS
	O'Neill *The Hairy Ape*		Sitwell *Facade*	
1923	Barry *You and I* Rice *The Adding Machine*	O'Casey *The Shadow of a Gunman* Shaw *St Joan*	Gershwin: 'Rhapsody in Blue'	Teapot Dome scandal *Time* magazine founded *Reader's Digest* founded Keynes *A Tract on Monetary Reform*
1924	Anderson & Stallings *What Price Glory?* Barry *The Youngest* Howard *They Knew What They Wanted* Kaufman & Connelly *Beggar on Horseback* Kaufman & Ferber *Minick* O'Neill *All God's Chillun Got Wings*	Coward *The Vortex* O'Casey *Juno and the Paycock*	Dickinson *Complete Poems* Forster *A Passage to India* Hemingway *In Our Time* Melville *Billy Budd*	Coolidge elected President First winter olympics J. Edgar Hoover named head of FBI

	O'Neill *Desire Under the Elms* O'Neill *Welded* Selwyn & Goulding *Dancing Mothers*			
1925	Kaufman *The Butter and Egg Man* Kelly *Craig's Wife* Kennedy & Hawthorne *Mrs. Partridge Presents* O'Neill *The Fountain*	Coward *Hay Fever*	Dos Passos *Manhattan Transfer* Dreiser *An American Tragedy* Fitzgerald *The Great Gatsby* Lewis *Arrowsmith* Pound *Cantos* Woolf *Mrs. Dalloway* film: 'Battleship Potemkin'	Scopes Trial (on teaching of evolution) Hitler *Mein Kampf*

DATE	AMERICAN DRAMA	WORLD DRAMA	OTHER LITERATURE AND ART	HISTORICAL/CULTURAL EVENTS
1926	Howard *The Silver Cord* O'Neill *The Great God Brown*	O'Casey *The Plough and the Stars*	Faulkner *Soldiers' Pay* Hemingway *The Sun Also Rises* Lawrence *The Plumed Serpent* Milne *Winnie the Pooh* film: 'Metropolis'	Hirohito becomes emperor of Japan General Strike in England Goddard launches first rocket T.E. Lawrence *Seven Pillars of Wisdom*
1927	Barry *Paris Bound* Heyward *Porgy* Kaufman & Ferber *The Royal Family* Kern & Hammerstein *Show Boat*	Brecht *Man is Man*	Lewis *Elmer Gantry* Marquis *archy and mehitabel* Wilder *The Bridge of San Luis Rey* Woolf *To the Lighthouse* film: 'The Jazz Singer' (first talkie) Stravinsky: 'Oedipus Rex'	Lindbergh makes first solo transatlantic flight Sacco and Vanzetti executed Babe Ruth hits 60 home runs

1928	Barry *Holiday*	Brecht *The Threepenny Opera*	Benet *John Brown's Body*	Hoover elected President Fleming discovers penicillin First Mickey Mouse cartoon Mead *Coming of Age in Samoa*
	Hecht & MacArthur *The Front Page*		Huxley *Point Counterpoint*	
	O'Neill *Lazarus Laughed*		Lawrence *Lady Chatterley's Lover*	
	O'Neill *Marco Millions*		Woolf *Orlando*	
	O'Neill *Strange Interlude*		Gershwin: 'An American in Paris'	
1929	Kaufman & Lardner *June Moon*	Giraudoux *Amphitryon 38*	Aiken *Selected Poems*	Stock market crash triggers decade-long Depression
	O'Neill *Dynamo*	Mayakovsky *The Bedbug*	Faulkner *The Sound and the Fury*	
	Rice *Street Scene*		Hemingway *A Farewell to Arms*	
			Lewis *Dodsworth*	
			Remarque *All Quiet on the Western Front*	
			Wolfe *Look Homeward, Angel*	

DATE	AMERICAN DRAMA	WORLD DRAMA	OTHER LITERATURE AND ART	HISTORICAL/CULTURAL EVENTS
			Woolf *A Room of One's Own*	
1930	Anderson *Elizabeth the Queen* Connelly *The Green Pastures* Glaspell *Alison's House* Kaufman & Hart *Once in a Lifetime* Weitzenkorn *Five Star Final*	Brecht *The Measures Taken* Coward *Private Lives*	Crane *The Bridge* Dos Passos *42nd Parallel* Faulkner *As I Lay Dying* Frost *Collected Poems* Hammett *The Maltese Falcon* Wood: 'American Gothic'	Haile Selassie becomes Emperor of Ethiopia Constantinople renamed Istanbul Planet Pluto discovered Freud *Civilization and its Discontents*
1931	Behrman *Brief Moment* O'Neill *Mourning Becomes Electra*	Giraudoux *Judith* Lorca *The Love of Don Perlimplin and Belisa in the Garden*	Buck *The Good Earth* Faulkner *Sanctuary* Dali: 'Persistence of Memory' film: 'Frankenstein'	Empire State Building completed 'Star Spangled Banner' made national anthem

1932	Barry *The Animal Kingdom*	Anouilh *Thieves' Carnival*	Dos Passos *1919*	Roosevelt elected President
	Behrman *Biography*	Priestley *Dangerous Corner*	Eliot *Selected Essays*	
	Hecht & MacArthur *Twentieth Century*	Shaw *Too True to be Good*	Faulkner *Light in August*	
			Huxley *Brave New World*	
1933	M. Anderson *Mary of Scotland*	Coward *Design for Living*	Crane *Collected Poems*	Hitler appointed Chancellor of Germany Prohibition repealed
	Hagan *One Sunday Afternoon*	Lorca *Blood Wedding*	Orwell *Down and Out in Paris and London*	
	Kingsley *Men in White*		Stein *Autobiography of Alice B. Toklas*	
	O'Neill *Ah, Wilderness!*		West *Miss Lonelyhearts*	
	O'Neill *Days Without End*		Yeats *Collected Poems*	
	Rice *We the People*			

DATE	AMERICAN DRAMA	WORLD DRAMA	OTHER LITERATURE AND ART	HISTORICAL/CULTURAL EVENTS
1934	M. Anderson *Valley Forge*	Lorca *Yenna*	Auden *Poems*	Dionne quintuplets born
	Behrman *Rain From Heaven*		Fitzgerald *Tender is the Night*	
	Hellman *The Children's Hour*		Graves *I, Claudius*	
	Kaufman & Hart *Merrily We Roll Along*		Miller *Tropic of Cancer*	
	Raphaelson *Accent on Youth*		W.C.Williams *Collected Poems*	
	Thomas *No More Ladies*			
1935	M. Anderson *Winterset*	Coward *Tonight at 8:30*	Farrell *Studs Lonigan*	Social Security Act
	Kingsley *Dead End*	Eliot *Murder in the Cathedral*	Moore *Selected Poems*	
	Odets *Waiting for Lefty*	Giraudoux *The Trojan War Will Not Take Place*	Steinbeck *Tortilla Flat*	
	Odets *Awake and Sing!*		Gershwin: 'Porgy and Bess'	

	Odets *Till the Day I Die*			
	Odets *Paradise Lost*			
	Sherwood *The Petrified Forest*			
	Spewack *Boy Meets Girl*			
1936	M. Anderson *The Wingless Victory*	Lorca *The House of Bernarda Alba*	Dos Passos *The Big Money*	Spanish Civil War begins George V dies; Edward VIII abdicates; George VI king O'Neill wins Nobel Prize Ford Foundation established *Life* magazine founded Carnegie *How To Win Friends and Influence People*
	Behrman *End of Summer*	Rattigan *French Without Tears*	Eliot *Collected Poems*	
	Boothe *The Women*		Faulkner *Absalom, Absalom!*	
	Ferber & Kaufman *Stage Door*		Mitchell *Gone With the Wind*	
	Kaufman & Hart *You Can't Take It With You*			
	Sherwood *Idiot's Delight*			

DATE	AMERICAN DRAMA	WORLD DRAMA	OTHER LITERATURE AND ART	HISTORICAL/CULTURAL EVENTS
1937	M. Anderson *The Masque of Kings* M. Anderson *High Tor* M. Anderson *The Star-Wagon* Odets *Golden Boy*	Giraudoux *Electra* Priestley *Time and the Conways*	Hemingway *To Have and Have Not* Steinbeck *Of Mice and Men* Picasso: 'Guernica' film: 'Snow White'	Japan invades China Dirigible 'Hindenburg' burns Golden Gate Bridge opens
1938	Conkle *Prologue to Glory* Goldsmith *What a Life* Odets *Rocket to the Moon* Osborn *On Borrowed Time* Sherwood *Abe Lincoln in Illinois*		cummings *Collected Poems* Hemingway *The Fifth Column* Isherwood *Goodbye to Berlin*	Germany annexes Austria, occupies Sudetenland House Unamerican Activities Committee formed

	Wilder *Our Town*			
	Wilder *The Merchant of Yonkers*			
1939	M. Anderson *Key Largo*	Brecht *Galileo*	Frost *Collected Poems*	Second World War begins as Germany invades Poland New York World's Fair
	Barry *Philadelphia Story*	Eliot *The Family Reunion*	Joyce *Finnegans Wake*	
	Behrman *No Time for Comedy*	Giraudoux *Ondine*	Steinbeck *The Grapes of Wrath*	
	Hellman *The Little Foxes*		West *The Day of the Locust*	
	Kaufman & Hart *The Man Who Came to Dinner*		films: 'Gone With the Wind', 'The Wizard of Oz'	
	Lindsay & Crouse *Life With Father*			
	Osborn *Morning's at Seven*			
	Saroyan *The Time of Your Life*			

DATE	AMERICAN DRAMA	WORLD DRAMA	OTHER LITERATURE AND ART	HISTORICAL/CULTURAL EVENTS
1940	Fields & Chodorov *My Sister Eileen*	E. Williams *The Corn is Green*	cummings *Fifty Poems*	Germany conquers most of western Europe
	Kaufman & Hart *George Washington Slept Here*		Faulkner *The Hamlet*	
	Sherwood *There Shall Be No Night*		Hemingway *For Whom the Bell Tolls*	
	Williams *Battle of Angels*		Koestler *Darkness at Noon*	
			Van Doren *Collected Poems*	
			Wright *Native Son*	
1941	M. Anderson *Candle in the Wind*	Brecht *Arturo Ui*	Fitzgerald *The Last Tycoon*	Japanese attack Pearl Harbour; US enters war
	Chodorov & Fields *Junior Miss* *Hellman* *Watch on the Rhine*	Brecht *The Good Woman of Sezuan*	Millay *Collected Sonnets*	
		Brecht *Mother Courage*	film: 'Citizen Kane'	
	Kesselring *Arsenic and Old Lace*	Coward *Blithe Spirit*		

1942	M. Anderson *The Eve of St. Mark*	Coward *Present Laughter*	Camus *The Stranger*	Fermi splits the atom
	Fields *The Doughgirls*	O'Casey *Red Roses for Me*	Eliot *Four Quartets*	Internment of Japanese–Americans
	James *Winter Soldiers*		Faulkner *Go Down, Moses*	
	Steinbeck *The Moon is Down*		Copland: 'Rodeo'	
	Wilder *The Skin of Our Teeth*			
1943	Gow & d'Usseau *Tomorrow the World*	Sartre *The Flies*	Saroyan *The Human Comedy*	Allied invasion of Italy
	Herbert *Kiss and Tell*			
	Kingsley *The Patriots*			
	Van Druten *The Voice of the Turtle*			
	Rodgers & Hammerstein *Oklahoma!*			

DATE	AMERICAN DRAMA	WORLD DRAMA	OTHER LITERATURE AND ART	HISTORICAL/CULTURAL EVENTS
1944	Behrman *Jacobowsky and the Colonel*	Anouilh *Antigone*	Bellow *The Dangling Man*	D-Day, allied invasion of Europe
	Chase *Harvey*	Camus *Caligula*	Borges *Ficciones*	
	Hellman *The Searching Wind*		Copland: 'Appalachian Spring'	
	Krasna *Dear Ruth*			
	Marquand & Kaufman *The Late George Apley*			
	Miller *The Man Who Had All the Luck*			
	Osborn *A Bell for Adano*			
	Van Druten *I Remember Mama*			
1945	Laurents *Home of the Brave*	Anouilh *Romeo and Jeanette*	Orwell *Animal Farm*	Roosevelt dies; Truman becomes President

	Patrick *The Hasty Heart*	Giraudoux *The Madwoman of Chaillot*	Steinbeck *Cannery Row*	US drops atom bombs on Japan Second World War ends
	Rice *Dream Girl*	Sartre *No Exit*	Waugh *Brideshead Revisited*	
	Williams *The Glass Menagerie*		Wright *Black Boy*	
			Britten: 'Peter Grimes'	
1946	M. Anderson *Joan of Lorraine*	Anouilh *Medea*	Warren *All the King's Men*	United Nations founded Spock *Baby and Child Care*
	Hellman *Another Part of the Forest*	De Filippo *Filumena Marturano*		
	Kanin *Born Yesterday*	Fry *A Phoenix Too Frequent*		
	O'Neill *The Iceman Cometh*	Rattigan *The Winslow Boy*		
1947	Goetz *The Heiress*	Genet *The Maids*	Camus *The Plague*	India granted independence Dead Sea Scrolls discovered Transistor invented CIA formed
	Haines *Command Decision*		Capote *Other Voices, Other Rooms*	

DATE	AMERICAN DRAMA	WORLD DRAMA	OTHER LITERATURE AND ART	HISTORICAL/CULTURAL EVENTS
	Krasna *John Loves Mary*		Frank *The Diary of Anne Frank*	
	Miller *All My Sons*		Michener *Tales of the South Pacific*	
	O'Neill *A Moon for the Misbegotten*			
	Williams *A Streetcar Named Desire*			
1948	M. Anderson *Anne of the Thousand Days*	Brecht *The Caucasian Chalk Circle*	Faulkner *Intruder in the Dust*	Ghandi assassinated Truman elected President Israel founded Berlin airlift Kinsey *Sexual Behavior in the Human Male*
	Heggen & Logan *Mr. Roberts*	Fry *The Lady's Not For Burning*	Mailer *The Naked and the Dead*	
	Lindsay & Crouse *Life With Mother*	Rattigan *The Browning Version*	Paton *Cry, the Beloved Country*	
	McEnroe *The Silver Whistle*			
	Williams *Summer and Smoke*			

1949	Kingsley *Detective Story*	Genet *Deathwatch*	Orwell *1984*	Republic of Ireland founded Communist revolution in China
	Miller *Death of a Salesman*	O'Casey *Cock-a-Doodle Dandy*	film: 'The Third Man'	NATO Treaty signed
	Odets *The Big Knife*			
	Spewack *Two Blind Mice*			
	Rodgers & Hammerstein *South Pacific*			
1950	Archibald *The Innocents*	Eliot *The Cocktail Party*	Hemingway *Across the River and Into the Trees*	Korean War begins
	Inge *Come Back, Little Sheba*	Fry *Venus Observed*	Williams *The Roman Spring of Mrs Stone*	
	McCullers *The Member of the Wedding* Miller *An Enemy of the People*	Ionesco *The Bald Soprano*	Pollock 'Lavender Mist'	
	Odets *The Country Girl*			
1951	M. Anderson *Barefoot in Athens*	Ionesco *The Lesson*	Faulkner *Requium for a Nun*	2nd Amendment limits President to two terms

DATE	AMERICAN DRAMA	WORLD DRAMA	OTHER LITERATURE AND ART	HISTORICAL/CULTURAL EVENTS
	Bevin & Trzcinski *Stalag 17*		Frost *Complete Poems*	
	Hellman *The Autumn Garden*		Jones *From Here to Eternity*	
	Williams *The Rose Tattoo*		Moore *Collected Poems*	
			Salinger *The Catcher in the Rye*	
			Sandburg *Complete Poems*	
1952	Axelrod *The Seven Year Itch*	Anouilh *Waltz of the Toreadors*	Capote *The Grass Harp*	George VI dies; Elizabeth II queen Eisenhower elected President
	Behrman *Jane*	Christie *The Mousetrap*	Ellison *Invisible Man*	Revised Standard Version of Bible published First H-bomb test
	Laurents *The Time of the Cuckoo*	Ionesco *The Chairs*	Hemingway *The Old Man and the Sea*	Peale *The Power of Positive Thinking*
		Rattigan *The Deep Blue Sea*	Malamud *The Natural*	
			Steinbeck *East of Eden*	

			film: 'High Noon'	
1953	R. Anderson *Tea and Sympathy*	Anouilh *The Lark*	Bellow *The Adventures of Augie March*	Korean War ends Hillary and Tenzing climb Mt. Everest *Playboy* magazine founded Kinsey *Sexual Behavior in the Human Female*
	Inge *Picnic*	Beckett *Waiting for Godot*	Fleming *Casino Royale*	
	Miller *The Crucible*	Rattigan *The Sleeping Prince*	Salinger *Nine Stories*	
	Patrick *The Teahouse of the August Moon*			
	Teichmann & Kaufman *The Solid Gold Cadillac*			
	Williams *Camino Real*			
1954	M. Anderson *Bad Seed*	Behan *The Quare Fellow*	Amis *Lucky Jim*	Supreme Court, in Brown v Board of Education, bars school segregation Salk introduces anti-polio vaccine
	Hayes *The Girl on the Via Flaminia*	Eliot *The Confidential Clerk*	Faulkner *A Fable*	
	Odets *The Flowering Peach*	Ionesco *Amédée*	Golding *Lord of the Flies*	
	Wilder *The Matchmaker*	Rattigan *Separate Tables*	Stevens *Collected Poems*	

DATE	AMERICAN DRAMA	WORLD DRAMA	OTHER LITERATURE AND ART	HISTORICAL/CULTURAL EVENTS
			Thomas Under Milk Wood Britten: 'The Turn of the Screw' film: 'On The Waterfront'	
1955	Axelrod *Will Success Spoil Rock Hunter?* Goodrich & Hackett *The Diary of Anne Frank* Inge *Bus Stop* Levin *No Time for Sergeants* Marchant *The Desk Set* Miller *A View From the Bridge* Miller *A Memory of Two Mondays*		Donleavy *The Ginger Man* Mailer *Deer Park* Nabokov *Lolita* O'Hara *Ten North Frederick*	Montgomery, Alabama bus boycott energizes Civil Rights movement, produces Martin Luther King as leader Rock and roll music appears McDonald's founded

	Williams *Cat on a Hot Tin Roof*			
1956	Chayefsky *Middle of the Night*	Anouilh *Poor Bitos*	Bellow *Sieze the Day*	Suez Canal crisis Soviet invasion of Hungary Churchill *History of the English-Speaking Peoples* Kennedy *Profiles in Courage* Mills *The Power Elite* Whyte *The Organization Man*
	Lawrence & Lee *Auntie Mame*	Durrenmatt *The Visit*	Ginsberg *Howl*	
	O'Neill *Long Day's Journey into Night*	Genet *The Balcony*	Moore: 'The Ballad of Baby Doe'	
	Lerner & Loewe *My Fair Lady*	Osborne *Look Back in Anger*	films: 'Around the World in Eighty Days', 'The Seventh Seal'	
1957	Frings *Look Homeward, Angel*	Beckett *Endgame*	Cheever *The Wapshot Chronicle*	USSR launches first space satellite
	Inge *The Dark at the Top of the Stairs*	Osborne *The Entertainer*	Kerouac *On the Road*	
	Laurents *A Clearing in the Woods*	Pinter *The Room*	Rand *Atlas Shrugged*	
	Vidal *Visit to a Small Planet*		Seuss *The Cat in the Hat*	
	Williams *Orpheus Descending*			
	Bernstein & Sondheim *West Side Story*			

DATE	AMERICAN DRAMA	WORLD DRAMA	OTHER LITERATURE AND ART	HISTORICAL/CULTURAL EVENTS
1958	MacLeish *J.B.*	Arden *Live Like Pigs*	Capote *Breakfast at Tiffany's*	Alaska becomes 49th state
	O'Neill *A Touch of the Poet*	Beckett *Krapp's Last Tape*	Ferlinghetti *A Coney Island of the Mind*	US launches first satellite
	Schary *Sunrise at Campobello*	Behan *The Hostage*	O'Hara *From the Terrace*	European Common Market formed
	Stevens *The Marriage-Go-Round*	Delaney *A Taste of Honey*	Pasternak *Dr. Zhivago*	Galbraith *The Affluent Society*
	Taylor & Skinner *The Pleasure of His Company*	Frisch *The Firebugs*		
	Williams *Suddenly Last Summer*	Genet *The Blacks*		
		Pinter *The Birthday Party*		
		P. Shaffer *Five Finger Exercise*		
		Wesker *Chicken Soup With Barley*		

1959	Albee *The Zoo Story*	Anouilh *Becket*	Bellow *Henderson the Rain King*	Hawaii becomes fiftieth state
				Fidel Castro becomes Premier of Cuba
	R. Anderson *Silent Night, Lonely Night*	Arden *Serjeant Musgrave's Dance*	Burroughs *Naked Lunch*	
	Chayefsky *The Tenth Man*	Ionesco *Rhinoceros*	Grass *The Tin Drum*	
	Gelber *The Connection*	Wesker *Roots*	Mailer *Advertisements for Myself*	
	Gibson *The Miracle Worker*		Roth *Goodbye, Columbus*	
	Hansberry *A Raisin in the Sun*		films: 'La Dolce Vita', 'Ben Hur'	
	Williams *Sweet Bird of Youth*			
1960	Albee *Fam and Yam*	Bolt *A Man for All Seasons*	Barth *The Sot-Weed Factor*	Kennedy elected President
				Laser invented
				Goodman *Growing Up Absurd*
	Albee *The Sandbox*	Pinter *The Caretaker*	Lee *To Kill a Mockingbird*	
	Hellman *Toys in the Attic*	Pinter *The Dumb Waiter*	*Sillitoe* *The Loneliness of the Long Distance Runner*	
	Williams *Period of Adjustment*	Rattigan *Ross*	Updike *Rabbit, Run*	

DATE	AMERICAN DRAMA	WORLD DRAMA	OTHER LITERATURE AND ART	HISTORICAL/CULTURAL EVENTS
		Sartre *The Condemned of Altona*	film: 'Psycho'	
		Wesker *I'm Talking About Jerusalem*		
1961	Albee *The American Dream*	Beckett *Happy Days*	Ginsberg *Kaddish*	Peace Corps founded
	Albee *Bartleby*	Fugard *The Blood Knot*	Heller *Catch-22*	Berlin Wall erected
	Albee *The Death of Bessie Smith*	Jellicoe *The Knack*	Salinger *Franny and Zooey*	Bay of Pigs, failed invasion of Cuba
	Kerr *Mary Mary*	Osborne *Luther*	Steinbeck *The Winter of Our Discontent*	
	Simon *Come Blow Your Horn*	Pinter *The Collection*		
	Williams *The Night of the Iguana*	Pinter *A Slight Ache*		
		Wesker *The Kitchen*		
		Whiting *The Devils*		

1962	Albee *Who's Afraid of Virginia Woolf?* Kingsley *Night Life* Kopit *Oh Dad, Poor Dad, . . .*	P. Shaffer *The Private Ear and The Public Eye* Wesker *Chips With Everything*	Faulkner *The Reivers* Kesey *One Flew Over the Cuckoo's Nest* Nabokov *Pale Fire* Porter *Ship of Fools* Steinbeck *Travels With Charlie* Britten: 'War Requiem'	Cuban missile crisis Thalidomide produces deformed children Pop Art appears Carson *Silent Spring*
1963	Albee *The Ballad of the Sad Cafe* Brown *The Brig* Hellman *My Mother, My Father and Me* Simon *Barefoot in the Park* Williams *The Milk Train Doesn't Stop Here Anymore*	Beckett *Play* Hochhuth *The Deputy* Ionesco *Exit the King* Pinter *The Dwarfs* Pinter *The Lover*	Baldwin *The Fire Next Time* LeCarre *The Spy Who Came In From the Cold* Plath *The Bell Jar* Pynchon *V*	Kennedy assassinated; Johnson becomes President Civil Rights March on Washington; King's 'I have a dream' speech

DATE	AMERICAN DRAMA	WORLD DRAMA	OTHER LITERATURE AND ART	HISTORICAL/CULTURAL EVENTS
	L. Wilson *So Long at the Fair*		Vonnegut *Cat's Cradle*	
	L. Wilson *The Bottle Harp*		film: 'Dr. Strangelove'	
1964	Albee *Tiny Alice*	Arden *Armstrong's Last Goodnight*	Bellow *Herzog*	Johnson elected President
	Behrman *But For Whom Charlie*	Orton *Entertaining Mr. Sloane*	Golding *The Spire*	Beginnings of US involvement in Vietnam War
	Hansberry *The Sign in Sidney Brustein's Window*	Osborne *Inadmissable Evidence*	Hemingway *A Moveable Feast*	Beginnings of several summers of race riots in various US cities
	Jones *Dutchman*	*P. Shaffer* *Royal Hunt of the Sun*		Major Civil Rights legislation passed
	Jones *The Slave*	Weiss *'Marat/Sade'*		Beatles popular
	Jones *The Toilet*			McLuhan *Understanding Media*
	Manhoff *The Owl and the Pussycat*			

	Miller *After the Fall*			
	Miller *Incident at Vichy*			
	O'Neill *Hughie*			
	Shepard *Cowboys*			
	Williams *Eccentricities of a Nightingale*			
	L. Wilson *Home Free!*			
	L. Wilson *The Madness of Lady Bright*			
1965	Bullins *Clara's Old Man*	Bond *Saved*	Capote *In Cold Blood*	Civil Rights march in Selma, Alabama
	Burrows *Cactus Flower*	Fugard *Hello and Goodbye*	Mailer *An American Dream*	
	Shepard *Chicago*	Marcus *The Killing of Sister George*	Plath *Ariel*	
	Shepard *Icarus's Mother*	Orton *Loot*	films: 'Dr. Zhivago.' 'The Sound of Music'	

DATE	AMERICAN DRAMA	WORLD DRAMA	OTHER LITERATURE AND ART	HISTORICAL/CULTURAL EVENTS
	Simon *The Odd Couple*	Osborne *A Patriot for Me*		
	Terry *Calm Down Mother*	Pinter *The Homecoming*		
	Terry *Keep Tightly Closed in a Clean Dry Place*	P. Shaffer *Black Comedy*		
	Ward *Day of Absence*			
	Ward *Happy Ending*			
	L. Wilson *Balm in Gilead*			
	L. Wilson *Days Ahead*			
	L. Wilson *Ludlow Fair*			
	L. Wilson *This is the Rill Speaking*			
	Zindel *The Effects of Gamma Rays . . . .*			

1966	Albee *Breakfast at Tiffany's*	Barth *Giles Goat-Boy*	US and USSR land unmanned spacecraft on moon
	Albee *A Delicate Balance*	Greene *The Comedians*	
	Albee *Malcolm*	Malamud *The Fixer*	
	Shepard *Fourteen Hundred Thousand*	Pynchon *The Crying of Lot 49*	
	Shepard *Red Cross*		
	Simon *The Star-Spangled Girl*		
	Terry *Viet Rock*		
	van Itallie *America Hurrah*		
	L. Wilson *The Rimers of Eldritch*		
	Williams *The Gnadiges Fraulein*		
	Williams *The Mutilated*		

DATE	AMERICAN DRAMA	WORLD DRAMA	OTHER LITERATURE AND ART	HISTORICAL/CULTURAL EVENTS
1967	Albee *Everything in the Garden*	Ayckbourn *Relatively Speaking*	Brautigan *Trout Fishing in America*	Six-day Arab-Israeli War First human heart transplant
	R. Anderson *You Know I Can't Hear You When the Water's Running*	Nichols *A Day in the Death of Joe Egg*	Mailer *Why Are We in Vietnam?*	
	Friedman *Scuba Duba*	Stoppard *Rosencrantz and Guildenstern Are Dead*	Styron *The Confessions of Nat Turner*	
	Garson *MacBird*		Wilder *The Eighth Day*	
	Horovitz *Line*		film: 'The Graduate'	
	O'Neill *More Stately Mansions*			
	Shepard *La Tourista*			
	Shepard *Melodrama Play*			
	Williams *The Two-Character Play*			

1968	Albee *Box-Mao-Box*	Ayckbourn *How the Other Half Loves*	Dickey *Poems*	Soviet invasion of Czechoslovakia Assassinations of Robert Kennedy and Martin Luther King Demonstrations at Democratic Convention in Chicago bring anti-war movement to a head Student riots throughout Europe Nixon elected President
	Bullins *The Electronic Nigger*	Barnes *The Ruling Class*	Mailer *The Armies of the Night*	
	Bullins *Goin' a Buffalo*	Bond *Narrow Road to the Deep North*	Portis *True Grit*	
	Bullins *In the Wine Time*	Osborne *A Hotel in Amsterdam*	Spark *The Prime of Miss Jean Brodie*	
	Crowley *The Boys in the Band*			
	Gurney *The Golden Fleece*			
	Horovitz *The Indian Wants the Bronx*			
	Kopit *Indians*			
	Miller *The Price*			
	Simon *Plaza Suite*			
	van Itallie *The Serpent*			

DATE	AMERICAN DRAMA	WORLD DRAMA	OTHER LITERATURE AND ART	HISTORICAL/CULTURAL EVENTS
	Williams *Kingdom of Earth*			
1969	Elder *Ceremonies in Dark Old Men* Gordone *No Place to be Somebody* McNally *Next* May *Adaptation* Russell *Five on the Black Hand Side* Shepard *The Holy Ghostly* Shepard *The Unseen Hand* Simon *The Last of the Red-Hot Lovers*	Fugard *Boesman and Lena* Nichols *The National Health* Orton *What the Butler Saw* Pinter *Landscape* Pinter *Silence* Storey *In Celebration*	Puzo *The Godfather* Roth *Portnoy's Complaint* Vonnegut *Slaughterhouse Five* film: 'Easy Rider'	First manned moon landing Woodstock rock festival

	Williams *In the Bar of a Tokyo Hotel* Williams *Out Cry* MacDermot, Ragni & Rado *Hair*			
1970	Friedman *Steambath* Shepard *Operation Sidewinder* Shepard *Shaved Splits* Simon *The Gingerbread Lady* L. Wilson *Lemon Sky*	Hampton *The Philanthropist* A. Shaffer *Sleuth* Storey *Home* Wesker *The Friends*	Angelou *I Know Why the Caged Bird Sings* Bellow *Mr Sammler's Planet* Dickey *Deliverance* Hemingway *Islands in the Stream* Olson *Collected Poems* Pound *Cantos*	Anti-war demonstrations on college campuses; students killed at Kent State University
1971	Albee *All Over* Bullins *In New England Winter*	Bond *Lear* Gray *Butley*	Segal *Love Story* Updike *Rabbit Redux*	Eighteen-year-olds given the vote

DATE	AMERICAN DRAMA	WORLD DRAMA	OTHER LITERATURE AND ART	HISTORICAL/CULTURAL EVENTS
	Guare *The House of Blue Leaves*	Osborne *West of Suez*		
	McNally *Where Has Tommy Flowers Gone?*	Pinter *Old Times*		
	Rabe *The Basic Training of Pavlo Hummel*	Storey *The Changing Room*		
	Rabe *Sticks and Bones*			
	Shepard *Cowboy Mouth*			
	Shepard *Back Bog Beast Bait*			
	Shepard *Mad Dog Blues*			
	Simon *The Prisoner of Second Avenue*			
	L. Wilson *The Great Nebula in Orion*			

1972	Mamet *The Duck Variations* Miller *The Creation of the World and Other Business* J. Miller *That Championship Season* Randall *6 Rms Riv Vu* Shepard *The Tooth of Crime* Simon *The Sunshine Boys* Walker *The River Niger* L. Wilson *Ikke, Ikke, Nye, Nye, Nye* Williams *Small Craft Warnings*	Ayckbourn *Absurd Person Singular* Fugard *Sizwe Banzi is Dead* Osborne *A Sense of Detachment* Stoppard *Jumpers* Wesker *The Old Ones*	film: 'The Godfather'	US ends draft, begins all-volunteer army Break-in at Democratic headquarters begins Watergate scandal
1973	Jones *The Last Meeting of the Knights of the White Magnolia*	Ayckbourn *The Norman Conquests*	Jong *Fear of Flying*	Vietnam War ends Arab oil embargo triggers worldwide energy crisis

DATE	AMERICAN DRAMA	WORLD DRAMA	OTHER LITERATURE AND ART	HISTORICAL/CULTURAL EVENTS
	Medoff *When You Comin' Back, Red Ryder?* Patrick *Kennedy's Children* Simon *The Good Doctor* L. Wilson *The Hot l Baltimore*	Beckett *Not I* Hampton *Savages* P. Shaffer *Equus*	Pynchon *Gravity's Rainbow* Vidal *Burr* Vonnegut *Breakfast of Champions* Wilder *Theophilus North*	Supreme Court legalizes abortion Watergate scandal spreads
1974	Jones *Lu Ann Hampton Laverty Oberlander* Jones *The Oldest Living Graduate* Mamet *Sexual Perversity in Chicago* Moore *The Sea Horse* Rabe *In the Boom Boom Room*	Ayckbourn *Absent Friends* Barnes *The Bewitched* Brenton *The Churchill Play* Stoppard *Travesties*		Nixon resigns US Presidency in Watergate scandal; Ford becomes president

	Shepard *Action*			
	Shepard *Geography of a Horse Dreamer*			
	Simon *God's Favorite*			
1975	Albee *Seascape*	Ayckbourn *Bedroom Farce*	Adams *Watership Down*	Vietnam War ends
	Bullins *The Taking of Miss Janie*	Berkoff *East*	Bellow *Humboldt's Gift*	
	Horovitz *The Primary English Class*	Gray *Otherwise Engaged*	Doctorow *Ragtime*	
	Shange *for colored girls . . . .*	Griffith *Comedians*		
	Shepard *Killer's Head*	Pinter *No Man's Land*		
	Slade *Same Time, Next Year*			
	L. Wilson *The Mound Builders*			
	Williams *The Red Devil Battery Sign*			

DATE	AMERICAN DRAMA	WORLD DRAMA	OTHER LITERATURE AND ART	HISTORICAL/CULTURAL EVENTS
	Hamlisch, Kirkwood, Dante & Kleban *A Chorus Line*			
1976	Albee *Counting the Ways*	Beckett *That Time*	Haley *Roots*	Carter elected President US Bicentennial
	Albee *Listening*	Brenton *Weapons of Happiness*	Vidal *1876*	
	Mamet *A Life in the Theater*	Osborne *Watch It Come Down*	Warren *Selected Poems*	
	Rabe *Streamers*	Stoppard *Dirty Linen*	film: 'Rocky'	
	Shepard *Angel City*	Wesker *The Merchant*		
	Shepard *Suicide in* B^b			
	Simon *California Suite*			
	L. Wilson *Seranading Louie*			

1977	Kopit *Wings*	Nichols *Privates on Parade*	McCullough *The Thorn Birds*
	Mamet *American Buffalo*	Rattigan *Cause Célèbre*	Morrison *Song of Solomon*
	Mamet *The Water Engine*	Stoppard *Every Good Boy Deserves Favour*	films: 'Annie Hall,' 'Star Wars'
	Mamet *The Woods*		
	Miller *The Archbishop's Ceiling*		
	Norman *Getting Out*		
	Shepard *Curse of the Starving Class*		
	Simon *Chapter Two*		
	Wasserstein *Uncommon Women and Others*		
	Williams *Vieux Carré*		

DATE	AMERICAN DRAMA	WORLD DRAMA	OTHER LITERATURE AND ART	HISTORICAL/CULTURAL EVENTS
1978	Durang *A History of the American Film* Norman *Third and Oak* Shepard *Buried Child* Shepard *Seduced* Valdez *Zoot Suit* L. Wilson *5th of July*	Ayckbourn *Joking Apart* Bond *The Woman* Hare *Plenty* Pinter *Betrayal* Stoppard *Night and Day*	Irving *The World According to Garp*	First 'test-tube baby' born in England
1979	Durang *Sister Mary Ignatius Explains It All For You* Henley *Crimes of the Heart* Mamet *Prairie du Chien*	Ayckbourn *Sisterly Feelings* Churchill *Cloud Nine* P. Shaffer *Amadeus*	Heller *Good as Gold* Mailer *The Executioner's Song* Roth *The Ghost Writer*	US establishes diplomatic relations with China Soviet invasion of Afghanistan Fundamentalist revolution in Iran

1979	Medoff *Children of a Lesser God*	Stoppard *Dogg's Hamlet, Cahoot's Macbeth*	Styron *Sophie's Choice*	
	Williams *A Lovely Sunday for Creve Coeur*			
	S. Williams *Home*			
	L. Wilson *Talley's Folly*			
1980	Albee *The Lady from Dubuque*	Ayckbourn *Season's Greetings*	Toole *A Confederacy of Dunces*	Reagan elected President
	Fuller *Zooman and the Sign*	Brenton *The Romans in Britain*	Welty *Collected Stories*	
	Hwang *FOB*	Pinter *The Hothouse*		
	Mamet *Lakeboat*	Russell *Educating Rita*		
	Mastrosimone *The Woolgatherer*			
	Miller *The American Clock*			
	Shepard *True West*			

DATE	AMERICAN DRAMA	WORLD DRAMA	OTHER LITERATURE AND ART	HISTORICAL/CULTURAL EVENTS
	Simon *I Ought To Be In Pictures*			
	Williams *Clothes for a Summer Hotel*			
1981	Albee *Lolita*	Beckett *Catastrophe*	Helprin *Ellis Island*	AIDS identified as world health threat
	Durang *Beyond Therapy*	Beckett *Ohio Impromptu*	Irving *The Hotel New Hampshire*	
	Fierstein *The Torch Song Trilogy*	Beckett *Rockaby*	Morrison *Tar Baby*	
	Fuller *A Soldier's Play*	Bond *Restoration*	Plath *Collected Poems*	
	Gurney *The Dining Room*	Gray *Quartermaine's Terms*	Updike *Rabbit is Rich*	
	Henley *The Miss Firecracker Contest*			
	Hwang *The Dance and the Railroad*			

	McKay *Sea Marks* Shue *The Nerd* Simon *Fools*			
1982	Albee *The Man Who Had Three Arms* Mamet *Edmond* Miller *Elegy for a Lady* Miller *Some Kind of Love Story* Norman *'night, Mother* L. Wilson *Angels Fall*	Churchill *Top Girls* Hare *A Map of the World* Pinter *Other Places*	Walker *The Color Purple*	Falklands War between Britain and Argentina

DATE	AMERICAN DRAMA	WORLD DRAMA	OTHER LITERATURE AND ART	HISTORICAL/CULTURAL EVENTS
1983	Mamet *Glengarry Glen Ross*	Stoppard *The Real Thing*		
	Norman *The Holdup*			
	Shepard *Fool for Love*			
	Shue *The Foreigner*			
	Simon *Brighton Beach Memoirs*			
	Wasserstein *Isn't It Romantic?*			
1984	Norman *Traveler in the Dark*	Ayckbourn *A Chorus of Disapproval*	Ginsberg *Collected Poems*	
	Rabe *Hurlyburly*	Frayn *Benefactors*		
	Shanley *Danny and the Deep Blue Sea*	Gray *The Common Pursuit*		
	A. Wilson *Ma Rainey's Black Bottom*	Pinter *One for the Road*		

1985	Durang *The Marriage of Bette and Boo*	Barnes *Red Noses*	Spender *Collected Poems*	
	Hoffman *As Is* Kramer *The Normal Heart*	Bond *The War Plays*		
	McNally *It's Only a Play*	Brenton & Hare *Pravda*		
	Mamet *The Shawl*	Carrière *Mahabharata*		
	Shepard *A Lie of the Mind*			
	Simon *Biloxi Blues*			
	L. Wilson *Talley & Son*			
1986	Gurney *The Perfect Party*	Ayckbourn *Woman in Mind*	Capote *Answered Prayers*	Accident at Soviet nuclear plant, Chernobyl US air attack on Libya in response to terrorism
	Simon *Broadway Bound*			

DATE	AMERICAN DRAMA	WORLD DRAMA	OTHER LITERATURE AND ART	HISTORICAL/CULTURAL EVENTS
1987	Harling *Steel Magnolias*	Ayckbourn *A Small Family Business*	Morrison *Beloved*	Stock Market drops 23% in one day
	Henley *The Debutante Ball*	P. Shaffer *Lettice and Lovage*		
	McNally *Frankie and Johnny in the Clair de Lune*			
	Miller *Clara*			
	Miller *I Can't Remember Anything*			
	Uhry *Driving Miss Daisy*			
	A. Wilson *Fences*			
	L. Wilson *Burn This*			

1988	Gurney *The Cocktail Hour*	Wertenbaker *Our Country's Good*	Larkin *Collected Poems*	Bush elected President USSR begins major restructuring of government and economy
	Gurney *Love Letters*		Rushdie *The Satanic Verses*	
	Hwang *M. Butterfly*		Wolfe *The Bonfire of the Vanities*	
	Mamet *Speed-the-Plow*			
	Wasserstein *The Heidi Chronicles*			
	A. Wilson *Joe Turner's Come and Gone*			
1989	Henley *Abundance*			Soviet domination of Eastern Europe ends
	McNally *The Lisbon Traviata*			Berlin Wall down US invades Panama
	Simon *Rumors*			
1990	Guare *Six Degrees of Separation*			East and West Germany united
	A. Wilson *The Piano Lesson*			

DATE	AMERICAN DRAMA	WORLD DRAMA	OTHER LITERATURE AND ART	HISTORICAL/CULTURAL EVENTS
1991	McNally *Lips Together, Teeth Apart* Miller *The Ride Down Mt. Morgan* Shepard *The War in Heaven* Simon *Lost in Yonkers*			Soviet Union dissolved 'Desert Storm' war in Iraq

General Bibliographies

Note: *Each section is arranged alphabetically. Place of publication is New York City unless otherwise noted.*

(i) Bibliographies and reference works

(Note: Many major dramatists are the subjects of individual bibliographies or reference guides; see the biographical section.)

Adelman, Irving and Dworkin, Rita — *Modern Drama: A Checklist of Critical Literature on Twentieth Century Plays* (Metuchen, NJ, 1967). (Covers world drama, focusing on books and popular periodicals.)

Arata, Esther Spring and Rotoli, Nicholas John — *Black American Playwrights 1800 to the Present: A Bibliography* (Metuchen, NJ, 1976). (Primary and secondary listings; see supplementary volume below.)

Arata, Esther Spring — *More Black American Playwrights: A Bibliography* (Metuchen, NJ, 1978). (Supplement to the previous volume.)

Bonin, Jane F. — *Prize-Winning American Drama: A Bibliography and Descriptive Guide* (Metuchen, NJ, 1973). (Plot summaries and brief lists of reviews of winners of Pulitzer, Tony, Obie and other awards 1917–70.)

Bordman, Gerald — *The Oxford Companion to the American Theatre* (1984). (Encyclopedia-style entries for authors, plays, etc.)

Bronner, Edwin *The Encyclopedia of the American Theatre 1900–1975* (1980). (Entries for all Broadway and many other plays, with brief plot, cast list and other data.)

Carpenter, Charles A. (ed). *Modern Drama: Scholarship and Criticism 1966–1980* (Toronto, 1986). (Extensive bibliographies arranged by dramatist.)

Coleman, Arthur and Tyler, Gary R. *Drama Criticism: A Checklist of Interpretations Since 1940 of English and American Plays* (Denver, 1966). (Includes journal articles and book sections.)

Coven, Brenda *American Women Dramatists of the Twentieth Century: A Bibliography* (Metuchen, NJ, 1982). (Primary and secondary listings.)

Eddleman, Floyd Eugene (ed.) *American Drama Criticism: Interpretations 1890–1977* (Hamden, CT, 1979). (Primarily theatre reviews; this volume supplants an earlier edition edited by Helen H. Palmer. See supplement below.)

Eddleman, Floyd Eugene (ed.) *American Drama Criticism: Supplement I* (Hamden, CT, 1984).

French, William P. (ed.) *Afro-American Poetry and Drama 1760–1975: A Guide to Information Sources* (Detroit, 1979). (Primary bibliographies.)

Harris, Richard H. *Modern Drama in America and England, 1950–1970: A Guide to Information Sources* (Detroit, 1982). (Primary and secondary bibliographies.)

Hatch, James V. and Abdullah, OMANii *Black Playwrights 1823–1977: An Annotated Bibliography of Plays* (1977). (Primary only.)

Helbing, Terry (ed.) *Gay Theatre Alliance Directory of Gay Plays* (1980). (Brief descriptions and plot summaries.)

King, Kimball *Ten Modern American Playwrights: An Annotated Bibliography* (1982). (Covers Albee, Baraka, Bullins, Gelber, Kopit, Mamet, Rabe, Shepard, Simon, L. Wilson with detailed and annotated primary and secondary listings.)

Kirkpatrick, D. L. (ed.) *Contemporary Dramatists*, 4th edition (London, 1988). (Covers world drama with primary bibliographies and brief critical essays.)

Kolin, Philip C. (ed.) *American Playwrights Since 1945: A Guide to Scholarship, Criticism and Performance* (1989). (Primary and secondary bibliographies and summary essays on plays and criticism, for 40 dramatists from Williams and Miller through August Wilson.)

Leiter, Samuel L. (ed.) *The Encyclopedia of the New York Stage: 1920–1930*, 2 vols (Westport, CT, 1985); *1930–1940* (Westport, 1989). (Alphabetic listing of plays with plot outlines, excerpts from reviews, etc.)

Long, E. Hudson *American Drama From Its Beginnings to the Present* (1970). (Brief bibliographies arranged by subject and by dramatist.)

Lovell, John Jr *Digests of Great American Plays* (1961). (Detailed plot summaries of plays from 1766 to 1959.)

Mantle, Burns and subsequent editors *The Best Plays of 1919–1920* [and subsequent years]. (Annual volume with abridged texts of ten plays, essays summarizing the New York season, and complete production details – casts, etc. – for Broadway and, in later volumes, Off-Broadway and regional theatres. Edited in turn by Mantle, 1919–46; John Chapman, 1947–51; Louis Kronenberger, 1952–60; Henry Hewes, 1961–3; Otis L. Guernsey Jr, 1964–84; Guernsey and Jeffrey Sweet, 1985–.)

Millett, Fred B. *Contemporary American Authors* (1940). (Biographies, primary and secondary bibliographies; good coverage of minor writers.)

Petersen, Bernard L. Jr *Contemporary Black American Playwrights and Their Plays: A Biographical Directory and Dramatic Index* (1988). (Biographies and primary bibliographies.)

Rigdon, Walter *The Biographical Encyclopedia and Who's Who of the American Theatre* (1966). (Who's-Who-style entries for dramatists, actors, producers, etc.)

Salem, James M. *A Guide to Critical Reviews, Volume I: American Drama 1909–1982* (London, 1984). (Listings of theatre reviews.)

Seller, Maxine Schwartz (ed.) *Ethnic Theatre in the United States* (Westport, CT, 1981). (Essays and extensive bibliographies on Armenian–American, Black, Yiddish, etc., theatre.)

Stein, Rita and Rickert, Friedhelm (eds) *Major Modern Dramatists*, 2 vols (1984). (Brief excerpts from reviews, books, etc., for each author.)

Woll, Allen *Dictionary of the Black Theatre* (Westport, CT. 1983) (Casts, plots, reviews of plays 1898–1981).

(ii) General drama histories

Bigsby, C.W.E. *A Critical Introduction to Twentieth-Century American Drama*. Vol. One: *1900–1940* (1982); Vol. Two: *Williams/Miller/Albee* (1984); Vol. Three: *Beyond Broadway* (1986). (Extensive critical analysis of major plays and playwrights.)

Dickinson, Thomas H.	*Playwrights of the New American Theater* (1925). (Covers MacKaye through O'Neill; though obviously dated, offers a valuable contemporary view, particularly of minor writers.)
Downer, Alan S. (ed.)	*American Drama and Its Critics: A Collection of Critical Essays* (Chicago, 1965). (Seventeen essays on writers from Herne to Albee.)
Downer, Alan S.	*Fifty Years of American Drama 1900–1950* (Chicago, 1951). (A short but useful overview and summary.)
Dukore, Bernard F.	*American Dramatists 1918–1945* (1984). (A short history with some analyses.)
Gould, Jean	*Modern American Playwrights* (1966) (Biographical sketches of writers from Rice to Albee.)
Krutch, Joseph Wood	*The American Drama Since 1918*, revised edition (1957). (First published in 1939, valuable for its contemporary viewpoint.)
Lewis, Allan	*American Plays and Playwrights of the Contemporary Theatre*, revised edition (1970). (Chapters on major writers from O'Neill to Albee.)
Mantle, Burns	*American Playwrights of Today* (1929). (Brief biographical sketches and analyses of plays, particularly valuable for minor figures. See next item.)
Mantle, Burns	*Contemporary American Playwrights* (1938) (Follow-up to the previous volume, covering major and minor writers of the 1930s.)
Meserve, Walter J.	*An Outline History of American Drama* (Totowa, NJ, 1965). (Useful brief review, covering from pre-revolutionary writers to early Albee.)
Moses, Montrose J.	*The American Dramatist* (Boston 1925) (From the beginnings to early O'Neill; though obviously dated, valuable for its contemporary view of early twentieth-century writers.)
Moses, Montrose J. and Brown, John Mason (eds)	*The American Theatre As Seen By Its Critics, 1752–1934* (1934). (Reprints contemporary reviews of major plays.)
Quinn, Arthur Hobson	*A History of the American Drama from the Civil War to the Present Day*, revised edition (1936). (First published 1923; like the Dickinson and Moses books it remains valuable for its contemporary view and its coverage of minor writers.)
Slide, Anthony (ed.)	*Selected Theatre Criticism*, three volumes: 1900–1919; 1920–1930; 1931–1950 (Metuchen, NJ, 1985–6). (Reprints contemporary reviews of major and minor plays.)

Wilson, Garff B. — *Three Hundred Years of American Drama and Theatre* (Englewood Cliffs, NJ, 1973). (A readable general history textbook.)

(iii) Special period studies

(Note: For books focusing on the early years of the century, see the older volumes in the preceding section.)

Bigsby, C.W.E. — *Confrontation and Commitment: A Study of Contemporary American Drama 1959–1966* (Missouri, 1968). (Focus on political and social critics: Miller, Albee, Jones, Hansberry, etc.)

Cohn, Ruby — *New American Dramatists 1960–1980* (1982) (Primarily Off- and Off Off-Broadway writers, but also Neil Simon.)

Goldstein, Malcolm — *The Political Stage: American Drama and Theater of the Great Depression* (1974). (A history of the left-wing theatre with analysis of major and minor political plays.)

Hughes, Catharine — *American Playwrights 1945–75* (London, 1976). (A brief introduction to Williams, Miller, Inge, Albee, Simon, Shepard and Rabe.)

Kirby, Michael (ed.) — *The New Theatre* (1974). (Descriptions and texts of some experimental productions of the 1960s and 1970s.)

Marranca, Bonnie and Dasgupta, Gautam, (eds) — *American Playwrights: A Critical Survey* (1981). (Critical essays on 18 Off- and Off Off-Broadway writers from Gelber to Shepard.)

Rabkin, Gerald — *Drama and Commitment: Politics in the American Theatre of the Thirties* (Bloomington, IN, 1964). (On politically committed theatres and writers: Odets, Rice, Anderson, etc.)

Weales, Gerald — *American Drama Since World War II* (1962) (Covers the 1945–60 period.)

Weales, Gerald — *The Jumping Off Place: American Drama in the 1960s* (1969). (Williams, Miller, Albee and many secondary Off-Broadway writers.)

(iv) Books focusing on special topics

Abramson, Doris E. — *Negro Playwrights in the American Theatre, 1925–1959* (1969). (Particularly valuable on secondary writers.)

Adler, Thomas — *Mirror on the Stage: The Pulitzer Play as an Approach to American Drama* (West Lafayette, IN, 1987). (Analysis of recurring themes in Pulitzer Prize plays.)

Broussard, Louis — *American Drama: Contemporary Allegory from Eugene O'Neill to Tennessee Williams* (Norman, OK, 1962). (On expressionism and other non-realistic styles.)

Laufe, Abe — *Anatomy of a Hit: Long-run Plays on Broadway from 1900 to the Present Day* (1966). (Descriptions and analyses of popular successes.)

Murphy, Brenda — *American Realism and American Drama, 1880–1940* (Cambridge, 1987). (Traces the rise of domestic realism as the dominant dramatic mode.)

Sanders, Leslie Catherine — *The Development of Black Theater in America* (Baton Rouge, LA, 1988). (Analysis of black drama with focus on Langston Hughes, LeRoi Jones and Ed Bullins.)

Schlueter, June (ed.) — *Modern American Drama: The Female Canon* (Rutherford, NJ, 1990). (Essays on women dramatists from Glaspell to Norman.)

Sievers, W. David — *Freud on Broadway: A History of Psychoanalysis and the American Drama* (1955). (Plays about psychology or built on Freudian characterizations.)

(v) Theatre histories

Atkinson, Brooks — *Broadway*, revised edition (1974). (A readable and authoritative history 1900–74.)

Berkowitz, Gerald M. — *New Broadways: Theatre Across America 1950–1980* (Totowa, NJ, 1982). (On the rise of Off- and Off Off-Broadway, and of regional and non-commercial theatre.)

Brown, John Mason — *Upstage: The American Theatre in Performance* (1930). (Essays on key actors, critics, directors, playwrights, etc.)

Chinoy, Helen Krich and Jenkins, Linda Walsh (eds) — *Woman in American Theatre* (1981). (Articles on actresses, playwrights, directors, theatre companies, etc.)

Clurman, Harold — *The Fervent Years* (1945). (A memoir of the Group Theatre in the 1930s.)

Durham, Weldon B. (ed.) — *American Theatre Companies, 1888–1930* (Westport, CT, 1987) and *American Theatre Companies, 1931–1986* (Westport, CT, 1989). (Descriptions, histories and performance records.)

Flanagan, Hallie — *The History of the Federal Theatre* (1940). (Memoir of the government-sponsored company of the 1930s, by its director.)

Goldman, William — *The Season* (1969). (Close study of a single Broadway season, 1967–68, by an insightful critic.)

Gottfried, Martin — *A Theatre Divided: The Postwar American Stage* (Boston, 1967). (Premise: Broadway retreated into shallow popular fare, leaving serious drama to other venues.)

Guernsey, Otis L. Jr — *Curtain Times: The New York Theater 1965–1987* (1987). (Reprints his Introductions to the annual *Best Plays* volumes, excellent contemporary accounts of the New York theatre.)

Himelstein, Morgan Y. — *Drama Was a Weapon: The Left-Wing Theatre in New York 1929–1941* (New Brunswick, NJ, 1963). (Focus more on personalities and theatre companies than plays.)

Hughes, Glenn — *A History of the American Theatre 1700–1950* (1951). (Focus on managers, actors, etc., more than plays.)

Little, Stuart W. — *Off-Broadway: The Prophetic Theater* (1972). (History 1950–70.)

Little, Stuart W. and Cantor, Arthur — *The Playmakers* (1970). (A primer on how Broadway works, with chapters on producers, critics, theatre owners, etc.)

Mordden, Ethan — *The American Theatre* (1981) (Anecdotal history placing Broadway in cultural and historical perspective.)

Novick, Julius — *Beyond Broadway* (1968). (On the beginnings of regional theatre.)

Poggi, Jack — *Theater in America: The Impact of Economic Forces 1870–1967* (Ithaca, NY, 1968). (On the fall of regional theatre, the dominance of Broadway, the rise of Off-Broadway and the reappearance of regional theatre, explained in economic terms.)

Sainer, Arthur	*The Radical Theatre Notebook* (1975). (Histories and descriptions of experimental theatres of the 1960s and 1970s.)
Sarlós, Robert Károly	*Jig Cook and the Provincetown Players* (Amherst, MA, 1982). (History of the company that produced O'Neill's first plays.)
Shank, Theodore	*American Alternative Theatre* (London, 1982). (Experimental companies of the 1960s and 1970s.)
Taubman, Howard	*The Making of the American Theatre* (1965). (A readable general history.)
Waldau, Roy S.	*Vintage Years of the Theatre Guild* (Cleveland, OH, 1972). (Early history of the important Broadway producers.)
Williams, Jay	*Stage Left* (1974). (Memoir of left-wing companies of the 1930s.)
Williams, Mance	*Black Theater in the 1960s and 1970s* (Westport, CT, 1985). (Accounts of companies, theatres, playwrights, etc.)
Zeigler, Joseph Wesley	*Regional Theatre* (Minneapolis, 1973). (On the growth of resident companies around the country from 1947 onwards.)

Individual Authors

ALBEE, Edward (1928–), born in Virginia and adopted as an infant by Reed and Frances Albee, members of the vaudeville theatre-owning family. Albee was raised in luxury, attending a series of boarding schools and spending three semesters at Trinity College in Connecticut (1946–47). He moved to New York City in 1950, living in relative poverty while working in a variety of jobs including Western Union messenger (1955–58). His first play, *The Zoo Story*, was produced in Berlin in 1959 and Off-Broadway in 1960. It was followed by *The Death of Bessie Smith* (Berlin 1960, New York 1961), *The Sandbox* (1960), *Fam and Yam* (1960) and *The American Dream* (1961). With *Who's Afraid of Virginia Woolf?* (1962) Albee became the first Off-Broadway dramatist to move to Broadway, where most of his subsequent plays were produced: *Tiny Alice* (1964), *A Delicate Balance* (Pulitzer Prize 1966), *Box and Quotations from Mao Tse-Tung* (1968), *All Over* (1971), *Seascape* (Pulitzer 1975), *Listening* and *Counting the Ways* (1977), *The Lady from Dubuque* (1980), *The Man Who Had Three Arms* (1982), *Finding the Sun* (1982), *Walking* (1984), *Marriage Play* (1986). Albee also wrote several adaptations: of Melville's *Bartelby* (1961), McCullers's *The Ballad of the Sad Cafe* (1963), Purdy's *Malcolm* (1966), Capote's *Breakfast at Tiffany's* (1966), Cooper's *Everything in the Garden* (1967), Nabokov's *Lolita* (1984); with the partial exception of the McCullers, all were critical and commercial failures. During the 1960s Albee and his producers used profits from . . . *Virginia Woolf?* to subsidize a playwrights' workshop that developed and produced the plays of many younger Off-Broadway writers.

See: Amacher, Richard E. and Margaret Rule, *Edward Albee at Home and Abroad* (1973) (Bibliography of criticism).
Amacher, Richard E., *Edward Albee*, revised edition (Boston, 1982).
Bigsby, C.W.E., *Edward Albee: Bibliography, Biography, Playography* (London, 1980).
Bigsby, C.W.E., *Albee* (Edinburgh, 1969).
Bloom, Harold, (ed.), *Edward Albee: Modern Critical Views* (1987).
De La Fuente, Patricia, (ed.), *Edward Albee: Planned Wilderness; Interviews, Essays and Bibliography* (Edinburg, TX, 1980).
Giantvalley, Scott, *Edward Albee: A Reference Guide* (Boston, 1987).
Green, Charles Lee, *Edward Albee: An Annotated Bibliography, 1968–77* (1980).
Kolin, Philip C. and J. Madison Davis (eds), *Critical Essays on Edward Albee* (Boston, 1986).
Tyce, Richard, *Edward Albee: A Bibliography* (Metuchen, NJ, 1986).

ANDERSON, Maxwell (1888–1959), born in Atlantic, Pennsylvania, son of a Baptist minister, and raised in Iowa and North Dakota. With degrees from the University of North Dakota (BA 1911) and Stanford (MA 1914), he taught at Whittier College before becoming a newspaper writer in San Francisco and New York. His first produced play, *White Desert* (1923) is a verse tragedy about North Dakota farmers; his first success, *What Price Glory?* (1924, with Laurence Stallings) a realistic comedy-drama about men at war. A series of unsuccessful plays followed in the 1920s until he returned to verse with *Elizabeth the Queen* (1930), *Night over Taos* (1932), *Mary of Scotland* (1933), *Valley Forge* (1934), *Winterset* (1935), *The Masque of Kings* (1936), *The Wingless Victory* (1936), *High Tor* (1937), and *Key Largo* (1939). Interspersed with these were the prose plays *Both Your Houses* (Pulitzer Prize, 1933), *The Star-Wagon* (1937), *Candle in the Wind* (1941), *The Eve of St Mark* (1942) and *Storm Operation* (1944); and the musical *Knickerbocker Holiday* (1938). Of Anderson's post-war plays, only *Joan of Lorraine* (1946), *Anne of the Thousand Days* (1948), the musical *Lost in the Stars* (1949) and *Bad Seed* (1954) were successful. Anderson also wrote eight films, including *All Quiet on the Western Front* (1930), *Rain* (1932), *Death Takes a Holiday* (1934) and *The Wrong Man* (1957); and a volume of poetry, *You Who Have Dreams* (1925).

Dramatist in America: Letters of Maxwell Anderson 1912–56 (Chapel Hill, NC, 1958).
The Essence of Tragedy (1939).
Off Broadway: Essays about the Theatre (1947).
Shivers, Alfred S., *The Life of Maxwell Anderson* (Briarcliff Manor, NY, 1983).

See: Bailey, Mabel Driscoll, *Maxwell Anderson: The Playwright as Prophet* (1957).
Clark, Barrett H., *Maxwell Anderson: The Man and His Plays* (1933).
Klink, William, *Maxwell Anderson and S. N. Behrman: A Reference Guide* (Boston, 1977).
Shivers, Alfred S., *Maxwell Anderson* (Boston, 1976).
Shivers, Alfred S., *Maxwell Anderson: An Annotated Bibliography* (Metuchen, NJ, 1985).

ANDERSON, Robert Woodruff (1917–), born in New York City, attended Philips Exeter Academy and Harvard (BA 1939, MA 1940). Served as Navy intelligence officer 1942–46; *Come Marching Home*, written while in the Navy, was his first produced play. From 1945 to 1953 Anderson adapted such plays as *The Petrified Forest* and *The Glass Menagerie* for radio and television, while several of his own plays were produced outside New York. His first critical success, *All Summer Long* (Washington, 1953) led to his Broadway debut and best-known play, *Tea and Sympathy* (1953). Others are *Silent Night, Lonely Night* (1959), *The Days Between* (Dallas, 1965), *You Know I Can't Hear You When the Water's Running* (1967), *I Never Sang for my Father* (1968), *Solitaire/Double Solitaire* (1971) and *Free and Clear* (1983). Anderson also wrote several films, including *Tea and Sympathy* (1956), *The Nun's Story* (1959), *The Sand Pebbles* (1966) and *I Never Sang . . .* (1970); as well as two novels, *After* (1973) and *Getting Up and Going Home* (1978). He is co-editor of a series of textbooks, *Elements of Literature* (1988).

See: Adler, Thomas P., *Robert Anderson* (Boston, 1978).

BARAKA, Amiri. *See* Jones, LeRoi.

BARRY, Philip (1896–1949), born in Rochester, NY, educated in Catholic schools, Yale (BA 1919) and Harvard (1919–21). His first play, written at Harvard, is *A Punch for Judy* (1921); his first professional production *You and I* (1923). Barry's strength was in sophisticated social comedy, as in *The Youngest* (1924), *Paris Bound* (1927), and particularly *Holiday* (1928) and *The Philadelphia Story* (1939). He was more interested, however, in writing symbolic and expressionistic plays on religious and spiritual subjects, such as *John* (1927), *Hotel Universe* (1930) and *Here Come the Clowns* (1938); most of these were commercial and critical failures. His novel *War in Heaven* appeared in 1938; his last play, *Second Threshold*, was revised by Robert E. Sherwood and produced posthumously in 1951.

See: Hamm, Gerald, *The Drama of Philip Barry* (Philadelphia, 1948).
Roppolo, Joseph P., *Philip Barry* (1965).

BEHRMAN, S[amuel] N[athan] (1893–1973), born in Worcester, MA, educated at Clark College, Harvard (BA 1916) and Columbia (MA 1918). Work as a freelance writer and as press agent for Broadway shows led to a couple of unsuccessful plays in the early 1920s. With his first success, *The Second Man* (1927), he found his niche in urbane, sophisticated social comedy frequently involving the Shavian discussion of political and social issues. Notable plays are *Meteor* (1929), *Biography* (1932), *End of Summer* (1936), *Amphitryon 38* (1937), *No Time for Comedy* (1939), *The Pirate* (1942), *Jacobowsky and the Colonel* (1944), and his last play *But For Whom Charlie* (1964). Behrman wrote about thirty films including *Queen Christina* (1933), *A Tale of Two Cities* (1935), *Waterloo Bridge* (1940) and *Quo Vadis* (1951); two novels, *The Suspended Drawing Room* (1965) and *The Burning Glass* (1969); and a biography, *Portrait of Max* (1960).

The Worcester Account (1954) (memoir).
People in a Diary: A Memoir (1972).

See: Klink, William, *Maxwell Anderson and S.N. Behrman: A Reference Guide* (Boston, 1977).
Klink, William, *S. N. Behrman: The Major Plays* (Amsterdam, 1978).
Reed, K. T., *S. N. Behrman* (Boston, 1975).

BULLINS, Ed (1935–), born in Philadelphia, dropped out of high school to join the Navy 1952–55. In the 1960s Bullins was active in the Black Power movement, serving as Culture Minister of the Black Panther Party until 1967, when he moved to New York as playwright-in-residence at the New Lafayette Theatre and other Off-Broadway companies. His plays fall into two general groups: revolutionary and consciousness-raising pieces such as *How Do You Do?* (1965), *The Theme is Blackness* (1966), *The Gentleman Caller* (1969) and *Death List* (1970); and less overtly political depictions of black life, such as *The Electronic Nigger* (1968), *Goin' a Buffalo* (1968), *In the Wine Time* (1968) *In New England Winter* (1971) and *The Taking of Miss Janie* (1975). He has taught drama, playwriting or black studies at such universities as Fordham, Columbia and Berkeley; edited the magazine *Black Theatre*; and published a novel, a book of short stories, a volume of poetry, and two anthologies of plays by black writers.

See: King, Kimball, *Ten Modern American Playwrights: An Annotated Bibliography* (1982).

CHAYEFSKY, Paddy (1923–81), born Sidney Chayefsky in New York City to a Russian immigrant family; BA City College 1943. Served in the Army, where he wrote the musical *No T.O. for Love* (1945). Of Chayefsky's six stage plays only *Middle of the Night* (1956) and *The Tenth Man* (1959), both effective comedy-dramas of small people finding small hope, are noteworthy; his major works were written for television – *Marty* (1953), *The Bachelor Party* (1953), *The Catered Affair* (1955) – and film – *Marty* (1955), *The Americanization of Emily* (1964), *Network* (1976), *Altered States* (1979).

See: Clum, John M., *Paddy Chayefsky* (Boston, 1976).

CONNELLY, Marc (1890–80), born in McKeesport, PA, to a theatrical family, and wrote for newspapers in Pittsburgh and New York, and later for *The New Yorker*. His plays fall into two groups, a series of successful farces written with George S. Kaufman in the 1920s, and later works written alone or in collaboration with others. The Kaufman comedies include *Dulcy* (1921), *To the Ladies!* (1922), *Merton of the Movies* (1922) and *Beggar on Horseback* (1924); the only success among the post-Kaufman plays is *The Green Pastures* (Pulitzer Prize, 1930). Connelly wrote a number of films, including *Captains Courageous* (1937) and *I Married a Witch* (1942); a radio play, *The Mole on Lincoln's Cheek* (1941); and a novel, *A Souvenir from Qam* (1965).

Voices Off-Stage: A Book of Memoirs (1968).

See: Nolan, Paul T., *Marc Connelly* (Boston, 1969).

CROTHERS, Rachel (1878–1958), born in Bloomington, Illinois, daughter of two physicians. After attending Illinois Normal School, she became an actress, eventually training and teaching at the Stanhope-Wheatcroft School of Acting in New York. After some plays written for her students, her first professional production was the one-act comedy *The Rector* (1902). More than twenty-five full-length plays followed, most of them built on conventional plot situations, but with an unusual awareness of psychology or moral issues, particularly relating to women. Such plays as *The Three of Us* (1906), *A Man's World* (1909), *He and She* (1911) and *Mary the Third* (1923) dramatize the constraints on women in a male-dominated world; while *Let Us Be Gay* (1929), *As Husbands Go* (1931), *When Ladies Meet* (1932) and *Susan and God* (1938) explore the mutual obligations of married couples, in the context of social comedy. A series of lectures was published as *The Art of Playwriting* (1928).

See: Gottlieb, Lois C., *Rachel Crothers* (1979).

GELBER, Jack (1932–), born in Chicago, (BA University of Illinois 1953). *The Connection* (1959), one of the first original plays produced Off-Broadway, is his most significant work; of his ten subsequent plays only *The Apple* (1961), *The Cuban Thing*, (1968) and *Jack Gelber's New Play: Rehearsal* (1976) have attracted much attention. Gelber has been a successful director of his plays and others, and has taught at City College, Columbia and Brooklyn College, where he is Professor of Drama. He has two novels, *On Ice* (1964) and *Rehearsal* (1976).

See: Gale, Steven, 'Jack Gelber: An Annotated Bibliography', *Bulletin of Bibliography* 44 (1987), 102–10.

GLASPELL, Susan (1882–1948), born in Davenport, IA, attended Drake and Chicago Universities, and worked as political reporter for an Iowa newspaper.

Married George Cram Cook 1913 and joined the circle of leftist intellectuals that formed the Provincetown Players. Plays include *Suppressed Desires* (1915), a parody of Freud written with Cook; and *Trifles* (1916), *The Outside* (1917), *Inheritors* (1921), *The Verge* (1921) and *Alison's House* (Pulitzer Prize 1930), most of which deal with women alienated from society and self. Glaspell also wrote ten novels, two volumes of short stories, and a biography of her husband.

See: Waterman, Arthur E., *Susan Glaspell* (1966).

GUARE, John (1938–), born in New York City, educated at Georgetown (BA 1961) and Yale Drama School (MFA 1963). After some early plays produced at Yale and Off Off-Broadway, Guare made his mark with *Muzeeka* (1967), the musical *Two Gentlemen of Verona* (1971) and particularly *The House of Blue Leaves* (1971). Subsequent plays were generally poorly received, though *Landscape of the Body* (1977), *Marco Polo Sings a Solo* (1977) and *Bosoms and Neglect* (1979) are worth attention, and *Six Degrees of Separation* (1990) is an effective social satire. Guare wrote the films *Taking Off* (1971) and *Atlantic City* (1981).

HANSBERRY, Lorraine (1930–65), born in Chicago, daughter of a prosperous black businessman who challenged convention by moving into a white neighbourhood in 1938. Educated at the University of Wisconsin, the Art Institute of Chicago and Roosevelt College, and worked as journalist and editor. *A Raisin in the Sun* (1959), about a black family moving to a white neighbourhood, is her best play; the unsuccessful *The Sign in Sidney Brustein's Window* (1964), about a white liberal, was produced as she was dying of cancer. Her husband Howard Nemiroff adapted and edited her uncompleted works as *To Be Young, Gifted, and Black* (1969), *Les Blancs* (1970), *The Drinking Gourd* (1972) and *What Use are Flowers?* (1972).

See: Cheney, Anne, *Lorraine Hansberry* (Boston, 1984).
Kaiser, Ernest and Robert Nemiroff, 'A Lorraine Hansberry Bibliography', *Freedomways* 19 (1979), 285–304.

HART, Moss (1904–1961), born in New York City and educated in city schools. Early work in amateur theatre and as social director at a summer resort led to playwriting; a producer teamed him with George S. Kaufman, and the pair wrote some of the most successful comedies of the 1930s, Hart generally providing the story and Kaufman the structure and jokes: *Once in a Lifetime* (1929), *Merrily We Roll Along* (1934), *You Can't Take It With You* (Pulitzer Prize 1936), *The Man Who Came to Dinner* (1939) and *George Washington Slept Here* (1940); they also wrote the books for several musicals, including *As Thousands Cheer* (1933), *I'd Rather Be Right* (1937) and *Lady in the Dark* (1941). On his own Hart wrote some serious plays – *Winged Victory* (1943), *Christopher Blake* (1946) and *The Climate of Eden* (1948) – and the comedy *Light Up the Sky* (1948); he also wrote several films, including *Gentlemen's Agreement* (1947) and *A Star is Born* (1954), and directed the musical *My Fair Lady* (1956).

Act One: An Autobiography (1959).

HECHT, Ben (1894–1964) and MACARTHUR, Charles (1895–1956), former newspapermen who collaborated on two classic comedies. Hecht was born in New York and raised in Wisconsin; MacArthur was born and raised in

Scranton, PA. Both worked on Chicago newspapers from about 1910 through the early 1920s before moving to New York. Each had written a couple of plays with others before teaming for the high-energy, wisecracking *The Front Page* (1928) and *The Twentieth Century* (1932). They also wrote the musical *Jumbo* (1935) and, together or separately, such films as *Scarface* (1932), *Design for Living* (1933), *Rasputin and the Empress* (1933), *Wuthering Heights* (1939) and *Gunga Din* (1939). Each wrote several lesser plays alone or with other collaborators, and each published several volumes of reportage – Hecht, *1001 Afternoons in Chicago* (1922); MacArthur, *A Bug's-Eye View of the War* (1919). Hecht also wrote several novels.

Hecht, *A Child of the Century* (1954) (autobiography).
Hecht, *Charlie* (1957) (biography of MacArthur).
Hecht, *Gaily, Gaily* (1963) (autobiography).
Fetherling, D., *The Five Lives of Ben Hecht* (Toronto, 1977).

HELLMAN, Lillian (1905–84), born in New Orleans and raised there and in New York City; studied at NYU and Columbia. Worked as book editor and reviewer, play reader for Broadway producers and script reader in Hollywood. After a failed marriage to Arthur Kober she began a lifelong relationship with mystery writer Dashiell Hammett, who encouraged her writing. Major plays include *The Children's Hour* (1934), *The Little Foxes* (1939), *Watch on the Rhine* (1941), *Another Part of the Forest* (1946), *The Lark* (1955), the musical *Candide* (1956) and *Toys in the Attic*, 1960). Films include *Dead End* (1937) and *The Chase* (1966); she also edited *The Selected Letters of Anton Chekhov* (1955) and *The Big Knockover* (1966) a collection of Hammett's stories. The historical accuracy of her three volumes of memoirs (see below) has been questioned.

An Unfinished Woman (1969).
Pentimento (1973).
Scoundrel Time (1976).
Rollyson, Carl, *Lillian Hellman: Her Legend and Her Legacy* (1988).
Wright, William, *Lillian Hellman: The Image, The Woman* (1986).

See: Bills, Steven H., *Lillian Hellman: An Annotated Bibliography* (1979).
Estrin, M. W., *Lillian Hellman: A Reference Guide* (Boston, 1980).
Falk, Doris V., *Lillian Hellman* (1978).
Holmin, Lorena R., *The Dramatic Works of Lillian Hellman* (Austin, 1973).
Lederer, Katherine, *Lillian Hellman* (Boston, 1979).
Moody, Richard, *Lillian Hellman, Playwright* (1972).
Riordan, Mary M., *Lillian Hellman, A Bibliography* (1980).

HENLEY, Beth (1952–), born in Jackson, MS, daughter of a state politician; set out to be an actress, studying at Southern Methodist University (BFA 1974) and the University of Illinois, and acting in Dallas and Los Angeles. *Crimes of the Heart* was entered by a friend in the annual competition of the Actors Theatre of Louisville; it won and, after a Louisville production in 1979 and several other regional productions, went to Broadway and a Pulitzer Prize in 1981. Other plays are *The Wake of Jamey Foster* (1982), *The Miss Firecracker Contest* (1984), *The Debutante Ball* (1985), *The Lucky Spot* (1987) and *Abundance* (1989); she also wrote the film *Nobody's Fool* (1986) and the film versions of *Crimes* (1986) and *Miss Firecracker* (1989).

HERNE, James A. (1839–1901), born in West Troy, New York, and educated in local schools. He became an actor, with seasons in Troy, Baltimore, Washington (where he delivered the opening address at the inauguration of Ford's Theatre), Philadelphia and on tour of Canada, eventually establishing himself in San Francisco and later New York. Like most other actor-dramatists he began writing to provide good roles for himself and for his second wife, actress Katherine Corcoran, adapting such works as *Oliver Twist* and *Rip Van Winkle*, sometimes in collaboration with producer-dramatist David Belasco. His plays – *Hearts of Oak* (1879), *The Minute Men* (1886), *Drifting Apart* (1888), *Margaret Fleming* (1890), *Shore Acres* (1892), *Griffith Davenport* (1899) and *Sag Harbor* (1899) – are generally realistic, though not without the melodramatic flourishes of nineteenth-century drama. *Margaret Fleming* is usually cited as the first modern American play, for its serious treatment of complex moral issues.

See: Perry, J., *James A. Herne – The American Ibsen* (Chicago, 1978).

HOWARD, Sidney (1891–1939), born in Oakland, CA, studied at Berkeley (AB 1915) and Harvard (MA 1916). After service as ambulance driver and pilot in the First World War, he joined *Life* magazine, becoming its literary editor in 1922. Half of his twenty-eight produced plays are journeyman translations or adaptations – e.g. *The Last Night of Don Juan* (1925), *Dodsworth* (1934). Among his original plays are *They Knew What They Wanted* (Pulitzer Prize 1924), *Lucky Sam McCarver* (1925), *Ned McCobb's Daughter* (1926), *The Silver Cord* (1926) and *Alien Corn* (1933). He also wrote eleven films, including *Gone With the Wind* (1939); a book of short stories; and two volumes of non-fiction.

See: White, Sidney H., *Sidney Howard* (Boston, 1977).

INGE, William (1913–73), born in Independence, Kansas to a family much like that in his *The Dark at the Top of the Stairs* – distant father, over-protective mother. Degrees from the University of Kansas (BA 1935) and George Peabody College (MA 1938) led to positions teaching English at Stephens College and Washington University, and reviewing theatre for a St Louis newspaper. A meeting in that capacity with Tennessee Williams encouraged his play writing; his first production was *Further Off From Heaven* (an early version of *Dark . . .*) in Dallas, 1947. Inge's next four plays, all about little people making little compromises to achieve little happinesses, were Broadway successes: *Come Back, Little Sheba* (1950), *Picnic* (Pulitzer Prize, 1953), *Bus Stop* (1955), *The Dark at the Top of the Stairs* (1957); he also wrote the films *Splendor in the Grass* (1961) and *All Fall Down* (1962). Unfortunately none of his subsequent plays was a success, nor were his novels *Good Luck, Miss Wyckoff* (1970) and *My Son Is a Splendid Driver* (1971); and depression and alcoholism led to his suicide.

Voss, Ralph F., *A Life of William Inge* (Lawrence, KS, 1989).
See: McClure, Arthur F., *William Inge: A Bibliography* (1982).
Shuman, Robert B., *William Inge* (Boston, 1989).

JONES, LeRoi (BARAKA, Amiri) (1934–) born in Newark, NJ, educated at Howard University (BA 1954), served in the Air Force 1954–57. Edited the literary magazine *Yugen* (1958–63) while writing music criticism and teaching creative writing at the New School, Columbia and SUNY Buffalo; he would later teach at San Francisco State, Yale, George Washington, Rutgers and SUNY Stony Brook, where he is Professor of African Studies. Early plays *A Good Girl is Hard to Find* (1958) and *Dante* (1961) preceded his first successes *Dutchman, The Baptism, The Slave* and *The Toilet* (all 1964). Soon afterwards Jones left the literary world to devote himself to social and political activism in the black community of Newark; in 1968 he took the name Amiri Baraka. Subsequent plays, such as *A Black Mass* (1966), *Slave Ship* (1967) and *Arm Yourself, or Harm Yourself!* (1967) are consciousness-raising or agitprop pieces designed for inner city black audiences. Most of his publications since 1968 have been political – e.g. *Raise Race Rays Raze* (1971), a collection of essays – though he has edited collections of black literature and published volumes of fiction and poetry. The convention of retaining the name Jones for the works written before 1968 is one Baraka himself follows.

Autobiography of LeRoi Jones/Amiri Baraka (1984).

See: Benston, Kimberly W., *Baraka* (New Haven, 1976).
Brown, Lloyd W., *Amiri Baraka* (Boston, 1980).
Dace, Letitia, *LeRoi Jones (Imamu Amiri Baraka): A Checklist of Works By and About Him* (London, 1971).
Hudson, Theodore R., *From LeRoi Jones to Amiri Baraka* (Durham, NC, 1973).
King, Kimball, *Ten Modern American Playwrights: An Annotated Bibliography* (1982).
Lacey, Henry C., *To Raise, Destroy, and Create: The Poetry, Drama and Fiction of Amamu Amiri Baraka* (Troy, NY, 1981).
Sollors, Werner, *Amiri Baraka/LeRoi Jones* (1978).

KAUFMAN, George S[imon] (1889–1961), born in Pittsburgh, briefly studied law and held a variety of jobs until his freelance contributions to newspaper columnists led to his own columns in Washington and New York papers and to the position of drama critic for the *New York Times* 1917–30. From *Someone in the House* (1918) he wrote more than fifty plays, virtually all comedies, and most in collaboration with others; generally the partner supplied the plot and Kaufman the jokes and comic structure. His major collaborations were with Marc Connelly – *Dulcy* (1921), *To the Ladies!* (1922), *Merton of the Movies* (1922), *Helen of Troy, New York* (1923) and *Beggar on Horseback* (1924) – Edna Ferber – *Minick* (1924), *The Royal Family* (1927), *Dinner at Eight* (1932), *Stage Door* (1936) and others – and Moss Hart – *Once in a Lifetime* (1929), *You Can't Take It With You* (Pulitzer Prize 1936), *The Man Who Came to Dinner* (1939), *George Washington Slept Here* (1940) and others. Other successes include *The Butter and Egg Man* (1924), *The Solid Gold Cadillac* (1953, with Howard Teichmann), and the musicals (with various collaborators) *The Cocoanuts* (1925), *Animal Crackers* (1928), *The Band Wagon* (1931), *Of Thee I Sing* (Pulitzer 1931), *I'd Rather Be Right*, (1937) and *Silk Stockings* (1955). Kaufman also wrote several films, notably *A Night at the Opera* (1935), and directed many of his own plays and others.

Hart, Moss, *Act One* (1959) (Hart's autobiography, with much about Kaufman).
Meredith, Scott, *George S. Kaufman and his Friends* (1974).
Teichmann, Howard, *George S. Kaufman: An Intimate Portrait* (1977).

See: Goldstein, Malcolm, *George S. Kaufman: His Life, His Theatre* (1979).
Pollack, Rhoda-Gale, *George S. Kaufman* (Boston, 1988).

KINGSLEY, Sidney (1906–), born Sidney Kirshner in New York City; he began writing after college (Cornell, 1928) and brief careers as actor and script reader. His strongest plays are fairly conventional melodramas in intensely realistic slice-of-life settings: *Men in White* (Pulitzer Prize 1933), *Dead End* (1935), *The World We Make* (1939), *Detective Story* (1949). *The Patriots* (1943) about the rivalry of Hamilton and Jefferson, and *Darkness at Noon* (1951), from Koestler's novel, were also successful.

KOPIT, Arthur (1937–), born in New York City, raised on Long Island, began writing plays at Harvard (BA 1959). *Oh Dad, Poor Dad, Mamma's Hung You in the Closet and I'm Feeling So Sad* (1962) was one of the first original plays to be an Off-Broadway hit, but none of his subsequent plays succeeded until *Indians* (1968) and then again until *Wings* (1978). Among the others *The Day the Whores Came Out to Play Tennis* (1965) and *End of the World* (1984) are worth attention, as is his version of Ibsen's *Ghosts* (1982). He also wrote the book for the musical *Nine* (1982) and a television (1990) and stage (1991) version of *The Phantom of the Opera*.

See: Auerbach, Doris, *Sam Shepard, Arthur Kopit and the Off-Broadway Theatre* (Boston, 1982).
King, Kimball, *Ten Modern American Playwrights: An Annotated Bibliography* (1982).

MACARTHUR, Charles. *See* HECHT, Ben.

MACKAYE, Percy Wallace (1875–1956), born in New York City, son of playwright-director Steele MacKaye; attended Harvard (BA 1897), toured Europe, and taught at a private school in New York until devoting himself full time to writing in 1904. Early plays were verse comedies – *The Canterbury Pilgrims* (1903) – and tragedies – *Fenris the Wolf* (1905), *Jeanne d'Arc* (1906), *Sappho and Phaon* (1907), but he also wrote romantic prose comedy – *The Scarecrow* (1908) – political and social satire – *Mater* (1908), *Anti-Matrimony* (1910) – and problem plays – *Tomorrow* (1913). Much of his work, particularly in the second decade of the century, was in large-scale pageant plays, with casts numbering in the thousands: *St Louis* (1914), *Sanctuary, a Bird Masque* (1913), *Washington, the Man Who Made Us* (1916), *The Evergreen Tree* (1917), *The Roll Call* (1918), among others. MacKaye subsequently taught at Miami University, Ohio, and published several volumes of essays and lectures on theatre, along with a biography of his father and two collections of poetry, generally celebratory odes. His last major work, *The Mystery of Hamlet* (1949) was a tetralogy of verse plays set in the years before the opening of Shakespeare's version.

MCNALLY, Terrence (1939–), born in St Petersburg, FL, and raised in Corpus Christi, TX. After college (Columbia, 1960) worked as stage manager at the Actors Studio (1960-61), travelling tutor for John Steinbeck's family (1961-62), film critic for *Seventh Art* magazine (1963-65) and assistant editor of the Columbia alumni magazine (1965-66). His early plays, among them *And Things That Go Bump in the Night* (1964), *Next* (1968), *Botticelli* (1968), *Sweet Eros* (1968) and *Where Has Tommy Flowers Gone?* (1971), are generally anti-war or social criticism pieces in the spirit of the times, though a strong farcical sense is apparent in *Next* and *Noon* (1968) that prepared for the apolitical farce *The Ritz* (1975). Few of his plays of the next decade are noteworthy until the comic *It's Only a Play* (1986), the romantic *Frankie and*

Johnny in the Clair de Lune (1988); and the character studies *The Lisbon Traviata* (1989) and *Lips Together, Teeth Apart* (1991). McNally wrote the musical *The Rink* (1984), the film *Frankie and Johnny* (1991), and several short plays for television.

MAMET, David (1947-), born in Chicago, attended private schools and Goddard College (BA 1969), studied acting in New York and taught at Marlboro College (1969–70) and Goddard (1971–73). A theatre company he directed at Goddard regrouped as the St Nicholas Players in Chicago in 1974, and thereafter most of Mamet's plays were first staged by various Chicago theatres. Plays include *Lakeboat* (1970), *The Duck Variations* (1972), *Sexual Perversity in Chicago* (1974), *American Buffalo* (1975), *The Water Engine* (1977), *A Life in the Theatre* (1977), *Edmond* (1982), *Glengarry Glen Ross* (Pulitzer Prize 1984), *Speed-the-Plow* (1987), and translations of *The Cherry Orchard* (1987) and *Uncle Vanya* (1989). Mamet wrote the films *The Postman Always Rings Twice* (1979), *The Verdict* (1980), *The Untouchables* (1986), *House of Games* (1986), *Things Change* (1987), *We're No Angels* (1987) and *Homicide* (1991); two collections of essays – *Writing in Restaurants* (1986) and *Some Freaks* (1989) – and two children's books – *Warm and Cold* (1985) and *The Owl* (with Lindsay Crouse, 1987).

See: Bigsby, C.W.E., *David Mamet* (1985).
Carroll, Dennis, *David Mamet* (London, 1987).
Davis, J. Madison and John Coleman, 'David Mamet: A Classified Bibliography', *Studies in American Drama 1945- Present* 1 (1986), 83–101.

MILLER, Arthur (1915–), born in New York City, son of a prosperous clothing manufacturer who lost his money just before the onset of the Depression, an event alluded to in *A Memory of Two Mondays* (1955), *After the Fall* (1964), *The Price* (1968) and *The American Clock* (1980). Miller worked at various menial jobs before attending the University of Michigan (BA 1938), where his first plays won several prizes. During the 1940s he wrote for radio and films; his first professional stage play, *The Man Who Had All the Luck*, failed in 1944. It was followed by the successful *All My Sons* (1947) and his best play *Death of a Salesman* (Pulitzer Prize 1949). Miller was of the generation whose youthful politics endangered them in the anti-communist hysteria of the 1950s; he dealt with this subject in his version of Ibsen's *An Enemy of the People* (1950) and in *The Crucible* (1953). Blacklisted in 1955 and convicted of contempt of Congress for not naming names in 1957 (the conviction was reversed on appeal), Miller wrote only one other play during the decade, *A View from the Bridge* (1956). His first marriage (to Mary Grace Slattery, 1940-55) ending in divorce, Miller married actress Marilyn Monroe in 1956, and spent the next six years aiding her career; he wrote the film *The Misfits* (1961) for her. After their divorce in 1961 he married photo-journalist Inge Morath; they have collaborated on several books of photography and reportage. Plays of the next decade include *After the Fall, Incident at Vichy* (1964), *The Price* and *The Creation of the World and Other Business* (1972). Another gap, during which Miller was active in international writers' organizations, ended with *The Archbishop's Ceiling* (1977), followed by *The American Clock* (1980), the double bill *Elegy for a Lady* and *Some Kind of Love Story* (1982), the double bill *I Can't Remember Anything* and *Clara* (1987), and *The Ride Down Mt Morgan* (1991). Other works include a novel, *Focus* (1945); a children's book, *Jane's Blanket* (1963); a volume of short stories, *I Don't*

Need You Anymore (1967), the television film *Playing for Time* (1980), and a film adaptation of *Some Kind of Love Story*, *Everybody Wins* (1990).

The Theatre Essays of Arthur Miller (1978).
Salesman in Beijing (1984). (An account of directing the play in China.)
Timebends (1987). (A memoir.)
Roudane, Matthew, (ed.), *Conversations with Arthur Miller* (Jackson, MS, 1988). (Collects 37 interviews from five decades.)

See: Carson, Neil, *Arthur Miller* (London, 1982).
Ferres, John H., *Arthur Miller: A Reference Guide* (Boston, 1979).
Hayashi, Tetsumaro, *An Index to Arthur Miller Criticism* (Metuchen, NJ, 1976).
Hayman, Ronald, *Arthur Miller* (1970).
Jensen, George, *Arthur Miller: A Bibliographical Checklist* (Columbia, SC, 1976).
Moss, Leonard, *Arthur Miller* (Boston, 1980).
Nelson, Benjamin, *Arthur Miller: Portrait of a Playwright* (1970).
Schleuter, June and James K. Flanagan, *Arthur Miller* (1987).
Welland, Dennis, *Arthur Miller: The Playwright* (London, 1983).

NORMAN, Marsha (1947–), born Marsha Williams in Louisville, KY, studied at Agnes Scott College (BA 1969) and University of Louisville (MAT 1971). Married Michael Norman 1969 (divorced 1974; married Dann C. Byck Jr 1978) and kept his name. Taught disturbed adolescents at Kentucky Central State Hospital and gifted children at the Brown School; created the children's page of the Louisville *Times* 1976. Her first play, *Getting Out* (1977), was a winner in the Actors Theatre of Louisville competition, leading to productions there and elsewhere. Subsequent plays include *Third and Oak* (1978), *Circus Valentine* (1979), *The Hold-up* (1980), the Pulitzer Prize winning *'night, Mother* (1983), *Traveler in the Dark* (1984) and the book and lyrics for the musical *The Secret Garden* (1991). She has also written two television plays and a novel, *The Fortune Teller* (1987).

See: Wolfe, Irmgard H., 'Marsha Norman: A Bibliography', *Studies in American Drama 1945–Present* 3 (1988).

ODETS, Clifford (1906–63), born in Philadelphia, son of a printer who later prospered, though Odets always considered himself a worker's son. Raised in New York City, he quit high school to become a radio and stage actor, joining the newly formed Group Theatre in 1930. His first play, *Waiting for Lefty* (1935), was first done by the New Theatre League, another leftist company, then staged by the Group Theatre, as were his next several plays, almost all reflections of the Depression through domestic melodrama: *Awake and Sing* (1935), *Till the Day I Die* (1935), *Paradise Lost* (1935), *Golden Boy* (1937), *Rocket to the Moon* (1938). Subsequent plays – *Night Music* (1940), *Clash by Night* (1941), *The Russian People* (1942), *The Big Knife* (1949), *The Flowering Peach* (1954) – were generally unsuccessful, though *The Country Girl* (1950) proved an effective dissection of marital politics. Odets worked briefly in Hollywood in the 1930s (*The General Died at Dawn*, 1936) and more extensively in the next two decades – his films include *None But the Lonely Heart* (1944), *The Sweet Smell of Success* (1957) and *The Story on Page One* (1960) – and remained ambivalent about easy money's effect on his work.

The Time is Ripe: The 1940 Journal of Clifford Odets (1989).

See: Brenman-Gibson, M., *Clifford Odets, American Playwright* (1981).
Cantor, Harold, *Clifford Odets, Playwright-Poet* (Metuchen, NJ, 1978).
Clurman, Harold, *The Fervent Years* (1945). (Memoir of the Group Theatre, with discussion of Odets.)
Mendelsohn, Michael J., *Clifford Odets, Humane Dramatist* (Deland, FL, 1969).
Murray, Edward, *Clifford Odets: The Thirties and After* (1968).
Shuman, R. Baird, *Clifford Odets* (1962).
Weales, Gerald, *Odets the Playwright* (1985).

O'NEILL, Eugene (1888–1953), born in New York City, youngest son of popular actor James O'Neill, educated in Catholic schools, with a year at Princeton (1906–07) and later at Harvard (1914–15). A period of rootlessness and dissipation, briefly interrupted by marriage, menial jobs in New York and Latin America, and two seasons touring with his father, ended with a sanatarium stay for tuberculosis in 1912–13, when he read heavily and decided to be a writer. Several short plays written 1913–16 remain unproduced or unpublished; his first production, *Bound East for Cardiff* (1916) was the result of his association with the young radicals and artists who made up the Provincetown Players. Other short plays followed, in Provincetown and New York, leading to his Broadway debut with *Beyond the Horizon* (1920). His plays of the 1920s are marked by experiments in form, virtually no two written in the same style: *The Emperor Jones* (1920), *Anna Christie* (1921), *The Hairy Ape* (1922), *All God's Chillun Got Wings* (1924), *Desire Under the Elms* (1924), *The Fountain* (1925), *The Great God Brown* (1926), *Marco Millions* (1928), *Strange Interlude* (1928), *Lazarus Laughed* (1928), *Dynamo* (1929). Only three plays appeared in the 1930s – *Mourning Becomes Electra* (1931), *Ah, Wilderness!* (1933) and *Days Without End* (1933) – and only two more during his lifetime – *The Iceman Cometh* (1946) and *A Moon for the Misbegotten* (1947) – as O'Neill withdrew into depression and a degenerative nerve disease. At his death it was revealed that he had continued writing, and *Long Day's Journey into Night* (1956) is now considered his masterpiece; other posthumous plays are *A Touch of the Poet* (1957), *Hughie* (1958) and the incomplete *More Stately Mansions* (1967). O'Neill was married three times, to Kathleen Jenkins 1909–12, Agnes Boulton 1918–29 and Carlotta Monterey 1929–53. *Beyond the Horizon, Anna Christie, Strange Interlude* and *Long Day's Journey* won Pulitzer Prizes; and O'Neill was awarded the Nobel Prize for Literature in 1936.

Work Diary 1924–43 (New Haven, 1981).
'As Ever, Gene': The Letters of Eugene O'Neill to George Jean Nathan (Rutherford, NJ, 1987).
'Love and Admiration and Respect': The O'Neill–Commins Correspondence (Durham, NC, 1986).
Selected Letters of Eugene O'Neill (New Haven, 1988).
'The Theatre We Worked For': The Letters of Eugene O'Neill to Kenneth Macgowan (New Haven, 1982).
Gelb, Arthur and Barbara, *O'Neill* (1973).
Sheaffer, Louis, *O'Neill: Son and Playwright* (Boston, 1968).
Sheaffer, Louis, *O'Neill: Son and Artist* (Boston, 1973).

See: Atkinson, Jennifer McCabe, *Eugene O'Neill: A Descriptive Bibliography* (Pittsburgh, 1974).

Berlin, Normand, *Eugene O'Neill* (1988).
Bogard, Travis, *Contour in Time: The Plays of Eugene O'Neill* (1972).
Cargill, Oscar *et al.* (eds), *O'Neill and His Plays: Four Decades of Criticism* (1961). (Collection of reviews and analyses.)
Carpenter, Frederick, *Eugene O'Neill* (Boston, 1979)
Floyd, Virginia (ed.), *Eugene O'Neill: A World View* (1979). (Collection of foreign criticism.)
Griffin, Ernest (ed.), *Eugene O'Neill: A Collection of Criticism* (1976).
Miller, Jordan Y., *Eugene O'Neill and the American Critic: A Bibliographical Checklist* (Hamden, CT, 1973).
Ranald, Margaret Loftus, *The Eugene O'Neill Companion* (Westport, CT, 1984). (Encyclopedia-style entries for plays, characters, family, etc.)
Reaver, J. Russell, *An O'Neill Concordance*, 3 vols (Detroit, 1969).

RABE, David (1940–), born in Dubuque, Iowa, educated in Catholic schools, Loras College (BA 1962) and Villanova. Drafted in 1965, he spent one of his two army years in Vietnam, returning to Villanova for an MA in theatre 1968. He worked as a journalist 1969-70 and taught at Villanova 1970-72. His reputation rests on three powerful plays about the Vietnam War, *The Basic Training of Pavlo Hummel* (1971), *Sticks and Bones* (1971) and *Streamers* (1976). Other plays – *The Orphan* (1973), *In the Boom Boom Room* (1974), *Goose and Tomtom* (1982) and *Hurlyburly* (1984) – are less effective. He wrote the films *I'm Dancing as Fast as I Can* (1982), *Streamers* (1983) and *Casualties of War* (1989).

See: Kolin, Philip C., *David Rabe: A Stage History and Primary and Secondary Bibliography* (1988).

RICE, Elmer (1892–1967), born Elmer Reizenstein in New York City, he dropped out of high school but studied law while working in a law office, and was admitted to the bar 1913. He turned immediately to writing, taking the name Rice; his first play, *On Trial* (1914) contains the first use of flashbacks in American drama. The rest of his more than forty produced plays fall into three groups: conventional and generally minor melodramas and comedies such as *Cock Robin* (with Philip Barry 1928), *Counsellor-at-Law* (1931) and *Love Among the Ruins* (1963); plays with a strong social or political content, such as *Street Scene* (Pulitzer Prize 1929), *We, the People* (1933) and *Flight to the West* (1940); and a few experiments in style and form, notably the expressionistic *The Adding Machine* (1923) and *Dream Girl* (1945). Rice directed the New York branch of the Federal Theatre Project 1935-36, helping to create the 'Living Newspaper' style. He directed all his own plays after 1929, and wrote several films and two novels.

The Living Theatre (1959).
Minority Report: An Autobiography (1963).

See: Durham, Frank, *Elmer Rice* (Boston, 1970).
Hogan, Robert, *The Independence of Elmer Rice* (Carbondale, IL, 1965).

SHEPARD, Sam (1943–), born Samuel Shepard Rogers in Fort Sheridan, Illinois, and raised in Duarte, California. After one year of junior college he moved to New York City in 1963, supporting himself with menial jobs and as a rock musician. Shepard was one of the first Off Off-Broadway writers to attract attention, first with Theatre Genesis, then with the Open Theatre; he later became playwright-in-residence for San Francisco's Magic Theatre. He wrote more than 25 short plays in his first decade, many of them unpublished. Among the most significant are *Cowboys* (1964), *Chicago* (1965), *Icarus's Mother* (1965), *Red Cross* (1966), *Forensic and the Navigators* (1967), *The Unseen Hand* (1969) and *Cowboy Mouth* (1971). Longer plays from this period include *Operation Sidewinder* (1970), *The Tooth of Crime* (1972) and *Geography of a Horse Dreamer* (1974). Shepard's plays after 1975 are generally more realistic in style, and generally more powerful than his early work: *Curse of the Starving Class* (1977), *Buried Child* (Pulitzer Prize 1978), *True West* (1980), *Fool for Love* (1983), *A Lie of the Mind* (1985) and *The War in Heaven* (1991). Shepard has written for films, and has a second career as film actor (*Days of Heaven*, *The Right Stuff*, *Country*, *Fool for Love*, *Crimes of the Heart*, etc.). His non-dramatic writing includes two volumes of stories and verse, *Hawk Moon* (1972) and *Motel Chronicles* (1982), and an impressionistic report on a rock concert tour, *Rolling Thunder Logbook* (1978).

Oumano, Ellen, *Sam Shepard: The Life and Work of an American Dreamer* (1986).
Shewey, Don, *Sam Shepard* (1985).

See: Auerbach, Doris, *Sam Shepard, Arthur Kopit, and the Off-Broadway Theatre* (Boston, 1982).
Hart, Lynda, *Sam Shepard's Metaphorical Stages* (Westport, 1987).
King, Kimball, *Ten Modern American Playwrights: An Annotated Bibliography*, (1982).
Marranca, Bonnie (ed.), *American Dreams: The Imagination of Sam Shepard* (1981).
Mottram, Ron, *Inner Landscapes: The Theatre of Sam Shepard* (Columbia, MO, 1984).

SHERWOOD, Robert E. (1896–1955), born in New Rochelle, NY, his education at Harvard (BA 1918) interrupted for service in the Canadian army. Wrote for *Vanity Fair* (1919) and *Life* (1920–28). Sherwood could write romance – *Waterloo Bridge* (1930) – and light comedy – *Reunion in Vienna* (1931) – but his best plays have a liberal social or political edge: *The Petrified Forest* (1934), *Idiot's Delight* (Pulitzer Prize 1936), *Abe Lincoln in Illinois* (Pulitzer 1938), *There Shall Be No Night* (Pulitzer 1940). Sherwood was active in theatrical and political circles, serving as President of the Dramatists Guild and later of ANTA; during the Second World War he held several government posts. He wrote twelve films, including *The Scarlet Pimpernel* (1935), *Rebecca* (1940) and *The Best Years of Our Lives* (1946); and a Pulitzer Prize biography, *Roosevelt and Hopkins* (1948).

Brown, John Mason, *The Worlds of Robert E. Sherwood* (1965).
Brown, John Mason, *The Ordeal of a Playwright* (1970).

See: Meserve, Walter J., *Robert E. Sherwood, Reluctant Moralist* (1970).
Shuman, R. Baird, *Robert E. Sherwood* (1964).

SIMON, Neil (1927–), born in New York as Marvin Neil Simon, attended New York schools and NYU, and served in the Air Force in Colorado during the Second World War. With his older brother Danny, he wrote comic sketches and scripts for radio and television 1948–52, notably the Sid Caesar *Your Show of Shows* and the Phil Silvers *Sgt Bilko* series. The brothers contributed sketches to two Broadway revues, *Catch a Star!* (1955) and *New Faces of 1956* (1956). Neil Simon's first stage comedy, *Come Blow Your Horn* (1960) began an unprecedented string of successes; for the next three decades there was hardly a season without one or as many as four Simon comedies on Broadway: *Barefoot in the Park* (1963), *The Odd Couple* (1965), *The Star-Spangled Girl* (1966), *Plaza Suite* (1968), *Last of the Red Hot Lovers* (1969), *The Prisoner of Second Avenue* (1971), *The Sunshine Boys* (1972), *The Good Doctor* (1973), *God's Favorite* (1974), *California Suite* (1976), *Chapter Two* (1977), *I Ought To Be in Pictures* (1980), *Fools* (1981), *Rumors* (1988). Simon's first attempt at a serious play, *The Gingerbread Lady* (1970), was unsuccessful, but his later trilogy of autobiographical comedies *Brighton Beach Memoirs* (1983), *Biloxi Blues* (1985) and *Broadway Bound* (1986) had a realism and emotional depth hitherto missing from his work, as did *Lost in Yonkers* (Pulitzer Prize 1991). Simon wrote the books for the musicals *Little Me* (1962), *Sweet Charity* (1966), *Promises, Promises* (1968) and *They're Playing Our Song* (1979). He has written more than twenty films, including adaptations of many of his own plays and *After the Fox* (1966), *The Out-of-Towners* (1970), *The Heartbreak Kid* (1972), *The Goodbye Girl* (1977), *The Cheap Detective* (1978) and *The Lonely Guy* (1984). He is known as a play doctor, and is reported to have edited or rewritten many scripts by others without credit.

See: Johnson, Robert K., *Neil Simon* (Boston, 1983).
King, Kimball, *Ten Modern American Playwrights: An Annotated Bibliography* (1982).
McGovern, Edythe M., *Neil Simon: A Critical Study* (Van Nuys, CA, 1978).

VAN ITALLIE, Jean-Claude (1936–), born in Brussels, came to America 1940, educated at Harvard (BA 1958). He studied acting in New York, and edited the *Transatlantic Review* 1963–68. During the same period he was playwright-in-residence at the Open Theatre, an experimental company specializing in improvisations and group creations; many of his plays were actually developed in collaboration with the actors. Among his early plays the three eventually combined as *America Hurrah* (1966) are the most notable; later *The Serpent* (1969) proved a particularly impressive group creation. While continuing to write original plays, among them *Bag Lady* (1979) and *The Tibetan Book of the Dead* (1983), van Itallie produced a series of well-regarded translation-adaptations: *The Sea Gull* (1973), *The Cherry Orchard* (1977), *The Three Sisters* (1979), *Uncle Vanya* (1983) and *The Balcony* (1986). He wrote several television documentaries in the 1960s, and has taught theatre or playwriting at Yale, the New School, the Naropa Institute, Amhurst, Princeton, NYU, Colorado and Columbia.

See: Brittain, Michael J., 'A Checklist of Jean-Claude van Itallie 1961–72', *Serif* 9 (1972), 75-77.
Pasolli, Robert, *A Book on the Open Theatre* (1970).

WASSERSTEIN, Wendy (1950–), born in Brooklyn, raised there and in Manhattan, attending private schools and Mount Holyoke (BA 1971), City University of New York (MA in creative writing 1973) and Yale Drama School (MFA 1976). Her first plays were written at Yale and produced there or Off Off-Broadway: *Any Woman Can't* (1973), *Montpelier Pazazz* (a musical, with David Hollister, 1976), *When Dinah Shore Ruled the Earth* (revue, with Christopher Durang, 1977). Wasserstein found her subject, the experience of her own generation of women, in *Uncommon Women and Others* (1977), continuing the exploration in *Isn't It Romantic* (1981) and *The Heidi Chronicles* (Pulitzer Prize 1987). Along with some minor plays, Wasserstein has also written for television, is contributing editor of *New York Woman* magazine, and published a book of essays, *Bachelor Girls* (1990).

WILDER, Thornton (1897–1975), born in Madison, Wisconsin, the survivor of a pair of twins. In 1906 his father became US Consul General in Hong Kong; the family lived there and in Shanghai between stays in California. Wilder studied at Oberlin and Yale (AB 1920), and taught at a New Jersey boys' school (1921-28) and the University of Chicago (1930-36). He made his name as a novelist with *The Cabala* (1926) and *The Bridge of San Luis Rey* (Pulitzer Prize 1928), and continued to write fiction throughout his life: *The Woman of Andros* (1930), *Heaven's My Destination* (1935), *The Ides of March* (1948), *The Eighth Day* (1967), *Theophilus North* (1973). A collection of early short plays was published in 1928 as *The Angel That Troubled the Waters*; it was followed by another collection, *The Long Christmas Dinner* (1931). After a few minor works came *Our Town* (Pulitzer Prize 1938), *The Merchant of Yonkers* (1939), *The Skin of Our Teeth* (Pulitzer 1943), *The Matchmaker* (1954), *The Alcestiad* (1955) and *Plays for Bleeker Street* (1962). He wrote the films *Our Town* (1940) and *Shadow of a Doubt* (1943), served as Lieutenant-Colonel in Army Air Intelligence in the Second World War, and was Norton Professor of Poetry at Harvard 1950-51.

Goldstone, Richard H., *Thornton Wilder* (1975).
Harrison, Gilbert A., *The Enthusiast: A Life of Thornton Wilder* (1983).

See: Burbank, Rex J., *Thornton Wilder* (Boston, 1978).
Castronovo, D., *Thornton Wilder* (1986).
Goldstein, Malcolm, *The Art of Thornton Wilder* (Lincoln, NB, 1965).
Goldstone, Richard H. and G. Anderson, *Thornton Wilder: An Annotated Bibliography* (1982).
Haberman, Donald, *The Plays of Thornton Wilder* (Middletown, CT, 1967).
Papajewski, Helmut, *Thornton Wilder* (1968).
Simon, Linda, *Thornton Wilder* (Garden City NY, 1979).

WILLIAMS, Tennessee (1911–83), born in Columbus, Mississippi, to a minister's daughter and a travelling salesman; his parents would appear both as characters and as conflicting moral energies in his work. The family moved to St Louis 1919, and Williams attended the University of Missouri 1929-32, but finances forced his withdrawal to work in various jobs, including a shoe warehouse. He briefly attended Washington University, writing his first plays there, and ultimately graduated from Iowa (1938). In 1937 his sister Rose was diagnosed as schizophrenic, probably falsely, and subjected to a lobotomy; Williams would support her for the rest of his life. In 1939 he went to New Orleans hoping to find work with the Federal Writers Project, but found bohemian poverty instead. It was at this time that he took the name Tennessee, submitting some early plays to a national contest, where they came to the attention of agent Audrey Wood, who would manage most of his career. His first professional production, *Battle of Angels* (1940), closed out of town; a contract writing films for MGM gave him time to write his first success, *The Glass Menagerie* (1945). This was followed by his best play, *A Streetcar Named Desire* (Pulitzer Prize 1947), and a string of plays that, along with those of

Arthur Miller, dominated the next decade: *Summer and Smoke* (1948), *The Rose Tattoo* (1951), *Camino Real* (1953), *Cat on a Hot Tin Roof* (Pulitzer 1955), *Orpheus Descending* (1957), *Suddenly Last Summer* (1958), *Sweet Bird of Youth* (1959), *Period of Adjustment* (1960), and *The Night of the Iguana* (1961). In 1963 the death of Frank Merlo, his long-time lover, triggered a depression which would be compounded by alcoholism and drug dependency, and Williams never again regained the artistic control of his early work, though he continued to write almost compulsively for two decades. Of the later plays *The Milk Train Doesn't Stop Here Anymore* (1963), *Small Craft Warnings* (1972), *Out Cry* (1973) and *Vieux Carré* (1977) show the clearest remnants of his lost talent. Williams also wrote two volumes of poetry, two novels (*The Roman Spring of Mrs Stone*, 1950; *Moise and the World of Reason*, 1976), five volumes of short stories (ultimately assembled in *Collected Stories* 1985), and the film versions of several of his plays.

Memoirs (1975).
Where I Live (1978) (theatrical and personal essays).
Tennessee Williams's Letters to Donald Windham (1976).
Leavitt, Richard Freeman, (ed.), *The World of Tennessee Williams* (1978., (A biography through montage of photos, documents and text.)
Spoto, Donald, *The Kindness of Strangers: The Life of Tennessee Williams* (Boston, 1985).

See: Boxill, Roger, *Tennessee Williams* (1987).
Devlin, Albert J., (ed), *Conversations With Tennessee Williams* (Jackson, MS, 1986). (36 interviews from 1940 to 1981.)
Falk, Signi, *Tennessee Williams* (Boston, 1978).
Fayard, Jeanne, *Tennessee Williams* (Paris, 1972).
Gunn, Drewey Wayne, *Tennessee Williams: A Bibliography* (Metuchen, NJ, 1980).
Londré, Felicia Hardison, *Tennessee Williams* (1979).
McCann, John S., *The Critical Reputation of Tennessee Williams: A Reference Guide* (Boston, 1983).
Phillips, Gene D., *The Films of Tennessee Williams* (London, 1980).
Stanton, Stephen S. (ed.), *Tennessee Williams: A Collection of Critical Essays* (Englewood Cliffs, NJ, 1977).
Tharpe, Jac (ed.), *Tennessee Williams: A Tribute* (Jackson, MS, 1977). (Critical essays.)

WILSON, August (1945–), born in a black slum in Pittsburgh, son of a white father who had little to do with the family; Wilson was raised by his mother and stepfather. He quit school at 15 and held various odd jobs. In the 1960s Wilson became active in the Black Power movement, founding the Black Horizons Theatre Company in Pittsburgh in 1968. In 1978 he moved to St Paul, writing educational scripts for the Science Museum of Minnesota and working with the Minneapolis Playwrights Center. After some early plays he committed himself to the project of depicting the black American experience in the twentieth century, one play for each decade. Four plays have appeared so far: *Ma Rainey's Black Bottom* (1984), *Fences* (Pulitzer Prize 1987), *Joe Turner's Come and Gone* (1988) and *The Piano Lesson* (Pulitzer Prize 1990). Typically, Wilson's plays go through several years of development involving workshop stagings and preliminary productions before their final versions.

WILSON, Lanford (1937–), born in Lebanon, Missouri, and raised, after his parents' divorce in 1942, in Springfield and Ozark, Missouri. Briefly attended Southwest Missouri State College and San Diego State College. Worked in Chicago as a graphic artist for an advertising agency 1956–62; moved to New York 1962 and discovered the Caffe Cino, one of the first Off Off-Broadway theatres. His first plays, including *So Long at the Fair* (1963), *Home Free!* (1964) and *The Madness of Lady Bright* (1964), were staged at the Caffe Cino; others from this period include *Balm in Gilead* (1965), *The Rimers of Eldrich* (1966) and *The Gingham Dog* (1968). In 1969 he joined director Marshall W. Mason and others in forming the Circle Theatre, later the Circle Repertory Company; most of his subsequent plays would be written for specific actors and directed by Mason. Major plays include *The Hot l Baltimore* (1973), *The Mound Builders* (1975), *5th of July* (1978), *Talley's Folly* (Pulitzer Prize 1979), *Angels Fall* (1983) and *Burn This* (1987). Wilson also wrote the libretto for an operatic version of Williams's *Summer and Smoke* (1971), an adaptation of *The Three Sisters* (1984) and some television plays.

See: Barnett, Gene, *Lanford Wilson* (Boston, 1987).
King, Kimball, *Ten Modern American Playwrights: An Annotated Bibliography* (1982).
Ryzuk, Mary S., *The Circle Repertory Company* (Ames, IA, 1989).

Index